Formulating For Sun

Formulating For Sun

ISBN 978-1-932633-15-3

Allured
Global Information Leader

Allured Publishing Corporation
362 South Schmale Road, Carol Stream, IL 60188-2787 USA
Tel: 630/653-2155 Fax: 630/653-2192
E-mail: books@allured.com

Table of Contents

CHAPTER 5 – STABILITY OF FILTERS

CHAPTER 6 – PHOTOPROTECTION

CHAPTER 7 – FORMULATING

CHAPTER 8 – SKIN DELIVERY

CHAPTER 9 – TESTING

CHAPTER 10 – PHOTOAGING

Introduction

More than ever, consumers are becoming increasingly educated and astutely aware of the damage that skin suffers at the hands of sun overexposure. For most, preventing skin cancer followed closely by the desire to stave off premature aging, are reasons enough to take precautions when outdoors.

With the popularity of multifunctional products, it is no surprise that sunscreen ingredients are making appearances in every conceivable form of daily makeup, in skin care lines and in bug repellant. Even hair care products tout their ultraviolet (UV)-protection properties.

Experts are well aware that exposure to ultraviolet radiation (UVR), either from sunlight or by artificial sources, contributes heavily to skin cancer. However, today's consumer still equates the bronzed glow that tanning provides with youth, vitality and good health. Even with all the data in, this seems to be a standard of beauty that never can be totally eradicated—no matter what the cost. Consumers seek this "fix" not only from the outdoors, but also from either indoor tanning facilities or from cosmetic self-tanners.

Since the reality dictates that sun worshipping in any form still is "in," formulators need to compile as much knowledge as possible and continue to court innovation and efficacy in this segment of personal care.

Cosmetics & Toiletries magazine offers the cosmetic chemist a compendium of information on creating sun care products in this ***Formulating for Sun*** compilation. Pulling from its most popular articles written by world-renown experts, this book delivers the same quality and international appeal as the magazine.

Formulating for Sun covers the gamut of information for sun industry insiders: from *Beginning Cosmetic Chemistry* gurus Randy Schueller and Perry Romanowski discussing the basics of SPF, to sun experts like Christopher Vaughn, Craig Bonda and Ken Klein discussing testing, formulating and sun filters.

This book spans a wide range of topics such as UV damage in hair and skin, outdoor tanning and self-tanners, the importance of filters, forms of delivery and treatments in photoaging, offering a comprehensive look at the formulation aspect of sun care.

Anyone designing products for sun protection, tanning or sunless tanning should find answers here to many troublesome formulation questions encountered on a day-to-day basis.

The challenges are many; so, often, are the answers.

The articles have been edited from their original publication for style consistency, but otherwise remain mostly unchanged from their original publication in *C&T* magazine.

The ABCs of SPFs: An Introduction to Sun Protection Products

Keywords: sunscreens, UV, photoaging, PABA, sun block, sunburn

This article discusses the biological effects of sunlight, the chemistry of common UV absorbers and factors to consider when formulating and testing sun-care products.

Sunscreens are a special class of personal care products containing active ingredients that can absorb ultraviolet (UV) radiation to shield skin from the damaging effects of the sun. These products are classified as OTC drugs and are regulated by the Food and Drug Administration (FDA) in the US.

UV Light and Its Effects

Sunlight is composed of many wavelengths spread across the electromagnetic spectrum. As this solar radiation passes through Earth's atmosphere, some of these wavelengths are filtered out. The remaining radiation reaches the Earth as UV and infrared light. UV light is of particular concern because it can interact with skin cells and cause a variety of damaging effects. It consists of wavelengths grouped into the following three categories.

- UVC light ranges from 200-290 nm. This range has the lowest wavelength and consequently the highest energy because wavelength and energy are inversely related. Almost all light in this range is filtered out by the atmosphere, although a small amount is produced by arc welders and certain tanning lights.
- UVB light ranges from 280-320 nm and is called the burning or erythemal region because it penetrates through the stratum corneum (SC) and the epidermis. This type of light causes most of the skin damage that is immediately apparent, like reddening, which is known as erythema or sunburn.
- UVA light is composed of wavelengths from 320-360 nm. This range has the lowest energy, but it penetrates deeper into the skin and interacts with more skin structures. Low doses of UVA can penetrate into the dermis, where it can stimulate the production of melanin, the pigment responsible for tanning and protecting the skin from further damage.

While UVB is responsible for most sunburns, high doses of UVA can also cause reddening. Furthermore, the amount of UVA that reaches the Earth's surface is higher than the amount of UVB. Higher doses of UVA can penetrate into skin structures and cause damage to structural components, such as the elastin and collagen matrix. Chronic sun exposure, particularly to the UVA range, also results in a condition known as UV-induced photoaging, which occurs when key support matrix elements are damaged by radiation. This is a cumulative process that contributes to wrinkles, sagging and other signs of aging. The impact of photoaging can easily be seen by comparing the difference in skin on the face to an area that is not exposed to constant sunlight, such as the buttocks. While this skin is physiologically identical to the face, it appears smoother because it is exposed to less light. Studies have shown intrinsic differences between photo-aged skin and intrinsically aged skin.

In the US, products that screen out this part of the spectrum can make anti-aging claims that are extremely popular now. These products do not just guard against sunburn and skin cancers, they also protect one's youthful appearance. In addition to dermal problems, there are also data suggesting UVA is one possible agent causing certain types of cataracts.

Hair Damage

In addition to its effects on skin, UV light has also been shown to cause two types of hair damage. When hair is exposed to enough UV light, it can become discolored. Brown hair tends to fade, while blond and red hair tends to yellow. This has been attributed to photo-oxidative bleaching processes and the photo-degradation of amino acids, such as cystine, tyrosine and tryptophan. The breakdown of amino acids is also responsible for the other types of hair damage induced by UV exposure, namely the reduction in tensile strength. Due to short- and long-term effects on health and beauty, it is desirable to protect skin and hair from UV light. Of course, the most direct solution is to shield them from the sun by physically blocking them with clothing, hats or umbrellas. However, this is not always practical or desirable. To protect exposed skin and hair from the damaging effects of sunlight, scientists have developed various chemicals that can absorb or block UV radiation.

The Chemistry of UV Absorbers

Certain chemical compounds are able to interact with UV light, defusing its damaging energy. The chemical bonds in these molecules can absorb UV light and remit or reabsorb it in a harmless form. These chemicals are typically aromatic compounds that have a carbonyl group. In general, they also have an electron releasing group, such as an amine or a methoxyl group in either the ortho- or para-position on the aromatic ring. When UV light strikes one of these molecules it causes a photochemical excitation; the molecule is stimulated to a higher energy level. When the molecule returns to its original energy state, the excess absorbed energy is emitted as light with a different energy state. Most sunscreens emit energy in the infrared region and may contribute a minuscule heating effect to the skin. Others emit in the blue range of visible light. In fact, products that use these types of compounds give the skin a slight bluish cast. Each sunscreen molecule can repeat this absorption-emission cycle multiple times before it decays.

A variety of compounds have been developed that have a molecular structure capable of UV absorption. These absorbers can be formulated in appropriate vehicles that can be conveniently applied to exposed skin to protect cells from interaction with the radiation. Common UV absorbers include organic compounds, such as para-amino benzoic acid (PABA) and its esters (salicylates cinnamates, benzophenones, anthanilates, dibenzoylmethanes and camphor derivatives. In addition, inorganic sun blocking materials, such as titanium dioxide, are commonly used in sunscreen formulations.

Key Sunscreen Chemicals

PABA and its derivatives: PABA is an aromatic ring with amino and carboxylic acid groups that has been used in sunscreen products since the 1950s. PABA is an effective sunscreen, but because of the way its functional groups are positioned (the parastructure), the molecule is prone to oxidation and discoloration. It also tends to form a crystalline structure that can make the material hard to work with in cosmetic formulations. Furthermore, in recent years, concerns have been raised about the safety of the material. Scientists have attempted to overcome some of these drawbacks by esterifying PABA with other materials. For example, combining PABA with glycerin yields glyceryl PABA that is more water-soluble than the parent compound. Despite these improvements, PABA-based sunscreens have gained a negative reputation with certain consumers, and many formulators avoid using them. In fact, many products are now labeled "PABA-free."

Salicylates: Salicylates were the first sunscreen chemicals to be widely used in commercial preparations, and they continue to be popular today. Benzyl salicylate, octyl salicylate and homomenthyl salicylate are among the most popular. These molecules all contain functional groups attached to the ortho position and are able to create internal hydrogen bonds. Because they absorb UV radiation around 300-310 nm, they are ideal for use as UVB screens. Although salicylates are less-effective UV absorbers than other sunscreens, they have an excellent safety profile and can be readily incorporated into cosmetic vehicles. Due to their active groups interacting with each other, they are less likely to react with the skin or other ingredients.

Cinnamates: Cinnamates, particularly benzyl cinnamate, were used as early sunscreen compounds. They contain an extra unsaturated group conjugated to both the aromatic ring and the carboxylic acid group. This structure gives these molecules good UV absorbance in the 305 nm range. Currently, octyl methoxycinnamate is preferred because it is insoluble in water. This makes it ideal for use in waterproof products.

Benzophenones: Benzophenones are the only UV absorbers belonging to the aroma ticketone category. This unique structure allows them to absorb UV light beyond 320 nm. Unfortunately, these materials are solids that are difficult to handle and hard to incorporate into cosmetic products.

Physical blockers: Other materials, such as titanium dioxide and zinc oxide, do not absorb UV light but are capable of reflecting light, thereby preventing it from reaching the skin. These sunblockers are the most effective way to chemically shield the skin, but they tend to leave a white film that can be aesthetically unpleasing. A recent innovation in sunblock technology has created micronized versions of these molecules whose particles are reduced in size to the point where they do not reflect visible light, but are able to scatter UV light. These micronized versions are a significant improvement because they do not leave a noticeable film on the skin.

Sun-Care Formulation

If a sunscreen ingredient is not properly incorporated into a cosmetic formulation, it will not matter how well that ingredient blocks or scatters UV radiation. The type of formulation depends on many factors, such as the desired product form and function, the type of sunscreen ingredients to be used, the aesthetic characteristics and other factors like cost and manufacturing capabilities. Although these factors are common to most formulating efforts, sunscreen ingredients also have their own special considerations related to effectiveness, stability and regulatory issues. In all cases, formulations that claim to be active against UV-induced damage must provide a uniform sunscreen film on the skin or hair to be effective. Ideally, this film will be thick yet invisible. For products designed to be used while swimming, the film must be waterproof. Also, care must be taken to minimize the ingredient interaction between the sunscreen active and the formulation. In some cases, the improper mix of ingredients can result in a significant reduction in the effectiveness of the sunscreen active. A variety of formulation vehicles that meet these requirements are available. The most common include oils, emulsions, gels, sticks, aerosols and ointments.

Oils: The simplest way to apply a sunscreen to the skin is by incorporating it into an oil-based product. In fact, oils were one of the earliest forms of sunscreen products. Evidence suggests that the ancient Greeks used mixtures of sand and oil to prevent sunburn as early as 400 B.C. The sand could effectively scatter the UV light while the oil allowed it to be distributed evenly on the body. Today, oils are still used as formulation vehicles for sunscreen actives. This type of formulation has various advantages and disadvantages. Because most sunscreen actives are lipophilic, oils can be manufactured easily. In fact, they are typically produced in a single phase. Additionally, oils can be spread readily on the skin and do not rinse off. Oils also have excellent product-stability profiles.

Unfortunately, oil-based products have some significant drawbacks. For instance, their rapid spreadability results in a thin, transparent film. This film has a lowered amount of active sunscreen, thus reducing the product's overall sunblocking effectiveness. In general, oils have higher costs associated with them because no water is present and more active sunscreen ingredients are needed. Aesthetically, oils can feel greasy and leave an unappealing coated feel. There are also stability issues related to putting an oil-based sunscreen product in certain types of plastic packaging. Products that use this type of formulation are tanning oils. These products, designed to be worn while sunbathing, typically provide minimal UV protection and are used merely to prevent serious burning.

Emulsions: Emulsions are the most commonly used delivery vehicle for sunscreen actives. Made up of a mixture of oils and water and stabilized by an emulsifying ingredient, emulsions are used to produce both cream and lotion products. While emulsion technology has been known for centuries it was not until the 1920s, with the introduction of the first organic sunscreens, that they could be used as sunscreen delivery vehicles. They are one of the best delivery forms because they are the most versatile. Unlike oils, they can be used to create a product providing an excellent sunscreen film on the skin and still have an elegant feel. Additionally, emulsions are relatively inexpensive, especially compared to oil formulations.

One common drawback to the use of emulsions is their difficulty in stabilization. Over time, any emulsion product will destabilize and separate into its component oil and water phases. However, with careful formulation efforts and the right emulsifiers, an adequately stable system can be made. From a manufacturing standpoint, emulsions are more difficult to produce than oil-based products. Also, emulsions are limited in the amount of protection they can provide. Because UV active ingredients are typically lipophilic, increasing the amount of the oil phase is the only way to produce products that protect better and last longer. However, the more oil in a system, the more difficult it is to stabilize. Products that use emulsions include sunscreen creams and lotions. These products are designed to be spread on the skin, leaving a film that is effective against sunburns. If formulated appropriately, they can be made to last many hours and be waterproof.

Gels: Gels provide an alternative to typical emulsion-based sunscreen products. They are a thick, clear product, with an elegant look that appeals to consumers. There are various types of formulations that can produce this effect, including aqueous and hydroalcoholic gels and microemulsions. Aqueous and hydroalcoholic gels are produced by incorporating acrylic polymers in the system for thickening. For aqueous gels, water-soluble sunscreens are typically required to keep the formula clear. More lipophilic actives can be used with hydroalcoholic gels because alcohol is included to keep the system clear. While microemulsions are more technical emulsions, they have significantly smaller particles making them transparent and they appear more gel-like than emulsion-like.

The primary advantage gels have over emulsions is their clarity and unique look. However, significant drawbacks limit their use. For example, they are typically limited to non-water-resistant applications because they produce water-soluble films. Additionally, these films are not uniform, so they may not provide the expected protection. The thickeners used in certain gel products can leave a sticky feel on the skin. In general, microemulsion gels tend to be more irritating on the skin than standard emulsions. Also, hydroalcoholic gels can negatively interact with certain types of plastic packaging. In general, gel products cost more than emulsions, and they can be more difficult to manufacture. Despite these drawbacks, gels do have their place as sunscreen delivery vehicles–perhaps the most promising in hair care. Many manufacturers have recently introduced styling gels with sunscreens incorporated in them. It is hoped that these products will reduce hair damage induced by UV exposure.

Sticks: Sticks are a solid form that has only recently been introduced as a vehicle for sunscreens. They are typically made up of waxes and oils thickened with petrolatum. They work well for sunscreens because they are stable and compatible with the ingredients. They are also more water-resistant than any other delivery vehicle. Unfortunately, they often have a greasy feel and are generally more expensive and difficult to produce. Sticks are used for lip or facial products. For these applications they are useful because they can be applied easily to a small, specific area.

Aerosols: Aerosols represent another type of formulation vehicle for UV active ingredients. Recently, aerosol mousses have gained popularity. Mousses are essentially emulsion lotions that are put in sealed containers and have propellant added to them. When the actuator is depressed, the product is released as a foam. Because UV active materials can be readily added to the oil phase, mousses make a convenient form for sunscreen products. They are easy to use and have good aesthetic characteristics. Because they are pressurized, they are not necessarily suitable for products that will be left out in the sun. Other types of aerosols can be used for sunscreen delivery, however they have significant drawbacks. One problem is that they are often oil-based, so they are expensive and have lower SPF ratings. Also, they produce a discontinuous film on the skin, reducing effectiveness. However, they are a novel and convenient product form that may find greater acceptance in the future.

Ointments: The last significant category of delivery vehicles are ointments. These are oily products based on thickened mineral oil formulas. Their primary advantage over emulsion-based products is that they are more water-resistant and may be able to provide more protection. Aesthetically, they are generally not desirable.

Testing Sun-Care Products

One of the primary measures for products designed to protect against the damaging effects of UV light is the SPF. Simply put, a product's SPF is equal to the ratio of the time it takes to burn if wearing the product versus the time it takes to burn without any product. For example, suppose a product has an SPF rating of 8. This means product users could stay in the sun about eight times longer and get the same amount of sunburning than they would have if they were not wearing any protection. The SPF value can be affected by a variety of factors including the amount and type of sunscreen active, the type of delivery vehicle and the other ingredients in the formulation. By increasing the amount of sunscreen active, sunscreen marketers have found that they can increase the SPF rating. According to the recently published FDA monograph on sunscreen drug products, a maximum numerical SPF rating of 30 has been established. However, because the data are not clear whether higher SPF values are possible, the FDA now allows the use of the claim 30 plus for certain products.

Obtaining a Rating

To obtain any SPF rating, specific types of tests have been developed. These include both biological and chemical evaluations. A variety of methods are possible for determining the SPF of a formulation. One biological evaluation involves the use of volunteers. In this test, the lower back of a non-sunburned volunteer is subjected to UV light until a minimum amount of erythema or redness develops. Light-proof barriers are placed around a measured section of the skin so clear margins are visible. The amount of energy required to produce this effect is determined and provides a baseline for the SPF calculation. The next day, the sunscreen formula is evaluated. It is first put on the volunteer's skin and allowed to dry. The skin is then irradiated with UV light at the estimated SPF value. A preliminary value for this number is obtained by using a spectrophotometric absorption estimation. The amount of UV light required is recorded and the SPF is determined. In addition to SPF, other characteristics of the formulation can be tested. For example, the water resistance of sunscreens can be determined.

To determine the water-resistance capability of a sunscreen, in vitro testing can be done. One method involves treating a hairless mouse with the product and immersing it in a pool of water. The amount of sunscreen that is leached off the animal is then measured. Another more appropriate method using in vivo human testing has been proposed by the FDA. In this method, sunscreen is applied to an area of the body measuring at least 50 square centimeters. The static SPF is determined by various methods. The test product is reapplied and allowed to dry for 20 min. The subjects are then allowed to be active in an indoor pool for a period of 20 min, dry and rest for 20 min and then get back into pool for another 20 min. For water-resistance claims, a cumulative period of 40 min in the water is required. For waterproof claims, a cumulative time of 80 min in the water is required. The SPF is determined again and if it maintains the value found initially, the claim is allowed. Because sunscreens are an OTC drug in the US, stability testing is required to establish an accurate expiration date. According to the regulations, if a product is stable and maintains its SPF rating after three months of stability testing, a one-year expiration date is required. If an entire year of testing is done, a three-year expiration date is allowed. If the product maintains its SPF rating after three full years of testing, no expiration date is required.

Future of Sunscreens

The FDA has issued a final OTC monograph that determines which materials can be legally used in sunscreens in the US. Even with a final monograph, however, research into UV absorbers must continue. As the ozone layer continues to decline and more damaging radiation reaches the Earth's surface, more effective sunscreen molecules will be required to protect the public. The challenge will be for regulatory concerns to keep pace with scientific discoveries to ensure the public has safe and effective products to choose from.

—**Randy Schueller and Perry Romanowski,** *Alberto Culver, Melrose Park, Illinois USA*

Additional Reading

1. J Lowe, *Sunscreens Development, Evaluation, and Regulatory Aspects*, N.&N Shaath (eds),

Marcel Dekker, New York (1990)
2. W Stevenson, Chemistry Saves Sun Worshipers, *Today's Chemist at Work,* 10 47-52 (1998)
3. P&G, Gourley, *Protect Your Life in the Sun*, Highlight Publishing Albuquerque (1993)
4. Formulation of Sun Protection Emulsions with Enhanced SPF Response. *Cosmet Toil* 6 55-64 (1997)
5. Code Federal Regulations Parts 310, 352, 700, and 740. Sunscreen Drug Products for Over the Counter Human Use; Final Monograph (May 21, 1999)

The South Beach Sunscreen Survey 2001

Keywords: sunscreen, UVA, UVB, SPF, skin type, sunburn resistance

A scientific survey of sunbathers on Miami's South Beach in July 2001 provides new information needed to answer the questions, "How much sunscreen SPF protection do I need?" and "What are consumers really using?"

This study evaluates the level of risk and protection existing under a "worst case" situation in which the UV exposure approaches 30 Minimal Erythemal Dose (MED).

How Much Sun Protection Do I Need?

It is remarkable that we could find no references to published research designed to determine the maximum level of sun protection justified under worst-case conditions within the territorial United States. The best references we found were photometric UV studies conducted at various US locations, and brand surveys derived from samplings of warehouse invoices from major chain stores.

Ours is the first study we know of that combines observation of exposed subjects with UVB dosimetry in a severe location. Though this may appear as a significant scientific void, it is not unusual; in many fields of science such voids exist with regularity (see sidebar).

Since the development of Robinson-Berger UV meters, many thorough studies of solar irradiance have been conducted[4] in various locations across the United States. These studies have clearly identified sites of extreme irradiance, such as Florida, Arizona and Hawaii. However, in the two studies reported here, eight years apart, we accurately measured 50% higher levels of UVB irradiance than were considered possible when the regulations controlling sunscreens were proposed.

Collecting UV irradiance data at severe locations can provide information to support rational market (and regulatory) decisions regarding the delivery of UV protection. Irradiance data becomes meaningful when it is combined with human exposure data, including the sensitivity (skin types) of exposed individuals, the time spent in exposure by actual subjects, and the level of protection that they are using. Only then can we answer the burning question: "How much protection do I need?"

Filling in the Gaps

Research gaps exist in many scientific specialties. For example, physicists only recently claim to understand the mechanics of the most common physical force we encounter: gravity.[1] But, as yet, there have been no reliable or successful attempts at the modification of gravitational force.

Until recently, no trained biologist had ever observed a living specimen of one of Earth's largest creatures. Only sailors and fishermen had observed the giant squid in a live state until about five years ago.

We chemists will find it difficult to locate a basic textbook that explains mechanically why some things mix[2] and others don't. Recent guidance in this area, however, may be found through the tables of solubility parameters (molecular stickiness) listed in the *Cosmetic Bench Reference*.[3]

This is indeed the paramount question in the sun-care industry. This is the first question any consumer wants answered. It is the primary question every sun-care marketer needs to address, and for the regulators it is a question that frames the limitations to be posed upon the market.

The goal of the survey described here was to measure and record both the exposure and probable injury to sunbathers caused by solar UV, in a worst-case situation. In 1993 we (Raketty and Vaughan) conceived that we could determine the exact UV exposure of the sunbathers and estimate their level of sunburn injury relatively precisely if we did the following:

1. Measure the incident UV (in MED values)
2. Simultaneously record the levels of UV protection (SPF) being used by an exposed population
3. Rate each sunbather's skin type
4. Measure each sunbather's time of exposure.

Then we could calculate the sunburn (UVB) injury according to the following formula:

$$\text{MEDs Absorbed} = \frac{\text{Exposure MEDs x Adjusting Factor for skin type}}{\text{SPF protection worn by the sunbather}}$$

The Adjusting Factor compensates for the fact that MED varies with skin type. According to Pathak's study on irradiation of unexposed buttock skin,[5] it takes approximately 30% more time (or radiation) to burn Type III (medium) skin than Type II (light) skin, and it takes an additional 30% more time (or radiation) to burn Type III skin than Type I (very light) skin. Thus, a sunbather exposed to 15 MEDs with light skin (30% more sensitive) wearing SPF 22 will have received slightly less than enough UVB to produce a minimal (just visibly pink) sunburn, as follows:

$$\text{MEDs Absorbed} = \frac{15 \times 1.3}{22} = 0.89 \text{ MEDs}$$

The practical response to the question "How much sun protection do I need?" has initially been limited by the amount of sun protection available (see sidebar), as well as the consumer's perception of its effectiveness.

Surveying Sunbathers

The 1993 survey: The purpose of the first South Beach Sunscreen Survey (completed in1993) was to assist the regulators in determining the maximum level of UV protection to be permitted on a product label. The survey results were submitted to the US Food and Drug Administration (FDA)[25] during the one-year comment period following the publication of the Tentative Final Sunscreen Drug Monograph (TFM).[26]

After reviewing the initial 62-person survey, the FDA responded that the sampling was too small to extrapolate results onto the population of the United States. The FDA assessment is reasonably supported by statistical evaluation of the sample size. However, because no other scientific evidence was available, we are told this study effectively supported the FDA's eventual decision to consider higher SPF labels because it, for the first time, defined the need for higher levels of UV protection than permitted by the currently delayed Final Sunscreen Monograph.

The 2001 survey: Our follow-up 2001 South Beach Sunscreen survey was almost four times larger than the first, consisting of 208 subjects, and it introduced a new evaluation technique designed to produce predictions of UV damage with a statistically significant level of confidence.

The results of the 2001 survey support the conclusions of the original survey, with respect to the amount of solar UV radiation to which sunbathers are subjected, and the amount of protection that they must use under worst-case conditions, to avoid more than one MED of sunburn.

The 2001 survey also measured UVA exposure; for the first time we have quantitative, field-generated scientific evidence that may support the need for – and the level of – UVA protection in extended-wear (or mid-day) sun-care products designed to protect against UV exposure in an extreme environment. Although we surveyed sunbathers, we clearly expect the results to be transferable to those who must be exposed daily to solar radiation, as a result of their occupations or avocations.

Survey Methods

The first South Beach Sunscreen survey was conducted on July 3, 1993, and submitted to the FDA approximately six months later. The FDA's response was included in the Final Sunscreen Drug Monograph published in 1999. Our follow-up survey was conducted July 30, 2001. Both surveys were performed on bright sunny days.

The weather in Miami, Florida, is not typically bright and sunny in June or July. In fact, the National Oceanic and Atmospheric Administration's 30-year weather records show that Miami has the distinction of leading the nation in precipitation, peaking in June,[27] with an average (mean) of 9.3 inches that month. The 2001 survey was delayed more than 20 days by rain, and slightly lower total irradiance was recorded as a result.

The Development of Sun Protection

Anyone asked for a practical response to the question "How much sun protection do I need?" must begin with a precise sun protection scale to express greater or lesser values. In 1975, the concept of SPF (Sun Protection Factor) was introduced by Sayre et al. at Coppertone.[6] Shortly thereafter, researchers in the irradiance area[7] concluded that an SPF of approximately 15 provided maximum protection, although at the time SPFs of up to 22 were feasible. The authors suspect that this assessment was based on the limited technical-level competence at that time, and also on a general belief that UVB doses above 15 MEDs per day were highly unusual. The first sunscreen monograph (proposed rule) of 1978 cites an example of 19 MEDs recorded atop Mauna Loa volcano in Hawaii. Currently the American Cancer Society recommends[8] a protection level of SPF 15 as a minimum.

Figure 1 shows the progression of increasing levels of SPF offered by the major marketers with respect to time. The maximum available formula potency in 1975 was SPF 15.

Because formula levels of sunscreening ingredients are limited by regulation, subsequent improved product benefits came as a result of the discovery of various synergisms.[9-12] Interactions between various ingredient components, which act to increase the UV blocking effect of the active ingredients, have resulted in continual product claim improvements.[13-20] The scientific progression of sunscreen synergism and efficiency will not be addressed in this article; it is covered by other articles and conferences.[21] However it is important to point out that researchers in the 1970s knew that adding common UV absorbers, such as padimate O, octyl salicylate, homosalate, or octyl methoxycinnamate, to mineral oil would provide a maximum SPF of 8 no matter how much UV absorber was added.[22] The discovery of the synergism of 3% oxybenzone and 7% padimate to provide SPF 15 protection was a remarkable advance[23,24] at the time. Recently one major brand has introduced SPF 70 protection.

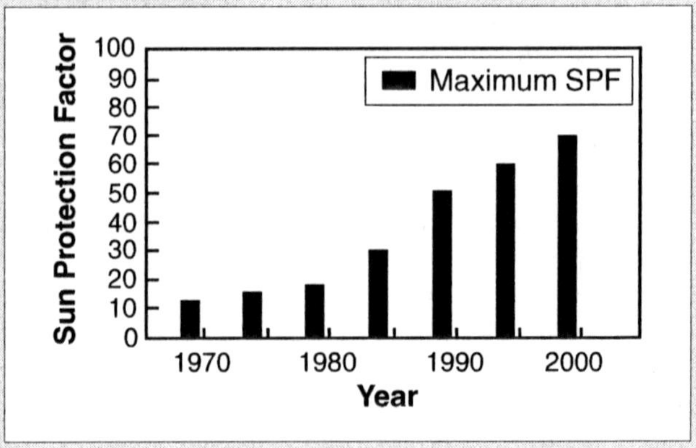

Figure 1. Maximum sun protection available from major brands

Survey forms: The seven survey questions were designed for simplicity and directness. Each subject's brand preference and skin type awareness were the only expected fallout data, beyond the intended UV damage data.

The 2001 survey form (Figure 2) was changed to include room for more than one sunscreen product, and a question on UVA awareness was added.

Subjects: The subjects of this survey were primarily vacationers (approximately 80%). They were of all six skin types, although we only included the five lightest skin types based on normal accepted testing protocols. The nationalities encountered on Miami Beach were diverse, with between 30% and 40% being international tourists. We were surprised that the majority of the foreigners were using American products, despite differences in European and domestic SPF scales. The profile of subjects for both surveys was very similar (Table 1).

Skin types: Our interviewers were trained in use of the Harvard six-level skin typing methodology (Table 2),[28] and they were instructed to use it on each subject before asking the subjects to assess their own skin type (Table 1). Thus, two assessments of skin type were made in each case. We used the ratings by the trained evaluators (experts) for the sunburn projection data.

We were surprised that 29% of the 1993 survey population could not accurately identify their personal skin type. That proportion shrank to 5.3% in the 2001 survey. Of course, in light of the changes in self-protective behavior we observed, it would be reasonable to suppose the same population has an improved awareness of many of the factors involved in sun protection, especially where failure to learn could be associated with levels of personal discomfort proportional to the size of one's error.

Table 1. Profile of subjects

		1993		2001	
		N	%	N	%
Gender	Male	31	50.0	94	45.4
	Female	31	50.0	114	54.6
	Total	62	100.0	208	100.0
Average age	Male	32.2		33.0	
	Female	28.5		28.9	
	Mean	30.3		30.8	
Skin types	(I) Very light	4	6.5	10	4.8
	(II) Light	19	30.7	48	23.1
	(III) Medium	26	41.9	111	53.4
	(IV) Dark	12	19.3	34	16.3
	(V) Very dark	1	1.6	5	2.4
	Total	62	100.0	208	100.0

Table 2. A Harvard-based skin typing chart used in the survey[36]

Hair	Complexion	Freckles	Sun reaction	Tanning	Phototype
red	very fair	+++	always burns	never tans	I
blond	fair	++	often burns	tans lightly	II
blond or light brown	fair to medium	+ to 0	sometimes tans	tans progressively	III
brown	olive	0	rarely burns	tans easily	IV
brown to black	dark	0	rarely burns	tans deeply	V
black	very dark	0	never burns	tans deeply	VI

Awareness of skin type and sensitivity is indeed a major tool in personal control of UV-induced skin damage from over exposure. Numerous health and welfare organizations are using charts like the one in Table 2 (used courtesy of L'Oréal) to teach skin-type awareness. Clinical test data[5] predicts that a difference of approximately 30% in UV sensitivity separates each category.

Worst case UV exposure on South Beach: Three freshly calibrated Robinson Berger Type UV dosimeters[a] were used to monitor UVA and UVB during the 2001 survey, while only one UVB dosimeter was available for the initial survey in 1993. Nevertheless, similar levels of UVB exposure were recorded during both surveys: irradiation approaching 30 MEDs. 30 MEDs comprise enough radiation to give a Type III (medium) UV sensitivity subject 30 sunburn doses. Four doses will induce a painful bright red burn, while eight will result in blistering. We recognize that these conditions present a significant health risk.

Table 3 shows hour-by-hour irradiation recorded at the same place on two days eight years apart. We initially recorded the UV irradiation in MED units, but Table 3 also presents MEDs adjusted for more sensitive Type I and Type II skin.

Our "worst-case" scenario is based on Type I skin subjects who are exposed to the extreme UV irradiation, near the longest day (and highest sun angle) of the year and at the southern-most major beach. Based on Pathak's irradiation study,[5] the erythemal effect on Type II and Type I skin is an approximate 30% increase in sensitivity between each skin type. Therefore, 28.1 MEDs for Type III skin, recorded in 1993 converts to 47.5 MEDs for very light (Type I) skin, while the 2001 survey exposed Type I subjects to 49.5 MEDs.

This is the portion of the study that scientifically addresses a real consumer need for protection up to SPF 50 for persons with light skin exposed for 6 to 8 hours on a sunny June or July day at our nation's southern-most major beach, in Miami. This, we believe, constitutes the worst-case scenario for sunscreens. Under these exposure conditions, even our suggested remedy may not prevent sunburn! An SPF 50 only serves to limit the UVB injury – for those wearing a heavy (2 mg/cm^2) layer of it – to approximately one sunburning UVB dose.

[a] Solar Light Co., Philadelphia, Pennsylvania USA

SPF CONSULTING LABS, INC.
1425 SW 1st Court
Pompano Beach, FL 33069

SOUTH BEACH SUNSCREEN SURVEY II

Date: July 30th, 2001

Subject: *MALE*_____ *FEMALE*____ *AGE:*_________ *est?*

SKIN RATING by INTERVIEWER:

VERY LIGHT   LIGHT   MEDIUM   DARK   DEEP DARK

QUESTIONS:

1. **WHAT SKIN TYPE DO YOU HAVE ?**

   VERY LIGHT   LIGHT   MEDIUM   DARK   DEEP DARK

2. **ARE YOU USING SUNSCREEN?**   YES   NO

3. **HOW LONG DO YOU EXPECT TO BE IN THE SUN?**

   Hours: 1   2   3   4   5   6   7   8

4. **WHAT SPF IS IT ?**   ____________   ____________

5. **WHAT BRAND IS IT ?**   Fill in below

   1. BANANABOAT ____________
   2. COPPERTONE ____________
   3. NEUTROGENA ____________
   4. STORE BRAND ____________
   5. HAWAIIAN TROP. ____________
   6. COPPERTONE SPORT ____________
   7. NO AD ____________
   8. BAIN DE SOLEIL ____________
   9. OTHER ____________

6. **ARE YOU SATISFIED WITH IT?**   YES   NO

7. **DOES IT PROVIDE UVA PROTECTION ?**   YES   NO   Don't Know

Interviewer: C. Vaughan   S. Porter   Jim Gilbert   Linda Dyer   ____________

Figure 2. The survey form and questions

According to our count, 23.6% of the sunbathers in this survey used SPF 45 or higher. Our calculations, accounting for skin type, exposure time, and SPF, project that none of these sunbathers went home sunburned. This observation, under extreme conditions, serves to substantiate the erythemal protection value of products in the SPF 30 to 50 category.

Results

The most alarming finding of the first (1993) survey was that more than 40% of the sunbathers surveyed went home with more than 4 MED, which *usually* results in a painful and damaging sunburn. By 2001 this proportion had declined to 21.2%, and the vast majority of the 21.2% projected to be sunburned were among the unprotected group, who used no product, or who applied SPF zero (actually SPF 1, or less) tanning products.

The recent survey (2001) disclosed that Miami beach-goers who used sunscreen had addressed the sunburn problem with remarkably higher SPF protection and significantly reduced exposure time, such that only 3.0% of the sunscreen users received a painful sunburn.

Table 3. UVB radiation recorded in June on South Beach, Miami

Time	Hours elapsed	Irradiation rate (MED/h)	Cumulative dose (MED)	Cumulative dose adjusted for light skin (MED)	Cumulative dose adjusted for very light skin (MED)	Irradiation (%)
7/3/93						
09:00 am	0	0.88	0.00	0.00	0.00	0.00
10:00 am	1	2.20	3.80	4.94	6.42	13.5
11:00 am	2	4.32	9.59	12.47	16.21	34.1
12:00 pm	3	4.11	14.39	18.71	24.32	51.1
01:00 pm	4	4.71	18.50	24.05	31.27	65.7
02:00 pm	5	3.77	22.32	29.02	37.72	79.3
03:00 pm	6	2.74	25.06	32.58	42.35	89.1
04:00 pm	7	1.71	26.77	34.80	45.24	95.1
05:00 pm	8	1.37	28.14	36.58	47.56	100.0
7/30/01						
09:00 am	0	0.63	0.00	0.00	0.00	0.00
10:00 am	1	1.45	0.89	1.16	1.50	3.0
11:00 am	2	4.80	3.44	4.47	5.81	11.7
12:00 pm	3	6.39	8.40	10.92	14.20	28.7
01:00 pm	4	7.12	15.60	20.28	26.36	53.2
02:00 pm	5	5.52	20.53	26.69	34.70	70.1
03:00 pm	6	4.30	24.76	32.19	41.84	84.5
04:00 pm	7	3.70	27.95	36.34	47.24	95.4
05:00 pm	8	1.74	29.30	38.09	49.52	100.0

Table 4. Unprotected sunbathing

	1993	2001
Subjects not using sun product – bare skin	12.9%	14.9%
Subjects using non-protecting product (SPF 0)	6.4%	5.8%
Percent unprotected subjects	19.3%	20.7%
Count of unprotected subjects	12	43
Tanning oils (SPF 0) used by women	3	9
Tanning oils (SPF 0) used by men	1	3
Unprotected sunbathers – Type I/II skin	25%	30.2%
Unprotected sunbathers – Type III/IV skin	75%	69.8%
Unprotected sunbathers – male	33%	49%
Unprotected sunbathers – female	67%	51%

Table 5. Protected sunbathing

Skin type	Population portions		Exposure		SPFs		Protection (Avg SPF/h)	
	1993	2001	1993	2001	1993	2001	1993	2001
Type I	6.5%	4.8%	5.50 h	2.60 h	15.7	27.8	2.9	10.7
Type II	30.6%	23.1%	4.74	3.36	8.5	20.8	1.8	6.2
Type III	41.9%	53.4%	4.04	3.34	8.1	21.9	2.0	6.6
Type IV	19.4%	16.4%	4.83	2.97	9.4	16.5	1.9	5.5
Type V	1.6%	2.4%	3.00	2.20	0.0	10.2	0.0	4.6
Mean			4.48 h	3.22 h	9.8	24.1	2.2	6.0

Table 6. Multiple product usage

	1993	2001
Users of multiple products	22.6%	13.6%
Percent of women who use two products	38.7%	18.4%
Percent of men who use two products	6.4%	7.5%

Unprotected sunbathing: The 2001 follow-up survey revealed no significant change in the 20% of beach-goers who chose unprotected sunbathing. This group is divided between the "used nothing" subjects and the subjects who used SPF zero or tanning oil. Together, this group was responsible for 90.7% of the severe sunburns over 4 MEDs projected for the study group.

Table 4 shows that the greater number of subjects in the 2001 survey did not alter the 4:1 ratio of sunscreen users vs. unprotected subjects, nor did it affect the ratio of tanning oils usage by women over men (3:1). The unprotected sunbathers were predominantly among the more pigmented subjects. In data not shown in Table 4, the portion of unprotected subjects who didn't use any sun product at all slightly favored males in both surveys (58% and 62%).

Protected sunbathing: The mean SPF of all products surveyed in 2001 was 24.1, vs. 9.8 in 1993. That is a 246% change. Because some subjects used two products (which were averaged), the averaged mean SPFs were 20.1 and 7.3, which is a 275% change. Such a substantial change in the SPF preference levels disclosed by the survey subjects may result in reducing the level of UV injury experienced by the sunbathers in Miami. This is a very encouraging discovery.

We further evaluated the data with respect to sunburn sensitivity as indicated by skin type. The results are shown in Table 5. The mean increase in SPFs and decrease in exposure time combine to document a 272% increase in sunburn protection being used by sunbathers in the field (i.e., on the beach) under severe conditions.

Multiple product usage: During the 1993 survey we discovered that an unexpectedly large number (38.7%) of the female subjects were using two different sunscreen products: one for the face, and one for the body. Generally the multiple product users chose a higher SPF on the face and a lower SPF on the body. Our 2001 sample population has increased their SPF preference and reduced their use of multiple products, as shown in Table 6. It appears that products once used only for the face may now be applied all over.

The widespread usage of two different sun products in 1993 was a surprise to us, and it caused us to modify our survey form before the 2001 study to allow room for multiple responses to the brand and SPF questions.

Brand data: The brand usage data showed a larger than expected component of small brands and store brands. This component varied markedly with the market shares generated by the major surveying organizations, however the order of brand predominance did generally correspond with the commercial surveys.

The top brands in 1993 were Coppertone, BananaBoat, Nothing, Panama Jack, Hawaiian Tropic, Vaseline.

In 2001 the top brands were BananaBoat, Coppertone, Nothing, Hawaiian Tropic, Australian Gold, No Ad.

Exposure times and protection: The mean exposure time spent by survey subjects on the beach was 4.5 hours in 1993. In 2001 this exposure had declined to 3.2 hours, while the mean SPF of products used by the subjects on Miami's South Beach rose from 7.3 to 20.1 during the 8-year interval.

The protection and exposure times were combined with the incident UV measured during the survey. With the exception of the all day (6+ hour) sunbathers, we were unable to record the exact time position, during the day, of the exposure span of each individual subject. Therefore the exposure times may be matched with both the maximum and minimum possible UV dose to provide exposure values that may be treated statistically. For example, 1993 subjects who were exposed for 5 hours

Table 7. Cumulative exposure to UVB over selected exposure periods in June on South Beach, Miami

Time	Hours elapsed	Dose at 2 h (MED)	Dose at 3 h (MED)	Dose at 4 h (MED)	Dose at 5 h (MED)	Dose at 6 h (MED)	Dose at 7 h (MED)	Dose at 8 h (MED)
7/3/93								
09:00 am	0							
10:00 am	1	9.6	14.4					
11:00 am	2	10.6	14.7	18.5	22.3			
12:00 pm	3	8.9	12.7	18.5	21.3	25.1	26.8	
01:00 pm	4	7.9	10.7	15.5	17.2	23.0	24.3	28.1
02:00 pm	5	6.6	8.3	12.4	13.8	18.6		
03:00 pm	6	4.5	5.8	9.6				
04:00 pm	7	3.1						
05:00 pm	8							
Max		10.6	14.7	18.5	22.3	25.1	26.8	28.1
Mean		7.3	11.1	14.9	18.6	22.2	25.6	28.1
Min		3.1	5.8	9.6	13.8	18.6	24.3	28.1
Variance		7.5	8.9	8.9	8.5	6.5	2.5	0.0
% Variance vs mean		103%	80%	60%	46%	29%	10%	0%
Number of exposures		5	10	18	11	11	0	6
7/30/01								
09:00 am	0							
10:00 am	1	3.4	8.4					
11:00 am	2	7.5	14.7	15.6	20.5			
12:00 pm	3	12.2	17.1	19.6	23.9	24.8	28.0	
01:00 pm	4	12.1	16.4	21.3	24.5	27.1	28.4	29.3
02:00 pm	5	9.2	12.4	19.6	20.9	25.9		
03:00 pm	6	7.4	8.8	13.7				
04:00 pm	7	4.5						
05:00 pm	8							
Max		12.2	17.1	21.3	24.5	27.1	28.4	29.3
Mean		8.1	12.9	18.0	22.5	25.9	28.2	29.3
Min		3.4	8.4	13.7	20.5	24.8	28.0	29.3
Variance		8.8	8.7	7.6	4.0	2.3	0.4	0.0
% Variance vs mean		109%	67%	42%	18%	9%	1%	0%
Number of exposures		58	39	50	12	7	2	8

were clearly exposed to a UV dosage that could not be lower than 13.8 MEDs, because that was the smallest UVB dose recorded in any 5-hour span during the survey. Likewise they could not have been exposed to greater than 22.3 MEDs. In the 2001 survey, 5 hours exposure provided between 20.5 and 24.5 MEDs, with 100% probability.

Table 7 shows the range of UVB exposures possible during all time spans during the day, as well as the most likely (mean) exposure for a given time span. Table 7 also shows the variance (max dose minus min dose) for each time span as well as the percentage of variance versus the mean. This percentage defines the precision of the sunburn estimate for each individual subject. As can be seen from Table 7, which shows the centralized intensity of the UV (from 11 a.m. to 2 p.m.), the range of possible MEDs declines with increased exposure.

The increased certainty of the higher exposure values provides a high level of confidence for all overexposure conclusions based on this survey. This is because data variances decrease as exposure periods get larger. Thus, the data from the high-exposure subjects, who were the focus of all questions investigating overexposure in the 2001 survey, showed mean deviations of less than 18%, which is low enough to yield projections of high precision.

Discussion

Other contributing factors: Many other arguments still remain, affecting the determination of what is an appropriate level of protection. Our study only provides a scientific benchmark from which those arguments may begin.

For example, Vorhees[29] has determined that as little as 0.25 MED will trigger the release of matrix metalloproteinases, such as collagenases and elastases, which may result in wrinkling and photoaging. Elias[30] has reported that a similar level of exposure depletes the skin's retinol (vitamin A), which is not normally replenished for four to five days.

Other researchers have claimed that very few sunscreen users apply the amount of sunscreen needed (2 mg/cm^2) to provide the label SPF.[31] Indeed, Yankell[32] long ago reported that UV absorbers may be absorbed through the skin, resulting in a loss of protection during wear. Lorenzetti[33] in 1975 reported that formula components such as proteins can greatly affect penetration of sunscreens. Recently, Bronough and Yourick[34] identified hydrolysis by esterase enzymes found in the skin as a pathway for UV absorber loss. Conversely, Stockdale[35] has suggested that resinous materials such as PVP can trap sunscreens in a matrix on the skin surface to prevent penetration.

For this study, and in our calculations, we have decided to use the most common, but possibly not the best, benchmark of UVB damage. That benchmark is 1 MED, the UV exposure that gives the first slight redness (erythema) to exposed Type III human skin.

Sunburn resistance: It was not our goal to study UVB exposure in unprotected subjects. However, we found this cohort to be involved in most of the major intended conclusions, either as sources of anomalies, or because they introduced

Table 8. Mean UV exposures and erythemal damage projected (2001)

Hours Exposed	Average number of predicted burns
1	0.8
2	2.6
3	3.6
4	2.4
5	3.1
6*	25.2*
7	0.6
8	6.1
Mean	3.53

* 3/4 of this group were unprotected

data deviations that had to be investigated because they were so skewed. Because each investigation seemed to end up on the backs of the great unprotected, that group may be a very fruitful field to investigate. They are vastly overexposed to UV, they are the sales holdouts, and they are possibly the future cancer cohort. They do not respond to UV the way we vacationers do. (Approximately 80% of our subjects were vacationers.) *They are sunburn resistant.*

After we generated projected burn data (Table 8), we tried to find a relationship between hours exposed and the primary expected overexposure effect, erythema. The six-hour exposure group provided a severe anomaly. On further investigation we saw that this group consisted of mostly unprotected sunbathers who should have been placed in the hospital, according to our projections. In fact, most of these sunbathers (the hard-core tanners) were still "soaking up those rays" when we passed them again in the afternoon on the way to our cars.

For the 6-hour sun-exposed group we predicted an average of 25 sunburn doses per subject, with some receiving up to 39 MEDs. The unprotected group averaged 13.2 MEDs adjusted for skin type, split between the oil-users (Avg 21.2 MEDs) and the bare-skinners (Avg 10.1 MEDs). We estimate that this group most likely exhibits a natural protective response equivalent to an SPF of 4 to 8, because we did not observe the expected symptoms of overexposure on these sunbathers.

UVA exposure: The 2001 survey included a question designed to evaluate each subject's awareness of the UVA protection provided by the products the subject used. The sample population responses to this question were rather confused and highly inaccurate. Twelve percent admitted they didn't know about the UVA protection offered by their products, and an additional 26% claimed they were getting UVA protection, when it was not in the product! Only 47.7% of the products surveyed contained the recognized strong UVA absorbers titanium dioxide, zinc oxide or avobenzone, while 73.6% of the subjects thought that they were getting UVA protection. Curiously, amidst this confusion, 10 of the 12 SPF zero users accurately knew that they were getting no UVA protection.

We believe this situation emphasizes the need for better consumer education, and it supports the current FDA work designed to provide a uniform and accepted scale of UVA protection. Because we have found that the buying public has adopted higher SPFs, it is reasonable to conclude that UVA protection will be likewise adopted once the consumer understands the value.

We hope that this survey data may serve as a benchmark with which to measure the understanding and adoption of uniform protection measures when they become enacted. Since the discovery of UVA's role in triggering melanoma, the FDA has determined that claims for high levels of SPF should be accompanied by required minimal UVA protection. The development of these requirements is the current impediment to implementation of the Final Sunscreen Monograph. However, the FDA's work is progressing rapidly.

In the future, data from this study may help evaluate consumer acceptance of UVA benefits, as it has for UVB sunburn protection. We have seen in this study that consumer acceptance of SPF benefits initially lagged behind the permitted regulatory product classifications. In 1993 SPF 30 became the maximum SPF permitted by the FDA. In the survey that year, less than 2% of the products used by our subjects were above SPF 30. Today that technically illegal portion has risen to 23%, but the SPF maximum has remained at 30. A regulatory revision now will be retrospective rather than proactive. So, we will be curious to know if the rate of regulated UVA change will initially preceed the adoption rate of new products by the market as occurred with UVB (SPF) regulation.

Conclusion

We have twice measured and recorded an upper limit near 30 MEDs of UVB exposure risk within the United States to qualify our data for a "worst-case" analysis. Recording incident UV data is not new, but we have, for the first time, tied that data to exposure risk of sunbathers and recorded their response to this risk.

We have also analyzed the consumers' adaptation to the risk they are facing. Our surveys show that most consumers have adopted improved sunburn protection, but are, at best, confused about UVA protection.

We found that there still exists a hard-core 20% of the beach population who shun commercial UV protection and develop a natural UV resistance of their own.

Moreover we have presented data sufficient to evaluate the delay between the availability of new sunscreen technology and its eventual adoption by consumers almost 10 years later.

Finally, we discovered it was nice to get out of the lab and go to the beach once in a while!

—**Christopher D. Vaughan, Susan M. Porter, James A. Gilbert and Mary L. Posten,**
SPF Consulting Labs, Inc., Pompano Beach, Florida USA

Postscript on Sunscreens

We've seen explosive growth in sun care innovations in the last 25 years. The total number of issued sun care patents in the US rose from approximately 200 in 1975 to nearly 1,800 in 2000. Our research shows that for every sunscreen patent that expires today, 20 new ones are issued.

Chris Vaughan, S.P.F. Consulting Labs

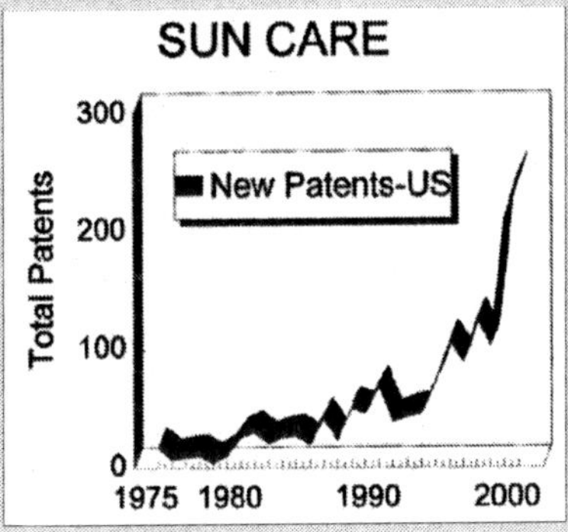

References

1. J Wambsganss, Gravity's kaleidoscope, *Scientific Am* 285(5) 64 (Nov 2001)
2. CD Vaughan, The solubility parameter: What is it? *Cosmet Toil* 106(11) 69-73 (1991)
3. *Cosmetic Bench Reference 2001*, Carol Stream, Illinois: Allured Publishing Corp (2001)
4. DHEW Report No. 76-1029, US Department of Health, Education and Welfare, National Institutes of Health (Nov 1975)
5. MA Pathak, Intrinsic photoprotection in human skin, Ch 5 in *Sunscreens*, NJ Lowe and NA Shaath, eds, New York: Marcel Dekker (1990)
6. Sunscreen drug products for OTC human drugs, Proposed Rule, *Fed Reg* 43(166) 38206-38269 (Aug 25, 1978)
7. R Goldemberg, *DCI* (Jun 1976) p 26
8. Am Cancer Soc, http//www.goaskalice. columbia.edu/0923.html
9. PS Lueng and ED Goddard, Proceedings of the 12th Congress of the IFSCC, Paris, France (1982)
10. US Pat 5340567, Sunscreen Compositions, Craig Vaughan (Aug 23, 1994)
11. US Pat 5658555, Photoprotective/cosmetic compositions comprising synergistic admixture, J-M Ascione (Aug 19, 1997)
12. US Pat 5914102, High SPF perspiration resistant sunscreen, K Fowler (Jun 22, 1999)
13. US Pat 6015548, High efficiency skin protection formulation with sunscreen agents and antioxidants, M Siddiqui (Jan 18, 2000)
14. US Pat 6030629, Photoprotective cosmetic/dermatological compositions comprising synergistic admixture of sunscreen compounds, I Hansenne (Feb 29, 2000)
15. US Pat 6048517, High SPF sunscreen formulations (over SPF 40), C Kaplan (Apr 11, 2000)
16. US Pat 6139855, Structured water in cosmetic compositions, G Cioca (Oct 31, 2000)
17. US Pat 6159456, Photoprotective/cosmetic compositions comprising camphorsulfonic acid and bisresorcinyltriazine sunscreens, D Candeau (Dec 12, 2000)
18. US Pat 6165451, Cosmetic screening composition, A Bringhen (Dec 26, 2000)
19. US Pat 6171579, Synergistically UV protecting triazine/silicone compositions, D Allard (Jan 9, 2001)
20. US Pat 6180090, Use of polysaccharide for improving light protection, H Gars-Berlag (Jan 30, 2001)
21. The 8th Florida Sunscreen Symposium, Coronado Resort, DisneyWorld, Orlando, Florida, April 24-28, 2002
22. CD Vaughan, The effect of inactive ingredients on UV absorption by sunscreen actives, 2nd Florida Sunscreen Symposium, Sanibel Island, Florida, Sept 1989

23. SI Katz, Relative effectiveness of selected sunscreens, *Arch Dermatol* 101 466-468 (1970)
24. RM Sayre, *Photochem Photobiol* 29 559 (1979)
25. Comment #C00282, *Fed Reg* 64(98) 27682 (May 21, 1999)
26. Sunscreen drug products for OTC human use, TFM, Proposed Rule, *Fed Reg* 58(90) 28194-28302 (May 12, 1993)
27. Monthly normal temperatures, precipitation, *World Almanac*, Mahwah: Funk & Wagnalls Corp (1996) p 180
28. TB Fitzpatrick, MA Pathak and JA Parrish, Protection of the human skin against the effects of the sunburn ultraviolet (290-320 nm), in *Sunlight and Man*, MA Pathak et al, eds, Tokyo: University of Tokyo Press (1974)
29. US Pat 5837224, Method of inhibiting photoaging of skin, JJ Vorhees (Nov 17, 1998)
30. P Elias and ML Williams, Retinoids, cancer and the skin, *Arch Dermatol* 117 160-180 (1981)
31. C Stenberg and O Larko, Sunscreen application and its importance for the Sun Protection Factor, *Arch Dermatol* 121 1400-1402 (1985)
32. SL Yankell, S Kemanic and MH Dolan, Sunscreen recovery studies in the Mexican hairless dog, *J Invest Dermatol* 55 31-33 (1970)
33. O Lorenzetti, J Broltralik, E Busby and B Fortenberry, The influence of protein vehicles on the penetrability of sunscreens, *J Soc Cosmet Chem* 26 593 (1975)
34. RL Bronough and JJ Yourick, Role of absorption in cosmetic development, *Cosmet Toil* 115(3) 47-52 (2000)
35. M Stockdale and D Roberts, Multifactorial influences on sun protection factors, Ch 27 in *Sunscreens*, NJ Lowe and NA Shaath, eds, New York: Marcel Dekker (1990)
36. http://www.loreal.com/sunprotectionorg/skin.asp

Sun Damage from Glycolic Acid Use

Keywords: AHA, protection, photodamage, glycolic acid, photosensitivity

Sun damage from products containing glycolic acid is a real concern to consumers.

Dermatologists have performed chemical peels for more than 200 years. However, dermatologists have generally discouraged daily or weekly peeling of the skin as a skin-care routine for rejuvenation. In spite of this, beauty therapists regularly give facial peels as salon treatments and consumers also use them as part of home care routines.

AHA Safety Issues

Two main reasons argue for discouraging frequent peels. First, barrier function might be compromised by the frequent removal of stratum corneum layers. Second, the esthetic effect of removing superficial stratum corneum is soon lost; the stratum corenum grows back within two weeks.

However, after two independent groups found the therapeutic usefulness of α-hydroxy acids (AHA), products containing AHA became popular in many dermatologic practices.[1,2] AHAs, especially glycolic acid, are widely used as moisturizers, keratolytics and cell activators for the treatment of xerosis, acne, photodamage and hyperpigmention.

Since 1989, the Food and Drug Administration has received more than 100 reports of adverse reactions in people using AHA products. Redness, swelling, burning and itching are among the most common complaints. The Cosmetic Ingredient Review (CIR) panel reviewed the submitted data and suggested that the use of glycolic acid and lactic acid as AHAs in consumer products should be limited to a pH greater than 3.5 and concentration lower than 10%.[3] Concern about AHA safety has also been raised in Europe.

Photosensitivity is another safety issue with AHA. Based on the increased formation of sunburn cells after once daily use of 10% AHA at pH 4.0, for 12 weeks, the CIR panel suggested the incorporation of sunscreen in, or concomitant use of sunscreen with, AHA products.[3] No change in sunburn cells was found after use of 10% AHA at pH 3.5 for four days in another study.

UVB was used in most of the studies. UVA was used to assess the phototoxic potential of 4% glycolic acid. Either erythema or sunburn-cell formation was used as the parameter indicating skin damage. However, UV light can exert other biologic influences on the skin, such as the tanning response. To evaluate the phototoxicity of glycolic acids in consumer products, glycolic acids with higher concentration should be used, to study what happens with the greater degree of exfoliation caused by the higher-concentration products.

Effect of Glycolic Acid on Pigmentation

Goals of the study: We performed a study to clarify two points:

- Whether a three-week use of glycolic acid hastens resolution of pre-existing light-induced pigmentation (caused by the initial irradiation) and
- Whether the skin tanned or burned more easily following AHA treatment.[4]

We also investigated whether racial differences in the tanning response exist.

Methods and materials: We used a 10% glycolic acid, buffered at pH 3.5 with sodium hydroxide, since this is the maximum glycolic acid concentration considered to be safe for use in cosmetic products. The placebo gel had a physiologic pH of 5.75. We formulated the placebo gel at a different pH than the test gel because acidity itself explains some of the effects of AHAs.

The test subjects were 6 Asian (5 male, 1 female) and 6 Caucasian (2 male, 4 female) adults. (Note: Because AHA-induced photosensitivity has not been correlated with sex differences, having different distributions of male and female subjects in the two skin groups should not present a problem.) The subjects received separate irradiations of UVB and UVA to both sides of the lower back. Then, during a 3-week period, they applied the AHA gel to one side of the back (including the irradiated area) and to the contra-lateral extensor forearms; they applied the placebo gel to the opposite sides. Because the experiment was performed in a double-blind fashion, neither subjects nor researchers knew which side received the AHA while the experiment was being run.

At the end of the 3 weeks, the subjects returned for colorimetric measurement[a] of remaining pigmentation and for additional irradiation to forearms and a second site on the back. We obtained additional colorimetric measurements immediately after the irradiation (immediate tanning), at two hours (persistent tanning), at one week (delayed tanning) and at one month (residual tanning) (see sidebar, "Tanning Types").

Results: On the areas pretreated with 10% glycolic acid, both races showed increased UVB-induced skin tanning on the forearm and the lower back. UVA also caused increased immediate, persistent and delayed tanning, but only on the extensor forearms of the Asian subjects. Treatment with 10% glycolic acid for 3 weeks after irradiation had no effect on pre-existing light-induced pigmentation. In both races, UVB induced more erythema on glycolic-acid treated skin, especially the forearms. On the back, Asian subjects showed a slight increase in erythema on the glycolic-acid treated side, but the increase was not statistically significant. We found no change in transepidermal water loss after application of the glycolic acid.

Discussion: Finding increased UVA-induced tanning only on exposed areas (in other words, finding that pre-exposed skin tended to tan more readily) might represent a pigmentary recall phenomenon.[9] Such an increased tanning ability may provide protection against further UV exposure.[10] However, because we found increased erythema with the increased tanning in our study, we think it more likely that the increased tanning we saw implied photodamage or photosensitivity following treatment with glycolic acid.

Tanning is now linked to UV-induced DNA damage and/or its repair. Most current literature suggests that tanning is also a type of photodamage.[11] Genetic differences in the results of our study might be explained in part by differences in the reflective function of the stratum corneum, which was reported to give more significant protection against longer UV light in moderate pigmented skin.[12,13] The differences measured between Asian and Caucasian subjects might also be due to differing amounts of recreational UV exposure, with resultant differences in the degree of pre-existing photodamage.

Glycolic Acid and Photodamage

AHAs are generally considered to be beneficial in the treatment of chronic photodamaged skin.[14,15] Their beneficial effects are ascribed to increased shedding of the stratum corneum and increased epidermal turnover. Melanin is removed in such a process, and a smoother skin surface is achieved. A resemblance of AHAs to ascorbic acid was also mentioned.[6] Researchers have demonstrated that glycolic

Tanning Types

Immediate tanning is a transient darkening of the skin occurring immediately after UVA exposure.[6] A migration of microtubules and melanosomes to melanocyte dendrites occurs. There is no increase in the number of melanocytes; the phenomenon results from the photochemical reaction (photooxidation) of preexisting melanin. Researchers remain divided on whether immediate tanning involves an increase in tyrosinase levels and the number of premature melanosomes.

Persistent tanning is the stable portion of immediate tanning, which is usually measured 1 to 2 h after UVA irradiation.[7] Immediate and persistent tanning have both been proposed as preferred methods for measuring the efficacy of UVA sunscreens.[8] Even though the immediate and persistent tanning phenomena have not been shown to represent photodamage by themselves, the intensity of tanning measured does reflect the amount of UVA reaching the viable epidermis.

Delayed tanning, on the other hand, is caused by an increase in the activity and number of melanocytes. Both UVB and UVA are involved, and delayed tanning becomes visible 48 to 72 h after the irradiation.

acid can stimulate new collagen formation in the papillary dermis and help repair photodamaged skin.

Glycolic acid may have antioxidant activity and has also appeared to serve as a photoprotective agent.[16,17] When glycolic acid was applied 4 h after UV irradiation, four times a day, Perricone and DiNardo measured a decrease in erythema, but observed increased pigmentation.[16,17] However, glycolic acid has not been shown to inhibit tyrosinase activity, and not all kinds of melanogenesis carry the same biologic meaning.

The seemingly contradictory results regarding glycolic acid and photodamage are due to different study design and assessment goals. Glycolic acid may prevent the secondary effects after the UV irradiation. However, the innate photoprotective function of the skin is impaired after glycolic acid use. The increased photosensitivity after use of glycolic acids might be due to thinning of the stratum corneum or might be due to a smoother, less light-scattering skin surface.

Skin's barrier function, as judged by the transepidermal water loss, seems to be unaffected when the glycolic acid is limited to 10% and pH no lower than 3.5. However, thinning of the stratum corneum is a consistant feature during the course of glycolic acid use.[18] Hence, more UV light can thus get into the skin, and sun damage from the use of glycolic acid in consumer products is a real concern.

Most glycolic acid products now use some form of sunscreen in their formulation. However, our studies suggest the inclusion of UVA as well as UVB sunscreen will minimize the potential sun damage from the use of glycolic acid.

—**Tsen-Fang Tsai, MD,** *Department of Dermatology, National Taiwan University Hospital, Taipei, Taiwan*

References

1. EJ Van Scott and RJ Yu, *Arch Dermatol* 110 586-590 (1974)
2. JD Middleton, *J Soc Cosmet Chem* 25 519-534 (1974)
3. Cosmetic Ingredient Reviews, in *Tentative Report on Glycolic Acid and Lactic Acid,* Washington DC: The Cosmetic, Toiletry, and Fragrance Association (1997) pp 179-184
4. TF Tsai, BH Paul, SH Jee and HI Maibach, *J Am Acad Dermatol* 43 238-243 (2000)
5. JC Seitz and CG Whitmore, *Dermatologica* 177 70-75 (1988)
6. C Routaboul, A Denis and A Vinche, *Eur J Dermatol* 9 95-99 (1999)
7. A Chardon, D Moyal and C Hourseau, *Sunscreens: Development, Evaluation, and Regulatory Aspects.* 2nd ed, NJ Lowe, NA Shaath and MA Pathak, eds, New York: Marcel Dekker (1997) pp 559-581
8. K AlaiKaidbey and W Gange, *J Am Acad Dermatol* 16 346-353 (1987)
9. LA Applegate, C Scaletta, G Treina, RE Mascotto, A Fourtanier and E Frenk, *Dermatology* 194 41-49 (1997)
10. JM Sheehan, CS Potten and AR Young, *Photochem Photobiol* 68 588-592 (1998)
11. BA Gilchrest, HY Park, MS Eller and M Yaar, *Photochem Photobiol* 63 1-10 (1996)
12. A Kawada, *Photodermatology* 3 327-333 (1986)
13. A Cader and J Jankowski, *Health Phys* 74 169-172 (1998)
14. CM Ditre et al, *J Am Acad Dermatol* 34 187-195 (1996)
15. N Newman, A Newman, LS Moy, R Babapour, AG Harris and RL Moy, *Dermatol Surg* 22 455-460 (1996)
16. NV Perricone and JC DiNardo, *Dermatol Surg* 22 435-437 (1996)
17. NV Perricone, *J Geriatr Dermatol* 1 101-104 (1993)
18. JC DiNardo, GL Grove and LS Moy, *Dermatol Surg* 22 421-424 (1996)

UV Damage on Human Hair

Keywords: UV damages, human hair, UV filters, tensile strength, oxidation, deposition

A comparison study of 10 UV filters

In the past few years, several groups started to study how UV damages human hair.[2,4,7,8] However, only a few papers describe the effect of UV light on human hair or quantify the photodamage. We conducted preliminary studies with virgin blonde Caucasian and virgin black Asian hair to develop a protocol for a study comparing different UV filters.

Preliminary Study

Color: Changes in color are measurable as differences in color (DE) and differences in lightness (DL), according to CIE LAB. The bleaching of black hair is detectable after 300 h of exposure to simulated sunlight (real time, not accelerated) and continues to increase up to 600 h. In blonde hair, the situation is complex due to two phenomena. During the first 300 h, yellowing is the dominant occurence; bleaching becomes more significant from 300 to 1,200 h.

Mechanical tests: We measured single-hair friction force (root to tip and against scales), which increased by 200 to 300 percent when virgin hair was exposed to simulated sunlight for 300 h. It continued to steadily increase up to 1,200 h. These findings paralleled the results of our tests on both wet and dry combing, although these were less pronounced.

A trial run with UV filters revealed one complication: The product deposit on the hair surface may produce a "lubrication effect," providing additional friction because of microfine crystals and clusters. We verified the phenomenon by analyzing gold-splattered samples with a scanning electron microscope.

Tensile strength: We tested the hair's tensile strength using the F-15 test. In contrast to another study, we found the uncertainty tends to be larger than actual differences, making this type of analysis unreliable.[17]

FTIR: Using attenuated total internal reflection (ATR), Fourier Transform Infrared (FTIR) spectroscopy shows an increase of an absorption band at 1,040 cm^{-1}, which is attributed to S-O vibrations of cysteic acid, an oxidation product of cystine.[13,14] Relatively long exposure times (600 to 1,200 h) would be required to use this absorption increase as quantitative measurement.

Formulations Tested

The formulations used in this study were not created to please consumers but to meet the technical constraint that the listed filters would be soluble at the desired concentration. (Concentrations are indicated in brackets.)

UV filters

- 2-Hydroxy-4-methoxy-benzophenone (Benzo-3)
- 2-Hydroxy-4-methoxy-benzophenone-5-sulfonic acid (Benzo-4)
- 4-Butyl-4'-methoxy-dibenzoylmethane (BM-DBM)
- Experimental Polymeric Sunscreen (EPS)
- 2-Ethylhexyl-p-methoxycinnamate (OMC)

UVB filters

- Experimental Methoxy Cinnamate (EMC)
- 3-(4'-Methyl-Benzylidene) Camphor (MBC)
- 2-Ethylhexyl-p-N,N- dimethylaminobenzoate (O-Paba)
- 2-Phenylbenzimidazol-5-sulfonic acid (PISA)
- 2,4,6-Trianilin-(p-carbo-2'-ethylhexyl-1'-oxy)-1,3,5-triazin (O-Triazon)

Amino Acid Analysis

Sulfur-containing amino acids are sensitive to oxidation and, in particular, to the transformation of cystine to cysteic acid. Aromatic amino acids, such as tryptophan and tyrosine, are degraded by light.[5,12,15] We quantified the formation of cysteic acid and, for some series, the loss of cystine and tyrosine on the basis of the classical amino acid analysis method by Spackman et al.[14] We determined the amount of tryptophan colorimetrically according to a procedure developed originally by Cegarra[1] and modified for unpigmented wool[11] ("Tryptophan Analysis," location).

After 150 h of exposure, blonde hair showed substantial losses of tryptophan (25 to 30%), cystine (25%) and tyrosine (up to 80%) and an increase in cysteic acid (about 50%). Black hair was much less sensitive to UV-radiation; it took about twice as long to produce similar changes.

Substantivity: In a separate study, we analyzed the quantitative aspects of deposition of UV filters on hair following a procedure described by Unilever.[16] Although we only tested four formulations, we made two important observations:

- Deposition is strongly dependent on the formulation, up to ninefold.
- Differences among filters are even more important.

Based on these findings, we deliberately omitted the rinsing step after application in our experimental protocol.

Tryptophan Analysis

To analyze the content of tryptopahn in treated human hair, we put approximately 60 mg of the hair into a 50 mL volumetric flask along with 2 mL of an 18 N sulfuric acid solution, 1 mL of a p-dimethylaminobenzaldehyde solution in 10% sulfuric acid solution, and 11 mL of an 18 N sulfuric acid solution. We kept the flask for 2 h at 70°C.

After 2 h, the hydrolysate was cooled for 2 min in an ice bath. We then added 2 mL of a 0.01 M sodium nitrite solution and filled the flask with a 0.03 N sulfuric acid solution. The flask was kept for another 2 h at 60°C and then cooled to room temperature in a 2-min ice bath.

We then filtered and, after 15 min, measured[a] the solution at 585 nm. We compared our results against those of the control solution (the reaction mixture without p-dimethylaminobenzaldehyde). We also checked the stability of the formed dyestuff by measuring again after 30 and 45 min.

We calculated the tryptophan content of the hair samples by using a calibration curve obtained from solutions of λ-tryptophan in 0.03 N sulfuric acid solution.

[a]Shimadzu UV-160A Photometer

The Experiment

Method: We applied 1 mL of product ("Formulations," location) to 0.5 g of Caucasian blonde hair[a]. The product was brushed into the hair to evenly distribute it within 60 sec. The treated hair was then exposed to simulated sunlight[b] under controlled environmental conditions (Table 1-1).

After 30 h, the samples were washed. We wet each tress for 60 sec under running water. We then massaged 0.3 mL sodium laureth sulfate solution (15% w/v) into the fibers for 60 sec. The hair was then rinsed under running water for 3 min, dried at 50°C for 1 h and conditioned for more than 15 h at 20°C and a relative humidity of 65%.

We repeated this sequence of treatment, exposure and washing until each tress had been exposed to a cumulative of 150 resp. 180 h.

Analysis: For comparison purposes, we calculated the results of each test series by first normalizing the difference (Δ) between the highest and the lowest

Table 1-1. Environmental conditions

Relative humidity	70%
Temperature	23°C
UVB[b]	2 MED[a]/h±10%
UVA[b]	4 mW/cm^2±10%

[a]MED = Minimal Erythemal Dose
[b]Measured with a Berger instrument (Solar Light Co., Philadelphia, Pennsylvania, USA)

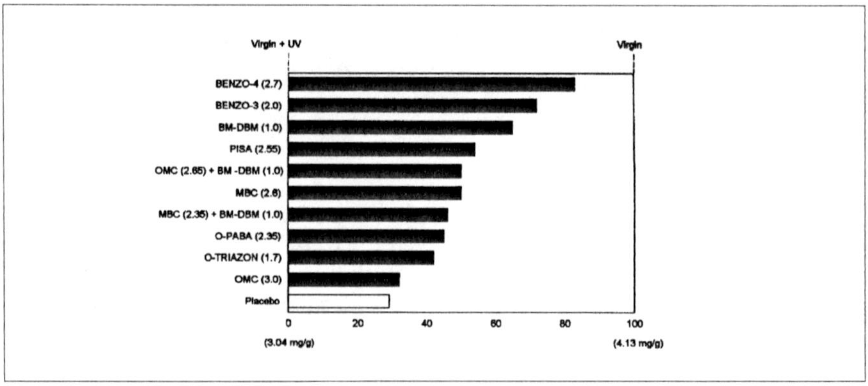

Figure 1-1. Tryptophan content of blonde Caucasian hair treated with a variable amount (%, w/w) of different UV filters in formula A and exposed 6 x 30 hours to simulated sunlight

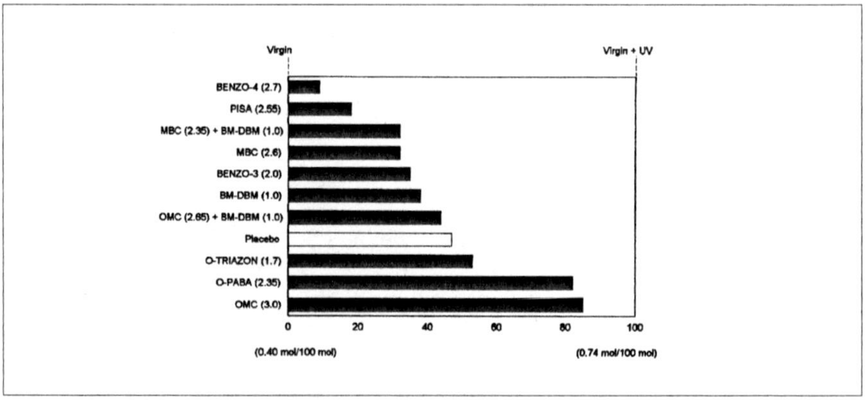

Figure 1-2. Cysteic acid content of blonde Caucasian hair treated with a variable amount (%, w/w) of different UV filters in formula A and exposed 6 x 30 hours to simulated sunlight

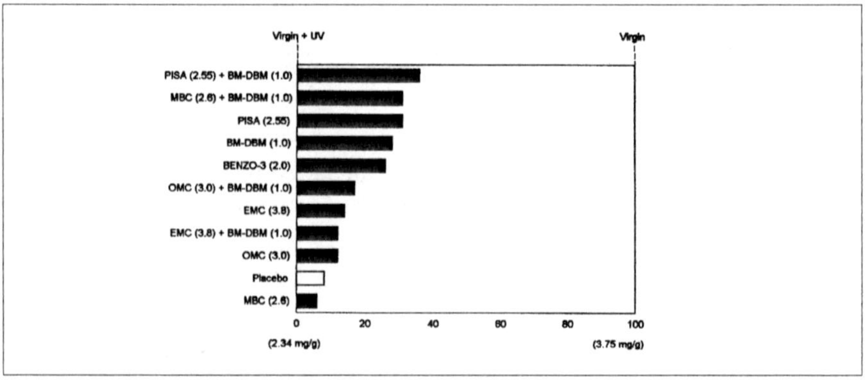

Figure 1-3. Tryptophan content of blonde Caucasian hair treated with a variable amount (%, w/w) of different UV filters in formula B and exposed 6 x 30 hours to simulated sunlight

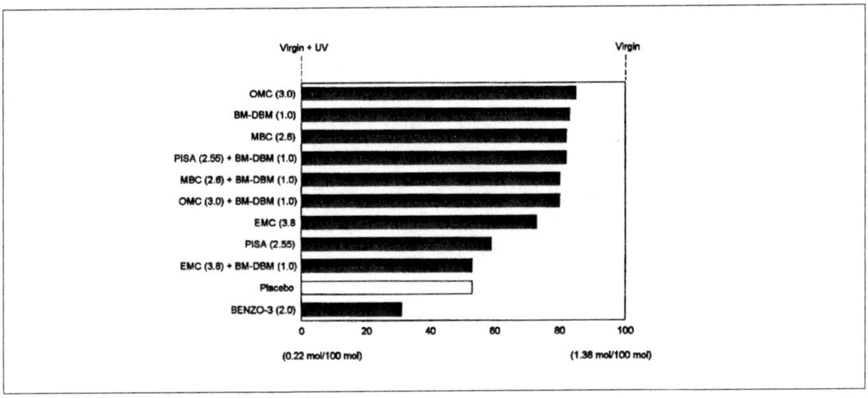

Figure 1-4. Tyrosine content of blonde Caucasian hair treated with a variable amount (%, w/w,) of different UV filters in formula B and exposed to 6 x 30 hours of simulated sunlight

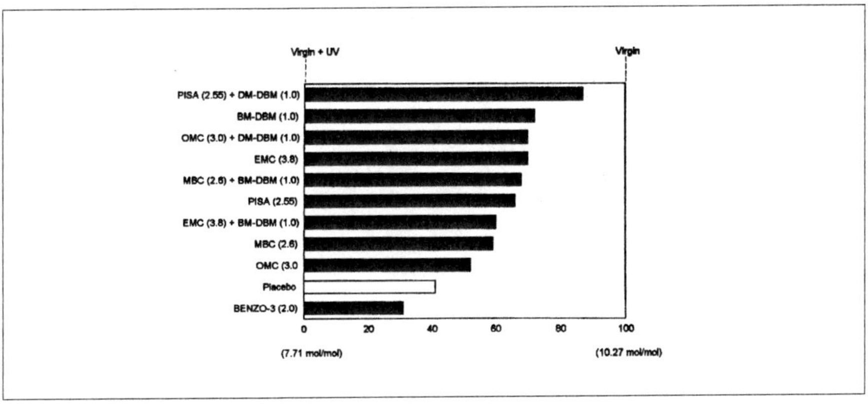

Figure 1-5. Cystine content of blonde Caucasian hair treated with a variable amount (%, w/w) of different UV filters in formula B and exposed to 6 x 30 hours of simulated sunlight

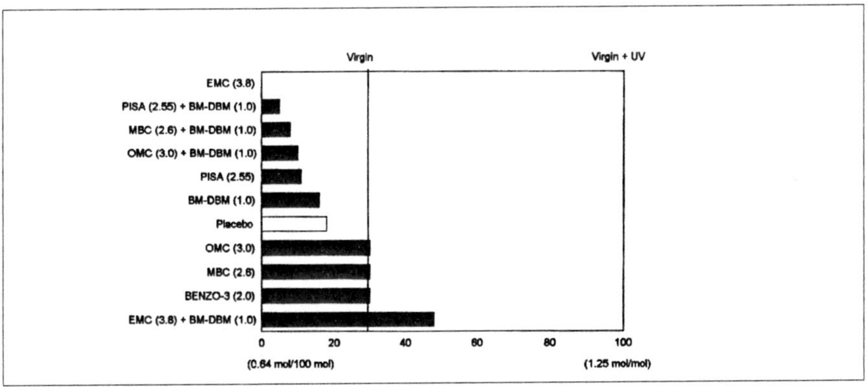

Figure 1-6. Cysteic acid content of blonde Caucasian hair treated with a variable amount (%, w/w) of different UV filters in formula B and exposed to 6 x 30 hours of simulated sunlight

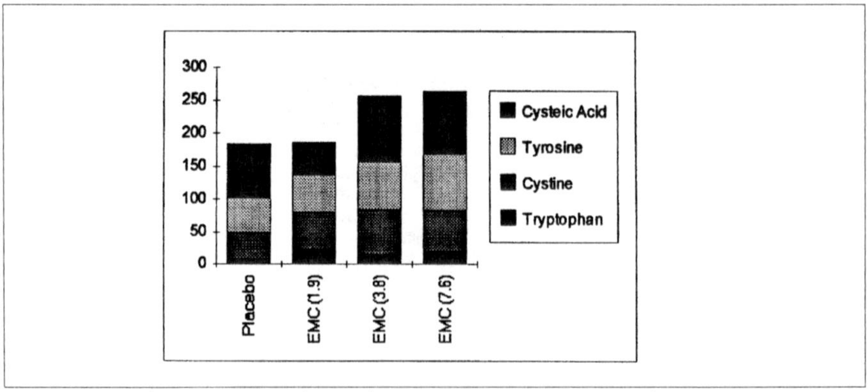

Figure 1-7. Cumulative protection of tryptophan, cystine, tyrosine, cysteic acid (maximum=400%) in blonde Caucasian hair by formula B with a variable content of filter EMC (%, w/w) exposed to 6 x 30 hours of simulated sunlight

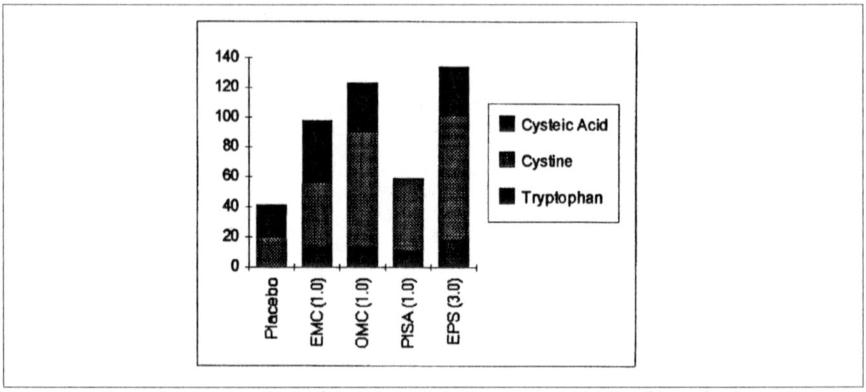

Figure 1-8. Cumulative protection of 3 amino acids (maximum=300%) of blonde Caucasian hair. Comparing 4 UV-B filters (%, w/w) in a clear shampoo (Formula C) exposed to 5 x 30 hours of simulated sunlight

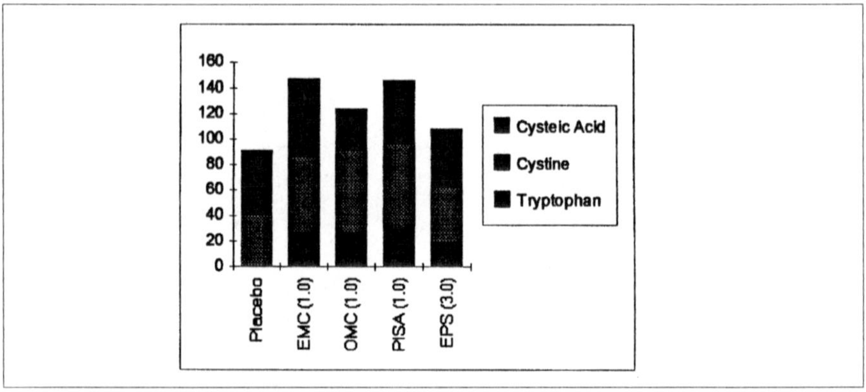

Figure 1-9. Cumulative protection of 3 amino acids (maximum=300%) of blonde Caucasian hair. Comparing of 4 UV-B filters (%, w/w) in a pearly shampoo (Formula D) exposed to 5 x 30 hours of simulated sunlight

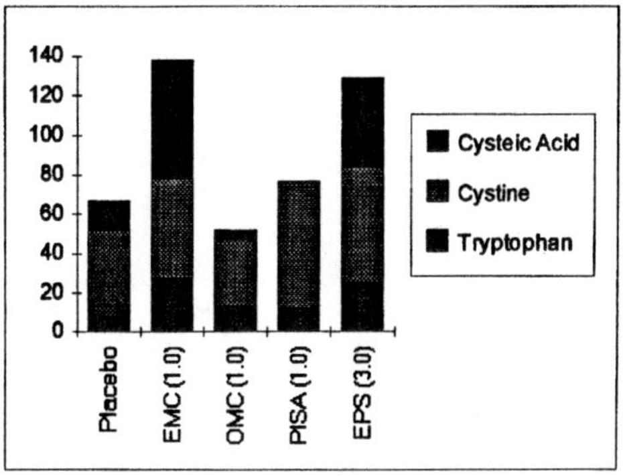

Figure 1-10. Cumulative protection of 3 amino acids (maximum=300%) of blonde Caucasian hair. Comparing 4 UV-B filters (%, w/w) in a protective cream (Formula E) exposed to 5 x 30 hours of simulated sunlight

values to 100%. All intermediate values were then calculated as a difference (δ) and expressed as a percentage (δ/Δ·100).

In most cases, Δ corresponds to the difference between the virgin samples (non-treated, non-exposed) and the control samples (non-treated, exposed), which is equivalent to irradiated virgin samples. Furthermore, we included in each series a placebo, samples treated with the vehicle and exposed to UV but without a filter in the formulation.

Results

The amino acids most susceptible to light damage are tryptophan, cystine and tyrosine. We report in Figures 1-1 through 1-6 the amount of each amino acid before and after exposure to simulated sunlight. (These are the values in brackets below the X axis.) Since these amino acids are degraded by UV, the higher the value, the better the protection. On the other hand, cysteic acid is a photoproduct. Therefore, the higher the value, the more pronounced the damage.

Figures 1-1 and 1-2 show the tryptophan and cysteic acid results from the test with the first series in which we treated blonde Caucasian hair with a fluid shampoo (Formula 1-1) containing different UV filters and exposed for 6 x 30 h to simulated solar UV radiation. The concentrations of the UVB filters (EMC, MBC, O-Paba, PISA and O-Triazon) were chosen to yield a similar optical density as E/Z equilibrated OMC at 3% (w/w). This is also true of the second series (Figures 1-3 to 1-6), which documents the results of four amino acid assays after treatment with Formula 1-2.

Figure 1-7 presents a comparison, in which the filter concentration varies. We applied EMC in Formula 1-2 at concentrations of 1.9%, 3.8% and 7.6% (w/w) to blonde Caucasian hair, which then was exposed to 6 x 30 h of simulated irradiation as before.

For the cumulative presentation, we converted the values for cysteic acid: [100 - (δ/Δ·100)].

Formula 1-1. Fluid emulsion

Ingredient	% w/w
UV filter	
Carbomer, 2% aqueous dispersion (Carbopol 981, BFGoodrich)	50.0
Triethanolamine	qs to pH approx 7.0
Water (aqua), deionized	qs to 100.0
C_{12}-C_{15} alkyl benzoates (Finsolv TN, Finetex)	10.0
Disodium EDTA (EDTA BD)	0.1
Phenoxyethanol (and) methylparaben (and) ethylparaben (and) propylparaben (and) butylparaben (Phenonip)	0.6

Formula 1-2. Shampoo

Ingredient	% w/w
UV filter	
Polysorbate (Tween 80)	5.0
Sodium laureth sulfate (Texapon N70)	30.0
Sodium lauroyl sarcosinate (Medialan LD)	25.0
Coco betaine (Dehyton AB 30)	5.0
Dimethicone copolyol (Silicone DC 190 Polyether)	1.5
Water (aqua), deionized	qs to 100.0
Sodium hydroxide, 10% solution	qs to pH approx. 7.5

Formula 1-3. Clear shampoo

Ingredient	% w/w
Ammonium lauryl sulfate (Texapon ALS)	50.0
Coco/oleamidopropyl	
Betaine (Murataine COB)	10.0
PEG-6 caprylic/capric	
Glycerides (Softigen 767)	10.0
Phenoxyethanol (and) methylparaben (and) ethylparaben (and) propylparaben (and) butylparaben (Phenonip)	0.6
UV filter	
Water (aqua), deionized	qs to 100.0

Formula 1-4. Pearly shampoo

Ingredient	% w/w
Coco betaine (Dehyton AB-30)	6.00
Sodium laureth sulfate & PEG-8 & socamide MEA & glycol distearate & glycerin (Texapon SG)	40.00
Polysorbate 80 & cetyl acetate & acetylated lanolin alcohol (Solulan 98)	0.20
PEG-150 distearate (Rewopal PEG-6000 DS)	1.00
Phenoxyethanol (and) methylparaben (and) ethylparaben (and) propylparaben (and) butylparaben (Phenonip)	0.60
Polysorbate 80 (Tween 80)	5.00
UV filter	
Dipropylene glycol	3.00
Water (Aqua), deionized	qs to 100.00
Tetrasodium EDTA (Sequestrene Na_4)	0.05
Panthenol (Panthenol)	0.60

Formula 1-5. Protective cream

Ingredient	% w/w
Behentrimonium chloride (Genamin KDM-F)	2.0
Cetyl phosphate acid (Amphisol A)	1.0
Mineral oil (Vaseline Oil)	2.0
Cetyl alcohol	2.0
Phenoxyethanol & methylparaben & ethylparaben & propylparaben & butylparaben (Phenonip)	0.6
UV filter	
Polysorbate 80 (Tween 80)	5.0
Deionized water	qs to 100.0
Disodium EDTA (EDTA BD)	0.10
Citric acid, 10% solution	qs

Table 1-2. Factors affecting UV damage to hair

Influence	Comment
Blonde vs. black hair	Preliminary work establishes that blonde hair is more susceptible to UV damage.[6,7]
UV wavelength, absorption characteristics	Results indicate that both UVB and UVA are important. Visible light is mostly responsible for photobleaching.[7]
Humidity, hydration	Water plays an important role.[3] As all experiments in this study were conducted under controlled conditions, this factor does not need to be taken into account.
Quantitative deposition	By brushing the product onto the hair and not rinsing afterwards, the total amount remains on the hair. Thus, our results are independent of quantitative deposition. This factor should not be neglected if uncontrolled.
Qualitative distribution (drops, clusters, crystals, film, etc.)	This is strongly formulation-dependent (solubility, spreading, affinity, etc.)
Lifetime of filter on hair	The filters used are primarily used in skin care (exposure time does not exceed 1 day). This is irrelevant since all exposure intervals in this study were constant (about 3 days).
Placebo effect	Hair treated with mineral oil is more resistant to UV damage than water-treated controls.[9] A similar effect is found here. In practically all cases, the placebo formulation without a filter provides some protection.

Finally, the last series focused on the influence of the formulation. The results can be found in Figures 1-8 to 1-10. We incorporated 4 UV filters (all of which absorb in the UVB range) into Formulas 1-3, 1-4 and 1-5. We reduced the concentrations of EMC, OMC and PISA to 1% (w/w) and that of the new EPS to 3% (w/w), taking into account the "polymeric ballast." The number of cycles of treatment, washing and exposure was reduced to 5, yielding a total exposure of 150 h. Due to troubles with our tyrosine determinations, we are confining our presentation to the results of our tryptophan, cystine and cysteic acid calculations.

Discussion

Figures 1-1 and 1-2 might lead you to a premature conclusion that water-soluble filters, such as Benzo-4 and PISA, are particularly suitable for hair protection and, given the high ranking of Benzo-4, Benzo-3 and BM-DBM, that UVA plays an important role in hair damage. This conclusion would reinforce that of Hoting et al.[7]

However, the second series (Figures 1-3 to 1-6) offer a different interpretation. With a different formulation, the overall protection is lower, particularly evident in the tryptophan values. Also, the ranking of the filters has changed.

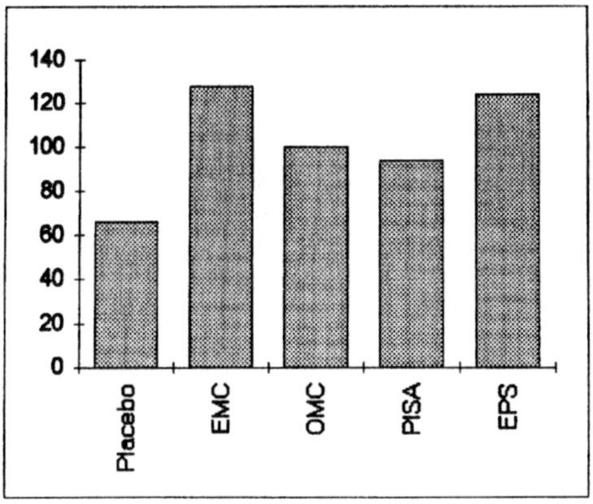

Figure 1-11. Average protective scores of 4 UV-B filters in 3 formulations

Interesting, too, are the results portrayed in Figure 1-7, suggesting there is an optimal concentration for a given filter in a given formulation.

All these results call for a careful consideration of all the factors that may govern UV damage and photoprotection of hair (Table 1-2).

Based on these latest results, we can hypothesize that qualitative deposition, which is strongly formulation dependent, is a key factor for successfully preventing UV-induced hair damage. Our final series (Figures 1-8 to 1-10) nicely confirms this. With each formulation, we obtained a clearly different ranking. This means that, to obtain maximum protection, a chemist must carefully select the most effective UV filter for a given formulation and thoroughly optimize a formulation for a given UV filter.

By comparing the average scores of the test filters (Figure 1-11), we found that all four do protect but that some have a higher likelihood of providing substantial protection. This is particularly true for the polymeric candidate (EPS). Perhaps the polymeric backbone helps to achieve an optimal qualitative deposition. Riedel et al found similar results after applying polymeric UV absorbers to wool.[10] Preliminary data on quantitative deposition yield relatively high scores for this material.

Efficacy studies taking into account both qualitative and quantitative factors remain to be done.

Conclusion

The present study confirms that UV damage to human hair is real, that both UVB and UVA play a role in this damage, and that blonde hair is more sensitive than black. It further demonstrates that amino acid analysis is a useful tool for quantifying UV damage. It also illustrates that several UV filters can effectively reduce UV damage. The choice of which filters to use as well as the careful optimization of the formulation are crucial for successfully protecting human hair from solar UV damage.

—**H. Gonzenbach, W. Johncock and K.-F. De Polo,** *Roche Vitamins and Fine Chemicals, Givaudan Roure Vernier-Geneva, Switzerland*
—**G. Blankenburg, J. Föhles, K. Schäfer and H. Höcker,** *Deutsches Wollforschungsinstitut an der RWTH, Aachen, Germany*

References

1. J Cegarra and J Gacen, *J Soc Dyers* Col 84, pp 216-220 (1968)
2. A Deflandre, JC Garson and F Leroy, IFSCC Congress (1990)
3. C Dubief, *Cosm & Toil* 107, pp 95-102 (1995)
4. M Giesen et al, IFSCC Congress (1990). Also: P Bernhardt et al, Intl J Cosm Sci 15, pp 181-199 (1993)
5. D Goddinger, K Schäfer and H Höcker, Wool Tech Sheep Breed 42(1), pp 83-89 (1994)
6. B Hollfelder et al, *Intl J Cosm Sci* 117, pp 87-89 (1995)
7. E Hoting, M Zimmermann and S Hilterhaus-Bong, *J Soc Cosm Chem* 46, pp 85-99 (1995)
8. S Nacht, Sunscreens and Hair, *Sunscreens* (Marcel-Dekker : NJ Lowe and NA Shaath, eds), pp 341-355 (1990)
9. CM Pande and J Jachowicz, *J Soc Cosm Chem* 44, pp 109-122 (1993)
10. JH Riedel and H Höcker, doctoral thesis, RWTH Aachen, (1992)
11. K Schäfer, IWTO-Report (1993). Also *J Soc Dyers Col* (in press)
12. K Schäfer et al, Textilberichte 74(3), pp 225-231, E103-E106 (1993)
13. U Schumacher-Hamedat, J Föhles and H Zahn, *Textilveredlung* 21, pp 121-125 (1986)
14. DH Spackman, WH Stein, and S Moore, *Anal Chem* 30, pp 1190-1206 (1958)
15. E Tolgyesi, *Cosm & Toil* 98, pp 29-33 (1983)

Patents

16. EP 0 386 898, priority GB 89 03 777 20 2 89. Also EP 0 573 229
17. WO 93 12763, priority

Wrinkles from UVA Exposure

Keywords: elastase, anti-inflammatory agents, wrinkles, honeysuckle flower, Engelhardtia chrysolepis, UVA

Research on mechanisms of wrinkle formation and methods of prevention

Recently, the increase in UV irradiation due to progressive depletion of the Earth's ozone layer has become a serious human health concern. In Japan, the frequency of skin cancer of persons living in sunny regions has showed a clear increasing trend from the 1970s to the 1990s. Solar keratosis, a precancerous condition, is also increasing rapidly.

Many UV-protective products are now available on the market. However, regardless of the amount of the product used, completely protecting against UV radiation under the intensity of the midsummer sun is difficult. In autumn, winter and spring, UV-protective cosmetics are rarely used, making adequate protection impossible.

UV radiation reaching the Earth's surface may increase substantially in the future. What is needed for protection may not only be sunscreen products with high SPF, but also anti-photoaging products, which can be used irrespective of region, season, lifestyle, gender and age.

Of the effects of UV irradiation on the Earth's surface, skin damage caused by UVB irradiation has been extensively investigated. However, the mechanisms of skin damage and photoaging caused by long-term exposure to UVA irradiation have not yet been sufficiently elucidated.

This study investigated the mechanism of skin sagging caused by long-term exposure to UVA radiation, in hope that better understanding of the mechanism will assist the development of new concepts in skin care and new products for the prevention of photoaging.

Chronic UVA and Sagging

We used hairless mice SKh:HR-1[a] as our animal model for chronic UVA damage to skin, modifying the method of Bissett et al.[1] At the start of experiments, the mice were 9 to 10 weeks old.

We produced UVA radiation by passing output from 8 lamps[b] through a 5 mm thick glass filter to cut out wavelengths below 320 nm. Animals housed individually in glass boxes with walls 2 mm thick were irradiated 5 times weekly for 6 months

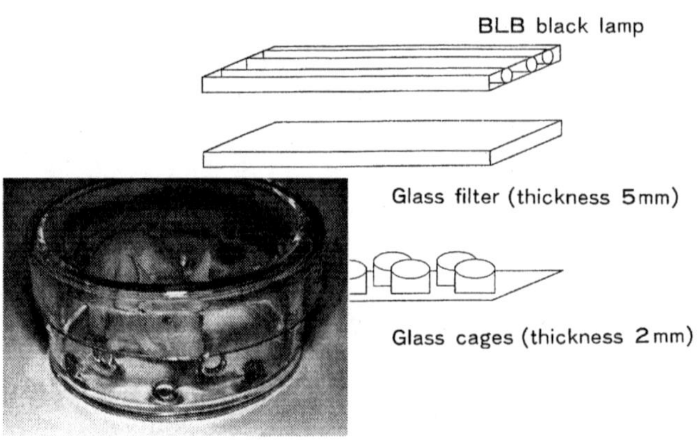

Figure 1. Experimental apparatus

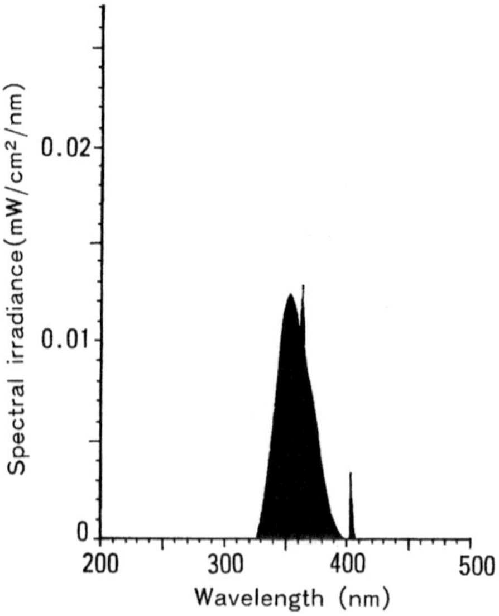

Figure 2. Spectrum of UVA light from the Toshiba BLB lamp passed through a 5 mm thick glass filter

with 27 J/cm^2 UVA radiation per daily exposure. The irradiation energy was determined using a radiometer.[c] The radiation spectrum through the glass boxes was determined using an spectroradiometer[d] (Figures 1 and 2).

Visual examination: The degree of sagging was graded on a scale of 0-3 (Table 1), according to the method of Bissett et al.[1]

After long-term exposure (24 weeks) to UVA radiation, hairless mice exhibited large, loose folds (sagging)[1] on the dorsal and lateral skin. At 3 months after the

[a]Charles River Laboratories, Boston MA, USA
[b]Toshiba-BLB, Toshiba, Tokyo, Japan

Table 1. Scale for grading sagging, 0 to 3

0 =	Pale pink color, fine striation (head to tail), no loose folds
1 =	Slight blanching, slight fine striation, no loose folds
2 =	Complete blanching, no fine striation, slight loose folds
3 =	Complete blanching, no fine striation, large loose folds

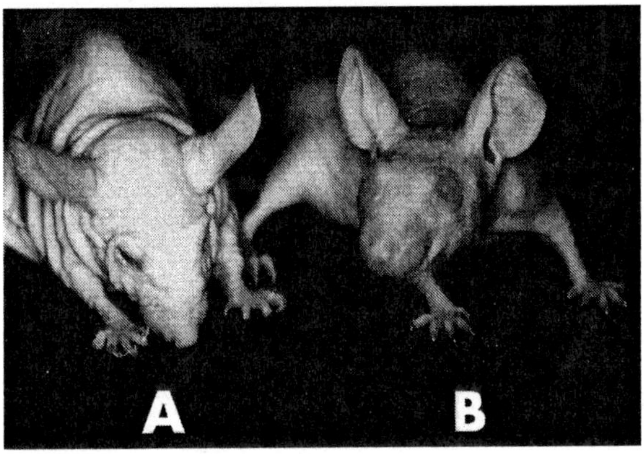

Figure 3. Visible changes at 24 weeks
A = UVA-irradiated hairless mouse
B = nonirradiated hairless mouse

first exposure, the dorsal skin became blanched and the skin texture became rough. After 6 months, there was complete blanching of the skin, nodular texture and loose folds (Figures 3, 4). Figure 5 shows the time-course for development of sagging.

Histological examination: On the last day of the experimental period, biopsies of the dorsal skin were performed on 5 animals per group. We fixed tissue specimens in formalin, embedded them in paraffin and sectioned to 5 µm in thickness. The sections were stained with H-E, van Gieson's and Luna stains, then examined for several parameters. One of the tissue specimens was reserved for transmission electron microscopy (TEM).

Histological and TEM findings after exposure to UVA for 24 weeks were difficult to analyze. Marked thickening of the epidermis and dermis was observed. In the dermis, an increase in mast cells and definite hypertrophic cysts were observed. In addition, partial infiltration into the dermis of inflammatory cells, including polymorphonuclear leukocytes (PMNs), was observed. In the upper layer of the dermis, there was a marked decrease in collagen fibers. As for elastin fibers, aggregations were intermingled with areas of partial absence (Figures 6 through 14).

[c]Topcon-UV radiometer 305/365 DII, Topcon, Tokyo, Japan
[d]Ushio spectroradiometer USR-20B, Ushio Denki, Yokohama, Japan

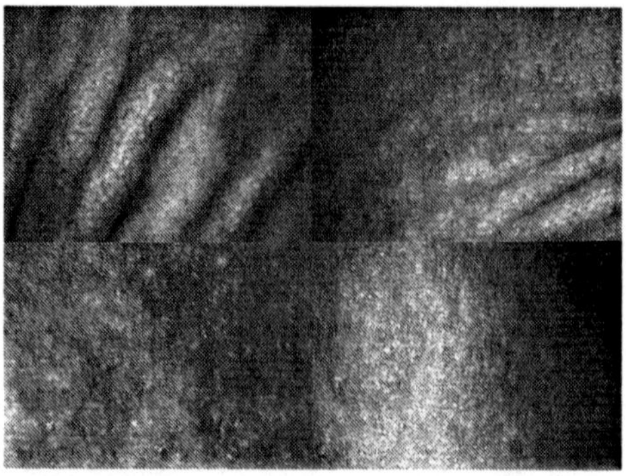

Figure 4. Visible changes at 24 weeks
a,b = UVA-irradiated hairless-mouse skin
c,d = nonirradiated hairless-mouse skin
a,c = Dorsal (back) skin
b,d = Abdominal skin

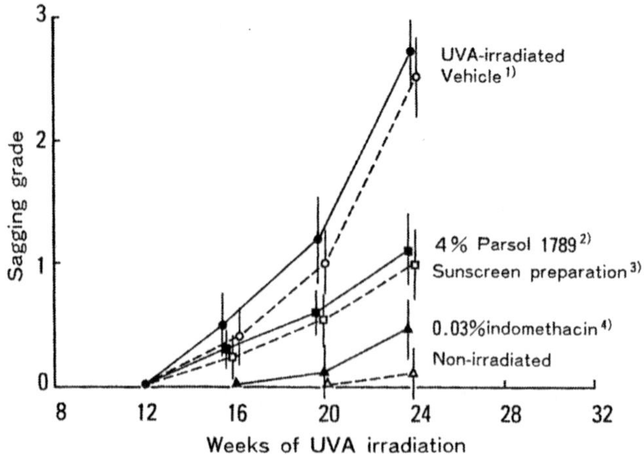

Figure 5. Effect of topical sunscreen preparation and indomethacin on the sagging of hairless-mouse skin caused by UVA irradiation
1) Control vehicle is 1/1 w/w propylene glycol and ethanol
2) Butyl methoyxydibenzoylmethane in carbitol
3) 5% Butyl methoyxydibenzoylmethane and 3% octyl methoxycinnamate
4) Nonirradiated control, treated with control vehicle

Determination of elastase activity: We removed the subcutaneous tissue from circular, dorsal-skin tissue specimens (2.1 cm in diameter) not used in the histological examinations. These were homogenized in a glass homogenizer with 2 ml of buffer (0.1 M HEPES [N-2-hydroxyethyl-piperazine-N'-2-ethanesulfonic acid] and 0.5 M

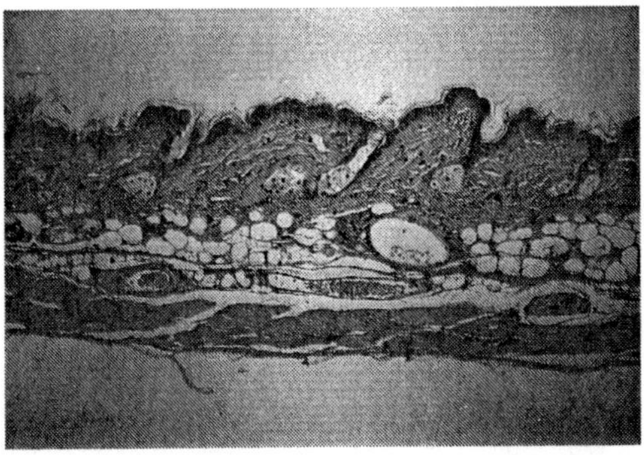

Figure 6. Intact (nonirradiated) skin; H-E stain, 400x

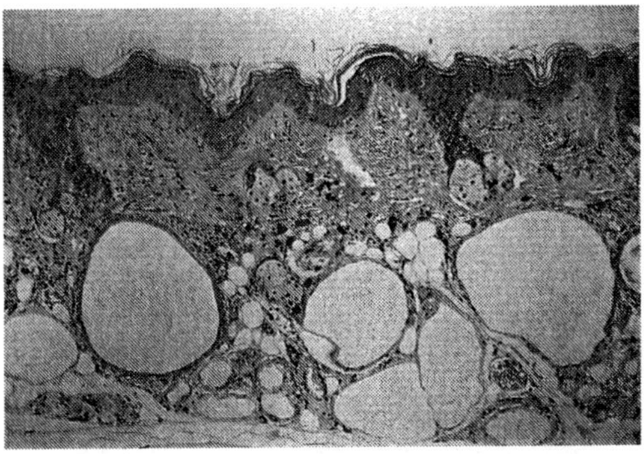

Figure 7. Thickening and hyperplasia of epidermis and dermis after 24 weeks UVA exposure; dermal mast cells increase in number while dermal cysts increase in both number and size. H-E stain, 400x

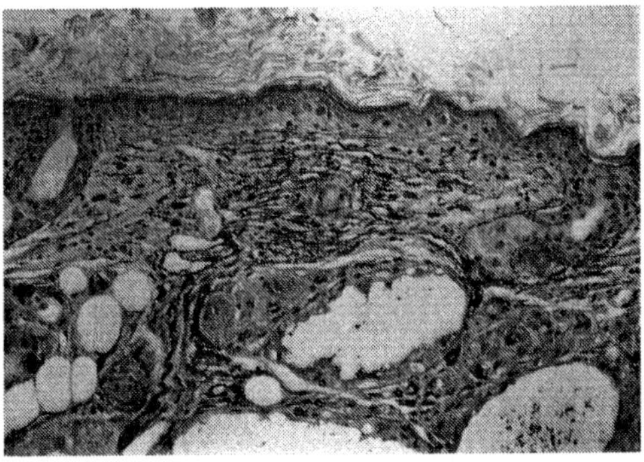

Figure 8. Increase of elastin fibers in focal area of UVA damage after 24 weeks UVA exposure; Luna stain, 400x

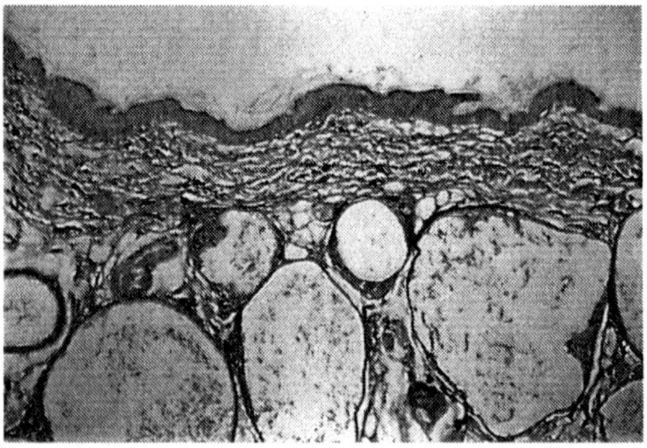

Figure 9. Damaged collagen fibers in the upper dermis after 24 weeks UVA exposure; Van Gieson stain, 400x

sodium chloride, pH 7.5). We centrifuged the homogenate at 15,000 rpm for 10 min. We assayed enzyme activity using 0.1 ml of supernatant and 1.4 ml of enzyme substrate solution at 37°C for 1 h. The amount of *p*-nitroaniline released was determined spectrophotometrically at 405 nm using a spectrophotometer[e] at 25°C.

The assay for elastase activity showed a significant increase in the activity in UVA-irradiated skin compared with non-irradiated skin (Figure 15).

Chronic UVA Exposure

Peroxidation of membrane lipids in the epidermis and dermis may be promoted by active oxygen and other free radicals produced by long-term exposure to UVA in hairless mice.[9] Skin damage and inflammation may be caused by the resultant peroxylipids.[10] Considering our histological and biochemical results, the infiltration into the dermal layer of cells that cause inflammation observed in UVA-irradiated hairless mice supports the idea that mild and persistent inflammation occurs in UVA-irradiated skin. Among the inflammatory cells infiltrated into the dermal layer, PMNs in particular release large amounts of elastase. The released elastase may attack the proteins of connective tissue, resulting in damages of elastin and collagen fibers, and finally causing sagging (Figure 16).

Histological and TEM examination of irradiated tissue specimens show partial absence and aggregation of elastin fibers in the dermis, supporting the above-stated hypothesis. Normally elastin fibers are not well developed in the skin of hairless mice, and a partial increase in elastin fibers occurred by a feed-back mechanism after degradation of elastin fibers by elastase, which collapsed to form aggregates.

Screening for Photoaging Inhibition

For mice receiving topical treatment of drugs, 50 µl of the test solution was applied to the dorsal skin surface 1 h prior to each exposure to UVA radiation. The

[e]Hitachi U-3210, Hitachi, Tokyo, Japan

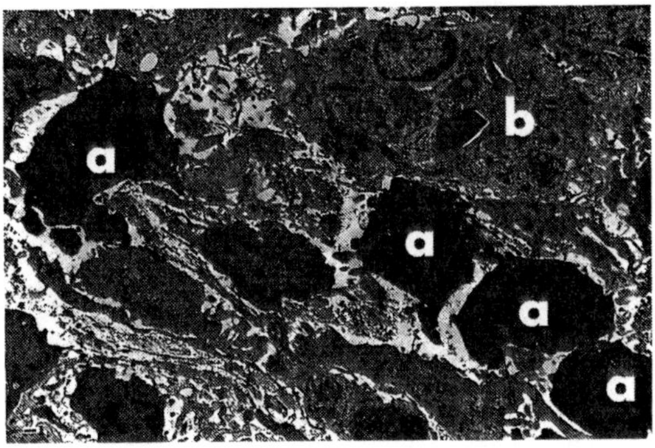

Figure 10. PMNs (a) and a macrophage (b) are seen in the dermis after 24 weeks of UVA exposure; TEM, scale bar = 1 μm

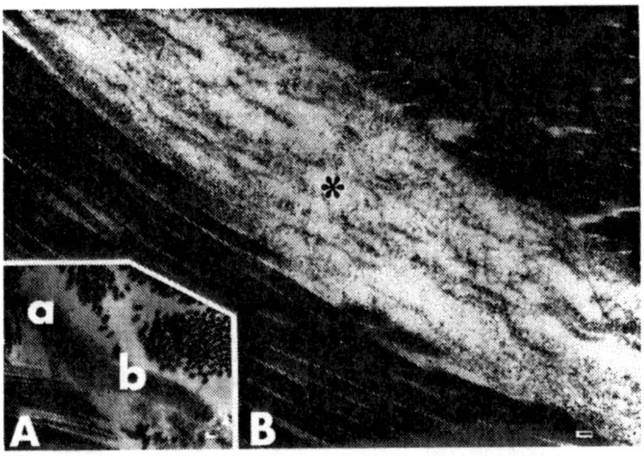

Figure 11. TEM of elastin fibers; A = normal fibers in intact skin, tannic acid positive elastic microfilaments (a) and tannic acid negative elastic microfilaments (b). B = damage after 24 week UVA exposure, tannic acid nagetive indistinct elastic microfilaments (*). Tannic acid positive elastic microfilaments disappeared. Tannic acid stain. Scale bar = 0.1 μm.

test solution was applied to the skin using an Eppendorf pipet and spread evenly over the entire dorsal skin surface using the flat part of the pipet tip.

For sunscreen products, the test solution was applied to the skin using a disposable syringe and spread evenly over the skin with a glass rod to achieve 2 mg/cm^2 coverage.

To investigate our hypothesis on the photoaging mechanism, it was necessary to determine which mechanism, antioxidation or anti-inflammation, might inhibit UVA-induced skin damage. We therefore included various antioxidants and anti-inflammatory agents in our photoaging inhibition study on hairless mice. In addition, we screened several Chinese homeopathic preparations for both antioxidative and anti-inflammatory activities

UV-induced erythema: We used male Hartley guinea pigs weighing 700 to 900 g, with 6 animals per group. Wide adhesive tape punctured with six 1.5 cm x

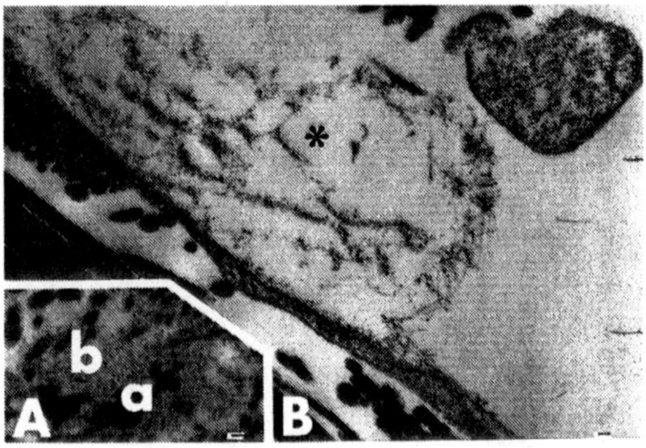

Figure 12. TEM of elastin fibers; A = normal fibers in intact skin, tannic acid positive elastic microfilaments (a) and tannic acid negative elastic microfilaments (b). B = damage after 24 week UVA exposure, tannic acid nagetive indistinct elastic microfilaments (*). Tannic acid positive elastic microfilaments disappeared. Tannic acid stain. Scale bar = 0.1 µm.

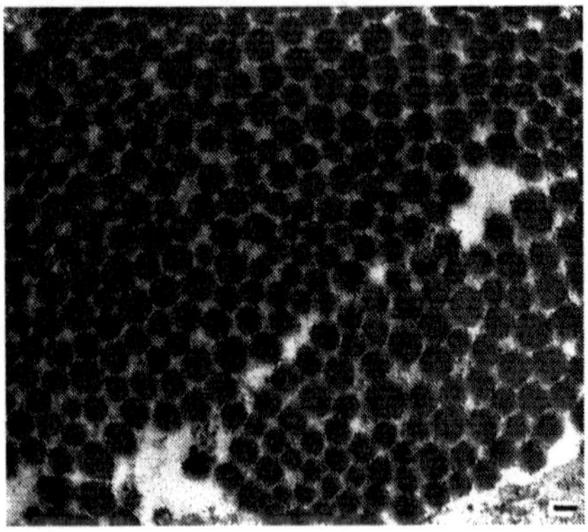

Figure 13. Intact (nonirradiated) skin, TEM, scale bar = 1 µm

1.5 cm holes (2 rows each with 3 holes) was applied on the shaved dorsal skin, and UV radiation was irradiated over the tape.[f] The amount of irradiation energy was controlled at 768 mJ/cm^2. After 24 h, a uniform erythematous reaction was observed at the 6 exposed sites. We administered 5 µl/site of the test drugs 5 times, at 1 h intervals beginning immediately after UV exposure. We evaluated the erythematous reaction at 24 h after exposure according to Table 2.

We calculated percent inhibition in erythema caused by each drug using Equation 1.

Equation 1.
% inhibition in erythema = $\{(Ev_c - Ev_d)/ Ev_c \} \times 100$.
Ev_c = Mean erythema value at the vehicle-treated site
Ev_d = Mean erythema value at the drug-treated site

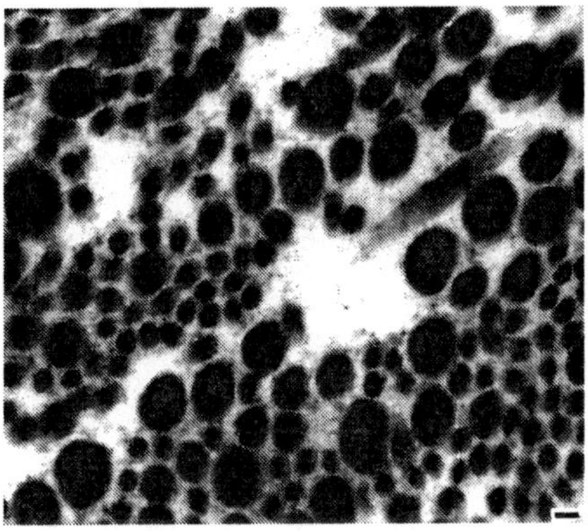

Figure 14. Skin after 24 weeks UVA irradiation, showing huge and small collagen fibers in the upper dermis, TEM, scale bar = 1 µm

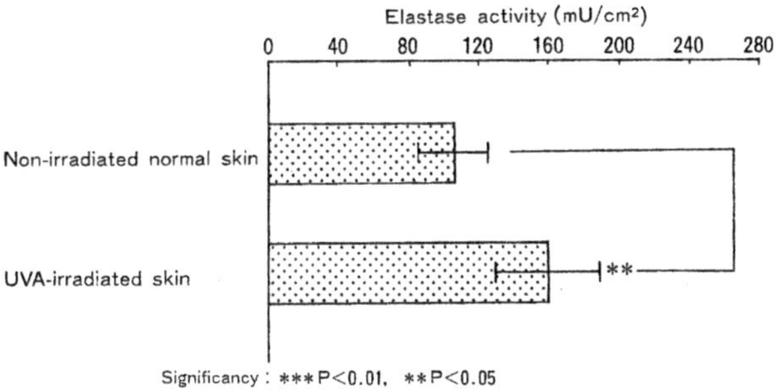

Figure 15. Effects of 24 weeks UVA irradiation on elastase activity in hairless mouse skin

Sunscreen preparations: Although a 4% solution of butyl methoxydibenzoylmethane[g] in carbitol and a sunscreen mixture containing 5% butyl methoyxydibenzoylmethane and 3% octyl methoxycinnamate absorb radiation, they only moderately inhibited the photoaging response (skin sagging) caused by long-term UVA irradiation in hairless mice (Figure 5).

Antioxidants: Ascorbic acid (0.3%) and α-tocopherol (0.3%) did not inhibit the photoaging response in hairless mice, as shown in Figure 17.

Anti-inflammatory agents: Dipotassium glycyrrhetic acid at 0.3% and 0.3% ε-aminocaproic acid each weakly inhibited the photoaging response in hairless mice, but there was no significant difference from the control group receiving the vehicle. The potent anti-inflammatory agents indomethacin (0.03%) and ibuprofen piconol (0.3%) markedly inhibited the photoaging response in hairless mice as shown in Figure 17. The skin of hairless mice treated with these potent anti-inflammatory agents showed that inflammatory cell infiltration into the dermis was markedly

[f]Toshiba FL30-SE lamp, Toshiba, Tokyo, Japan

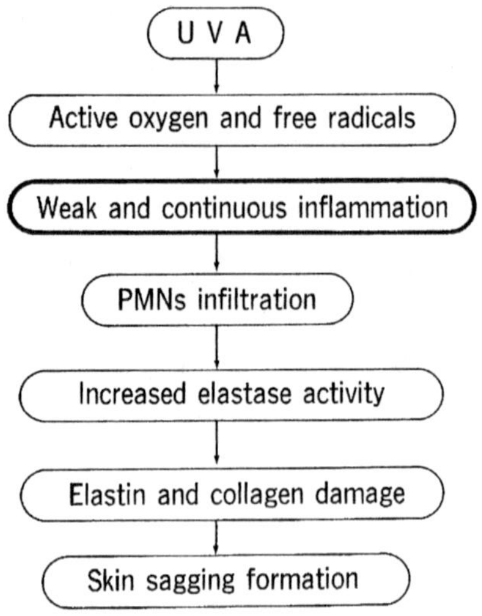

Figure 16. Hypothetical mechanism for development of sagging skin in chronically UVA-irradiated hairless-mouse skin

Table 2. Erythema scale, 0 to 3

0 =	No erythematous reaction is observed.
1 =	Slight erythema or erythema with vague border
2 =	Moderate erythema with well-defined border
3 =	Intense erythema with well-defined border (occasionally associated with edema)

inhibited. The inhibitory effect of these anti-inflammatory agents on the photoaging response in hairless mice was proportional to the anti-inflammatory potency of these agents (Figure 18).

New Anti-Inflammatory Substances

Because potent anti-inflammatory agents such as indomethacin have been reported to produce many adverse reactions,[8] they cannot be added to cosmetics for reasons of safety. We therefore screened various Chinese homeopathic preparations that were considered to have less tendency to cause adverse reactions. Among them, we selected the extracts of honeysuckle flower and *Engelhardtia chrysolepis* Hance, which showed moderate anti-inflammatory effects. These two Chinese homeopathic preparations markedly inhibited photoaging in hairless mice. They inhibited both infiltration of PMNs into the dermis and elastase activity in the skin of UVA-

[g]Parsol 1789, Hoffmann LaRoche, Basel, Switzerland

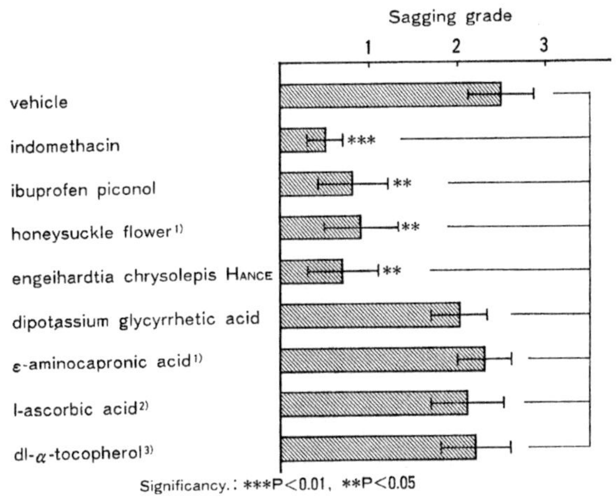

Figure 17. Inhibitory effects of various drugs on UVA-induced skin sagging in hairless mice; indomethacin was used at 0.03%, all other drugs at 0.3%. ***P < 0.01, **P < 0.05. Vehicle was propylene glycol and ethanol (1/1 w/w) except as marked:

1) Distilled water/ethanol 1/1
2) Distilled water/ethanol 3/2
3) Ethanol

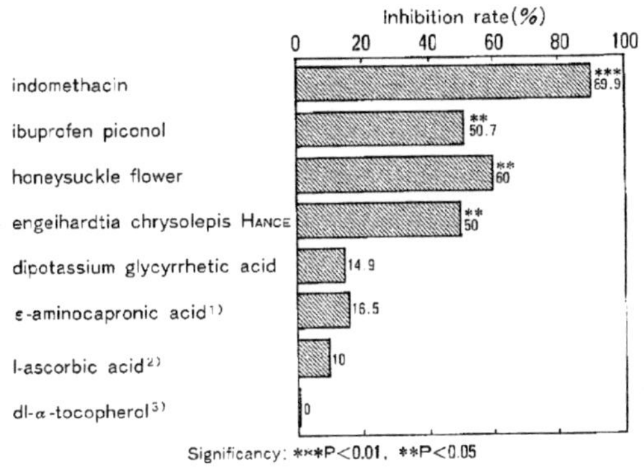

Figure 18. Inhibitory effects of various drugs on erythematous reactions in guinea pig dorsal skin; indomethacin was used at 0.03%, all other drugs at 0.3%. ***P < 0.01, **P < 0.05. Vehicle was propylene glycol and ethanol (1/1 w/w) except as marked:

1) Distilled water/ethanol 1/1
2) Distilled water/ethanol 3/2
3) Ethanol

irradiated hairless mice as can be seen in Figures 19, 20 and 21. The capacity for UV-absorbance of these two preparations was extremely low as compared with butyl methoxydibenzoylmethane, as shown in Figure 22.

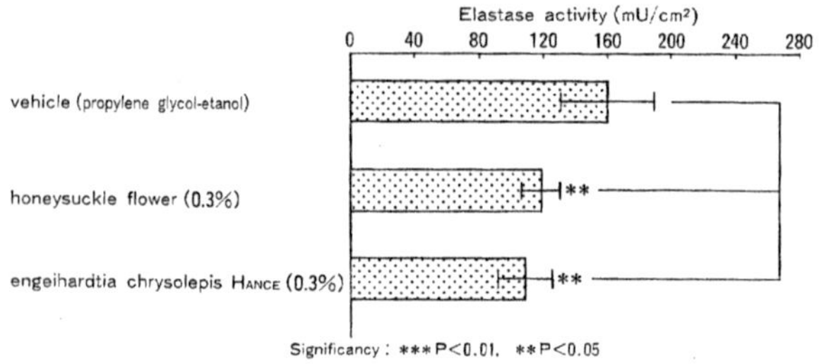

Figure 19. Effects of various drugs on elastase activity in UVA-irradiated hairless-mouse skin

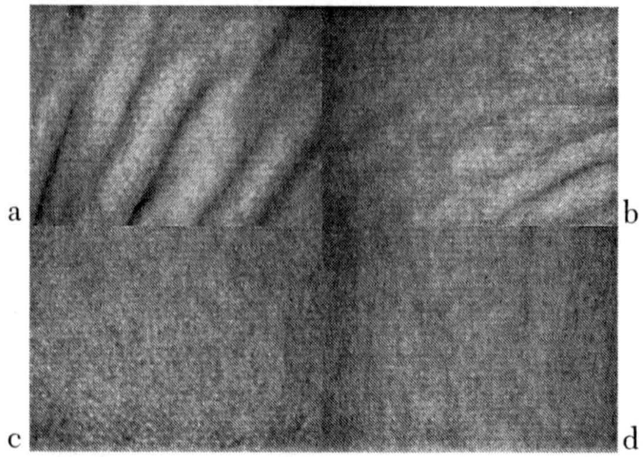

Figure 20. Visible changes at 24 weeks
a,b = Vehicle-treated UVA-irradiated hairless-mouse skin
c,d = UVA-irradiated hairless-mouse skin treated
a,c = Dorsal (back) skin
b,d = Abdominal skin

The usefulness of sunscreen preparations to prevent or reverse photoaging in hairless mice has been well documented.[5] Repair of collagen degradation, glycosaminoglycan deposition and elastin aggregation was observed with administration of high SPF sunscreen preparations to mice.[6] However, whether or not the sunscreen preparations may be truly useful and whether the usefulness may be altered by the vehicle remains unclear.

Recently, significant inhibition of elastin aggregation was reported in humans treated with UVA/UVB sunscreen preparations as compared with the vehicle. However, no significant differences were reported in inflammatory cell infiltration into the dermis, epidermal thickening and atrophy of keratinocytes between sunscreen-treated or vehicle-treated groups.[3]

In this study, the sunscreen preparation containing UVA absorbers moderately inhibited the photoaging response caused by long-term UVA irradiation. However, the skin of hairless mice treated with UVA absorbers showed some epidermal thickening, inflammatory cell infiltration into the dermis and partial elastin aggregation.

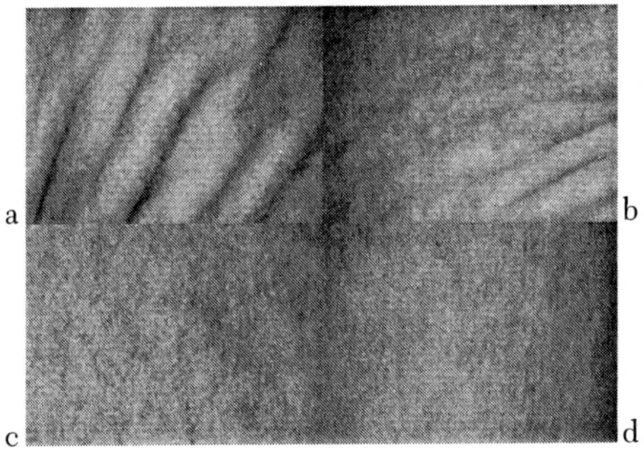

Figure 21. Visible changes at 24 weeks
a,b = Vehicle-treated UVA-irradiated hairless-mouse skin
c,d = UVA-irradiated hairless-mouse skin treated with
a,c = Dorsal (back) skin
b,d = Abdominal skin

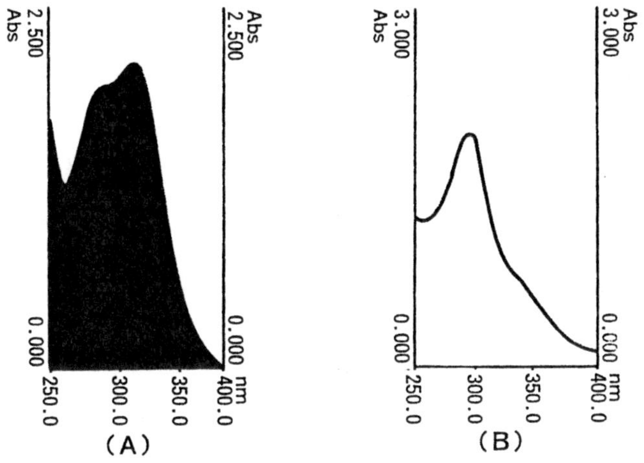

Figure 22. UV-absorbance spectra of honeysuckle flower extract (0.1%) and *Engelhardtia chysolepis* (0.01%); K = UV absorbance/sample conc. in g/l

	K	λ peak (nm)
Honeysuckle flower	2.1	323
Engelhardtia chrysolepis	19.2	290
Parsol 1789	111.0	357

Based on these results, we concluded that UVA-absorbing sunscreen preparations are somewhat, but not completely, effective in preventing photoaging.

Most commercially available sunscreen preparations that offer protection from UV radiation contain chemical absorbents and scattering agents. Our study results suggest that the use of sunscreen preparations alone may not be sufficient in preventing photoaging caused by UVA irradiation. Thus, the question arises: What should we use for the anti-photoaging cosmetics for the future?

We believe that anti-photoaging cosmetics should have a built-in anti-inflammatory effect and should be intended for daily use by all people, irrespective of geographical region, season, lifestyle, gender or age. It seems that skin care cosmetics for daily use containing the extracts of honeysuckle flower and *Engelhardtia chrysolepis* Hance may be useful for protecting the skin from photoaging.

Conclusions

1. The infiltration of PMNs into the dermis as observed in hairless mice submitted to long-term exposure to UVA radiation may play an important role in the photoaging response. The PMNs may release elastase which attack connective tissue proteins, resulting in damages of elastin and collagen fibers, and finally causing sagging.
2. Sunscreen preparations containing only UVA absorbents and scattering agents do not effectively inhibit the photoaging response to UVA irradiation.
3. The photoaging response was significantly inhibited by anti-inflammatory agents such as indomethacin, the inhibitory effect of which was proportional to the anti-inflammatory potency. However, anti-inflammatory agents such as indomethacin cannot be incorporated into cosmetics because of the possible adverse reactions. Therefore, we selected alternative anti-inflammatory agents — honeysuckle flower and *Engelhardtia chrysolepis* Hance extracts — because they showed the greatest anti-inflammatory effect among the Chinese homeopathic preparations tested. These two homeopathic preparations effectively inhibited the photoaging response and elastase activity.
4. The use of cosmetics having both anti-inflammatory effect and high safety level is a new anti-photoaging concept. Such cosmetics, when used on a daily basis, can prevent sagging skin caused by exposure to UVA radiation.

—Katsuhiro Motoyoshi, PhD, Yutaka Ota, Yuko Takuma and Masanori Takenouchi,
Pola R&D Laboratories, Pola Corporation, Yokohama, Japan

References

1. DL Bissett et al, *Photochem Photobiol* 46 367-378 (1987)
2. DL Bissett et al, *Photochem Photobiol* 50 763-769 (1989)
3. AS Boyd et al, *J Am Acad Dermatol* 33 941-946 (1995)
4. R Kasai et al, *Chem Pharm Bull* 36 (10) 4167-4170 (1988)
5. LH Kligman et al, *J Invest Dermatol* 78 181-189 (1982)
6. LH Kligman et al, *J Invest Dermatol* 81 98-102 (1983)
7. NJ Lowe et al, *J Invest Dermatol* 105 739-743 (1995)
8. Nippon Igaku Joho Center, Iryoyaku Nippon Iyakuhinshu, 10th ed, Yakugyo Jihosha, Tokyo (1986)
9. R Ogura et al, *J Invest Dermatol* 97 1044-1047 (1991)
10. H Ueda, *Active Oxygen & Free Radical* 3 291-297 (1992)

Ascorbic Acid and Its Derivatives in Cosmetic Formulations

Keywords: L-ascorbic acid, magnesium L-ascorbyl phosphate, magnesium ascorbate PCA, epidermis, skin penetration

The study evaluates skin penetration of L-ascorbic acid, magnesium L-ascorbyl phosphate and magnesium ascorbate PCA and their effects on the epidermis.

Ascorbic acid (vitamin C) and its derivatives have been widely used in cosmetic formulations. Ascorbic acid is a natural antioxidant that combats the reactive oxygen species that can cause damage to the lipids, proteins, and nucleic acids inside the cell, endangering tissue integrity.[6] The study of ascorbic acid in dermocosmetic formulations is important because of its suggested physiological functions in the skin. These include its inhibitory effect on melanogenesis,[3] free radical scavenger activity and anti-aging properties.[8]

The relationship between reactive oxygen species and melanomas, cutaneous photoaging and other disorders is well-known. These species are produced by sunlight associated with the presence of light absorbent powder components and molecular oxygen.[1,6]

Another property attributed to ascorbic acid is inhibition of the enzyme tyrosinase, which is the most important regulator of melanin pigment production. Topically applied ascorbic acid regulates production of melanin pigment.[2, 5, 9]

Derivatives

The use of ascorbic acid in cosmetic formulations is very difficult due to its low stability in aqueous solutions; oxidation easily occurs in formulations such as gels, cream-gels and o/w emulsions.[4, 7, 8] Therefore, scientists have looked for ascorbic acid derivatives that have an action similar to that of ascorbic acid but with better chemical stability and with comparable percutaneous penetration. As a result, magnesium L-ascorbyl phosphate and magnesium ascorbate PCA have been introduced on the world cosmetic market.

Magnesium L-ascorbyl phosphate was synthesized for higher stability than ascorbic acid when added to formulations with large aqueous content and pH above or equal to 7.[8] When penetrating the skin, magnesium L-ascorbyl phosphate releases free vitamin C by the action of phosphatase enzymes and then is able to exert its functions.[8]

Magnesium ascorbate PCA was synthesized for use in formulations with pH under 7 to permit its association with AHAs. This property promotes its percutaneous penetration and action.

The objectives of the present study were to comparatively evaluate the tissue changes and skin penetration occurring when formulations containing L-ascorbic acid, magnesium L-ascorbyl phosphate or magnesium ascorbate PCA are applied topically on guinea pig epidermis for a period of 15 days.

Epidermal Tissue Changes

Methods: We studied the effects of ascorbic acid and its derivatives on the guinea pig epidermis using a gel cream (Formula 1) on male guinea pigs weighing approximately 350 g.

The guinea pigs were shaved on the back. Five areas of 1.5 cm^2 each were used for the experiments. One untreated area was used for the control and the others for the application of gel cream formulation alone and gel cream formulations containing 1% ascorbic acid or its derivatives. The formulations were applied once a day for 15 days and biopsies were then obtained from each area using a dermatological punch. The material collected was analyzed by using histopathological, morphometric and stereological techniques.

Formula 1. Gel cream

Hydrogenated lecithin	0.8%
Squalane	4.0
Hydroxyethylcellulose	2.5
Propylene glycol	5.0
Methyl dibromoglutaronitrile (and) phenoxyethanol	0.02
Water (*aqua*), distilled	qs to 100 mL

Formula 2. O/W emulsion

Self-emulsifying base	7.0%
Mineral oil (*paraffinum liquidum*)	4.0
Propylene glycol	5.0
Methyl dibromoglutaronitrile (and) phenoxyethanol	0.02
Water (*aqua*), distilled	qs to 100 mL

Formula 3. Gel

Hydroxyethylcellulose	2.5%
Propylene glycol	5.0
Methyl dibromoglutaronitrile (and) phenoxyethanol	0.02
Water (*aqua*), distilled	qs to 100 mL

Results and discussion: Skin areas treated with the gel cream with or without the additives presented histologically a thickening of the basal, spinous and granulose layers of the epidermis compared to the untreated control (Figure 1 and Table 1). These layers also showed larger cells and nuclei (Table 2). This fact can be explained by the increased skin hydration promoted by these formulations, encouraging normal cell metabolism.

When we did the statistical analysis, our results concluded the addition of ascorbic acid or its derivatives to the vehicle increased the observed alterations. Ascorbic acid had the most visible action, followed by magnesium ascorbate PCA and magnesium L-ascorbyl phosphate, confirming the action of ascorbic acid and its derivatives on cell renewal. This may be due to the fact that ascorbic acid is an AHA in the lactone structure,[10] whose action on cell renewal has been extensively studied.[11] Ascorbic acid derivatives release it in the free form when it penetrates the cutaneous tissue, inducing the alterations observed.[8]

Skin Penetration

Methods: We studied the skin penetration of ascorbic acid and its derivatives by using three vehicles applied to harvested guinea pig skin in a glass Franz diffusion cell model. The three vehicles were an o/w emulsion (Formula 2), a gel cream

Table 1. Mean total thickness (μm) for the different epithelial layers of the guinea pig skin treated with gel cream containing 1% ascorbic acid or its derivatives

Layers	Control	Gel Cream	AA	MAP	MAPCA
Basal	5.25	9.0	14.0	12.5	12.5
Spinous	7.25	15.0	23.5	14.0	19.5
Granulose	4.25	9.5	11.0	11.0	13.0
Total	16.75	33.5	48.5	37.5	45.0

AA = L-ascorbic acid
MAP = magnesium L-ascorbyl phosphate
MAPCA = magnesium ascorbate PCA
Statistically significant $P< 0.01$. $n=5$. Mann Whitney test

Table 2. Mean volume (μm³) of the nuclei of cells in the different epithelial layers of the guinea pig skin treated with gel cream containing 1% ascorbic acid or its derivatives

Layers	Control	Gel Cream	AA	MAP	MAPCA
Basal	130.64	172.31	304.92	248.67	230.07
Spinous	162.07	276.34	349.56	310.41	257.36
Granulose	243.63	356.79	397.05	337.72	299.48

AA = L-ascorbic acid
MAP = magnesium L-ascorbyl phosphate
MAPCA = magnesium ascorbate PCA
Statistically significant $P< 0.05$. $n=5$. Mann Whitney test

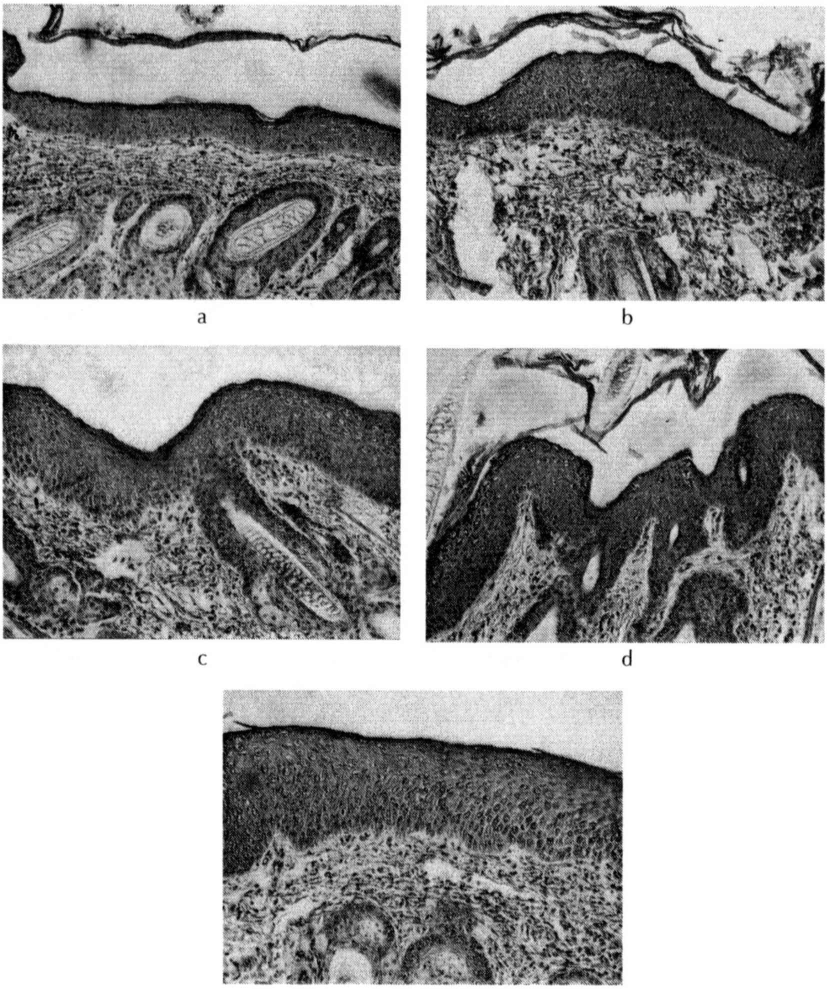

Figure 1. Photomicrograph of the skin from a guinea pig treated for 15 days. (a) control skin; (b) area treated with gel cream; (c) area treated with gel cream containing ascorbic acid; (d) area treated with gel cream containing magnesium L-ascorbyl phosphate; (e) area treated with gel cream containing magnesium ascorbate PCA. Note the thicker epithelium with larger and more numerous nuclei for all treated areas. (Magnification: x 200) (Stain: Hematoxylin and Eosin)

(Formula 1) and a gel (Formula 3). All formulations were supplemented with 1% ascorbic acid, magnesium ascorbyl phosphate or magnesium ascorbate PCA.

The receiver phase of the diffusion cell was phosphate buffer, pH 7.5. Samples of each test formulation (2 g) were applied to a second group of animals. Application was to the dorsal skin of female guinea pigs weighing 350 g. The receiver volume used for each cell was calibrated prior to use, and was constantly stirred with a

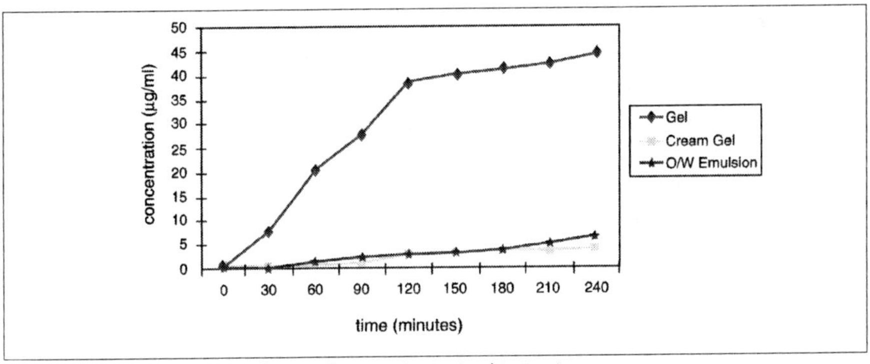

Figure 2. Penetration of ascorbic acid into guinea pig skin in vitro

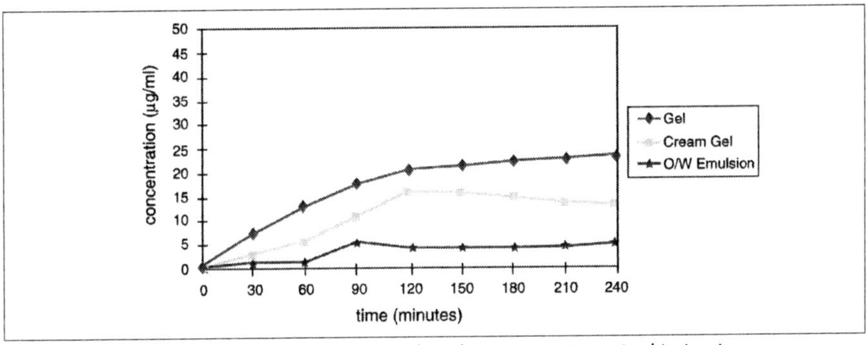

Figure 3. Penetration of magnesium ascorbyl phosphate into guinea pig skin in vitro

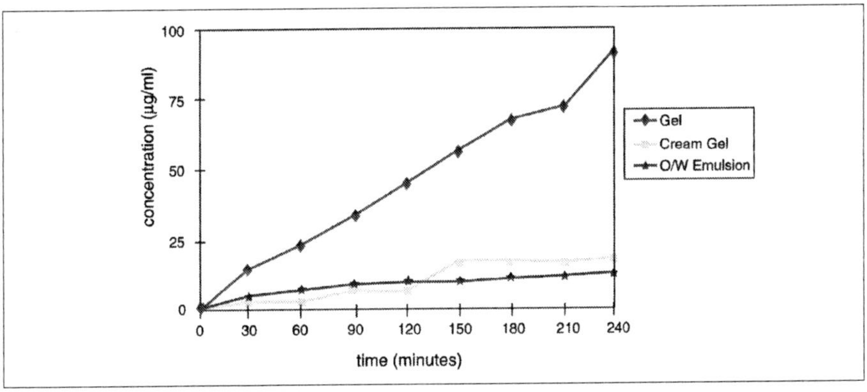

Figure 4. Penetration of magnesium ascorbate PCA into guinea pig skin in vitro

Teflon-coated bar magnet. A uniform 37°C receiver temperature was maintained with a jacket connected to a circulating bath. Portions were obtained at 0, 30, 60, 90, 120, 150 and 240 minutes. The amounts of ascorbic acid or its derivatives in the receiver phase were obtained by ultraviolet penetration analysis at 254 nm.

Results and discussion: As seen in Figures 2, 3 and 4, the magnesium ascorbate PCA was more readily absorbed by the skin than L-ascorbic acid or magnesium ascorbyl phosphate. The best vehicle for penetration into the skin was the gel formulation, followed by the emulsion for L-ascorbic acid and the gel cream for magnesium ascorbyl phosphate.

It is well-known that the polarity of a substance is important for its cutaneous penetration, and this was possibly a reason for the differences in penetration levels observed.

Some characteristics of the vehicle are necessary to facilitate the cutaneous penetration of the active principle added, such as ideal pH and viscosity. As previously mentioned, the gel was the best vehicle for penetration.

Formulating Considerations

Since L-ascorbic acid has several important properties for the prevention and treatment of skin aging, the development of formulations for skin care demands special care, such as the selection of vehicle and agents that will favor stability for the longest possible time. Confirmation of the biological activity of a topically applied active substance at the epithelial level is an essential factor for its use.

The cosmetics formulations, regardless of the active substances contained, must undergo rigorous quality control tests, in both the chemical-physical aspect and biological action.

The formulators must highlight the shelf life of formulations containing active substances subject to oxidation reactions, as in the case of vitamin C. These reactions cause product degradation, reducing its potential and resulting in skin damage through the presence of free radicals.

Conclusion

The results obtained under the present experimental conditions permit us to conclude that the magnesium ascorbate PCA was more readily absorbed by the skin than L-ascorbic acid and magnesium ascorbyl phosphate. The best vehicle for penetration into the skin was the gel formulation, followed by the emulsion for L-ascorbic acid and the gel cream for magnesium ascorbyl phosphate.

The gel cream formulation has a hydrating action on the epidermis of guinea pigs after 15 days of cutaneous treatment. It was also clear that the addition of vitamin C or its derivatives to this formulation intensified the intra- and extracellular hydration observed, a fact that favors normal cell metabolism.

—**Patricia M. B. G. Maia Campos and Gisele Mara Silva,** *Faculty of Pharmaceutical Sciences of Ribeirão Preto, University of São Paulo, Brazil*

References

1. D Darr and I Fridovich, *J Inv Derm* (Review) 102(5) 670 (1994)
2. K Iozumi, GE Hogansom, R Pennela, MA Everett and BB Fuller, *J Inv Derm* 100(6) 806 (1993)
3. CL Phillips, SB Combs and SR Pinell, *J Inv Derm* 103(2) 229 (1994)
4. SR Pinnell, *Rev Cosm Med Est* 3(4) 31 (1995)
5. G Prota, *J Inv Derm* 100(2) Sup 156-S (1993)
6. Y Shindo, E Witt and L Packer, *J Inv Derm* 100(3) 261 (1993)
7. GM Silva and PM Maia Campos, *XIX IFSCC Congress*, Acapulco, Mexico, 3 pp21-225 (1997)
8. M Tagawa, K Uji and Y Tabata, *XV IFSCC International Congress,* Yokohama, Japan, 3 p 399 (1988)
9. H Takashima, H Nomura, Y Imai and H Mima, *Am Perf Cosm* 86(7) 29 (1971)
10. R Hermitte, Pele envelhecida: retinóides e alfa-hidróxi ácidos, *Cosmet Toil Portugues* 5(5) 55-58 (1993)

Self-Tanners: Formulating with Dihydroxyacetone

Keywords: dihydroxyacetone, DHA, self-tanner, self-tanning, DHA dimers, Maillard reaction, DHA formulations

This paper discusses the background behind the development of self-tanning products using DHA. Its structure, chemistry, self-tanning mechanism and formulation guidelines are also discussed.

The majority of cosmetic formulations are designed to improve biological surfaces through some physical interaction. For example, shampoos help physically remove oil and dirt from hair; makeups physically change the color of skin. While these products significantly modify the appearance of their targeted surfaces, the effects are temporary because they do not chemically modify the surface.

There are, however, a few cosmetic product types that rely on chemical reactions with the biological surfaces to create more "permanent" effects. In hair care, these types include dyes, perms and relaxers. In skin care, the most common type is sunless tanners.

A well-rounded cosmetic chemist must be familiar with the chemical reaction involved in these types of cosmetics and the special formulating nuances that go along with creating them. In this article, the primary active for sunless tanners, DHA, is reviewed and described at length.

Background

The tanned look: The prevailing cultural aesthetic among fair-skinned people in many countries is to have tanned skin, which means skin with a light yellowish brown or even deeper brown color produced by exposure to the sun or to an artificial source of ultraviolet light. The tanning of the skin is produced as a result of the darkening of preformed melanin, accelerated formation of new melanin, and retention of melanin in the epidermis as a result of retardation of keratinization.

For those individuals who wish to achieve tanned skin, the most readily available means for doing so is by exposing their skin to natural sunlight. However, this method carries certain hazards. Chief among those hazards is the risk of sunburn, which is an injury to the skin produced by excessive exposure to ultraviolet rays. The injury is accompanied by erythema, tenderness and sometimes blistering.[1] Furthermore, excessive exposure to ultraviolet radiation is considered by the medical community to be a leading factor in skin cancer and premature aging.[2] So a completely safe natural tan is out of the question.

The solution to achieving a tanned look is to use self-tanners. They are available in hundreds of brands of creams, gels, mousses and sprays at drug stores, department stores and salons. They all work the same way. That is, they all have the same active ingredient: dihydroxyacetone (DHA), also known as 1,3-dihydroxypropanone, according to International Union of Pure and Applied Chemistry (IUPAC) nomenclature.

Discovery of DHA's browning effect: The browning effect of DHA was discovered by accident. In the mid-1950s, at Children's Hospital at the University of Cincinnati, Dr. Eva Wittgenstein was studying the effect of large oral doses of DHA in children who had glycogen storage disease. Sometimes the children vomited some of the sweet concentrated material, and it splashed on their skin. A few hours later, it was noticed that the children had brown spots on their skin where stray splashes hadn't been wiped off.

Wittgenstein was able to do something with her observation other than berating the staff for not getting those kids cleaned up. Curious, she prepared aqueous solutions of varying concentrations of DHA and was able to reproduce the pigmentation on her own skin. She was able to turn her skin brown.[3] This was the beginning of self-tanning products based on DHA. Additional information on this subject has recently been reviewed.[4,5]

Acceptability of DHA's cosmetic effect: The cosmetic acceptability of DHA products seems to have improved in recent years. More natural-appearing brown or golden hues are produced, as opposed to more off-color oranges observed with older formulations. The shades obtained may be more acceptable to medium-complexioned people than those with darker or fair complexions. Purer supplies of DHA, refining of DHA-containing vehicles to allow better penetration, recognition of the need for a lower pH, and more rapid color change with a lower concentration of DHA may have all contributed to the improved outcome.

The US Food and Drug Administration has approved DHA as a permanently listed colorant (21 *CFR* § 73.2150) exempt from certification for drugs and cosmetics. It may be used only in externally applied drugs and cosmetics intended solely or in part to give a color to the human body. A monograph for DHA has been included in the ninth supplement of the United States Pharmacoepia 23.

The European Economic Community's EEC Cosmetics Directive (76/768/EEC) places no limitations on DHA. The Japanese Cosmetic Ingredient Dictionary (JCID-II-80) includes DHA as a cosmetic ingredient.

Structure and Chemistry

Structure: DHA is a simple three-carbon sugar. It is nontoxic. In fact, it is an intermediate in carbohydrate metabolism in higher plants and animals, and is more rapidly metabolized than glucose in the body. Specifically, this three-carbon sugar is a physiological product (dihydroxyacetone monophosphate) of the body formed and utilized during glycolosis. DHA used in self-tanners is prepared mainly by fermentation of glycerol using *Gluconobacter oxydans*.[6]

DHA is a slightly hygroscopic material, which should be shipped and stored in tightly closed containers under refrigeration (<6°C). Degradation of DHA increases with higher temperature and higher pH.[7] DHA can be assayed using periodic acid by following the method described in the Official Monographs, USP 23.

In crystalline form, DHA is a mixture of one monomer and four dimers (Figure 1).[8] The dimeric forms with 1,4-dioxane structure predominate;[9] they result from intermolecular hemiketal formation. The dimeric DHA in the form of a cyclic ketal and trimeric DHA in the form of a bicyclic ketal have also been reported.[10,11] The monomer is formed by heating or melting dimeric DHA or by dissolving it in water. In aqueous solution, DHA occurs as a monomer that can gradually tautomerize into glyceraldehyde. The equilibrium shifts to DHA at acidic pH or to glyceraldehyde at alkaline pH.

Interconversion of DHA and glyceraldehyde was studied in different solvents and temperatures by FTIR spectroscopy.[12] Dissolution in water and increasing temperatures caused the dissociation of the dimeric forms of both compounds into monomers and the subsequent inter-conversion of DHA and glyceraldehyde. A five-membered ring form was predominant in aqueous solutions of the dissociated glyceraldehyde dimer, whereas a six-membered ring form was preferred in aqueous solutions of dissociated DHA dimer. DHA predominantly converted to the six-membered H-bonded conformation of glyceraldehyde when dissolved in water. This was attributed to the preferential formation of the trans- or E-enediol as an intermediate. Temperature-dependent spectra have indicated that increasing the temperature favored the formation of glyceraldehyde in the aqueous equilibrium mixtures of dimeric glyceraldehyde and DHA.

Chemistry: The site of action of DHA in the skin is the stratum corneum.[13] The first step of the self-tanning reaction is the conversion of DHA to pyruvaldehyde

The SPF Effect of DHA

We know that DHA can produce an artificial tan on human skin. We also know that the reaction product of DHA and the skin protein that produces the "tan" color is an effective sunscreen agent.[27-30] Experimental and clinical evidence show that skin that has been treated topically with 3% DHA solution overnight has a Sun Protection Factor (SPF) of at least 3 in the UVB region. Likewise, a photoprotection factor of 10 in the UVA region has been observed in skin treated topically with a 15% solution of DHA.

The advantage of this DHA-induced skin pigmentation is that it cannot be removed by perspiration, swimming or washing. It can only be removed by desquamation. It provides photoprotection both in the UVB and UVA regions when the regular sunscreen has been lost. However, one will still need plenty of sunscreen, sunglasses, and a big hat to remain protected.

This photoprotection property has been used in protecting uninvolved skin by DHA during psoralene-UVA (PUVA) treatment of psoriasis.[31] By preferentially protecting uninvolved skin, DHA allows higher UVA doses to be tolerated and delivered to the psoriatic plaques with acceleration in clearing.

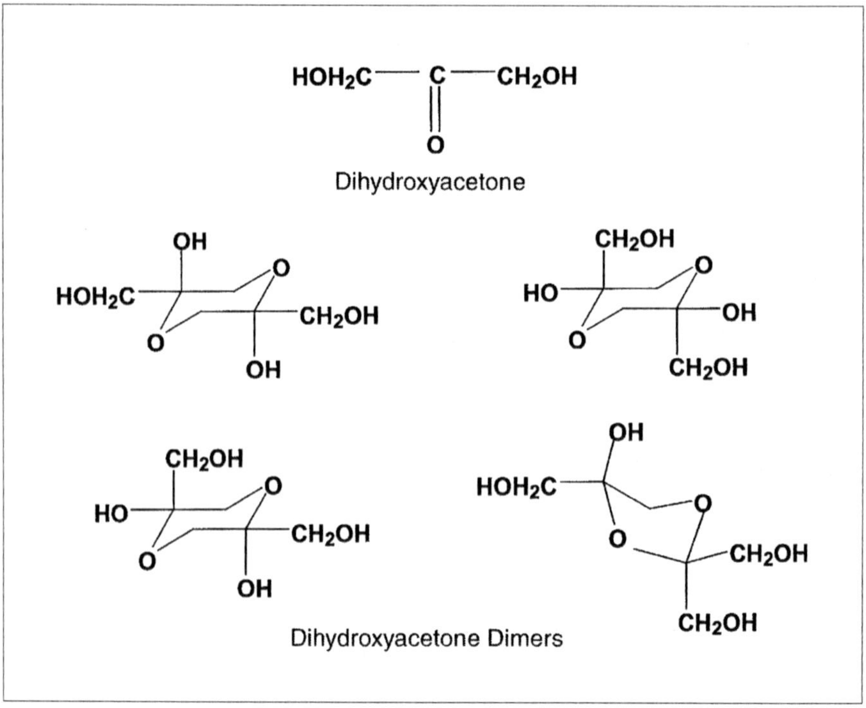

Figure 1. Structures of monomeric and dimeric dihydroxyacetone

with the elimination of water (Figure 2). Then the keto or aldehyde function reacts with the amine functionality of skin keratin to form an imine. Subsequent steps of this reaction are a complex chain of reactions and are not fully understood. However, it is well-known that the resulting products are cyclic and linear polymers having yellow or brown color.[14,15]

The self-tanning process takes place in the outer layers of the epidermis. Only the monomeric form undergoes the Maillard reaction that leads to tanning. In this process, named after Louis-Camille Maillard, who first described it in 1912, amino acids interact with sugars to create brown or golden brown compounds. The Maillard reaction is defined currently as the reaction of the amino group of amino acids, peptides, or proteins with the glycosidic hydroxyl group of sugars,[16] forming brown products referred to as melanoidins.

Melanoidins are polymeric compounds, which are linked by lysine side-chains to the proteins of the stratum corneum. Although the formation of melanoidins is different from that of melanin, some of their properties are similar, especially their absorption spectra.[17] Melanins consist of aromatic amino acids, originating mainly from tyrosine. Melanoidines, on the other hand, consist mainly of an aliphatic moiety with very few aromatic functions in the side chains.

pH plays a very important role in tanning. To achieve a uniform tan, the skin should be exfoliated to remove loose skin scales before applying DHA. The

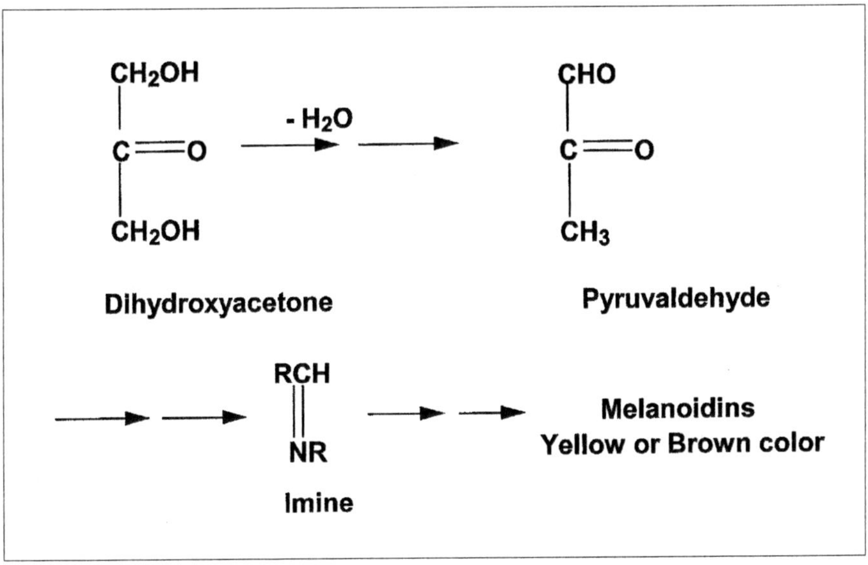

Figure 2. Self-tanning reaction of dihydroxyacetone

optimum pH for the Maillard reaction is between 5 and 6, which is the normal pH of healthy skin. When formulations with a higher or lower pH are applied, the skin buffer adjusts the pH on the skin to the optimal tanning pH. Tests showed that un-buffered formulations with a pH between 2 and 6 always lead to the same coloring results. Only when the pH was above 6 was the brown intensity reduced. However, the desired formulation pH for DHA is between 3 and 4.

Formula Types

Tanning products fall loosely into two categories: wash-offs and wear-offs. Wash-offs are cosmetic tanning products in cream or gel formulas, without DHA, that give the appearance of a tan, but wash-off in the shower at the end of the day. Wash-offs exist in a wide range of products available from all major cosmetic companies.

Wear-off formulations containing DHA are the true self-tanning products. They are available in creams, lotions, milks, gels or sprays. When these products are applied on the skin, they develop a tan over a matter of hours and wear off over a matter of days.

By far the most popular of all vehicles used for self-tanning products, the emulsion, offers a wide variety of options. DHA is most often formulated in o/w emulsions. Lotions are more popular than creams owing to their ease of spreadability on the skin and dispensability from bottles. Creams can lead to more intense tan than lotions because the applied film is thicker.

From an aesthetic viewpoint, emulsions are an elegant medium that can give the skin smooth silky feel without being greasy or tacky. They can

Formula 1. Self-Tanning Lotion (O/W)

A	Glycerol stearate (Arlatone 983 S, ICI)	1.50% w/w
	Ethoxylated fatty alcohol (Arlatone 985, ICI)	2.20
	Steareth-10 (Brij 76, ICI)	1.50
	Mineral oil *(paraffinum liquidum)*, medium viscosity	5.00
	Caprylic/Capric triglyceride (Miglyol 812, Huls)	5.00
B	Sorbitol (Sorbitol F liquid, Rona)	2.50
	Propylene glycol	2.50
	Preservative	qs
	Water *(aqua)*, demineralized	qs 100.00
C	Water *(aqua)*, deionized	5.00
	Dihydroxyacetone (Dihydroxyacetone, Rona)	5.00
		100.00

Procedure: Heat A to 75°C and B to 80°C. Add B slowly to A while stirring. Homogenize. Cool while stirring. Add C at 40°C.

Note: Viscosity 2,250 cps (Brookfield RV#3, 20 rpm @ 25°C)

accommodate a wide variety of raw materials. Sunscreens and other ingredients can be incorporated in the emulsion to make additional product claims.

Due to high water solubility of DHA, aqueous or aqueous-alcoholic lotions and gels can be prepared easily. By appropriate control of viscosity, spray lotions or gels can also be developed.

Formulation Guidelines

The content of DHA in tanning products depends on the desired browning intensity on the skin and is normally in the range of 4-8%. Depending on the type of formulation and skin type, a tanning effect appears on the skin in approximately 2-3 hours after use. During product storage, the pH of a DHA-containing formulation will drift over time to about 3-4. At this pH, DHA is quite stable. In order to ensure end-product stability, the following factors must be considered before developing new formulations containing DHA.

Temperature: Heating DHA above 40°C for a long period of time must be avoided as it causes rapid degradation of DHA (> 40% within three months). During manufacturing processes involving heating, as in the case of emulsions, DHA should not be added until the formulation has been cooled down to below 40°C. Products containing DHA should be stored in an opaque and resealable package.

pH: When DHA is incorporated in a formulation, the pH of the formulation drops to 3-4 within about two days from the date of preparation. In the past,

Formula 2. Self-Tanning Lotion

A	Water *(aqua)*, demineralized	39.10% w/w
	Propylene glycol	5.00
	Methylparaben	0.20
	Propylparaben	0.10
	Xanthan gum (Keltrol CGT, Calgon)	0.75
B	Glyceryl stearate (and) PEG-100 stearate (Arlacel 165, Uniqema)	1.20
	PEG-40 stearate (Myrj-52-S, Uniqema)	0.40
	Steareth-2 (Brij-72, Uniqema)	0.70
	Cetearyl alcohol (and) ceteareth-20 (Cosmowax J, Croda)	1.00
	Cetyl alcohol (Crodacol C-70, Croda)	1.50
	Cyclomethicone (Dow Corning 344 Fluid, Dow Corning)	5.00
	Dimethicone (Dow Corning 200 Fluid, 100 cst, Dow Corning)	0.50
	Isononyl isononanoate (Pelemol IN-2, Phoenix)	28.50
C	Citric acid, 50% soln	0.05
D	Water *(aqua)*, demineralized	10.00
	Dihydroxyacetone (Dihydroxyacetone, Rona)	5.00
E	Phenoxyethanol (Phenoxyethanol, Nipa)	1.00
		100.00

Procedure: Combine all ingredients of A except xanthan gum. Disperse xanthan gum into A with agitation. Heat A to 70-75°C. Combine B and heat to 70-75°C. Add B to A while stirring. Gently homogenize until 50°C. Cool down while stirring and add C at 40°C. Adjust pH to 5 with C. Add D at 40°C. Add E. Mix until uniform.

Note: Viscosity 4,000 cps (Brookfield RV#4, 20 rpm @ 25°C)

buffering was recommended to keep the pH at a level of 4-6. Recent investigations in our laboratories revealed that storage stability of DHA could be increased when the formulations are kept at pH 3-4. Buffering at a higher pH is counterproductive, as it enhances degradation of DHA thereby reducing its storage stability. pH 3, however, is not the optimal pH for the tanning reaction; skin buffers bring the applied formulation to the optimal tanning pH (5-6) and then the tanning occurs.

Buffers: Use of buffers to maintain the pH of a formulation above 4.5 is not recommended. The pH of the formulation may be adjusted to approximately 3-4 by using a small amount of citric acid or using acetate buffer because they do not affect DHA stability.[5]

Phosphate buffers are not recommended. At pH values above 7, brown-colored compounds are produced via isomerization and condensation reactions, and tanning efficacy is reduced.

Stability evaluation: The stability evaluation of self-tanning products is not unlike that required for other cosmetic products because DHA is not stable above

Formula 3. Self-Tanning Lotion

A	Water *(aqua)*, demineralized	58.80% w/w
	Propylene glycol	4.00
	Methylparaben	0.15
	Propylparaben	0.05
	Hydroxyethylcellulose (Natrosol 250 HHR, Aqualon)	0.25
	Xanthan gum (Keltrol CGT, Calgon)	0.25
B	Steareth-10 (and) steareth-7 (and) stearyl alcohol (Emulgator E2155, Goldschmidt)	2.00
	Glyceryl stearate (and) ceteth-20 (Teginacid H, Goldschmidt)	2.00
	Glyceryl stearate (Tegin M, Goldschmidt)	3.00
	Cetearyl octanoate (Luvitol EHO, BASF)	5.00
	Oleyl oleate (Schercemol OLO, Scher)	5.00
	Microcrystalline wax (White Microcrystalline wax 1275W, Frank B Ross)	1.00
	Capric/Caprylic triglyceride (Myritol 318, Cognis)	3.00
	Dimethicone (Dow Corning 200 Fluid, 100cst, Dow Corning)	0.50
C	Water *(aqua)*, demineralized	10.00
	Dihydroxyacetone (Dihydroxyacetone, Rona)	5.00
		100.00

Procedure: Combine water, propylene glycol, methylparaben and propylparaben from A. Disperse hydroxyethylcellulose and xanthan gum into A with agitation. Heat A to 70-75°C. Combine B and heat B to 70-75°C. Add B to A while stirring. Gently homogenize until 50°C. Cool down while stirring and add C at 40°C. Mix until uniform.

Note: Viscosity 5,200 cps (Brookfield RV#4, 20 rpm @ 25°C)

40°C. The finished formulation, in the production package, must be subjected to both high (40°C for 3 months or longer) and low (–10°C to 25°C, five freeze-thaw cycles) temperatures. Color, odor, viscosity, pH and DHA stability must be monitored to ensure a long-term storage stability of the final products. DHA content in cosmetic formulations can be determined by converting DHA to glycerol using glycerol dehydrogenase and NADH (reduced nicotinamide adenine dinucleotide) and measuring[a] the absorbance of NADH at λ_{max} 365 nm.

Selection of Ingredients

In order to have a stable DHA formulation, one must carefully select appropriate ingredients for each formula because there is always a potential for ingredient interaction. Certainly, all ingredients selected must be cosmetically acceptable, have a good safety record, be stable, and must not interfere with the efficacy of DHA in any way. The following are some comments on how to select ingredients.

Formula 4. O/W Self-Tanning Cream with UV Filter

A	Octyl methoxycinnamate (Eusolex[a] 2292, Rona)	5.00% w/w
	Benzophenone-3 (Eusolex 4360, Rona)	2.00
	Glyceryl stearate (and) steareth-25 (and) ceteth-20 (and) stearyl alcohol (Tego-Care 150, Goldschmidt)	8.00
	Cetearyl alcohol (Lanette O, Henkel)	1.50
	Stearoxy dimethicone (Abil Wax 2434, Goldschmidt)	1.60
	Cetearyl octanoate (Luvitol EHO, BASF)	5.00
	Isopropyl myristate (Emerest 2314, Henkel)	3.00
	Caprylic/Capric triglyceride (Miglyol 812 neutral oil, Huls America)	5.00
	Dimethicone (Dow Corning 200, 350 cs, Dow Corning)	0.50
B	Propylene glycol (1,2-Propanediol, BASF)	3.00
	Methylparaben	0.15
	Propylparaben	0.05
	Water *(aqua)*, deionized	50.20
C	Dihydroxyacetone (Dihydroxyacetone, Rona)	5.00
	Water *(aqua)*, deionized	10.00
		100.00

Procedure: Heat A to 75°C, B to 80°C. Add B slowly to A while stirring. Homogenize. Cool down while stirring and add C at 40°C.

Note: Viscosity 38,800 cps (Brookfield RVT, Sp. F, 10 rpm @ 25°C)

Note: SPF (by Diffey method) = 19.0; UVA PF = 9.5

[a] Eusolex is a registered trademark of Rona/EM Industries, Hawthorne, NY, USA.

Emulsifers: The use of nonionic emulsifiers is recommended over ionic emulsifiers because of improved stability of DHA in the formulations.

Emollients: Emollients represent one of the most important classes of emulsion components. They provide a silky skin feel on application.

There are many types of emollients: esters, waxes, fatty alcohols, mineral oils and silicone materials. Esters are the most common classes of emollients used in self-tanning products. Silicone-based oils have also enjoyed an increase in popularity in recent years. Dimethicone and cyclomethicones are the widely used materials of this type. All these emollients are compatible with DHA.

Thickeners: Thickening a formulation containing DHA, especially to produce a clear gel, is relatively difficult because many conventional cosmetic thickeners are not compatible with DHA. Scientists at Merck KGaA, Darmstadt, Germany screened a wide variety of thickeners in the early 1990s. The simple method used for this study called for a solution of 5% DHA and a thickener, storing at room

Formula 5. Sunless Tanning Spray-Lotion

A	Glyceryl stearate (and) ceteareth-20 (and) cetearyl alcohol (and) ceteareth-12 (and) cetyl palmitate (Emulgade SE, Henkel)	4.50% w/w
	Ceteareth-20 (Eumulgin B2, Henkel)	1.00
	Dicapryl ether (Cetiol OE, Henkel)	5.00
	Coco caprylate/caprate (Cetiol LC, Henkel)	5.00
B	Water *(aqua)*, demineralized	qs 55.00
C	Water *(aqua)*, demineralized	20.00
	Propylene glycol	2.50
	Dihydroxyacetone (Dihydroxyacetone, Rona)	6.00
	Propylene glycol (and) DMDM hydantoin (and) methylparaben (Paragon, McIntyre)	1.00
		100.00

Procedure: Combine A, stir and heat to 80-85°C. Heat B to 80-85°C. Slowly add A to B while stirring with a propeller mixer. Homogenize AB, allowing the mixture to cool. Combine C at room temperature by stirring. Add C to AB, when AB temperature has reached 30°C. Adjust pH to 3.5-4.0 with citric acid if needed.

Note: Viscosity < 100 cps (Brookfield RV#1, 50 rpm @ 25°C)

Emulsion Stability Freeze/Thaw – no separation after 5 cycles

Emulsion Stability 50°C – no separation after 4 weeks

temperature and at 40°C for three months, and analyzing for DHA content and solution colors.

Based on this and other studies, they determined that three good choices for thickeners are hydroxyethylcellulose, methycellulose and silica. Additionally, xanthan gum and polyquaternium-10 may also be used for thickening emulsions. Carbomers, sodium carboxymethylcellulose, PVM/MA decadiene crosspolymer, and magnesium aluminum silicate are not acceptable thickeners because they cause a rapid degradation of DHA at 40°C.

Moisturizers: The natural water content of stratum corneum is not sufficient to cause the tanning reaction. Therefore, water-free systems are not at all useful for formulating tanning products. Moisturizers, like sorbitol or propylene glycol, at a level of approximately 20% (w/w of formulation) help increase the tanning intensity.

Preservatives: Aqueous solution or emulsion of DHA is susceptible to microbial attack. For this reason, preservation of the final product as well as hygienically clean manufacturing and packaging procedures are particularly important. Parabens and phenoxyethanol, alone or in combination, are recommended preservatives.

Compounds containing nitrogen: Amines or other nitrogen-containing compounds should be avoided in the final formulation. These compounds include collagen, urea derivatives, amino acids and proteins.

The reactivity of DHA toward these nitrogen-containing compounds can lead to gradual breakdown of DHA and, thus, to a loss of efficacy in use. However, many of the available commercial formulations include amino acids. This combination gives a perceptual advantage; customers begin to see a tanning effect within about 45 minutes due to the reaction between DHA and the amino acids. Only glycine and histidine provide brown color. However, this tan is not substantive, and part of it is easily washed off.[5] In order to prevent the reaction from occurring before application, two suppliers have patented a two-compartment dispenser; the amino acids are in one compartment and the DHA in the other.[18]

Sunscreens: It must be noted that a tan achieved with DHA alone does not offer sun protection comparable to that resulting from natural melanin production during sunbathing. It is, however, possible to combine DHA with non-nitrogen-containing sunscreens, such as ethylhexyl methoxycinnamate, ethylhexyl salicylate, homosalate, benzophenone-3 or octocrylene. The combination delivers a product with additional sun protection effect. Inorganic sunscreens, such as titanium dioxide, zinc oxide, and other metal oxides, must be avoided because they induce rapid degradation of DHA.

Fragrance oils: The choice of a proper fragrance can enhance the aesthetics of any self-tanning formulation. However, their use level must be kept as low as possible because fragrance oils can be irritating. They can also discolor formulations or degrade DHA. Their breakdown products can be photosensitizers. Thus, great care must be taken in choosing them.

Tinting: To achieve a makeup effect with self-tanning products, we suggest combining DHA with organic compounds such as carotene, D&C Blue No. 4, staining dyes such as D&C Red No. 21, and other approved dyes. The use of iron oxides must be avoided, because DHA will degrade rapidly.

Quantification of Tan

The human eye is able to differentiate colors very easily, including the color obtained from self-tanning products. The challenges are to remember a color difference and to quantify it. Color standards like Pantone color cards can be useful, but the number of colors in such a system is very limited.

One successful method to evaluate the tan is chromometric measurement,[19,20] specifically, the L^* value (lightness), a^* value (indicates the color in red-green axis) and b^* value (indicates color in blue-yellow axis) system. With this method, L^* values for tanned skin were found to be too sensitive to different light conditions or pressure forces, whereas a^* and b^* remained quite stable.[21] From the a^* and b^* values, one can calculate two additional parameters, which are the hue angle (h) and the hue intensity (C):

$$h = \arctan (b^*/a^*)$$
$$C = \sqrt{a^{*2} + b^{*2}}$$

When the hue intensity (C) is higher, the tanning is more intense. The hue angle (h) for Caucasian skin remained in the yellow-red area. A good correlation between the visual impressions and the C or h values was demonstrated.[21] The eye still has a greater sensitivity than the chromometric method, but the latter has the advantage of quantifying the difference in tan color and intensity.

Formulations

An investigation of several self-tanning products on the market revealed that in some formulations, the content of DHA decreases considerably even when stored at room temperature. An elevated pH appears to be the main reason for DHA decomposition.

One can easily make a fairly good formulation if one follows the guidelines above regarding temperature, pH, buffers, stability evaluation, and selection of ingredients. Using those guidelines, Thekla Kurz of Merck KGaA, Darmstadt, Germany prepared a self-tanning o/w lotion (Formula 1) and studied its stability at room temperature storage for three and a half years. Only 10% of the DHA was lost (5% to 4.5%).

Several recent patents claim to have developed self-tanning compositions with improved stability. Estée Lauder scientists have developed a self-tanning composition using DHA and cyclodextrins.[22] The composition was found to be stable and have reduced odor associated with the reaction between DHA and the skin. They have also shown that an effective amount of a long-chain α-hydroxy acid provides enhanced stability of DHA in formulated products.[23]

Kurz et al. have developed a stable formulation that consists of DHA, metabisulfite salts and magnesium stearate.[24] The improved stability is achieved by the action of the metabisulfite salts scavenging formaldehyde released from DHA. Subsequent reaction products of formaldehyde, such as formic acid, are also reduced. Ordinarily, DHA and sodium metabisulfite cannot be combined because they interact to form viscous brown substances. However, the addition of a small amount of magnesium stearate forms a hydrophobic barrier that prevents the interaction. Procter & Gamble scientists have obtained a US patent on a similar composition.[25]

Pfizer scientists have developed an improved self-tanning composition that combines DHA with a polyethoxyglycol and a polyol.[26] Further improvements are claimed if a soluble diol is included, and also if the pH is adjusted to approximately 4.0 with a mild organic acid like sorbic acid. The claimed improvements are due to the equilibrium reaction between the ketone functionality of DHA and the polyols, forming a cyclic ketal that is more stable than the ketone. The cyclic ketal structure is stable as long as the pH of the composition remains between 3.5 and 4.5. This will not undergo the Maillard reaction necessary to promote the tanning reaction. However, when applied on skin, the higher pH of the skin surface converts the cyclic ketal to the monomeric DHA, and tanning results.

Formulas 2 through 5 show how DHA is used in different formula types currently on the market.

Conclusion

DHA, used as a self-tanning agent for more than 30 years, is a safe product for use instead of sun-induced tanning. We now have a much better understanding how DHA works with skin protein to provide an artificial tan. We also know how to formulate stable formulations with DHA by careful selection of ingredients, pH and storage conditions. Experimental and clinical evidence have shown DHA-containing product provides some photo-protection (for better photo-protection always use in combination with non-nitrogen-containing organic sunscreens). This photo-protective property of DHA has been utilized in the treatment of psoriasis.

—**Ratan K. Chaudhuri and Cristina Hwang,** *Rona/EM Industries, Inc., (Merck KgaA, Darmstadt, Germany) Hawthorne, NY USA*

References

1. D Meyers, IR Scott and NJ Lowe, in *Sunscreens: Development, Evaluation, and Regulatory Aspects*, LJ Lowe, NA Shaath and MA Pathak, eds, 2nd edition, New York: Marcel Decker (1997) p 101
2. LR Kligman and AM Kligman, in *Sunscreens: Development, Evaluation, and Regulatory Aspects*, LJ Lowe, NA Shaath and MA Pathak, eds, 2nd edition, New York: Marcel Decker (1997) p 117
3. E Wittgenstein and HK Berry, *Science* 132 894 (1960)
4. RK Chaudhuri, *The Chemistry and Manufacture of Cosmetics*, M Schlossman, ed, Carol Stream, Illinois: Allured Publishing (in print)
5. T Kurz, *Cosmet Toil* 109(11) 55-61 (1994)
6. US Pat 5,770,411, HL Ohrem and F Westmeier (1998)
7. C Aretz, R Buczys, K Buchholz and H Driller, *Eurocosmetics*, 7 32 (1999)
8. H Fuehrer, *Seifen-Oele-Fette-Wachse* 20 607 (1960)
9. T Kobayashi and H Higasi, *J Molec Struct* 35 85 (1976)
10. PA Levene and A Watt, *J Biolog Chem* 78 23 (1928)
11. AJ Showler and PA Darlet *Chem Rev* 67 427 (1967)
12. VA Yaylayan, S Harty-Majors and AA Ismail, *Carbohyd Res* 318 20 (1999)
13. L Goldman, J Barkoff and D Blaney, *J Invest Dermatol* 35 161 (1960)
14. M Angrik, *Chemie in Unserer Zeit* 14 149 (1980)
15. T Severin, *Lebensm Unters Forsch* 178 284 (1984)
16. GP Ellis, *Adv Carbohydr Chem* 14 63 (1959)
17. A Meybeck, *J Soc Cosmet Chem* 28 25 (1977)
18. EP 0547864A1, A Suares (1993)
19. RG Kuehni, CIELAB Color Difference and Lightness, Hue and Chroma Components for Objective Color Control, Technical Bulletin No 1, Detroit Color Control
20. RM Johnson, *Pigment Handbook*, vol III, TC Patton, ed, New York: John Wiley & Sons (1973) p 229
21. T Kurz, *Proceedings of International Federation of Society of Cosmetic Chemists* (1995) p 361
22. US Pat 5,514,367, PJ Lentini and JR Zecchino (1996)
23. US Pat 5,942,212, PJ Lentini, P Tchinnis and N Muizzuddin (1999)
24. DE 43 14 083 A1 T Kurz, S Hitzel, R Martin and R Emmert (1994); EP 0 622 070 B1 (1997); US Pat 5,656,262 (1997)
25. US Pat 5,514,437, CR Tanner and LR Robinson (1996)
26. EP 884045, JA Scott and EM Stroud (1998)

27. R Fusaro and JA Johnson, *Dermalogica* 150 346 (1975)
28. R Fusaro and JA Johnson, *Dermalogica* 172 53 and 58 (1987)
29. JA Johnson and R Fusaro, *Dermatology* 185 237 (1992)
30. RM Sayer, DL Torode, JA Johnson and RM Fusaro in *Melanin: Its Role in Human Photoprotection: A Melanin Symposium*, L Zeise, MR Chedekel and TB Fritzpatrick, eds, Overland Park, Kansas: Valdemar (1994) p 1
31. CR Taylor, C Kwangsuksith, J Wimberly, N Kollias and RR Anderson, *Arch Dermatol* 135 540 (1999)

Substantiating Claims for a Tanning Magnifier

Keywords: L-tyrosine, Tyrosinase, Melanogenesis, Melanin Biosynthesis, Individual Typological Angle, Potassium Caproyl Tyrosine

Tanning magnifiers, known in the United States as tanning accelerators, enhance the substrate for the production of melanin. One example is an innovative N-acyl derivative from L-tyrosine. This article describes a way to evaluate its pigmentation efficacy after UV irradiation. Also discussed is the regulatory status of tanning magnifiers.

Melanin biosynthesis occurs as the skin's defense reaction to UV rays. It includes both the preliminary pigmentation activated by UVA rays and the series of chain reactions that generate the real and longer duration pigmentation, defined as indirect.[1,2]

The melanins elaborated from melanocytes can be precisely defined as eumelanins, feomelanins and tricochromes. These three different types of pigments derive from a common forerunner, L-tyrosine, a nonessential amino acid synthesized in the body from phenylalanine.

L-tyrosine's functions are several. It is a precursor of neurotransmitters, such as norepinephrine and dopamine, both of which regulate mood, adrenalin (adrenocortical hormone) and thyroxine (thyroid hormone). Tyrosine deficiency has been also linked to hypothyroidism, low blood pressure, low body temperature and restless leg syndrome.

Because tyrosine binds unstable molecules that can potentially cause damage to the cells and tissues, it could also be considered a mild antioxidant.

L-Tyrosine is the starting material for melanin's biosynthesis, so it is the pigment responsible for hair and skin color.

Skin Pigmentation

Tyrosinase, an enzyme containing copper, catalyzes the initial stages of melanogenesis. More precisely, it catalyzes the hydroxylation of tyrosine to DOPA and a further oxidation of DOPA to DOPA-quinone. The next steps, which lead to the synthesis of the different types of melanins, occur spontaneously without an enzymatic catalyst.[2]

Melanin biosynthesis depends not only the activity of tyrosinase, but also the bioavailability of tyrosine. In fact, it is well known that in a biologic synthesis, increasing the concentration of the starting substance also increases the quantity of the synthesized molecule.

This is also true in the formation of melanin, where the UV rays stimulate the process of melanogenesis, reducing tyrosine in cells and at the same time leading to an insufficiency of tyrosine.[1]

The pigmentation of human skin could be the result of two different mechanisms: the melanin's biosynthesis at the level of melanocytes; and a simple reaction between skin proteins and specific reactive substances at the level of the stratum corneum.

Almost certainly, tanning is a complex reaction. One report delineated a number of the paracrine factors made by keratinocytes that stimulate tanning normally in the skin. These include alpha-melanocyte stimulating hormone (a-MSH), adrenocorticotropin hormone (ACTH), and endothelin 1, all of which stimulate melanogenesis; factors such as basic fibroblast growth factor (bFGF), which can increase the number of melanocytes in skin; and agents such as nerve growth factor (NGF), which can preserve melanocytes in skin that might otherwise be lost.[3]

A second category of agents that have been identified are molecules involved in the intracellular signal transduction pathways that lead to tanning, such as cyclic AMP, which mediates MSH-induced tanning; protein kinase C (PKC) beta, which activates tyrosinase and enhances melanogenesis; nitric oxide, which is released by diols and stimulates tanning. DNA fragments released during the course of repair of UV-induced DNA photoproducts might also enhance tanning, according to this same report.[3]

Artificial Tanners

There are several different types of topical products that produce an "artificial" tan without sunlight.

Bronzers: "Bronzers" are color additives that stain the skin without harming it. For example, there are combinations of synthetic certified colors. Others come from natural sources, such as walnut juice. Usually, soap and water will remove them easily. Among other products marketed as bronzers are tinted moisturizers and brush-on powders. These also produce a temporary effect, similar to other types of makeup.

Sunless tanners: "Tanning extenders," sometimes referred to as self-tanners or sunless tanners, also are color additives, but they work through chemical reaction with the proteins of the skin. In fact they are particularly reactive substances which, in contact with functional lipoproteic terminals of keratin in the skin, generate yellow-brown substances that give to the skin a tanned look. This last process is independent from the sun exposure, because the melanogenesis process is not started by UV rays. The most frequently used agent is dihydroxyacetone[4] (DHA), a simple sugar involved in plant and animal carbohydrate metabolism. DHA browns the skin through its interaction with the amino acid arginine that is found in surface skin cells.[5]

Sunless tanners also are viewed as camouflage products because they can be used to hide skin stains. It is important to emphasize that these products don't offer adequate sun protection, except for the cases in which sun filters are also present in the formulation.[6]

Tanning magnifiers: "Tanning magnifiers," or "tanning accelerators" as they are called in the United States, are substances that, when applied to the skin, will make melanin form faster than it normally forms in sunlight. They increase the substrate available for action by tyrosinase, the key enzyme in melanogenesis. The activity of tyrosinase increases the melanin biosynthesis but it decreases the bioavailability of tyrosine. As disclosed in a recent L'Oréal patent,18 n-acyl amino acid esters (in particular isopropyl N-lauroyl sarcosinate) are considered tanning magnifiers or enhancing agents because some compositions containing them (with other cosmetic ingredients) are able to increase the level of tanning or skin browning.

Regulatory Status of Tanning Magnifiers

In Europe: In the European Union, tanning magnifiers are viewed as cosmetics because they enhance the color of the skin. They are permitted as a cosmetic under the Cosmetic Directive Annex I.

In the United States: The United States Food and Drug Administration (FDA) has issued a Fact Sheet[7] discussing the supposed role of tyrosine in melanin production. The Fact Sheet defines tanning accelerators as "products marketed with claims that they speed up the skin's melanin production, generally by supplying doses of tyrosine (an amino acid) or its derivatives, perhaps in combination with other substances." Any product purporting to "accelerate the tanning process" or "stimulate the production of melanin" is claiming to affect the structure and function of the body and is therefore a drug, according to the definition of "drug" in the United States. The Fact Sheet observes that the agency is "not aware of any data demonstrating that tyrosine or its derivatives are effective in stimulating the production of melanin. Thus, any product containing tyrosine or its derivatives and claiming to accelerate the tanning process is an unapproved new drug."[7]

At Sinerga, it is belived that this position could be challenged because it is in opposition to the investigated and declared mechanism of actions of several recently introduced cosmetic ingredients that could regulate enzymes or biochemical processes.

Tyrosine and Derivatives

All these considerations led to the possibility of cosmetic products containing L-tyrosine, using the concept of tanning magnifier as an agent that is able to enhance the coloring of the skin. Thus, attempts to induce melanogenesis by L-tyrosine are based on the concept that the primary substrate for tyrosinase may be the limiting factor in melanogenesis.[8,9] Different experiences have proved that pure tyrosine applied topically does not have acceptable bioavailability, and it is also an irritant ingredient. In fact it is soluble only at pH >11.4, a value at which keratin destruction occurs. At lower pH, L-tyrosine crystallizes, because it can not be absorbed. In fact it has a poor solubility in water (0.05 g/dl at RT). Other experiments validate that minimum concentration of L-tyrosine in the applied product must be more than 0.4%, the level at which one first notices an increase of melanogenesis.[1,10]

Several derivatives have been created to enhance cutaneous absorption of L-tyrosine itself. Among these are L-tyrosine copper salt, N-acetyl-L-tyrosine,2 N-chloroacetyl-L-tyrosine, N-p-toluensulfonyl-L-tyrosine, L-tyrosine methylester hydrochloride and recently alpha-linoleoyl tyrosine.[11]

The release of L-tyrosine is mediated by enzymes activated by UV rays. This fact was demonstrated by tests utilizing water-soluble derivatives marked with tritium, formulated in a cream at 1%, and applied on the skin of rats. Radiography revealed the presence of tritium in the basal layer and in the dermis.[12]

The Sinerga R&D laboratories sought to increase the skin bioavailability of tyrosine. First it obtained caproyl tyrosinic acid via the condensation of tyrosine with capric acid, delivering the equivalent of approximately 10% pure tyrosine.[13] The condensation product is a water-soluble lipoaminoacid, in which capric acid, a C10 saturated fatty acid, is able to produce a hydrophilic molecule with excellent affinity for the skin. Two new forms were abtained from the condensation:

- The lipophilic form has the INCI name caproyl tyrosinic acid (and) glyceryl oleate (and) sorbitan isostearate[a].
- The hydrophilic form has the INCI name potassium caproyl tyrosine[b] (PCT) (Figure 1), an easy-to-use N-acyl derivative of tyrosine. It is a clear liquid that is completely water-soluble and compatible with traditional cosmetic ingredients, similar to other lipoaminoacids, oligopeptides and water-soluble amino acids. Its physical properties are described in Table 1.

The rest of this article describes how the skin tolerability and the tanning properties of PCT were substantiated.

Tolerability

The toxicological profile of PCT was obtained by an in vitro test with the aim of determining the ingredient's potential dermal irritation. This in vitro test, the Ir-

[a] Tyrosinol, a trade name of Sinerga SpA, Pero (Milan), Italy
[b] Tyrostan, a trade name of Sinerga SpA, Pero (Milan), Italy

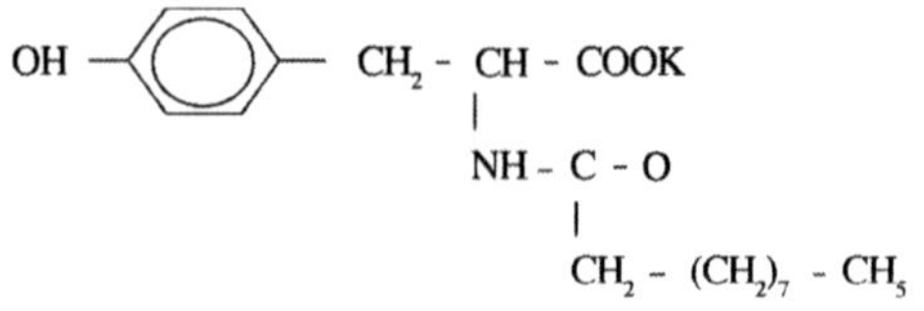

Figure 1. Potassium caproyl tyrosine

ritection Assay Systemc, is a reliable substitute for the patch test normally used for this characterization.

A standard concentration-dependent dose-response study was performed with the dermal Irritection test method. Results of the study indicated that the sample was classified as a non-irritating agent (score = 0.63).

Evaluation of the Pigmentation Efficacy After UV Irradiation

The aim of this study was to evaluate the skin pigmenting properties of a cosmetic product after exposure to UV radiation.[14,15] PCT was compared with a placebo and an untreated control area.

Materials and methods: PCT was formulated at a 5% concentration in an emulsion-gel, applied on the back of 12 volunteers (Fitzpatrick phototypes II, III, IV) for three consecutive weeks. UV irradiation was delivered using a solar simulator.

The effect on the skin was evaluated with chromameter up to three weeks after application.[13,16] The parameters (L^*, b^* parameters, ITA° value), which are sensitive indexes of change in pigmentation intensity, are taken by skin colorimetric measurements, before the product application and after several intervals. The Individual Typological Angle or ITA° value expresses the melanin index. It is calculated as a ratio obtained by complex calculations from L^* and b^* parameters. So any change in the ratio might indicate a significant change in colorimetric values.

The assessment was performed on three selected areas on the volunteer's back. Each area was treated with a different sample (active, placebo or untreated), with a non-occlusive patch. A fourth area of the back was exposed to UVA-UVB rays in order to determine the minimum erythemal dose on unprotected skin (MEDu). Baseline colorimetric measurements were taken on Day 1 (T0) before irradiation and application. Measurements were also taken

[c] Irritection Assay System, InVitro International, Irvine, California, USA. Irritection is a registered trademark of InVitro International.
[d] Multiport Solar UV Simulator Model 601, Solar Light Company, Philadelphia, Pennsylvania, USA

Table 1. Properties of potassium caproyl tyrosine

Aspect	clear liquid
Color	yellow
Odor	light
pH (c=10)	6.5–7.5
Solubility (water)	complete
Active substance (%)	28–32

Table 2. ITA° parameters (with standard deviation) for test areas treated with potassium caproyl tyrosine (PCT), a placebo or untreated control on the backs of human volunteers (averaged over n=12) following UV irradiation with 50% MEDu

Time	PCT	Placebo	Control
T0	23.86 (7.9)	22.93 (8.8)	22.02 (7.2)
Week 1			
T1	20.14 (9.2)	21.72 (8.9)	20.55 (7.4)
T2	20.08 (9.3)	23.48 (7.7)	20.68 (6.3)
T3	18.68 (8.1)	20.27 (8.6)	21.10 (7.4)
Week 2			
T4	19.91 (8.8)	21.49 (8.6)	21.88 (7.0)
T5	22.58 (9.8)	23.28 (8.4)	21.14 (7.1)
T6	19.98 (8.8)	20.33 (8.1)	21.33 (6.3)
T7	21.18 (8.1)	20.10 (8.7)	22.06 (7.6)
T8	20.33 (7.3)	21.03 (8.6)	21.88 (7.5)
Week 3			
T9	21.10 (6.9)	20.22 (6.6)	21.54 (5.5)
T10	21.65 (7.3)	20.45 (6.8)	22.30 (5.4)
T11	21.50 (7.0)	20.24 (7.6)	22.53 (6.3)
T12	22.31 (6.3)	20.95 (7.7)	22.73 (6.3)
T13	20.29 (6.3)	19.93 (7.2)	22.16 (5.9)
T14	18.60 (6.1)	20.33 (6.9)	20.84 (7.1)

before application during the first week (T1-T3), the second week (T4-T8) and the third week (T9-T14). UVA-UVB irradiation corresponding to 50% MEDu was given after any application of the non-occlusive patch of the product sample or the placebo.

Results: The pigmentation efficacy is shown by a decrease in ITA° values. Mean values and standard deviation were calculated for each set of values relating to the tested skin areas and the different time intervals. Table 2 shows ITA° mean values ($n = 12$) and standard deviation, calculated for each area at each measurement.

Variance analysis and Tukey test were carried out on the data to determine statistically significant differences among the set of values recorded at different times in the three areas.

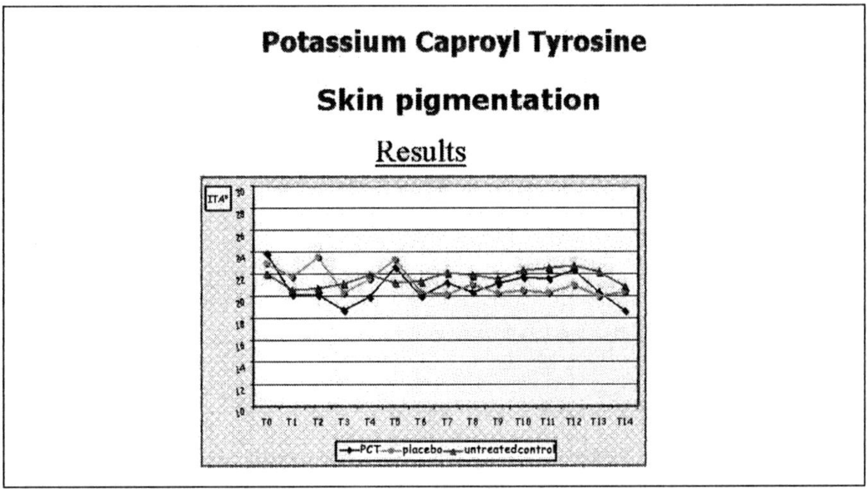

Figure 2. Skin pigmentation effects of potassium caproyl tyrosine (PCT), placebo and untreated control

- The area treated with the sample of PCT at 5% showed a highly significant decrease in ITA° values (for T0 versus T3, p=0.002; for T0 versus T14, p=0.002).
- The area treated with the sample of placebo did not show a significant decrease in ITA° values at any time.
- The untreated control area did not show a significant decrease in ITA° values at any time.

Discussion: When the results are graphed (Figure 2), it is possible to understand the long-lasting effect of PCT. The process of melanogenesis is activated when UV rays are supplied. When exogenous L-tyrosine (PCT) was supplied, endogenous L-tyrosine was activated too. Then melanogenesis occurs in two stages. The first is fast; the second is slow, but both need UV rays and copper to activate tyrosinase. So rapidly (T3) there is an increase of tanning, that is confirmed slowly (T14) too. Then a negative feedback of the process could be generated; meaning that cells don't produce L-tyrosine anymore because they realize that it already is present. That is an interpretation, not yet supported by in vitro results.

In the results, it can be noted[17] that the values recorded in the test areas treated with the two products are often lower than those recorded in the control area. That indicates a tendency toward a pigmentation increase. The difference of values is nevertheless not significant (p <0.05). PCT is significantly effective in increasing the skin tanning in comparison to the starting conditions.

A similar level of statistical significance could not be obtained in the comparison with the control areas, probably because of the high standard deviation.

[e] Chromameter CR-300 Minolta, Minolta GmbH, Germany. Chromameter is a registered trade name of Minolta GmbH.

Formula 1. Skin toning facial cream (O/W emulsion)	
Potassium palmitoyl hydrolyzed wheat protein, glyceryl stearate, cetearyl alcohol (Phytocream 2000, Sinerga)	5.00% w/w
Ethylhexyl ethylhexanoate (Dragoxat EH, Dragoco)	7.00
C12-15 Alkyl benzoate (Finsolv TN, Chemlink)	5.00
Dimethicone (SF 18-350, Sinerga)	0.50
Tocopherol, lecithin, ascorbyl palmitate, citric acid (Aperoxid TLA, Biochim)	0.50
Phenethyl alcohol, methylparaben, propylparaben, glycerin (Fenilight, Sinerga)	0.50
Water (*aqua*)	qs 100.00
Imidazolidinyl urea (Gram 1, Sinerga)	0.30
Disodium EDTA	0.10
Xanthan gum (Keltrol, Kelco)	0.25
Glycerin	2.00
Prunus amygdalus dulcis, hydrolyzed sweet almond protein, potassium palmitoyl hydrolyzed wheat protein (Mandorlat, Sinerga)	2.00
Water (*aqua*), algae (Seamollient, Sinerga)	2.00
Potassium caproyl tyrosine (Tyrostan, Sinerga)	5.00
Fragrance (*parfum*) (Sun 1667, Agieffe)	0.20

If baseline measurements are mathematically compared to final values it is possible to compare the tanning effects of PCT, placebo and control. The change in ITA° value from T0 to T14 is –5.3, –2.6 and –1.2 for PCT, placebo and control, respectively. Thus, application on volunteers of PCT for three weeks magnifies tanning or browning of the skin by about 50% in comparison to parallel application of placebo and about 77% in comparison to untreated areas.

Formulation Development

PCT is water-soluble, clear and easy to add to formulations where a tanning accelerator effect is requested, with significant efficacy, modulated in time. In formulating, the advice is to add it after the emulsifying phase at about 40°C and at any time during the cooling phase. To insure its stability, it is best to avoid its contact with strong oxidizing agents and alkaline solutions. Recommended percentage of use is 5%.

The cosmetic application as tanning magnifier has been developed in products for sun care during exposure, and for sun tan maintenance after exposure and its stability over time has been confirmed in several formulations in the form of cleansers, gels and emulsions.[13] An oil-in-water emulsion with PCT at 5% is shown in Formula 1.

Conclusions

Potassium caproyl tyrosine is a water-soluble, N-acyl derivative of tyrosine that increases the bioavailability of tyrosine, delivering the equivalent of approximately 10% pure tyrosine. Its effectiveness as a tanning magnifier was demonstrated with measurements of ITA° on irradiated human skin. Its benign toxicological profile was demonstrated in vitro with the Irritection assay.

Its use in cosmetic formulations, typically at 5%, has been demonstrated. However, in some markets, namely the United States, products making tanning accelerator claims are regulated as OTC drugs, not cosmetics, and would require a New Drug Application before they could be sold.

—**G. Guglielmini,** *Sinerga SpA, Pero (Milan), Italy*

References

1. DA Brown, Skin pigmentation enhancers, *J Photochem Photobiol B* 63(1–3) 148–161 (2001)
2. G Prota, Melanine e melanogenesi, *Cosmet Toil Ed Ital* 2 9–22 (1997)
3. BA Gilchrest and Dr Brown, speakers at a National Institutes of Health workshop on risks and benefits of exposure to ultraviolet radiation and tanning, Bethesda, Maryland, Sep 16-18, 1998. *www.niams.nih.gov/ne/reports/sci_wrk/1998/ulv.html* (Apr 2004)
4. N Muizzudin and D Kenneth, Tonality of suntan vs sunless tanning with dihydroxyacetone, *Skin Research and Technology* 6(4) 199–204 (2000)
5. Vari, *Chem Eng News* 78(26) (2000)
6. SB Levy, Tanning preparations, *Dermatol Clin* 18(4) 591–596 (2000)
7. *www.cfsan.fda.gov/~dms/cos-tan3.html* (Apr 8, 2002)
8. PP Agin et al, Tyrosine does not enhance tanning in pigmented hairless mice, *Photochem Photobiol* 37 559–564 (1983)
9. W Tur, Tanning accelerators, *Cosmet Toil* 105(12) 79–85 (1990)
10. C Jaworsky et al, Efficacy of tan accelerators, *J Am Acad Dermatol* 16 769–771 (1987)
11. A Slominski et al, L-tyrosine and tyrosinase as positive regulators of the subcellular apparatus of melanogenesis, *Pigment Cell Res* 2 109–116 (1989)
12. S Yehuda, Possible anti-Parkinson properties of N-(alpha-linoleyl) tyrosine, *Pharmacol Biochem Behav* 72(1–2) 7–11 (2002)
13. HS Zadeh, Tirosina biodisponibile e pigmentazione cutanea, *Cosmetic News* 146 307–311 (2002)
14. GH Pittel, Human test methods for determining SPFs, *Drug Cosm Ind* 9 24–32 (1988)
15. NJ Lowe et al, Sunscreens and phototesting, *Clin Dermatol* 3(7) 40–49 (1988)
16. C Elly, Eyetex: an in vitro method of predicting ocular safety, *Pharmacopeial Forum* 48154824 (1989)

17. G Guglielmini, Potassium caproyl tyrosine, *Cosmetic Technol* 6(2) 31–34 (2003)
18. US Pat 6,616,918, Self tanning composition containing N-acyl amino acid esters and a self tanning agent, D Candau, assigned to L'Oréal (Sep 2003)

Inhibitory Effects of Ramulus Mori Extracts on Melanogenesis

Keywords: ramulus mori, tyrosinase, melanogenesis

A compound from extracts of ramulus mori inhibits tyrosinase and melanin biosynthesis in melanocytes.

It has been observed that a local increase in melanin synthesis or uneven distribution of melanin can cause local hyperpigmentation or spots. Pigmentary disorders are caused by various factors, including inflammation, imbalance of hormones and genetic disorders.[1] Another factor is the recent increase in the harmful effects from UV radiation because of destruction of the ozone layer. Excessive exposure to UV radiation causes post-inflammatory pigmentation.

Most women want to avoid uneven skin pigmentation.[2,3] To satisfy this desire many cosmetic companies have been developing melanogenesis inhibitors and finding promising active agents for use in cosmetic preparations for skin whitening. In these preparations, inhibitors such as kojic acid,[4] arbutin,[5] ascorbic acid and licorice extracts[6] have been used as whiteners.

Plant extracts having an inhibitory effect on melanin formation may be a good choice as a cosmetic ingredient because they have relatively fewer side effects. Therefore, we screened 285 plant extracts for their inhibitory activity on tyrosinase, the enzyme that catalyzes the oxidation of the amino acid tyrosine in the formation of melanin.[7] We found potent tyrosinase inhibition activity in extracts of ramulus mori, a Latin medicinal name that means young twigs of *Morus alba* L.

In this article we report on the ramulus mori extract's active compound, its safety and its ability to inhibit tyrosinase activity and melanogenesis.

Ramulus Mori Compound

Morus alba L. is commonly found in many parts of East Asia. This deciduous tree, which reaches a height of approximately 15 m, grows in many parts of Korea at 100-1000 m above sea level. All parts of the tree—its leaves, root, fruit and branches—have been used in oriental herbal medicine, typically as a fever remedy, cold remedy and anti-hypertension agent.

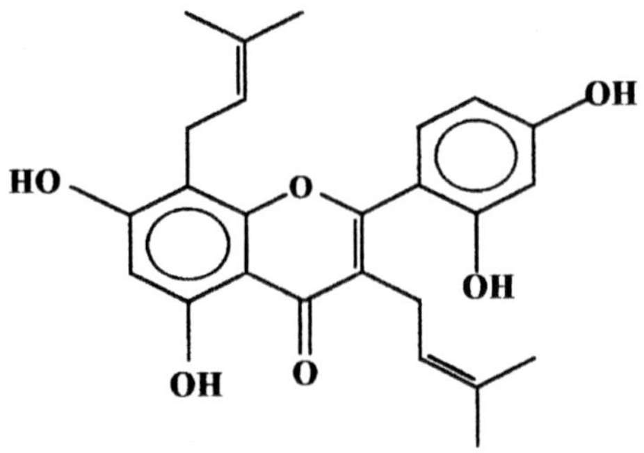

Figure 1. Structure of 2-(2,4-dihydroxyphenyl)-5,7-dihydroxy-3,8-bis(3-methyl-2-butenyl)-4H-1-benzopyran-4-one isolated from ramulus mori

Extraction and isolation: We extracted dried young twigs of *Morus alba* L. with 70% ethanolic aqueous solution. We removed the solvent by rotary evaporation. Then we purified the solid residue through solvent fractionation, column chromatography and recrystallization to obtain the tyrosinase inhibitor.

The ethanolic extracts were dissolved in water, then shaken with chloroform, ethyl acetate and butanol. We purified the ethyl acetate soluble fraction by silica gel and gel filtration[a] column chromatography. Then we isolated the tyrosinase inhibitor by chromatography.

Identification: We recrystallized the isolated tyrosinase inhibitor from a mixture of ether and benzene (1:1, v/v) to yield pale yellowish prisms with a melting point at 148-150°C. The UV λ_{max} (peak plateaus) of the compound in ethanol were 210 nm, 264 nm and 315 nm. Using infrared, mass spectrometry and nuclear magnetic resonance (see sidebar), we identified the compound as 2-(2,4-dihydroxyphenyl)-5,7-dihydroxy-3,8-bis(3-methyl-2-butenyl)-4H-1-benzopyran-4-one (Figure 1), which was first described by Nomura and Fukai.[8]

Tyrosinase Inhibition

Tyrosinase activity is generally determined by spectrophotometry. We followed the procedure described by Vanny et al.[9] For the assay, the test reaction mixture was prepared by adding 200 units of mushroom tyrosinase to 0.5 mL of a solution of the ramulus mori compound. Then that solution was added to 0.5 mL of 0.1 mg/mL L-tyrosine and 0.5 mL of 50 mM sodium phosphate buffer (pH 6.8). After incubating the test mixture for 10 min at 37°C, we measured tyrosinase activity by absorbance at 475 nm because DOPAchrome, the product of tyrosinase reaction, has maximum absorbance at that wavelength. We determined the effect of the test compound on tyrosinase inhibition by IC_{50}, the concentration of the compound at

[a] Sephadex LH-20, Sigma Chemical, St. Louis, Missouri USA

Table 1. Concentration (IC_{50}) required for selected tyrosinase inhibitors to reduce mushroom tyrosinase activity by 50%

Materials	IC_{50} (μg/mL)
Arbutin	65.2
Licorice extracts	12.88
Kojic acid	5.82
Ramulus mori extracts	12.48
Ramulus mori compound	0.507

Infrared, mass spectroscopic and nuclear magnetic resonance data on the tyrosinase-inhibiting compound in ramulus mori

IR[a]

λ_{max} (cm^{-1}) 3370, 1660, 1630, 1610, 1560

MS[b]

m/z 422, 407, 379, 367, 323

^{1}H-NMR[c]

δ (ppm) 1.43(3H, s, C_{11}-CH_3), 1.57(9H, s, C_{11}-CH_3 and C_{14}-CH_3x2), 3.12(2H, brd, J = 8Hz, C_9Hx2), 3.35 (2H, brd, J = 8 Hz, C_{12}-Hx2), 5.20(2H, m, C_{10} and C_{13}-H), 6.31(1H, s, C_6-H), 6.43(1H, dd, J = 2 and 8Hz, C_5'-H), 1.52(1H, d, J = 2 Hz, C_3'-H), 7.20(1H, d, J = 8 Hz, C_6'-H), 13.05(1H, s, OH)

^{13}C-NMR[d]

δ (ppm) 159.1(C-2), 119.6(C-3), 182.1(C-4), 103.4(C-4a), 155.1(C-5), 98.2(C-6), 161.9(C-7), 105.6(C-8), 160.2(C-8a), 123.4(C-9), 121.9 (C-10), 131.6(C-11), 25.6(C-12), 17.4(C-13), 21.2(C-14), 122.3(C-15), 130.9(C-16), 25.6(C-17), 17.4(C-18), 111.6(C-1'), 156.8(C-2'), 102.8(C-3'), 161.6(C-4'), 106.8(C-5'), 131.6(C-6')

[a] KBr is used as a dilutor. Instrumentation was a Nicolet 205 FT-IR Spectrophotometer, Nicolet Instrument Corp., Madison, Wisconsin, USA.
[b] Instrument was a VG Trio-2, VG Masslab, Atringham, UK.
[c] CD_3COCD_3 (acetone with deuterium substituted for hydrogen)
[d] DMSO-d_6 (dimethyl sulfoxide with deuterium substituted for hydrogen)

which half of the original tyrosinase activity is inhibited (Table 1).

We calculated the percent inhibition of tyrosinase activity as follows:

$$\% \text{ inhibition} = \frac{A - B}{A} \times 100$$

where A = absorbance at 475 nm without test sample,
and B = absorbance at 475 nm with test sample.

Our ramulus mori compound is a more potent tyrosinase inhibitor than kojic acid and arbutin. The compound shows a strong inhibitory effect on tyrosinase at very low concentration.

Inhibition of Melanogenesis

We examined the inhibition of melanogenesis in B-16 melanoma cells by a modification of the method of Oikawa and Nakayasu,[10] who developed a reproducible and precise method for quantifying melanin content. B-16 melanoma is a typical cell line for studies of melanogenesis.

We placed B-16 melanoma cells[b] in a 50 mL T-flask at a density of 5.76×10^6 cells/flask and cultured at 37°C in Dulbecco's modified eagle's medium (DMEM) containing 4.5 g/L of glucose, 10% fetal calf serum and 1% antibiotics.

After 48 h of cultivation, we replaced the medium with new DMEM medium containing ramulus mori compound of various concentrations.

After five days, we washed the cells with phosphate buffer saline and collected the cells by trypsinization and centrifugation. We separated melanin from the pellet of the cells using 5% trichloroaceticacid and dissolved the melanin in 1N NaOH solution.

We determined the melanin contents with an absorbance at 475 nm (Table 2). A standard curve for melanin determination was prepared using synthetic melanin[c].

Table 2. The effect of ramulus mori compound on melanogenesis in B-16 melanoma cells

Ramulus mori compound (µg/mL)	Melanin content (pg/cell)	Inhibition of melanogenesis (%)
Control	4.281 ± 0.172	0.00 ± 4.02
10	1.953 ± 0.493	54.38 ± 11.52
20	1.426 ± 0.398	66.69 ± 9.30
50	1.072 ± 0.281	74.96 ± 6.56
100	0.465 ± 0.305	89.13 ± 7.12

[b] ATCC CRL 6323, American Type Culture Collection, Virginia, USA

[c] Sigma Chemical

The cell number was determined with the coulter counter. Our ramulus mori compound showed an inhibitory effect on melanogenesis.

Safety Tests

Acute toxicity test: We investigated the potential toxicity of the ramulus mori compound. Using the guidelines of the Cosmetic, Toiletry and Fragrance Association,[11] we assessed the compound's acute oral toxicity in 60 rats and its acute dermal toxicity on 24 rabbits.

We did the acute toxicity examination 14 days after treatment. No deaths occurred and no abnormalities were detected in clinical findings from the rats to whom the compound was administered orally. No symptoms were detected in the rabbits to whom the compound was applied topically. We observed no changes of body weight in post mortems of sacrificed test animals (data not shown).

Skin irritation test: The primary skin irritation of the ramulus mori compound was investigated in six rabbits. We examined skin irritation and clinical signs for 24-72 h after a seven-day treatment using the Draize Primary Irritation Index.[12] We did not observe side effects such as erythema, scarring and edema.

Eye irritation test: The potential toxicity of the compound was evaluated according to Association Francaise de Normalization (AFNOR) guidelines in an eye irritation test on nine rabbits. We examined both eyes at 1 h and again at Days 1, 2, 3, 4 and 7 after treatment. We did not see any clinical signs.

Skin sensitization test: The sensitization potential of the compound was assessed in 33 guinea pigs by the methods of Magnusson and Kligman Maximization Test.[13] We observed no skin responses in test animals.

Human skin irritation test: We studied the potential of the compound to irritate human skin in 50 healthy female volunteers using a 24 h closed patch. No skin irritation occurred after application in 50 volunteers.

Conclusion

We found that ramulus mori extracts had a strong inhibitory activity against melanogenesis. We also found that the active compound was 2-(2,4-dihydroxyphenyl)-5,7-dihydroxy-3,8-bis(3-methyl-2-butenyl)-4H-1-benzopyran-4-one. The results of safety tests showed the compound to have no irritation or sensitization potential.

We suggest that the ramulus mori compound will be effective in treating uneven skin pigmentation.

—**Jeong-Ha Kim and Kang Tae Lee,** *Coreana Cosmetics Co. Ltd., Cheonan, Korea*

References

1. M Matsuda, T Tejima, T Suzuki and G Imokawa, Skin lighteners, *Cosmet Toil* 111(10) 65-77 (1996)
2. JM Pawelek, AK Chakraborty, MP Osber and JL Bolognia, Ultraviolet light and pigmentation of the skin, *Cosmet Toil* 107(11) 61-68 (1992)
3. G Imokawa, Melanocyte activation mechanism in UV pigmentation, *Fragrance J* (14) 38-48 (1995)

4. Y Mishima, S Hatta, Y Ohyama and M Inazu, Induction of melanogenesis suppression: cellular pharmacology and mode of differential action, *Pigment Cell Research* 1 367-374 (1988)
5. K Maeda and M Fukuda, In vitro effectiveness of several whitening cosmetic components in human melanocytes, *J Soc Cosmet Chem* 42 361-368 (1991)
6. T Ikeda and T Kambara, Melanogenesis inhibitory effect of oil-soluble licorice extract and its action mechanism, *Fragrance J* (14) 174-179 (1995)
7. KT Lee, BJ Kim, JH Kim, MY Heo and HP Kim, Biological screening of 100 plant extracts for cosmetic use(I): Inhibitory activities of tyrosinase and DOPA auto-oxidation, *Intl J Cos Sci* 19(4) 291-298 (1997)
8. T Nomura and T Fukai, On the structures of mulberrin, mulberrochromene, cyclomulberrin, and cyclomulberro-chromene, *Heterocycles* 12(10) 1289-1295 (1979)
9. A Vanni, D Gastaldi and G Giunata, Kinetic investigations on the double enzymic activity of the tyrosinase mushroom, *Annali Di Chimica* 80 35-60 (1990)
10. Oikawa and M Nakayasu, Quantitative measurement of melanin as tyrosinase equivalents and as weight of purified melanin, *Yale J Biol Med* 46 500-507 (1973)
11. *CTFA Safety Testing Guideline*, Washington, DC: The Cosmetic, Toiletry, and Fragrance Association (1991)
12. JH Draize, Dermal toxicity, in *Appraisal of the Safety of Chemicals in Food, Drugs, and Cosmetics*, Austin, Texas: Texas State Dept of Health (1959) pp 46-59
13. B Magnusson and AM Kligman, The identification of contact allergens by animal assay: the guinea-pig maximization test, *J Inv Derm* 52 268-276 (1969)

BEMT: An Efficient Broad-Spectrum UV Filter

Keywords: sunscreen, bis-ethylhexyloxyphenol methoxyphenyl triazine, BEMT, broad-spectrum UV protection, hydroxyphenyltriazine chemistry, persistent pigment darkening, UVA assessment

Efficient broad-spectrum UV protection is provided by bis-ethylhexyloxyphenol methoxyphenyl triazine (BEMT), an oil-soluble, photostable UV absorber that offers improved UVA protection and synergy with UVB filters.

Over the past decade the need for improved UV protection became apparent. UV radiation has adverse effects, including sunburn, photoaging and skin cancer. These effects can only be prevented by protecting against the whole range of UV radiation, including UVA.

On the other hand there is also a trend to limit the amount of "chemicals" on the skin. This means that very efficient UV absorbers should be available for the new requirement of broad UV protection. Bis-ethylhexyloxyphenol methoxyphenyl triazine[a] (BEMT) has been designed to fulfill this requirement. It is photo-stable, oil-soluble, very efficient and covers the UVB and UVA range.[1] Its physical and chemical properties are shown in Table 1. In the year 2000, European authorities added BEMT to the positive list of cosmetic UV absorbers.[2]

Design of BEMT

Due to their molecular structure, hydroxyphenyltriazines (HPTs) exhibit a UV spectrum with two distinctive absorption bands. This is due to the presence of two electronic transitions with strong dipole moments, both of which are polarized perpendicular to each other (Figure 1).

In order to obtain broad-spectrum absorbance, OH and O-alkyl substituents have to be introduced at the three phenyl groups and positioned as shown in Figure 1.

- UVA: Two ortho-OH groups are required for efficient energy dissipation via intramolecular hydrogen bridges. In order to obtain strong absorption in

---

[a] Tinosorb S, Ciba Specialty Chemicals Inc., Basel, Switzerland. Ciba and Tinosorb are registered trademarks of Ciba Specialty Chemicals.

Table 1. Physical and chemical properties of bis-ethylhexyloxyphenol methoxyphenyl triazine

Chemical Name	2,4-Bis-{[4-(2-ethyl-hexyloxy)-2-hydroxy]-phenyl}-6-(4-methoxy phenyl)-(1,3,5)-triazine
INCI name	Bis-Ethylhexyloxyphenol Methoxyphenyl Triazine (BEMT)
CAS no.	103597-45-1
Molecular formula	C38H49N3O5
Molecular mass	627.80 g/mol
Melting point	80°C
Photostability	Stable over 50 MED according to the method of Berset et al[3,5]
Appearance	Light yellow powder
Absorbance	E (1%, 1 cm) ≥ 790 (in isopropanol, 341 nm)
Use/concentration	Sunscreen products/max. 10%

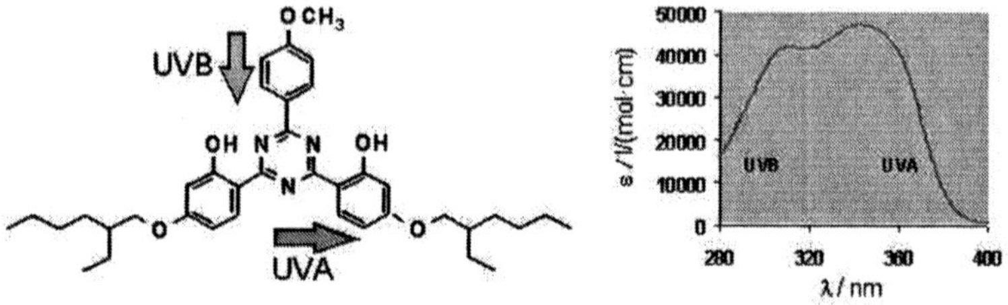

Figure 1. Molecular structure (left) and UV-absorption spectrum of BEMT. (Arrows indicate the polarization of the UVA and UVB transitions.)

the UVA, the para-positions of the two respective phenyl moieties should be substituted by O-alkyl, resulting in a bis-resorcinyl triazine chromophor.

- UVB: The remaining phenyl group attached to the triazine leads to UVB absorption. It can be demonstrated that maximal "full spectrum" performance is achieved with O-alkyl located in the para-position.[1]

Without solubilizing substituents, HPTs are nearly insoluble in cosmetic oils. They exhibit the typical properties of pigments (e.g., high melting points). In order to increase solubility in oil phases, the structure of the UV filter has been modified accordingly. The introduction of two 2-ethyl-hexyl groups as shown in Figure 1 results in the formation of BEMT, a new oil-soluble UV filter with true broad-spectrum performance.

In general the photostability of a UV filter depends on how well the molecule is able to release the absorbed energy to the environment in the form of heat.

Formula 1. O/W cream with 5% ethylhexyl methoxycinnamate and varying amounts of BEMT

A. Cetearyl alcohol/dicetyl phosphate/ceteth-10 phosphate (Crodafos CES, Croda GmbH)	4.00%w/w
C12-15 alkyl benzoate (Tegosoft TN, Goldschmidt)	6.0
Cetearyl octanoate/isopropyl myristate (Crodamol CAP, Croda GmbH)	3.50
Caprylic/Capric triglyceride (Myritiol 318, Cognis)	2.00
Caprylyl pyrrolidone (Surfadone LP-100, ISP)	1.50
Stearic acid	1.50
Glyceryl stearate (Tegin M, Goldschmidt)	2.50
Bis-ethylhexyloxyphenol methoxyphenyl triazine (Tinosorb S, Ciba)	x
Ethylhexyl methoxycinnamate (Tinosoft OMC, Ciba)	5.00
B. Water *(aqua)*	qs 100.00
C. Sodium acrylates copolymer (and) *Glycine soja* (and) PPG-1 trideceth-6 (Salcare AST, Ciba)	0.20
D. Diazolidinyl urea (and) iodopropynyl butylcarbamate (Germall Plus, ISP)	0.15
Propylene glycol	3.50
Water *(aqua)*	5.00
E. Sodium hydroxide, 30%	qs

Procedure: Heat A and B separately to 75°C. Pour B into A under increased stirring; add C; homogenize 10 sec at 10,000 rpm. Add D around 40°C. Cool under stirring. At RT, adjust the pH and homogenize (10 sec at 16,000 rpm).

BEMT contains two intramolecular hydrogen bonds, which enables excited state intramolecular proton transfer (phototautomerism) after photoexcitation. This results in rapid radiationless internal conversion that ensures that the UV radiation, efficiently absorbed by the filter, is almost quantitatively transformed into harmless vibrational energy. The entire phototautomeric cycle only lasts 10^{-12} seconds, leaving no time for undesirable side reactions, such as population of triplet states, formation of singlet oxygen or radical formation.[4] This explains the photostability of BEMT; in other words, more than 95% recovery of parent BEMT is observed analytically after UV irradiation of 50 minimal erythema doses (MED).[5]

Efficacy of BEMT in UVA Protection

Both in vitro and in vivo methods are used in the assessment of the UVA protection of a suncare formulation. UVA protection factors (PFA) of sunscreen formulations with ethylhexylmethoxycinnamate (EHMC) and different concentrations of BEMT (Formula 1) were measured using the method of persistent pigment darkening (PPD).[6,7] Irradiation of volunteers was performed with a UVA light source

(320 to 400 nm). Two hours after irradiation, the minimal pigmenting dose of protected (MPDp) and unprotected (MPDu) skin was evaluated. Each formulation was tested on 10 panelists. The results were expressed as PFAs in analogy to the SPF.

The reference sample of the Japanese UVA standard[8] was also measured for comparison, using the same procedure on 10 panelists. It contains 5% butylmethoxydi-benzoylmethane (BMBM) and 3% EHMC. In our tests, the UVA protection factor (PFA) of this reference sample was only 4.5 (in agreement with the standard, 4.3). However, non-irradiated BMBM has a higher extinction in the UVA range than BEMT. In fact the same PFA can already be achieved with only 1% BEMT (Figure 2). This clearly demonstrates the difference in photostability. PA+++ is the highest category of UVA protection according to the Japanese standard (where PA+, PA++ and PA+++ are increasingly good grades indicating protection against UVA radiation). We found that PA+++ is already reached with about 3% BEMT alone or 2% BEMT plus 5% EHMT. Herzog et al. have shown[6] that this synergistic effect with EHMC in the UVA range is not observed with BMBM, nor is it seen when BMBM is stabilized with octocrylene.

Formulating with BEMT

The cosmetic formulators have to design systems that will remain stable over the useful lifetime of the product.[9] This requires a good working knowledge about the characteristics of each part of an emulsion including the oil phase, the aqueous phase and the emulsifier.

Incorporation of BEMT in oil phase: The product form of BEMT is a very fine yellow powder with a melting point of 80°C. It is easily solubilized in most lipophilic emollients (Table 2), especially those with a high surface tension, because they present the lowest surface tension gradient (interfacial tension) between the oil and water phases.

The interfacial tension is used as a guide to the emollient's polarity; the more polar the oil, the lower its interfacial tension with water. With BEMT, the polarity of lipophilic oils used in the oil phase is an essential parameter to develop stable formulations.

The oil surface tension has a significant impact on the spreadability and therefore on the SPF of the formulation.[10] To create an uninterrupted and uniform film, an oil phase with high surface tension is required.

Handling of BEMT in a w/o system: Formulating a physico-chemically stable w/o emulsion requires a minimal quantity of nonpolar oils. On the other hand, BEMT solubilization requires a sufficient amount of polar oils. Therefore, the formulator has to compromise in order to guarantee a stable w/o emulsion with BEMT. A reasonable limit for the use of BEMT in w/o emulsions is 3%.

The stability of w/o emulsions that are built with mainly polar oils can be improved by adding some structure to the emulsion system.[11] For this purpose, one can use ethoxylated dodecylglycol copolymer, waxes and clays. One can also use polyglyceryl esters, an emulsifier category that builds liquid lamellar phases.

Combining BEMT with inorganic pigments: Formulators face a major challenge in stabilizing an emulsion containing significant quantities of particulate materials (such as microfine particles of TiO_2 or ZnO) while maintaining their

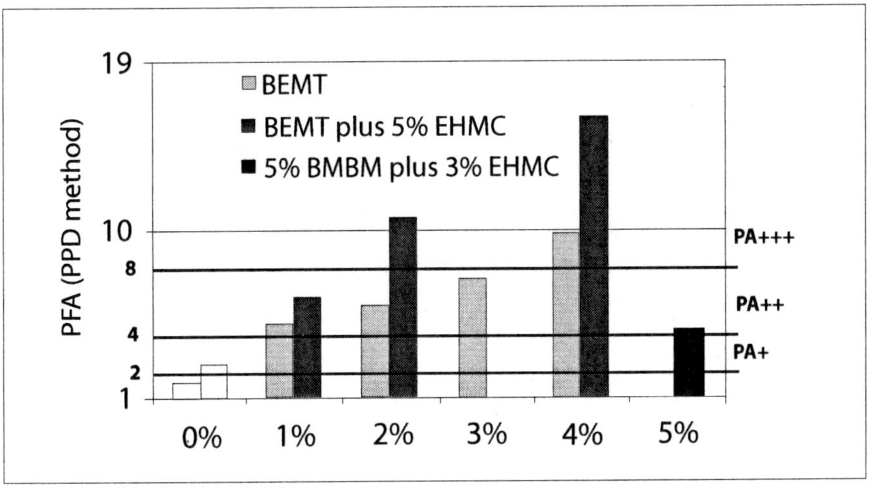

Figure 2. PFA of BEMT and BEMT/EHMC (PPD method)

Table 2. Solubility of BEMT in selected solvents

Solvent	Solubility
Caprylylpyrrolidone	20%
Ethylhexyl salicylate	20
Isodecyl salicylate	18
Ethylhexyl methoxycinnamate	17
C_{12}-C_{15} Alkylbenzoate	13
Butyloctyl salicylate	11
Diethylhexyl adipate	9
Diethylhexyl succinate	7
Isopropylmyristate	6
Hexyl laurate	6
Coco caprylate/caprate	5
Caprylic/Capric triglyceride	5
Isopropylpalmitate	5
Sesame oil	3
Mineral oil	<1

efficacy and controlling the whitening effect on skin.[12] To help meet this challenge, the combination of BEMT with TiO_2 offers several advantages:

- Lower levels of organic sunscreens in end-products
- Less whitening effect and more elegant end-products
- Synergistic effect on the sun protection efficacy

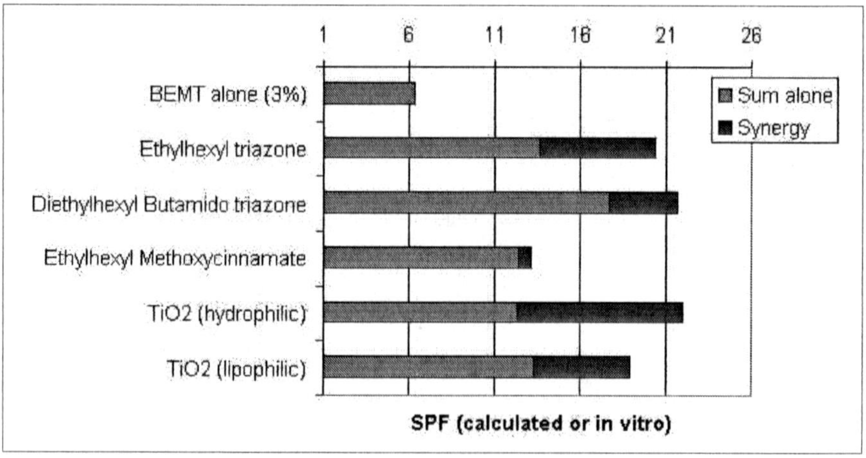

Figure 3. Synergy of BEMT with UVB filters

Synergy of BEMT with UVB Filters

Synergy is present when a particular combination of UV filters gives an SPF that is higher than the sum of the SPFs of each of the filters measured alone – assuming all SPFs are determined by the same method, such as that of COLIPA.[13]

We determined the synergy of 3% BEMT combined with 4% organic UVB filter, or 4% or 5% inorganic UVB filter (Figure 3). The SPF values of BEMT, the UVB absorbers and all combination pairs were calculated using a calibrated step film model[14] that was first proposed by O'Neill.[15] The calculations are in good agreement with in vivo data.

BEMT shows a very strong synergy with the two most efficient UVB filters (which are ethylhexyl triazone and diethylhexyl butamido triazone) and a slight synergy in combination with the most frequently used, but less efficient UVB filter (ethylhexyl methoxycinnnamate).

BEMT also shows a strong synergy with the inorganic UVB filter (TiO_2) that is used in most high SPF (e.g. 30+) sunscreens today.

The synergy was tested with o/w emulsion systems using a standard formula (Formula 2) built with polyglyceryl ester emulsifiers[b] at 4% and using the in vitro assessment with polymethylmethacrylate (PMMA) plates (1.2 mg/cm^2).

Two grades of TiO_2 were used together with 3% BEMT. We tested the hydrophilic grade[c] at 4% active ingredient and a lipophilic grade[d] at 5%.

One explanation for the synergy is that the scattering of light by physical sunscreens increases the optical path length through the film and increases the absorption of UV light by the organic filter (such as BEMT). As expected, the synergy

[b] Tego Care 450, Degussa Care Specialties, Essen, Germany. Tego is a registered trademark of Degussa. Care Specialties.
[c] Eusolex T-2000 from Merck, Darmstadt, Germany. The INCI name is water (*aqua*) (and) titanium dioxide (and) alumina (and) metaphosphate (and) phenoxyethanol (and) sodium methylparaben.
[d] UV-Titan M262 from Kemira Pigments Oy, Pori, Finland. The INCI name is titanium dioxide (and) alumina (and) dimethicone.

Formula 2. Sun protection emulsion for hydrophilic dispersion (A) or lipophilic dispersion (B)

	Ingredient	A (%w/w)	B (%w/w)
A.	Polyglyceryl-3 methylglucose distearate (Tego Care 450, Degussa)	4.00	4.00
	Caprylic/Capric triglyceride (Tegosoft CT, Degussa)	15.00	10.00
	Myristyl myristate (Tegosoft MM, Degussa)	1.00	1.00
	Dimethicone (Abil 350, Degussa)	0.50	0.50
	Mineral oil (*paraffinum liquidum*) (Paraffin, Merck)	1.00	1.00
	Bis-Ethylhexyloxyphenol methoxyphenyl triazine (Tinosorb S, Ciba)	3.00	3.00
	Titanium dioxide (and) alumina (and) dimethicone (UV Titan M262, Kemira)	-	5.90
	Cetyl dimethicone (Abil Wax 9840, Degussa)	0.50	0.50
B.	Water (*aqua*)	qs 100.00	qs 100.00
	Glycerin	3.00	3.00
	Xanthan gum (Rhodicare S, Rhodia)	0.30	0.30
	Aqua (and) titanium dioxide (and) alumina (and) metaphosphate (and) phenoxyethanol (and) sodium methylparaben (Eusolex T-Aqua, Merck)	12.50	-
C.	Phenoxyethanol (and) methylparaben (and) ethylparaben (and) butylparaben (and) propylparaben (and) isobutylparaben (Phenonip, Clariant)	0.70	0.70

Procedure: Heat A and B separately to 75°C under moderate stirring. Add A to B under stirring. Homogenize for 20 sec with Ultra Turrax at 11,000 rpm. Allow to cool to 40°C. Add C. Let cool to RT and adjust the pH value between 6.50 and 7.00.

with the lipophilic grade TiO_2 turned out to be lower than with the hydrophilic grade because in the latter case the filters are present in both the oil phase and the water phase.

We also checked the synergy using the Kull equation[14] (raw data BEMT 3% = SPF 7, TiO_2 5% = SPF 6, combined = SPF 18). The results again confirmed synergy.

Improved UVA Protection

One would expect the addition of BEMT into a sunscreen or day cream would improve UVA protection. To illustrate and quantify the improvement, we carried out some calculations with different formulations using the Ciba Sunscreen Simulator.[15-17] Table 3 shows the composition of these formulations. They all had similar SPFs, meaning similar UVB protection. But they differed significantly in

Table 3. "Experiments" on the Ciba Sunscreen Simulator[15-17]

Formula	F-1	F-2	F-3*	F-4*	F-5*
SPF	14.0	14.7	15.5	15.7	14.1
UVA Transmission 320-400 nm	100%	58%	60%	40%	26%
Ingredients					
EHMC	7%	7%	5%		
EHT**				2%	1%
BP-3	3%				
ZnO		3%			
BMBM				2%	
BEMT***			3%	3%	3%
MBBT***					3%
Total percentage of UV filter	10%	10%	8%	7%	7%

* These formulations cannot yet be used in the USA.
** Not yet approved in the USA; US application (TEA) under way.[18]
*** Does not yet have five years' non-US marketing experience that is required for TEA.[18]

the amount of UVA protection they offered. In spite of great differences, all these formulations could make "UVA" or "broadband" claims.

We also measured the amount of UVA (320-400 nm) transmitted through selected sunscreen films. The amount of UVA radiation still reaching the skin through a layer of sunscreen is best seen from the transmission curve of that sunscreen. The transmission spectrum was calculated using a 2-step film model (Figure 4).[15]

The sunscreen with the highest UVA transmission was F-1 with benzophenone-3. This area has arbitrarily been defined as 100%. Formula F-2 with ZnO reduces this UVA exposure through the sunscreen to below 60%. With formula F-5, using BEMT, and another new broad-spectrum filter called methylenebisbenzotriazolyl tetramethylbutylphenol (MBBT),[19,20] the UVA exposure is reduced to a quarter of the value achieved with the conventional formulation (F-1).

The new broad-spectrum UV filter BEMT provides a means to further improve the UVA protection as well as the overall efficacy of sunscreens. This is best illustrated by calculations of the transmission spectrum and confirmed by in vivo measurements. The residual UVA radiation still reaching the skin can be reduced to a quarter of the value of a conventional sunscreen.

The improved efficacy of BEMT in UVA protection is confirmed by the measurement of the in vivo PFA (UVA protection factor after the PPD method). With only 2% BEMT together with the UVB filter EHMC, the highest protection category of the Japanese Standard, PA+++, can be reached.

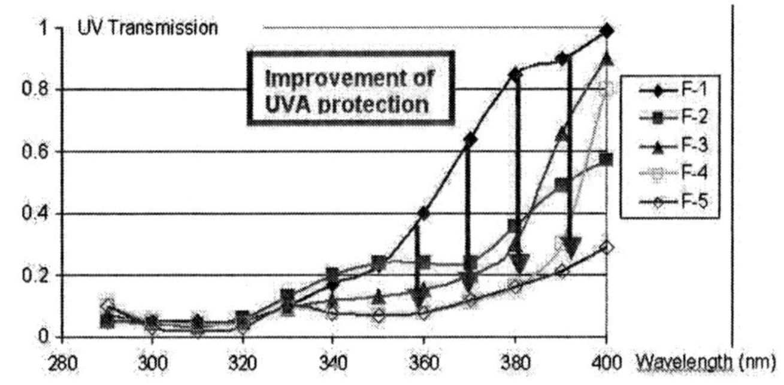

Figure 4. Improvement in UVA protection – shown as reduction in UVA transmission – due to new broad-spectrum filters, calculated by 2-step model.[15,16] Area below formula F-1 (EHMC plus benzophenone-3) was defined as 100%.

Conclusion

Efficient broad-spectrum UV protection is provided by bis-ethylhexyloxyphenol methoxyphenyl triazine (BEMT), an oil-soluble, photostable UV absorber. It is compatible with UVB filters and can thus cover all needs from daily care to special high SPF sunscreens, especially those providing improved UVA protection.

—**Sebastien Mongiat, Bernd Herzog, Cyrille Deshayes, Peter König and Uli Osterwalder,** *Ciba Specialty Chemicals, Basel, Switzerland*

References

1. D Hueglin, B Herzog, S Mongiat, Hydroxyphenyltriazines: A new generation of cosmetic UV filters with superior photoprotection, Oral presentation at 22nd IFSCC, Edinburgh, 23-26 September 2002
2. Twenty Fourth Commission Directive 2000/6/EC of 29 February 2000, Council Directive 76/768/EC, *Official Journal of the European Communities* L56/42 (Mar 1, 2000)
3. G Berset, H Gonzenbach, R Christ, R Martin, A Deflandre, RE Mascotto, JDR Jolley, W Lowell, R Pelzer and T Stiehm, *Int J Cosmet Sci* 18 167-177 (1996)
4. J-E Otterstedt, Photostability and molecular structure, *J Phys Chem* 58 5716-5725 (1973)
5. B Herzog and K Sommer, Investigations on photostability of UV-absorbers for cosmetic sunscreens, Proceedings (CD-ROM): 21st IFSCC Congress, Berlin, Poster P 60, 11-14 September 2000
6. B Herzog, S Mongiat, C Deshayes, M Neuhaus, K Sommer and A Mantler, In vivo and in vitro assessment of UVA-protection by sunscreen formulations containing either butyl methoxy dibenzoyl methane, methylene bis-benzotriazolyl tetramethylbutylphenol, or microfine ZnO, Oral presentation at 22nd IFSCC, Edinburgh, 23-26 September 2002
7. A Chardon, D Moyal and C Hourseau, Persistent pigment-darkening response as a method for evaluation of ultraviolet A protection assays, In *Sunscreens: Development, Evaluation, and Regulatory Aspects*, NJ Lowe, NA Shaath and MA Pathak, 2nd edn, New York: Marcel Dekker (1997) pp 559-582

8. *JCIA Measurement Standard for UVA Protection Efficacy*, Minato-Ku Tokyo: Japan Cosmetic Industry Association (1995) p 105
9. R Schueller and P Romanowski, Understanding emulsions, *Cosmet Toil* 113(9) 39-44 (1998)
10. GH Dahms, Choosing emollients and emulsifiers for sunscreen products, *Cosmet Toil* 109(11) 45-52 (1994)
11. JP Hewitt, The influence of emulsion structure on SPF in physical sunscreen formulations, Poster 22nd IFSCC Congress, Edinburgh 2002
12. B Innes, Nanotechnology and the cosmetic chemist, 22nd IFSCC Edinburgh, 23-26 September 2002
13. COLIPA sun protection factor test method, ref. 94/289, Brussels, Belgium: The European Cosmetic Toiletry and Perfumery Association (Oct 1994)
14. B Herzog, Prediction of sun protection factors by calculation of transmissions with a calibrated step film model, *J Cosmet Sci* 53(1) 11-26 (2002)
15. D Steinberg, Measuring synergy, *Cosmet Toil* 115(11) 59-62 (2000)
16. JJ O'Neill, Effect of film irregularities on sunscreen efficacy, *J Pharm Sci* 73 888-891 (1984)
17. Ciba Specialty Chemicals, The Ciba Sunscreen Simulator, www.cibasc.com/personalcare
18. S Onel, FDA finalizes rule that could expand OTC drug marketplace, *UPDATE Food and Drug Law, Regulations and Education*, issue 5 (Sep/Oct 2002) www.fdli.org
19. U Osterwalder, H Luther and B Herzog, UV-A protection with a new class of UV absorber, *Proceedings of the 47th SEPAWA Kongress*, Bad Durkheim: SEPAWA Vereinigung der Seifen-, Parfüm- und Waschmittelfachleute eV (2000) pp 153-164
20. U Osterwalder, H Luther and B Herzog, New Class of UV absorber with excellent performance in UV-A region, In *Cosmetics and Toiletries Manufacture Worldwide*, Hemel Hempstead, UK: Aston Publishing Group (2001) pp 153-158

Pigments as Photoprotectants

Keywords: pigments, UV radiation, sun products, ZnO, TiO_2, benzotriazol

This article gives an overview of two pigments (TiO_2 and ZnO), their use in photoprotection and their mechanism of action and ability to attenuate UV radiation. Finally, the organic pigments, the dominant members of a new family of photoprotectors, are described.

Pigments are solid particles and are insoluble in water and fatty matter. They may be classified into two groups, depending on their origin: mineral pigments, which are of mineral origin, and organic pigments, which are produced by organic sythesis.

Mineral pigments are inert and opaque powders that reflect and diffuse UV radiation and part of the visible light (Figure 1). Consequently, they provide broad-spectrum protection. The mineral pigments most often used for photoprotection are titanium dioxide (TiO_2) and zinc oxide (ZnO). Other minerals such as zirconium, cerium, talc and kaolin are less used because their capacity to attenuate UV radiation is low. This article will focus on the protectant properties of TiO_2 and ZnO.

TiO_2 and ZnO

Titanium is one of the 10 most abundant elements on Earth. It can be found as mineral rutile (93-97% titanium dioxide) or in conjunction with iron oxides in the ores such as ilmente (45-75% titanium dioxide).[2] It makes up 0.6% of the Earth's crust and can be found either in its pure state or combined with ferrous oxide.

Zinc is also found in large quantities on earth, although it is less abundant than titanium. Its principal ores are blend (sulfide), smithsonite (carbonate), calamine (silicate) and franklinite (zinc, manganese and iron oxide).[3] It can be found as zinc salt (carbonate, silica) or combined with other metals such as manganese or ferrous oxide.

The industrial sector consumes 3 million metric tons of TiO_2 annually. TiO_2 is used for its pigment properties and its whitening capacity (opacity in the visible light range). These properties present a real interest in application and are used in the pharmaceutical industry[4] as well as in the food, cosmetic and paint industries.

ZnO is used in the same industrial fields as TiO_2 and is conventionally used as a topical antiseptic in the pharmaceutical industry.[1]

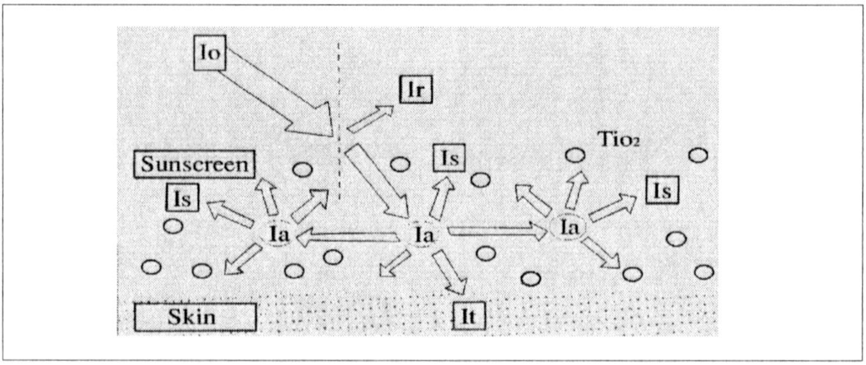

Figure 1. Attenuation of UV radiation by titanium dioxide[1]

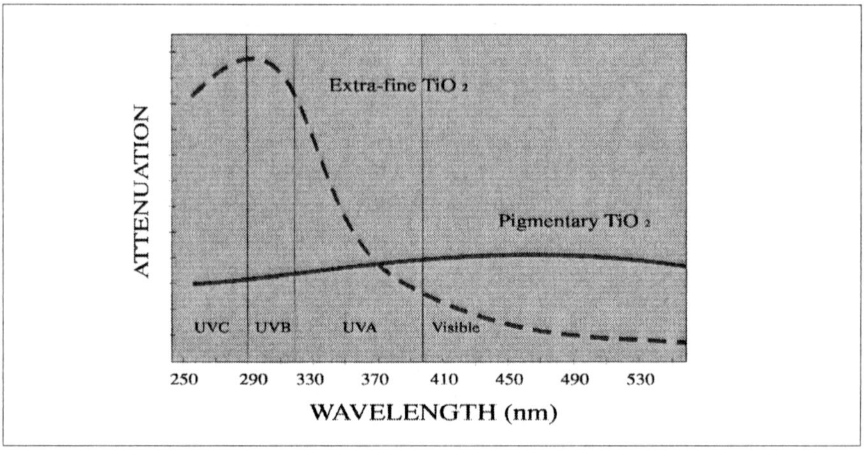

Figure 2. Attenuation of UV radiation and visible light by pigmentary and extra-fine titanium dioxide[5]

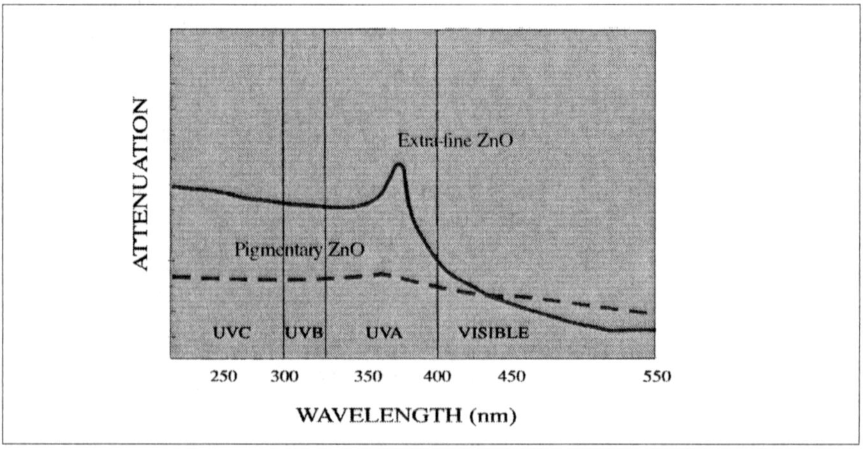

Figure 3. Attenuation of UV radiation and visible light by pigmentary and extra-fine zinc oxide[5]

As pigments, TiO_2 and ZnO are opaque to UVB and UVA radiation. Their use has nevertheless been limited in sun care products for cosmetic reasons: They make the skin look excessively white. In order for TiO_2 and ZnO to be used as photoprotectors in cosmetic products, they must be opaque to UVB radiation and UVA radiation, but they must also be as transparent as possible to visible light so they limit the appearance of whitening. Extra-fine mineral pigments with UV attenuation qualities (Figures 2 and 3) have been developed for this purpose.

Extra-fine TiO_2 may be obtained from hydrated titanium oxide by various methods. After a crystallization and washing phase, calcination and grinding (micronization) are performed, thus producing a micronized TiO_2 whose surface may eventually be treated.[6] The UV radiation attenuation qualities of mineral pigments depend on certain parameters that will directly impact their optical properties and thus affect their photoprotecting efficacy. Among these parameters are:

- The crystalline size and form
- The refractive index
- Coating and dispersion

Crystalline Size and Form

For a mineral pigment, the crystal's size is a fundamental determinant in filtration efficacy. The crystalline size of pigmentary TiO_2 is greater than 250 nm. The particle size of extra-fine TiO_2 ranges from 15 to 60 nm. This nanometric size allows maximum absorption spectrum in the region of short UVA and UVB rays (Figure 4). The efficacy of UV attenuation in proportion to the crystal's size may be determined by spectroscopy. With this tool, the crystalline size best adapted to each need may be chosen.

A smaller crystalline size means shorter attenuated wavelength (toward UVB) and greater transmittance of visible light (transparency). The larger the crystalline size is, the nearer the attenuated wavelength is to that of visible light, and the greater is the opacity and the appearance of whiteness on the skin.

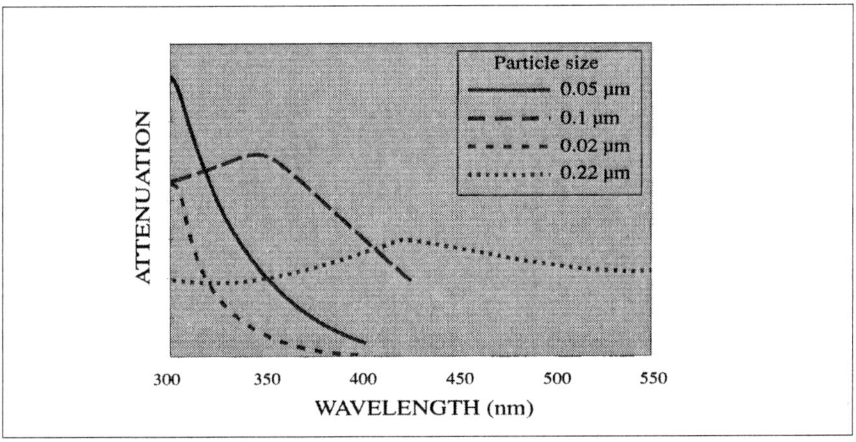

Figure 4. Attenuation of UV radiation according to the crystalline size of titanium dioxide[5]

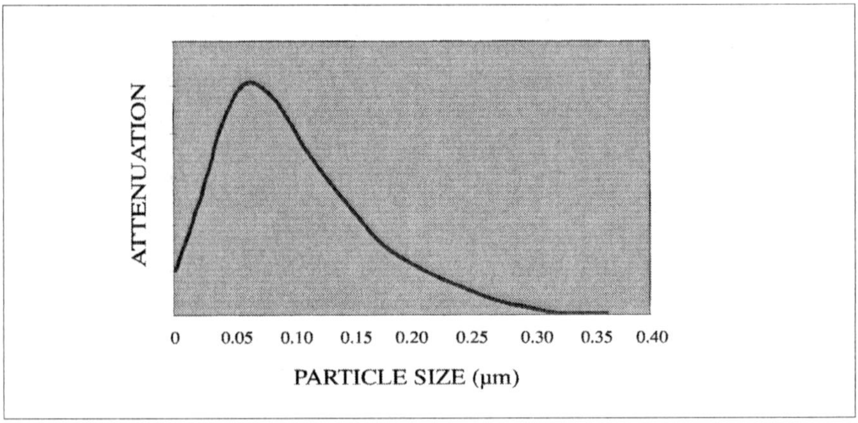

Figure 5. Attenuation of UVA radiation at 360 nm according to the particle size of zinc oxide[5]

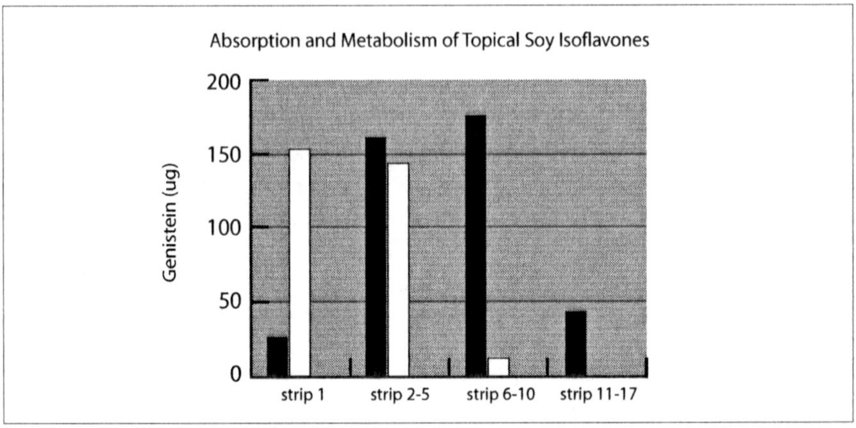

Figure 6. Attenuation of UV radiation depending on the quality of TiO_2 dispersion[7]

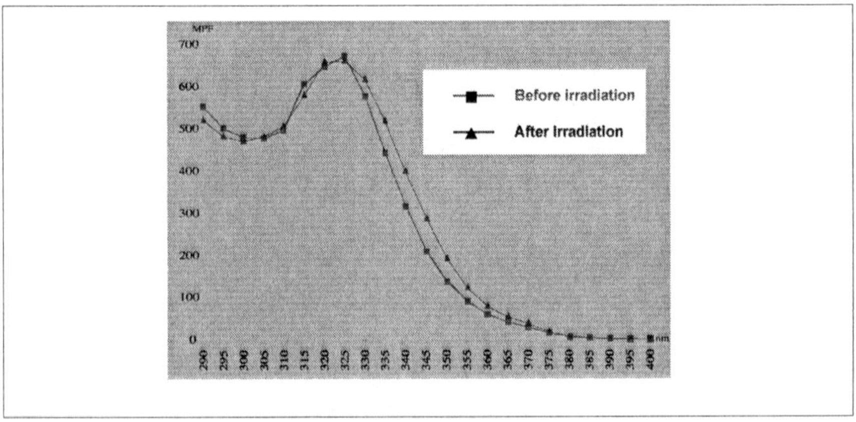

Figure 7. Absorption spectrum of a formula containing TiO_2 and ZnO before and after irradiation, 5 DEM (DEM=Minimal Erythemal Dose related to photostability)[8]

Titanium has various crystalline forms: brookite (very rare), anatase and rutile. Anatase and rutile crystals are used the most in photoprotection and have a tetragonal structure. The rutile form is the most abundant form and is very stable. Its refractive index and specific density are higher than those of anatase.

The industry has developed procedures similar to those used with TiO_2 to obtain extra-fine ZnO and optimize its filtration efficacy. Micronized ZnO has particle sizes that range from 40 to 100 nm. The highest efficacy of ZnO against UVA radiation is situated within this range (Figure 5). ZnO essentially attenuates short and long UVA radiation and complements TiO_2 attenuation, thus generating products with broad-spectrum protection.

Refractive Index

TiO_2 and ZnO have a refractive index of 2.7 and 2.01, respectively. A higher refractive index means a higher reflection of visible light. This explains why TiO_2, though its particle size is identical to that of ZnO, is more opaque (whitening) and photoprotective.

Coating and Dispersion

The quality of pigment dispersion within sun care products affects not only their cosmetic properties, but also their efficacy and performance as sun protectants (Figure 6). Proper pigment dispersion makes it easier to obtain the desired sun protection factor and does not whiten the skin as much.

The aim of proper TiO_2 and/or ZnO dispersion is to reach stable particle distribution over time and avoid particle re-agglomeration, which is mainly due to

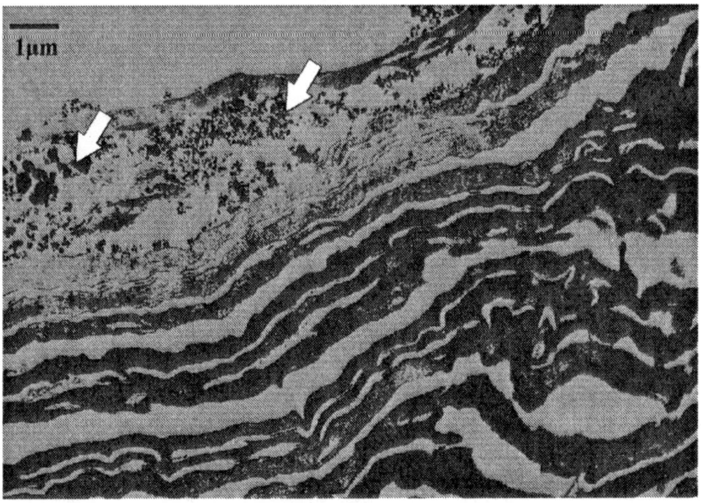

Figure 8. TEM image of TiO_2 localization in the upper layers of the stratum corneum[12]

Broad-Spectrum Protection from an Organic Pigment

Among the organic pigments is a new micronized organic pigment belonging to the benzotriazol family. This pigment is of great interest to photoprotection because it covers the whole UV spectrum.

Methylene bis-benzotriazolyl tetramethylbutyl-phenol is the INCI name of this new organic UV filter that acts by absorbing radiation as soluble organic filters do, and by diffusing radiation as mineral pigments do. Like mineral pigments, it has high photostability. It also has some of the mineral pigment properties, such as its particle size (approximately 200 nm), and it offers maximum filtration efficacy (Figure 9).

Methylene bis-benzotriazolyl tetramethylbutyl-phenol disperses within the sun care product, and its toxicological profile is excellent.

This new pigment from the benzotriazol family opens the way for a new generation of photoprotective agents.

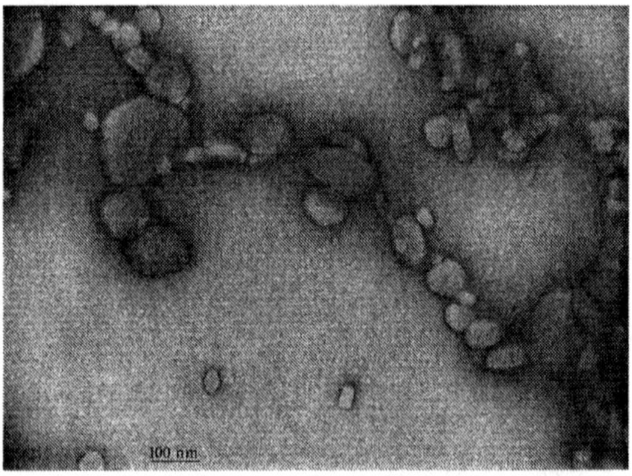

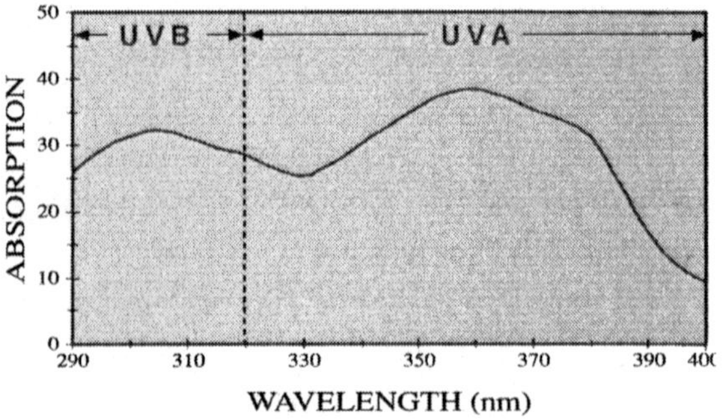

Figure 9. TEM image (top) and absorption spectrum (bottom) of methylene bis-benzotriazolyl tetramethylbutylphenol[17]

absorption phenomena and electric charges at the pigment's surface (zeta potential), which tend to attract pigments to each other and, consequently, form aggregates.

In order to avoid aggregation phenomena, the pigment's surface is coated. The primary particles of TiO_2 and ZnO are treated with organic and inorganic materials such as silicone, alumina, silica and stearic acid derivatives.

Surface treatments cancel surface charges and optimize pigment dispersion. It is then possible to choose the coating best adapted to the desired formula, thus offering the formulator a wider range of excipients. The excellent quality of mineral pigment coatings should guarantee the perfect photostability of the formulated products (Figure 7).

Toxicological Aspects

The excellent toxicological profile of TiO_2 and ZnO is due to their insignificant–considered by some to be nonexistent–transcutaneous passage.[9] Mineral pigments that do not dissolve in aqueous and organic solvents do not qualify for transcutaneous passage. In vivo studies on penetration have indeed shown that TiO_2 penetration is not detected in living skin tissues.[10]

Moreover, ex vivo studies with Franz cells have shown that TiO_2 is found in the upper layers of the stratum corneum.[11,12] Its localization was demonstrated by assaying TiO_2 in the stratum corneum using a stripping method associated with spectrophotometry. Its localization was also demonstrated by observations with a transmission electron microscope (Figure 8).

Studies on skin localization and eye tolerance after repeated applications (45 days) and on oral or parenteral toxicity did not show any anomaly.[13]

Lastly, epidemiological studies on workers in the titanium industry showed the safety of this substance with regard to respiratory diseases or skin intolerance.[14]

One can conclude that mineral pigments (TiO_2 and ZnO) are the most appropriate means of photoprotection for children and persons with sensitive skin. They are also reported to be the most effective means of reducing photosensitization risks.[15,16]

The qualities described in the article lead to the conclusion that TiO^2 and ZnO are seemingly the best way to obtain a safe and broad spectrum UV protective product.

—**Jose Ginestar,** *Pierre-Fabre, Center for Research, Castanet Tolosan, France*

References

1. NJ Lowe, NA Shaath and MA Pathak, *Sunscreens: development, evaluation and regulatory aspects*, Second Edition, revised and expanded, New York, Basel, Hong Kong, Marcel Dekker, Inc. (1997) pp 335-365
2. *European Pharmacopoeia*, 3rd edition, (2000) Direction de la Qualité du Médicament du Conseil de l'Europe, Strasbourg (France) Conseil de l'Europe
3. Technical report from Tioxide Specialties Limited, Peterlee, Durham, England
4. Technical report from Degussa France, Courbevoie, France
5. Technical report from Créations et Couleurs, presented at UV Harmony Seminar (Apr 8-9, 1999)

6. Unpublished correspondence from D Redoules and R Tarroux, Laboratoire de Dermochimie, IRPF, Vigoulet Auzil, France (March 1998)
7. RH Guy, JJ Hostynek, RS Hinz and CR Lorence, Metals and the skin: Topical effects and systemic absorption 36; 395-407 (1999)
8. J Lademann, HJ Weigmann, C Rickmeyer, H Barthelmes, H Schaefer, G Mueller and W Sterry, Penetration of titanium dioxide microparticles in a sunscreen formulation into the horny layer and the follicular orifice, *Skin Pharmacol Appl Skin Physiol*, 12 247-256 (1999)
9. F Pflucker and H Hohenberg, The outermost strateum corneum layer is an effective barrier against dermal uptake of topically applied micronized titanium dioxide, *Intl J Cosmet Sci* 21 399-411 (1999)
10. C Miquel, V Raufast, S Mouysset and A Mavon, Evaluation of titanium dioxide microparticles in a sunscreen formulation into the horny layer after UV radiation exposure, presented at 7th Perspectives in Percutaneous Penetration Conference (Apr 2000)
11. P Msika and F Boyer, Efficacité et sécurité de l'oxyde de titane ultrafin en photoprotection, *BEDC Cosmétologie* 237 (1993)
12. HS Tichy, Power block, SPC 37-42 (1992)
13. M Jeanmougin, JR Mancet, A Pons-Guiraud, G Laine and L Dubertret, Allergies et photoallergies de contact aux photoprotecteurs externes. Etudes sur six années, *Nouvelles Dermatologiques* 377 (1994)
14. S Shauder and H Ippen, Contact and photocontact sensitivity to sunscreens, *Contact Derm* 230 (1997)
15. J Ginestar, J Hemmerle and F Carrière, "Microscopic analysis and spectrophotometric evaluation of a new photoproction system (MPI-SORB) ," poster presented at the 13th Congress of Photobiology in San Francisco (July 1-6, 2000)
16. Technical report, Ciba Specialty Chemicals, Rueil Malmaison Cedex France (September 1998)

Safety and Efficacy of Microfine Titanium Dioxide

Keywords: microfine titanium dioxide, UVA, UVB, broad-spectrum protection, safety

Modern microfine TiO_2 products attenuate both UVA and UVB. The ingredient is practically chemically inert and "non-penetrating" and can be regarded a safe cosmetic raw material

The "healthy, sun-tanned appearance" is still a popularly held perception and, despite much information on the hazards of UV-induced skin diseases and cancers, there appears to be no lessening in the time spent sunbathing. It is a fact, however, that the cells of the human skin are sensitive toward and must be protected from both UVB (290-320 nm) and UVA (320-380 nm) radiation.

Alongside trying to educate people to the hazards of too much sun exposure, the cosmetics industry must focus on providing safe and effective formulations that offer maximum protection against these damaging properties of UV. An ideal sunscreen formulation should contain active ingredients that protect against radiation within the complete UV range. At the same time they should be stable against UV radiation themselves and, naturally, safe to human skin under irradiation.

Microfine titanium dioxide (TiO_2) complies with all these requirements. The Physical Sunscreens Manufacturers Association (PSMA) has spent many years producing and collating data that support this safety claim. This has led to titanium dioxide being approved as a UV filter for use in sunscreens in the major regulatory jurisdictions of the European Union (EU), the United States and Japan.

In the EU, following the passage of the Council's Cosmetic Directive 76/768/EEC of July 27, 1976, and as a part of the process of approximating the laws of the Member States relating to cosmetic products, the safety of titanium dioxide was evaluated. The EU's Scientific Committee for Cosmetics and Non-Food Products Intended for Consumers (SCCNFP) was asked to answer the following questions:

- Does the SCCNFP consider the safety profile of titanium dioxide to be sufficient to permit the listing of this material in part I of Annex VII?

- Does the SCCNFP concur with the proposed concentration limit of up to 25% or is a different figure recommended?
- Does the SCCNFP recommend additional requirements for its use in cosmetic products?

In October 2000 the answers came in a document called the SCCNFP Opinion Concerning Titanium Dioxide.[1] The SCCNFP found titanium dioxide safe for use in cosmetic products at a maximum concentration of 25% for the purpose of protecting the skin from certain harmful effects of UV radiation. That document and its references will be cited frequently in this article.

Absorption Spectrum

When its particle size is controlled, TiO_2 may draw out UV light either by scattering or through absorbance. While fine particle TiO_2 attenuates UVB mainly due to absorption, pigmentary TiO_2 does so mainly by scattering the light. Microfine TiO_2 is specifically designed for optimizing UV attenuation while minimizing the aesthetically adverse whitening effects of pigmentary TiO_2 on the skin. Typically, microfine TiO_2 offers protection between 290 nm and 350 nm (Figure 1).

Coatings

To achieve optimum UVA and UVB performance, it is crucial that microfine TiO_2 is ideally dispersed in the cosmetic formulation and, later, evenly spread on the skin. Some of today's sophisticated formulations that include microfine titanium dioxide are delivering smooth, light and easy application in cosmetic products. Uncoated TiO_2 is hydrophilic and can be inflexible in terms of formulation compatibility. It can also "drag" on the skin, giving it an unpleasant, dry feeling.

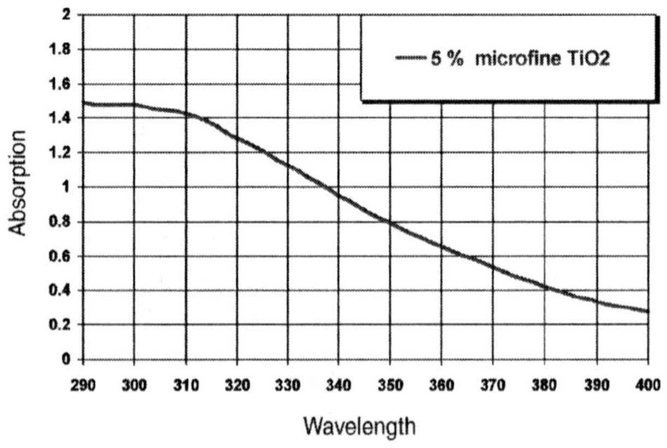

Figure 1. In vitro absorption of microfine TiO_2 (primary particle size < 20 nm). Formulation: Sun Milk

Scientists have developed a range of coatings in order to control the degree of hydrophilicity/lipophilicity and to provide the TiO_2 particles better compatibility with cosmetic formulations, which allows better dispersion in the oil or water phase of an emulsion. These coatings also aid in stabilizing the particles against agglomeration, thus delivering more uniform films and improving efficacy on the skin.

A further advantage of the coatings is that they reduce the photocatalytic activity of uncoated TiO_2. While this activity has been shown not to cause adverse effects on skin (as shown by the absence of any phototoxic, photomutagenic, photogenotoxic and photocarcinogenic effects), it could potentially lead to undesirable property changes in the cosmetic formulation (e.g., discoloration).

Typically, the coating materials selected are common cosmetic ingredients that are widely used for different purposes in a variety of cosmetic products. They contribute to the physio-chemical properties of the formulation in which they are used. The key purpose of the coatings is to improve dispersability, because they themselves are not UV absorbers.

The inorganic and organic coating materials used most on titanium dioxide make the surface of the particle more hydrophobic and allow better dispersion in the oil phase of an emulsion. This renders titanium dioxide more versatile for its use in different formulations. The following are commonly used coating materials found in commercial grades: alumina, silica, dimethicone, simethicone, glycerol, stearic acid and trimethoxycaprylylsilane.

The integrity of the coatings is important to retain improved dispersability, and the stability of coating materials on the TiO_2 particle has been clearly proven[2] with experiments using variations in pH, temperature and sheer force. The stability of the coating on the particle is also important for the technical properties of TiO_2-containing formulas (stability of emulsion, color, segregation of particles).

The product advantages provided by the coating technology have led to widespread usage of the various types of coated, microfine TiO_2 in modern sunscreen products in the United States, Europe and Japan. The ensuing high dispersability enables formulators to meet the current demand for high-SPF products (SPF 30+) with good skin compatibility.

Safety in Use

Titanium dioxide is generally chemically inert and, thus, will not undergo chemical reactions with the various constituents of the human skin. As a consequence, it is not attacked once applied to the skin, it does not cause skin irritation and sensitization, and it remains stable and efficient as a UV protector. Adverse reactions to microfine TiO_2 within suncreens have practically not been reported. The tests shown in Table 1 have been carried out over many years as part of developing the safety data for microfine TiO_2.

The titanium dioxide grades used in these tests had various coatings. The coatings did not influence the results, and the SCCNFP concluded in its opinion: "Many of the coating substances are already used as ingredients in cosmetics, and if they are acceptable in this role they should be acceptable as coatings for titanium dioxide."[3]

Effect on Living Dermal Cells

Extensive tests for percutaneous absorption, mostly in vivo, indicate that absorption does not occur under in-use conditions, with either coated or uncoated material. An impressive number of studies have been conducted using ex vivo (human) skin in Franz-type diffusion cells.[4-9] With this procedure, one can simultaneously assess the fraction remaining on the epidermal surface, the fraction penetrating into the layers of the epidermis, and the fraction that would fully penetrate the epidermis. As was expected, no penetration was detected. Titanium dioxide was found only in the top two to four layers of the stratum corneum, and thus did not reach living cells, even after repeated application.

These results were backed up by in vivo studies[10-13] with volunteers where consecutive skin layers were removed by tape stripping and analyzed under the (electron) microscope; again, no penetration beyond the upper layers of the stratum corneum was found.

Table 1. Safety tests of microfine TiO_2

Safety Parameter	Rating
Acute oral toxicity	Very low
Acute dermal toxicity	Low
Sub-chronic oral toxicity	Low
Long term feeding studies (rat/mouse)	No evidence of carcinogenesis
Skin irritation (animals/man)	Low or absent
Sensitization (animals/man)	Not found
Tests for mutagenicity	Non mutagenic
Tests for clastogenicity	Non clastogenic

Table 2. Microfine TiO_2 used for UV attenuation in cosmetic products: types and coatings

Type	Particle Size (nm)	Coating Material(s)	Coating %
Rutile/Anatase	21	SiO_2	<2.5
Rutile	14	Al_2O_3	8-11
		simethicone	1-3
Anatase	60	Al_2O_3	3-7
		SiO_2	12-18
Rutile	20	Al_2O_3	5-6
		dimethicone	1-4
Rutile	15	Al_2O_3	1-15
		stearic acid	1-13

Photocatalytic Properties

Titanium dioxide is photocatalytic in ultraviolet light. The biological relevance of this photoactivity is doubtful, given both the absence of dermal penetration and the fact that the coated preparations typically used as UV filters show significantly less photocatalytic activity than the uncoated preparations.[14-19]

In addition, the photocatalytic activity of titanium dioxide does not change or activate the coating materials[20,21] and the combination of TiO_2 particles and coatings does not show any photo-induced toxic effect.[39]

While it is understood that some photocatalytic activity in microfine TiO_2 itself may occur, none of the respective studies showed any photocatalytic activity under in-use conditions. Titanium dioxide did not show phototoxic activity in studies in vivo and in vitro, and has been shown to be non-photoirritant,[22,23] non-photosensitizing,[24-26] non-photomutagenic,[27-38] and non-photocarcinogenic.[40,41]

Phototoxic, Photoirritant and Photosensitive Properties

Titanium dioxide did not show phototoxic activity in studies in vivo and in vitro[22-25] and no photosensitization or photoirritation was observed. TiO_2 exists in three stable crystalline forms,[42] two of which – anatase and rutile – are commonly used in cosmetic products. Both are tetragonal crystalline structures, but rutile differs from anatase in structural properties, thus conferring higher refractivity and lower photoactivity.[43] The tests examined a large number of varieties of coated types of microfine TiO_2, including both rutile and anatase forms. TiO_2 types, particle sizes and coatings studied are shown in Table 2.

All of these have been shown to be nonphototoxic in the validated in vitro 3T3-NRU assay and nonphotoirritant in animals.[39] These findings are backed by the complete absence of any photosensitizing potential (Magnusson-Kligman tests under UV irradiation). All of these findings are also confirmed by the extensive experience of the use of the materials as UV filters.

Photomutagenic, Photogenotoxic and Photocarcinogenic Properties

All these types of coated microfine TiO_2 have consistently shown negative results in in vitro photomutagenicity tests (photo-Ames) and photogenotoxicity tests (photo-Chromosome-Aberration).[22-38,40,41]

One previously published article[44] mentions chromosome damage in vitro in the presence of UV light, although this was only evident at cytotoxic TiO_2 concentrations. Another article[45] reports oxidative damage when isolated DNA was treated with titanium dioxide in the presence of UV light. No DNA oxidation was seen, however, when treating intact cells. The relevance of these results is therefore doubtful.

An irradiation-promoting study[40] in mice using the tumor initiator DMBA and low-dose solar simulated UV radiation as promoter showed that a titanium dioxide formulation completely protected the animals from UV-induced tumor promotion and almost completely against carcinogenesis. The absence of phototoxicity in the 3T3-NRU assay and the fact that titanium dioxide particles do not penetrate the stratum corneum make it unlikely that it has a photocarcinogenic potential under in-use conditions. This is confirmed by years of experience from the use of the material as a UV filter.

Conclusion

Microfine TiO_2 is an active ingredient for sunscreen formulations. It protects against radiation within the complete UV range while it is, at the same time, stable against UV radiation itself and nontoxic to human skin under irradiation.

The safety and efficacy of microfine titanium dioxide has been the subject of numerous experimental and clinical studies and has been fully demonstrated. In the standard risk equation (i.e., Hazard × Exposure = Risk), both the hazard and the exposure have been minimized

As a consequence, microfine titanium dioxide has been officially registered in the United States and the EU. It is also recognized for use in cosmetics in Japan as a safe and effective ingredient for sunscreen products.

—**Rainer W. Christ,** *CIC GmbH, Frankfurt, Germany*

References

Note: Many of the following references are cited in Reference 1, which is the SCCNFP Opinion on Titanium Dioxide. The SCCNFP internal references are indicated here in brackets.

1. SCCNFP, Opinion concerning titanium dioxide (Oct 24, 2000). The full 39-page text is not publicly available, because it consists mainly of proprietary information from PSMA member companies. However, an Executive Summary is available at http://europa.eu.int/comm/food/fs/sc/sccp/out135_en.html.
2. Stability of coating materials on TiO_2 particles (Overview by PSMA, Jan 2000) [96]
3. SCCNFP, Opinion concerning titanium dioxide (Oct 24, 2000) Section 2.11 p 31
4. Degussa AG, The in-vitro percutaneous absorption through human abdominal epidermis of titanium dioxide from titanium dioxide T 805 formulation, a Degussa technical report US-IT No 94-0158-DGT (1996) [24]
5. R Christ, The in vitro percutaneous non-penetration of titanium dioxide from Eusolex TA and Eusolex TC formulations through human abdominal epidermis (Summarizing Report Nov 25, 1994), E Merck Project 154524, Report No 1 [25]
6. Jue-Chen Liu, Chung Ye-Tseng and F Viebrock, Percutaneaous absorption of titanium oxide, an internal publication from Sachtleben Chemie (1990) [28]
7. F Pflücker et al, The outermost stratum corneum layer is an effective barrier against dermal uptake of topically applied micronized titanium dioxide, *J Invest Dermatol* (1999) [132]
8. The in vitro percutaneous absorption through human abdominal epidermis of titanium dioxide, Inveresk Report No 14157 from Inveresk Research International (1996) [30]
9. I Castiel-Higounec et al, Demonstration of non-penetration of titanum dioxide contained in photo-protection products. An ex-vivo study conducted on human skin, Institut de Recherche Pierre Fabre (IRPF) [31]
10. R Christ, The in vivo percutaneous absorption of titanium dioxide from Eusolex TA and Eusolex TC formulations through human epidermis of the lower arm, E Merck Project 154524, Report No 2 (May 22, 1995) [27]
11. SR Spruce, Skin penetration with Tioveil formulations, Tioxide Internal Report (1993) [29]

12. W Sterry, Investigations of coated titanium dioxide, Humboldt Univesität zu Berlin, Medizinische Fakultät Charite Dermatologische Universitätsklinik und Poliklinik (May 1997) [62]
13. W Sterry, Investigations of alumina/silica coated titanium dioxide particles Tioveil AQ-N, Humboldt Univesität zu Berlin, Medizinische Fakultät Charite Dermatologische Universitätsklinik und Poliklinik (May 1997) [63]
14. RL Bickley and LT Hogg, The measurement of the specific intrinsic photoactivity of dispersed solids, a Tioxide internal publication (1994) [53]
15. J Braun, TiO2 Contribution to the durability and degradation of paint film, II: Prediction of catalytic activity, *J Coatings Tech* 62(785) 37-42 (1990) [54/1]
16. U Gesehues, Bedeckungsgrad und Photoactivität anorganisch nachbehandelter TiO2-Pigmente, *Farbe und Lack* 94 184-189 (1988) [54/2]
17. K Heikkila, The photocatalytic activity of titanium dioxide and a method for studying the same, a Masters Thesis of Helsinki University of Technology and internal research reports of the pigments development laboratory of Kemira Pigments (1991) [55]
18. VP Judin and VT Salonen, Correlation of crystal properties of ultrafine titanium dioxide with its performance as a physical UV filter, Oral presentation given at the SCC Annual Scientific Meeting, Dec 9-10, 1993, New York City [56]
19. J Leimbach, University of Regensburg, PhD Thesis (1995) [57]
20. H-J Driller, Coatings of titanium dioxide, SCCNFP Plenary Meeting (Feb 17, 1999) [72]
21. TA Egerton, Comments on photostability and photoactivity of titanium dioxide, including articles referenced in the comments (Sept 1999) [84]
22. Degussa AG, Acute dermal photoirritation study with titanium dioxide T 805 in albino rats, a Degussa technical report US-IT No. 92-0042-DGT (1992) [45]
23. Inveresk Research International, Determination of photoirritation potential in guinea pigs, a Tioxide UK Limited, Report No 6387 NP89/367 (1990) [46]
24. Rhone-Poulenc Chimie, TSG: Etude de phototoxicite chez le cobaye, Report 14899 (1996) [47]
25. Degussa AG, Photosensitization study with titanium dioxide T 805, A Degussa technical report US-IT-No. 92-0037-DGT (1992) [48]
26. Inveresk Research International, Determination of photosensitization potential in guinea pigs, a Tioxide UK Limited Report No 6388, NP89/367 (1990) [49]
27. D Utesch, TTO51 A, TTO51C and TioveilAQ: In vitro assessment for photomutagenicity in bacteria, Report Nos 40/110,112,113/93, Experiments T 13775/6/7 Institute of Toxicology, E Merck, Darmstadt (1993) [50,51,52]
28. M Ballantyne, Eusolex T-2000, PSMA-3, UV-TITAN M 160, UV-TITAN M 212, UV-TITAN M262 PSMA-5: Reverse Mutation in three histidine-requiring strains of *Salmonella typhimurium* and a tryptophan-requiring strain of *Escherichia coli* in the presence of ultra violet light, Covance Laboratories Report Nos 70/70-D5140 (Mar 1999) [73], 1770/1-D5140 (Mar 1999) [74], 520-D5140 (Oct 1999) [75], 520-D5140 (Oct 1999) [76], 520-D5140 (Oct 1999) [77], 1731/2-D5140 (Jul 1999) [78]
29. S Riley, PSMA-2, PSMA-3, PSMA-4, PSMA5, PSMA-6: Induction of chromosome aberrations in cultured Chinese hamster ovary (CHO) cells in the presence of ultra violet light, Covance Laboratories Report Nos 70/73-D5140 (Jul 1999) [79], 70/74-D5140 (Jul 1999) [80], 70/75-D5140 (Jul 1999) [81], 1731/1-D5140 (Jul 1999) [82], 346/3-D5140 (Jul 1999) [83]
30. M Ballantyne, X-200, ADFC, MT-100-V: Reverse mutation in three histidine-requiring strains of *Salmonella typhimurium* and a tryptophan-requiring strain of *Escherischia coli* in the presence of ultra violet light, Covance Laboratories Report Nos 520-D5140 (Jan 2000) [86], 520-D5140 (Jan 2000) [87], Report No XXX-D5140 (Jan 2000) [88]
31. J Whitwell, UV-Titan X-200, ADFC900524001: Induction of chromosome aberrations in cultured Chinese hamster ovary (CHO) cells in the presence of ultra violet light, Covance Laboratories Report Nos 520/32-D5140 (Dec 1999) [89], 520/34-D5140 (Dec 1999) [90]

32. M Ballantyne, Mirasun XC99/20: Reverse mutation in three histidine-requiring strains of *Salmonella typhimurium* and a tryptophan-requiring strain of *Escherischia coli* in the presence of ultra violet light, Covance Laboratories Report No 413/28-D5140 (Jan 2000) [97]
33. J Whitwell, Mirasun XC99/20: Induction of chromosome aberrations in cultured Chinese hamster ovary (CHO) cells in the presence of ultra violet light, Covance Laboratories Report No 413/27-D5140 (Feb 2000) [98]
34. Degussa AG, US-IT No 94-0014-DGM, Titanium Dioxide T 805: Reverse mutation in three histidine-requiring strains of *Salmonella typhimurium* and a tryptophan-requiring strain of *Escherischia coli* in the presence of ultra violet light (1998) [121]
35. Degussa AG, US-IT No 94-0015-DGM, Titanium Dioxide T 817: Reverse Mutation in three histidine-requiring strains of Salmonella typhimurium and a tryptophan-requiring strain of Escherischia coli in the presence of ultra violet light (1998) [122]
36. Degussa AG, US-IT No 94-0073-DGM, Titanium Dioxide T 805: Induction of chromosome aberrations in cultured Chinese hamster ovary (CHO) cells in the presence of ultra violet light (1999) [123]
37. Degussa AG, US-IT No 94-0073-DGM, Titanium Dioxide T 817: Induction of chromosome aberrations in cultured Chinese hamster ovary (CHO) cells in the presence of ultra violet light (1999) [124]
38. J Whitwell, Uncoated MT-100-TV: Induction of chromosome aberrations in cultured Chinese hamster ovary (CHO) cells in the presence of ultra violet light, Covance Laboratories Report No 413/18-D5140 (Jan 2000) [91]
39. WJW Pape, Preliminary data on phototoxicity of S 75 using the 3T3 NRU Phototoxicity test in vitro, January 27, 1999 [69]
40. R Bestak and G Halliday, Photochem Photobiol 64(1) 188-193 (1996) [58]
41. G Greenoak, A Torkamanzehi and MR Nearn, Reduction in tumour incidence by a sunscreen containing microfine titanium dioxide, Cosmetics, Aerosols in Australia 7(4) 12-17 (1993) [59]
42. NN Greenwood and A Earnshaw, *Chemistry of the Elements*, 2nd edition, Oxford: Pergamon Press (1993)
43. A Mills and S LeHunte, *J Photochem Photobiol A Chem* 108(1) (1997)
44. R Cai et al, Induction of cytotoxicity by photoexcited TiO2 particles, *Cancer Res* 52 2346-2348 (1992) [60]

Designing Broad-Spectrum UV Absorbers

Keywords: sunscreens, UVA radiation, UV filters, efficacy, safety, registration, patent freedom,

Design considerations for broad-spectrum UV filters include photostability, solubility, efficacy, safety, registration, patent freedom, and PPD performance. These criteria are described and then applied in the case of bis-ethyl-hexyloxyphenol methoxyphenyltri-azine (BEMT), a new broad-spectrum sunscreen active.

Overexposure of human skin by UV light leads to sunburn, an increased risk for skin cancer and also premature aging of the skin.[1] Although the role of UVA is not entirely known, it is now widely recognized that a certain UVA protection must be ensured in the development of sunscreens. More than 90% of ultraviolet energy received by the unprotected skin comes from the UVA range. Because there is no natural immediate warning sign similar to sunburn for UVB radiation, sunscreens with poor UVA protection may be transmitting vast amounts of UVA radiation onto the skin. This has even led to the assumption that sunscreens may contribute to rather than protect against skin cancer.[2]

The example in Table 1 illustrates the effect of three sunscreens with the same degree of sunburn protection of SPF 8, but different degrees of UVA protection. The examples were calculated by using the standard CIE solar spectrum[3] and a sunscreen simulatora.[4,5]

The UVA dose received during sun exposure varies largely, without causing any short-term effects. A person of photo-skin type II (fair skin, easily burned and rarely tanned) can expect mild sunburn after a sun exposure of about 15 minutes without protection. During this period that person receives a UVA dose of 5.5 J/cm2. With any SPF 8 sunscreen, the amount of time to get mild sunburn is, by definition, two hours for that person. The maximal UVA dose that could be received during that time would be eight times the UVA dose received with no protection after 15 minutes. Two hours with no protection would of course result in bad sunburn with painful blistering mainly due to the UVB radiation.

With an SPF 8 sunscreen, this sunburn is avoided for exposures up to two hours, but the example of three different SPF 8 sunscreens shows that the UVA dose received can vary significantly up to six times in the case of a "pure" UVB sunscreen. To ensure that not only sunburn is prevented but also the UVA dose is not excessive, photostable broad-spectrum or UVA/broad-spectrum filters are required.

---

[a]Sunscreen Stimulator is a product of Ciba Specialty Chemicals, Basel, Switzerland.

Table 1. Comparison of three SPF 8 sunscreens with different degrees of UVA protection

Protection type	None	UVB only	UVB plus little UVA	Broad-Spectrum
Sunscreen actives	UVB: None UVA: None	UVB: 5.9% EHMC UVA: None	UVB: 4.0% EHMC UVA: 1.6% Benzo-phenone-3	UVB: 0.9%EHMC UVB/UVA: 1.8% BEMT, 1.8% MBBT
UVA-Parameters:				
UVA/UVB ratio	(1.0)	0.25	0.35	0.85
Critical wavelength λ_c	(290)	339 nm	351	384
Australian UVA-Standard fulfilled:	No	No	No	Yes
Effect of UVR				
Mild Sunburn after	15 minutes	2 hours	2 hours	2 hours
UVA dose J/cm²	5.5	33	27	8.5

EHMC= ethylhexyl methoxycinnamateBEMT= Bis-ethylhexyloxyphenol methoxyphenyltriazine
MBBT= methylene bis-benzotriazolyl tetramethylbutyl phenol

Designing a New UV Filter

Any new UV filter must demonstrate efficient UV absorption, photostability, and solubility. UV filters for personal care applications must also demonstrate their performance in vitro and in vivo, and they must meet four additional requirements: efficacy, safety, registration and patent freedom.

Photostability: UV absorbers are widely used to protect polymers (e.g. plastics, fibers, coatings) against photo-degradation. Numerous investigations have demonstrated that photostability is of key importance to the filters in order to provide long-term protection of polymer substrates (e.g. no degradation after several years of Florida outdoor testing).

In general, the photostability of a UV filter depends on how well the molecule is able to release the absorbed energy to the environment in the form of heat rather than radiation. Also, in general, UV absorbers with an intramolecular hydrogen bond exhibit very efficient radiation-free energy transformation processes resulting in inherent photostability.

For polymer applications, the following UV stabilizer technologies have been developed over the years:

- Methyl salicylates (1960)
- o-hydroxybenzophenones (1965-1970)
- 2-(2-hydroxyaryl)-benzotriazoles (1975-1990)
- 2-(2-hydroxyaryl)-1,3,5-triazines (since 1995)

Presently, hydroxyphenyltriazines (HPTs) represent the most advanced class of UV absorbers for the photoprotection of all kinds of polymer substrates.[6-8] Their general structure is shown in Figure 1.

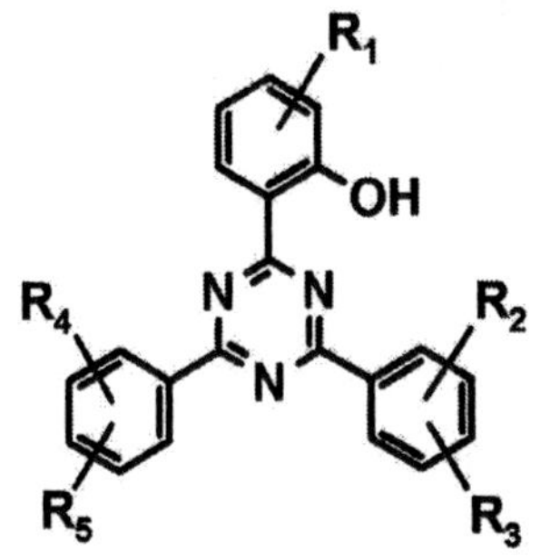

Figure 1. General structure of hydroxyphenyltriazines (HPTs)

Solubility: Without solubilizing substituents, HPTs are nearly insoluble in cosmetic oils. They exhibit the typical properties of pigments (e.g. high melting points). In order to increase solubility in oil phases, the structure of the UV filter has to be modified accordingly. An example in the case of BEMT is given later.

Efficacy: Besides efficient UV absorption, photostability and solubility as described above, there are other important parameters regarding efficacy to be considered. The UV absorber substance must be compatible with all other ingredients in a formulation; there should be no discoloration of skin and hair, no staining of textiles and no odor. For the water-resistant claim the UV absorber should be insoluble in water. And, last but not least, the UV filter should be economical in its use.

Safety: Sunscreen actives should have no adverse effect on humans and the environment. Although direct comparison with a new pharmaceutical drug is not

Table 2. Typical international safety dossier of a new sunscreen[8]
Acute oral and dermal toxicity
Dermal, ocular irritation, skin sensitization
Photoirritation, photosensitization
Subchronic oral and topical toxicity
Chronic toxicity
Fertility, early embryonic development
Embryofetal toxicity and peri-/post-natal toxicity
In vitro and in vivo percutaneous absorption
Topical and oral pharmacokinetic and metabolism
In vitro and in vivo genetic toxicity
Carcinogenicity
Photocarcinogenicity
Safety and efficacy in man

appropriate, the development of a new sunscreen active for global use is highly demanding. The toxicological studies required for a global registration are listed in Table 2.[9]

International registration: In order to exploit the full economic potential of a UV filter, UV absorber manufacturers are aiming for global registration. In Europe, South America, Asia and Africa, where sunscreens are labeled cosmetics, approval is possible within two years of filing. In Australia, Japan and the United States it takes longer.

Only recently in the U.S., a new procedure (TEA: Material Time and Material Extent Application) was introduced. After a minimum of five years foreign marketing experience in five countries, a new sunscreen active can be submitted for registration.[10] After the marketing experience, efficacy and safety data have to be submitted.

So far three UVB filters that are widely used outside the U.S. have received the status of "eligibile to enter the Sunscreen Monograph."[11] These three are: isoamyl p-methoxycinnamate (IMC) or Amiloxate (U.S. drug name); 4-methylbenzylidene camphor (MBC) or Enzacamene (U.S. drug name); and ethylhexyl triazone (EHT) or octyl triazone BEMT (U.S. drug name: Bemotrizinol). They will start the TEA process in 2005 when the five years of marketing experience will be available.

Patent freedom: Patent freedom means the free use of sunscreen actives by any sunscreen manufacturer, i.e. avoiding the infringement of any third-party patent rights.

Until about 10 years ago UV absorber manufacturers protected their inventions by simple substance patents that included the basic applications, e.g., "invention of a novel UV absorber for the incorporation in personal care formulations for the protection of skin and hair)."

In the mid 1990s important cosmetics manufacturers started to patent not only their specific technologies but also generic combinations of different ingredients. This strategy is aimed to keep competitors from using new technology that emerged on the market.[12] This limits the potential of the competitors, which is part of business. But it also is detrimental for the supplier who suddenly sees the potential of his new sunscreen active shrinking due to patent restrictions.

As a consequence the suppliers had to react and rethink the patenting strategy and the whole innovation process. As soon as the identity of a new ingredient becomes known, "all" measures have to be taken. These measures include publication of combinations of that novel ingredient with other sunscreen actives and other compounds such as emollients, emulsifiers or thickeners.

BEMT – An Example of a New UVA Filter

HPT structure suggested the use of HPT chemistry to design an efficient broad-spectrum sunscreen active. As a consequence of their molecular structure, HPTs exhibit a UV spectrum with two distinctive absorptions. This is due to the presence of two electronic transitions with strong dipole moments, both of which are polarized perpendicular to each other. In order to obtain broadband absorbance, OH and OR substituents at the three phenyl groups were introduced and positioned (as shown in Figure 2), resulting in the formation of bis-ethyl-hexylphenol methoxyphenyltriazine

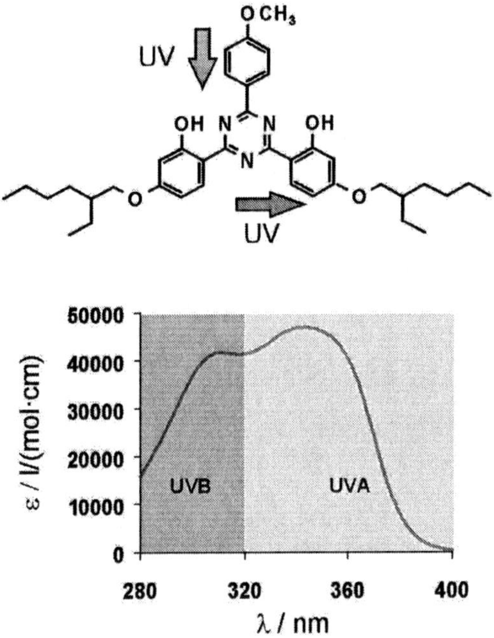

Figure 2. Molecular structure and UV-absorption spectrum of BEMT, measured in EtOH, E1,1 max = 820 (342 nm) (arrows indicate the polarization of the UVA and UVB transitions)

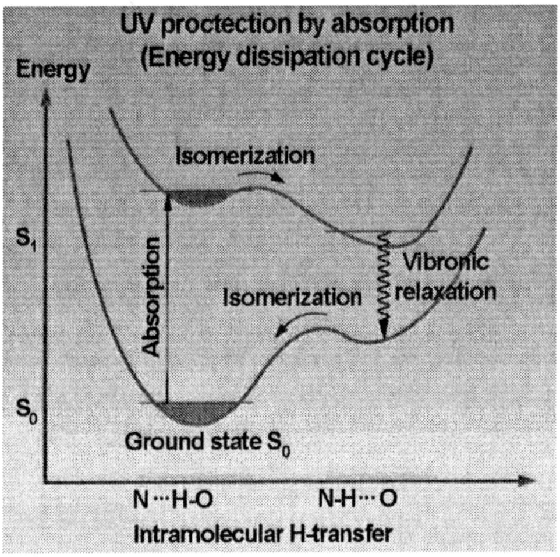

Figure 3. Model for the conversion of absorbed UV-energy to heat by extremely fast photo-tautomerism (approximately 10^{-12} seconds)

Gaining OTC Sunscreen Monograph Status via TEA

In January 2002, the FDA published the "foreign marketing rule," the so-called Material Time and Material Extent Application (TEA) process,[23] which opens all OTC drug monographs to foreign drugs or cosmetic ingredients under certain conditions. Here's how TEA works for a sunscreen ingredient.

1. The manufacturer submits evidence to the FDA showing that the ingredient has five years of continuous foreign marketing experience.
2. The FDA has approximately 120 days to determine eligibility and will issue a Federal Register Notice of Eligibility. This concludes the first phase (TEA1) of the "twostep" TEA process.
3. FDA's Notice of Eligibility triggers the second phase of the TEA process (TEA2), which requires that data on efficacy and safety be submitted to demonstrate that the ingredient and formulation can be generally regarded as safe and effective (GRAS/E). For example, in the case of Ciba's bemotrizinol under TEA2, Ciba will provide FDA with the necessary toxicological and efficacy studies, including a 2-year dermal carcinogenicity study and an 18-month photocarcinogenicity study to support the GRAS/E status.
4. FDA reviews the submitted efficacy and safety data. If they meet the requirements for GRAS/E status, FDA will make a determination of approval for inclusion of the ingredient into the OTC Sunscreen Monograph.

(BEMT), a new oil-soluble UV filter with true broad-spectrum performance. The two ortho-OH groups not only contribute to the broad-spectrum characteristics, but also enable efficient energy dissipation via intramolecular hydrogen bridges. Figure 3 illustrates the role of the intramolecular hydrogen bridge in the quantitative conversion of UV light to vibrational (heat) energy.[13]

In the United States, bemotrizinol is not yet approved as a sunscreen active ingredient by the United States Food and Drug Administration (FDA), however the ingredient will have five years of continuous marketing experience by March 2005 and can then seek approval via the TEA process (see sidebar).

Photostability: BEMT contains two intramolecular hydrogen bonds, which enable an excited state intramolecular proton transfer (photo-tautomerism) after photoexcitation. This results in rapid radiationless conversion ensuring that the UV radiation, efficiently absorbed by the filter, is almost quantitatively transformed into harmless vibrational (heat) energy. The entire photo-tautomeric cycle only lasts about 10-12 seconds, leaving no time for undesirable side reactions (e.g. population of triplet states, formation of singlet oxygen or formation of radicals). This explains the excellent photostability of BEMT (> 95% recovery of parent BEMT is observed analytically after UV irradiation of 50 minimal erythema doses, MED).[14]

Solubility: Solubility of BEMT in different cosmetic solvents is given in Table 3. The product form is a very fine yellow powder with a melting point of 80°C that is easily solubilized in most emollients.

Table 3. Solubilities of BEMT in cosmetic solvents

Caprylylpyrrolidone	20.0%
Ethylhexyl methoxycinnamate	17.0
C_{12}-C_{15} alkylbenzoate	13.0
Diethylhexylsuccinate	7.0
Isopropylmyristate	6.0
Hexyl laurate	6.0
Caprylic/capric triglyceride	5.0
Coco caprylate/caprate	5.0
Isopropylpalmitate	5.0
Sesame oil	3.0

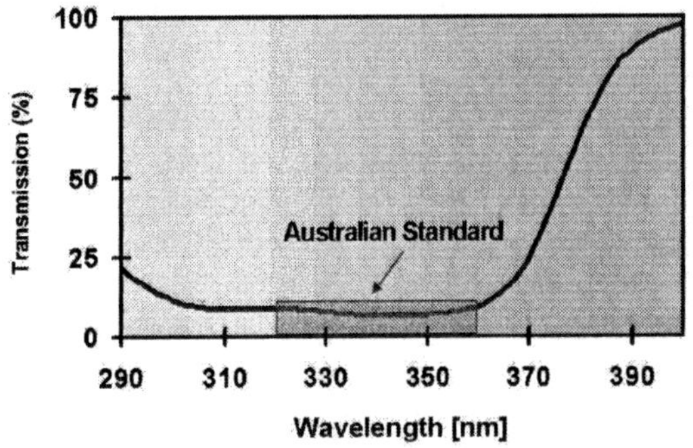

Figure 4. BEMT fulfilling the Australian UVA Standard

In vitro/In vivo performance: The performance of BEMT as a cosmetic UV filter has been assessed in in vitro and in vivo studies.[15] For rating UVA performance, the Australian standard is increasingly used. UVA protection is recognized when a sunscreen preparation transmits between 320 nm and 360 nm (in an 8 μm cuvette) less than 10% of the incoming light (Figure 4).[16]

In a comparative test with other oil soluble filters it was shown that BEMT exhibits the highest efficacy to satisfy the Australian Standard—only 1.9% of this broad spectrum filter is required in formulation.

In vivo assessment of UVA protection: The UVA protection factors (UVA-PF) obtained from in vivo Persistent Pigment Darkening (PPD) studies increases steadily, as expected, with the concentration of the UVA broad spectrum filter BEMT (Figure 5). Already low BEMT concentrations of 1-2% are sufficient for a PA++ protection, rated after the Japanese standard.[17] In combinations of BEMT and ethylhexyl methoxycinnamate (EHMC), a synergy regarding the PFA performance has been observed.

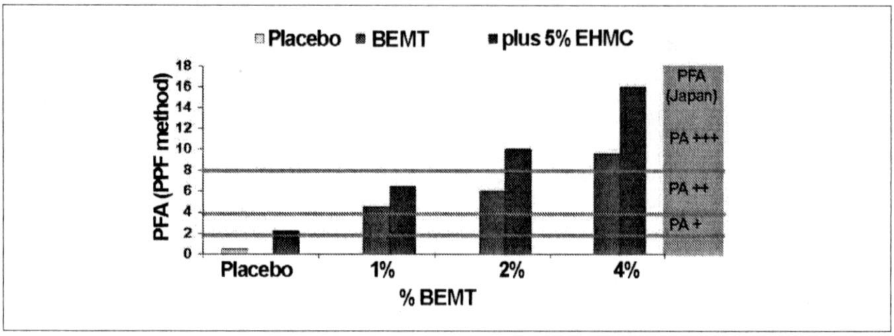

Figure 5. UVA protection with BEMT alone, and with EHMC (PPD method)

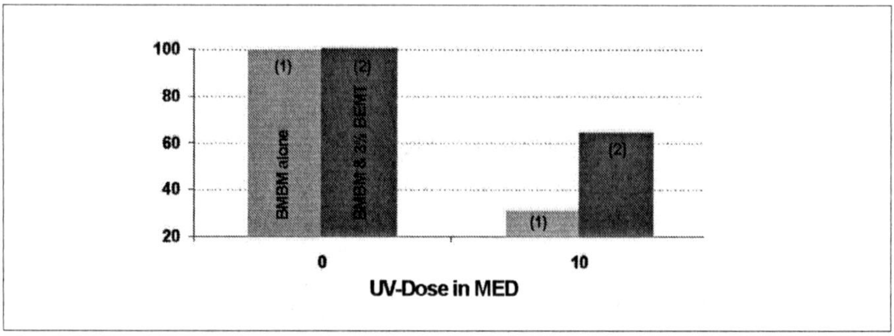

Figure 6. Stabilization of BMBM by BEMT

Due to its spectral performance, BEMT is able to boost photoprotection (SPF) when combined with conventional UV filters such as EHMC.[18] No adverse effect (incompatibility) has been observed in combinations of BEMT with other soluble filters or pigments. Moreover it has been observed that BEMT stabilizes photolabile sunscreens such as butylmethoxy-dibenzoylmethane (BMBM) and EHMC (Figure 6).[19]

Safety: A safety testing program was conducted with BEMT. According to applicable OECD/EC test guidelines, BEMT was GLP-compliant. The toxicological test results indicate no adverse effects for human use.

BEMT toxicology data have been reviewed by the European Commission Scientific Committee on Cosmetic Products and Non-Food Products Intended For Consumers (SCCNFP) and found to support an approved safe use level of up to 10% BEMT as a UV filter in leave-on and rinse-off cosmetics.

BEMT has no acute aquatic toxicity up to its maximum solubility and also no toxicity to microorganisms. The substance is not readily biodegradable but has an expected elimination of greater than 70%. BEMT has no bioaccumulation. The resulting Environmental Risk Assessment according to the EC "Technical Guidance Document" has a PEC/PNEC ratiob of < 0.001 in the environmental compartments water, sludge, sediment and soil. These favorable results indicate no adverse effects to the environment.

Patent freedom: In an attempt to protect both the filter and combinations of the filter with other ingredients, patents[20] have been issued for both BEMT and

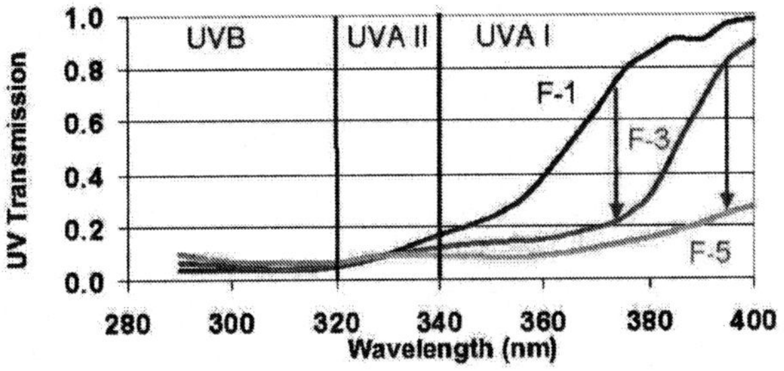

Figure 7. Improvement in UVA protection with new broad-spectrum filters

Formula	Composition	SPF(calc)	(relative %)
F-1	7% EHMC, 3% BP-3	14.0	100
F-3	5% EHMC, 3% BEMT	15.5	60
F-5	1% EHT, 3% MBBT, 3% BEMT	14.1	25

combinations thereof with commonly used cosmetic UV filters. Regarding BEMT there exist several specific third-party patent rights. However, according to our view there are no serious patent restrictions.

Conclusions

From the new broad-spectrum UV absorbers we expect better UVA coverage when incorporated into a sunscreen or day cream. To illustrate and quantify the improvement, we carried out some calculations with different formulations using a new sunscreen simulator.[4, 5]

Figure 7 shows the composition of three formulations with similar SPF (i.e. similar UVB protection) but different degrees of UVA protection.[21] The area below the sunscreen with the highest UVA transmission (320-400 nm) was Formula F-1 (with 3% benzophenone-3). We arbitrarily set that area as 100%. Formula F-3 (with BEMT) reduces this UVA exposure through the sunscreen already to below 60%. With formula F-5 (using BEMT and another new UVA/broad-spectrum filter such as methylenebisbenzotriazolyl tetramethyl-butylphenol or MBBT, U.S. drug name: Bisoctrizole[22]), the UVA exposure is reduced down to a quarter of the value achieved with the conventional formulation F-1. With modern broad-spectrum filters it is thus possible to achieve better UVA protection by using lower amounts of UV filter.

In spite of great differences, all these formulations could make "UVA" or "broadband" claims at the moment. What thus is needed is a standardized and harmonized method for the UVA assessment that discriminates between the vast differences in the degree of UVA protection.

—**Uli Osterwalder and Bernd Herzog,** *Ciba Specialty Chemicals, Basel, Switzerland*

References

1. LH Kligman, FJ Akin and AM Kligman, The contribution of UVA and UVB to connective tissue damage in hairless mice, *J Invest Dermatol* 3 215-227 (1985)
2. R Haywood, P Wardman, R Sanders and C Linge, Sunscreens inadequately protect against ultraviolet-A induced free radicals in skin: Implications for skin aging and melanoma?, *J Invest Dermatol* 121 862-868 (2003)
3. AF McKinlay and BL Diffey, A reference action spectrum for ultraviolet-induced erythema in human skin, *CIE Journal* 6 17-22 (1987)
4. B Herzog et al, The sunscreen simulator: A formulator's tool to predict SPF and UVA parameters, SÖFW 129(7) 25-36 (2003)
5. Ciba Sunscreen Simulator located at www.cibasc.com/personalcare, Ciba Specialty Chemicals, Basel, Switzerland, (September 2002)
6. JF Rabek, *Photostabilization of Polymers, Principles and Applications*, London: Elsevier Applied Science Publishers, London (1990)
7. F Gugumus, In: *Kunststoff-Additive*, R Gächter and H Müller, eds, Munich: C Hanser Verlag (1989)
8. F Waiblinger et al, Irradiation-dependent equilibrium between open and closed form of UV absorbers of the 2-(2-hydroxyphenyl)-1,3,5-triazine type, *Res Chem Intermed* 27 5-20 (2001)
9. G Nohynek and H Schaefer, Benefit and risk of organic ultraviolet filters, *Regulatory, Toxicology and Pharmacology*, 33 1-15 (2001)
10. Food and Drug Administration, Additional criteria and procedures for, classifying over-the-counter drugs as, generally recognized as safe and, effective and not misbranded, 21 CFR Part 330, [Docket No. 96N–0277], RIN 0910–AA01; *Federal Register* Vol 67, No 15 (Wednesday, Jan 23, 2002) Rules and Regulations, 3060-3076
11. Food and Drug Administration, Over-the-counter drug products; Safety and efficacy review; additional sunscreen ingredients, [Docket No. 2003N–0233], *Federal Register* Vol 68, No 133 (Friday, Jul 11, 2003) Notices, 41386-41387
12. M Rudolph, Specific UV filter combinations and their impact on sunscreen efficacy, Intl Sun Protection Conference, Commonwealth Institute, London (March 9-10, 1999)
13. JEA Otterstedt, Photostability and molecular structure, *J Chem Phys* 58(12) 5716-5725 (1973)
14. J Keck et al, Ultraviolet absorbers of the 2-(2-hydroxyaryl)-1,3,5-triazine class and their methoxy derivatives: Fluorescence spectroscopy and x-ray structure analysis, *J Phys Chem* B 102(36) 6975-6985 (1998)
15. B Herzog et al, In vivo and in vitro assessment of UVA-protection by sunscreen formulations containing either ZnO, butyl methoxy dibenzoyl methane, methylene bis-benzotriazolyl tetramethylbutylphenol, or bis-ethylhexyloxyphenol methoxyphenyl triazine, *Proceedings: 22nd ISFCC Congress*, Edinburgh, UK (2002)
16. Australian/New Zealand Standard, 15 / NZS 2604 (1993)
17. *JCIA Measurement Standard for UVA Protection Efficacy*, Japan Cosmetic Industry Association JCIA, 9-14, Toranomon 2-Chome, Minato-Ku Tokyo (1995) p 105
18. E Chatelain and B Gabard, Photostabilization of butyl methoxydibenzoylmethane (Avobenzone) and ethylhexyl methoxycinnamate by bis-ethylhexyloxyphenol methoxyphenyl triazine (Tinosorb S), a new UV broadband filter, *J Photochem Photobiol* 74 401-406 (2001)
19. B Herzog and K Sommer, Investigations on photostability of UV-absorbers for cosmetic sunscreens, *Proceedings (CD-ROM) 21st ISFCC Congress*, Berlin, Germany, Poster P60 (2000)
20. D. Hüglin, E. Borsos, H. Luther, B. Herzog, F. Bachmann, Bis-Resorcinyl-Triazine als UV-Absorber, EP 0 775 698 A1, Priority Date: 23 November 1995; Publication Date: 28 May 1997
21. S Mongiat et al, BEMT: An efficient broad-spectrum UV filter, *Cosmet Toil* 118(2) 47-54 (2003)

Boldine as a Sunscreen

Keywords: boldine, photostability, sunscreen, quantum yield, sun protection factor (SPF)

Boldine alkaloid, which occurs in the leaves and bark of the boldo (Peumus boldus Mol.) tree, exhibits structural features that confer to it UV absorption properties. It is also a very potent antioxidant compound. This article discusses boldine's photostability and photo-protector capacity.

UV light may produce harmful reactions on skin irrespective of its beneficial biological effects. Depending on the dose, time of irradiation, wavelength and type of exposed areas, UV light may cause a range of reactions, from mild skin burn, to skin cell DNA damage, to premature skin aging, and even to skin cancer (basal cell carcinoma, squamous cell carcinoma, epithelial melanoma).[1-4]

Free radicals are known to participate in either the cause or the development of most UV-induced skin lesions.[5] Although sunburn is a complex inflammatory reaction, free radicals, lipid peroxides and, secondarily, prostaglandins are known to play a major role in its course.[6]

Based on the oxidative hypothesis of UV-induced skin sunburn, substances displaying free-radical scavenging capacity have surfaced as potentially interesting photoprotective or photopreventive agents. Given the increasing recognition of the damaging role of free radicals in biological systems, much interest has arisen in recent years prompting the study of new substances with antioxidant activity.[7,8] Particular attention has been directed to the search and development of natural antioxidant compounds under the often — though not necessarily always — valid assumption that because of their natural origin, these substances could provide a source of new antioxidants substantially less toxic than most available synthetic compounds. Such a contention also arises from the increasing awareness and demonstration that many of the currently employed synthetic antioxidants are not sufficiently innocous.[9]

Studies conducted by our group on the therapeutic potential of new natural antioxidants have led us to recently focus on the characterization of boldine as a novel and potentially alternative antioxidant.[10]

Boldine, which is (S)-2,9-dihydroxy-1,10-dimethoxy-aporphine (Figure 1), occurs abundantly in the leaves and bark of boldo (*Peumus boldus* Mol.), a widely distributed native tree of Chile. Infusions of boldo leaves have been traditionally employed in folk medicine for their purported choleretic, diuretic, sedative and digestive stimulant properties.[11,12]

At low micromolar concentrations, boldine prevents both enzymatically and nonenzymatically induced damage to biological systems. In vitro, boldine inhibits

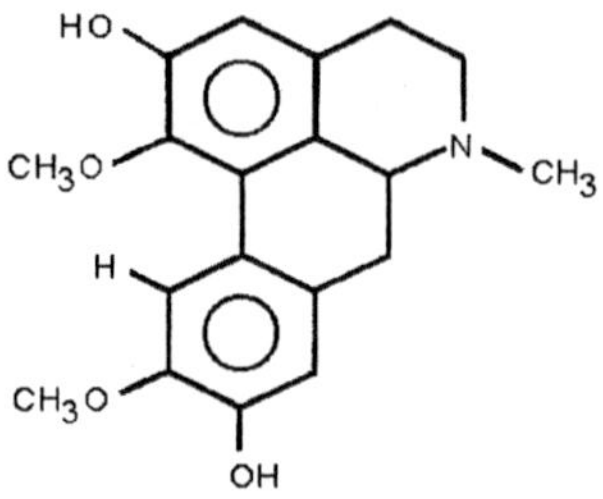

Figure 1. Boldine structure

the free-radical-mediated initiation and propagation of the peroxidative damage induced to various membrane types (such as liver homogenates, hepatic microsomes and ghost erythrocytes), and it blocks the free-radical-dependent lysis of red blood cells and intact hepatocytes.[11,13,14] In other ways, boldine behaves as a very potent antioxidative substance in biological systems. Undergoing peroxidative free-radical-mediated damage, the boldine molecule acts as an efficient hydroxyl radical scavenger.[14] The molecule is reported to be relatively innocuous.[10,15-17]

The present study deals with the following:

- The photostability of boldine in methanol and water.
- The photophysical behavior of boldine (singlet lifetimes, fluorescence yields).
- The photoprotective capacity of boldine evaluated by in vitro methods (solution dilution spectrophotometry) and in vivo methods (guinea pig skin treated with UVB).
- Evaluation of the skin temperature of both treated and untreated areas of guinea pigs, subjected to UV radiation.

The behavior of this compound was compared with that of two recognized sunscreens: homosalate and PABA.

Materials and Methods

Boldine was extracted from the bark of *Peumus boldus* Mol. and used as the hydrochloride.[18] The alkaloid was chromatographically pure, and its identity was established by usual spectroscopic methods.

Female Hartley albino guinea pigs, weighing between 400 and 500 g each, were used in all experiments. They were housed in a room with controlled temperature and humidity during the experiment, and fed with a commercial diet[a] and water ad libitum. Each animal's back was shaved and depilated[b] 24 h before starting the irradiation experiments.

We prepared two forms of the vehicle[c] (a moisturizing base cream)—one with boldine and the other with PABA.[d] For each time interval, we applied one of the vehicle forms to the backs of 10 guinea pigs, and the other vehicle form to the backs of 10 other guinea pigs. The application site for the test vehicle was a 4 cm^2 rectangle in an area of the back distal from the head. We applied the vehicle at a thickness of 20 μm through a micropipette at a dose of 2 mg/cm^2, as proposed by the US Food and Drug Administration. In an area of the back proximal to the head, all guinea pigs were treated with the vehicle cream alone as a control. Both the test vehicles and the control were applied 15 min before the animals were irradiated. This time was found to be sufficient for the cream to be absorbed. At the end of the irradiation period, the exposure sites were protected by covering the challenged areas with adequate cotton pads.

In vivo SPF determination: A bank of six lamps[e] providing a mean irradiance of 1.46 mW/cm^2 (maximal wavelength 310 nm) was used as UVB radiation source. Increasing irradiation doses (exposure times of 0, 5, 20, 35 and 50 min) were administered into the sunscreen-treated and untreated skin of the 100 animals. During the irradiation phase, each animal was immobilized within a 30 x 15 x 15 cm stainless chamber.

The minimal erythema dose (MED) was defined as the amount of energy (expressed in time) needed to produce erythema (pinkness response) perceptible to the normal human eye. Under these experimental conditions, the erythema was observed 4 h after the irradiation challenge was finished, and the MED was established as 5 min of exposure. To assess erythema, we used the following gradient scale:

Normal +1
Medium +2
Severe +3
Ulcerated +4

The in vivo SPF was determined for each individual as:

$$SPF = \frac{\text{MED of protected skin}}{\text{MED of unprotected skin}}$$

Equation 1.

Individual SPF values were used to determine a mean SPF for the preparation.

Determination of the skin temperature: Four hours after irradiation, the temperatures of the treated and untreated areas were measured by means of a thermocouple.[f]

In order to show the reproducibility of these experimental temperature results, each measurement was repeated six times and the values were statistically compared. Anova tests demonstrated that there is no statistically significant variation in the skin temperature of the skin with filter, as the irradiation time increases. The T Student test showed that, under conditions with a filter versus without, there are significant differences in the skin temperature at various irradiation times.

[a] Chaw, Champion S.A., Santiago, Chile
[b] Devellol, Laboratorio Farmoquímica del Pacífico, Santiago, Chile [c] Neutraderm, Galderma Laboratories, France, obtained from Laboratorio Alcon, Santiago, Chile
[d] Paba-Film-15 (containing octyl dimethyl PABA, benzophenone, propylene glycol and benzyl alcohol), Laboratorio Alcon
[e] Phillips TL 40 W/12, Eindhoven, The Netherlands

Table 1. Rate constant of boldine photodegradation (10^{-5} M) and quantum yield of consumption (Φ_c) in oxygen and nitrogen

	k x 10^3 (seg^{-1})	Φ_c x 10^2
oxygen	5.1±1.7	5.4
nitrogen	3.0±1.4	2.0

Table 2. SPF values of boldine (compared to SPF values of homosalate[19])

	In vivo[a]	In vitro[b]
boldine	7.9±0.89[c]	5.8±0.6[c]
homosalate	4.3±0.20	4.0±0.25

[a] 8 mg of boldine per 50 μL of vehicle per 4 cm^2 of skin surface
[b] 0.123 mg boldine per 5 mL universal solvent
[c] Standard deviation
n = 10 animals

Table 3. Temperature (°C) of guinea pig skin with and without boldine, subjected to UVB irradiation for selected time periods (min)

Irradiation time period	With boldine	Without boldine
0	27.33±0.921[a]	27.33±0.921[a]
5	27.94±0.907	29.21±0.953
20	27.44±1.888	30.00±1.808
35	28.40±1.656	30.49±1.685
50	28.63±1.029	31.22±1.439
	Sx=0.425	Sx=0.445
	F=1.810	F=11.16

[a] Standard deviation
n = 6 animals
Sx = Mean standard deviation
F = Statistic Test

In vitro SPF determination: We prepared fresh solutions of boldine to a final concentration of 0.123 mg per 5 ml using a universal solvent (12.5 g of methylene chloride, 37.5 g of cyclohexane and 50 g of isopropanol). We used solutions of identical concentration of homosalate (SPF = 4±0.25) as reference.[19]

Using a method described by Meybeck,[20] we assayed the in vitro photoprotector capacity of boldine by measuring the area over the curve for a transmittance spectra obtained after scanning the solution in the 280-320 nm range with a spectrophotometer.[g] The area was correlated with that of homosalate.

Fluorescence measurements: Absorption spectra were recorded with a spectrophotometer.[g] A spectrofluorimeter[h] measured fluorescence spectra. Excited state lifetime (τ_f) measurements were obtained by the phase modulation technique and appropriate equipment.[j] Fluorescence quantum yields were evaluated by comparison with those of anthracene in ethanol ($\Phi f = 0.27$).[21] All measurements were carried out at room temperature (20° ± 2°C).

Photostability studies: Photolyses were carried out employing a photochemical reactor.[k] Methanolic and aqueous solutions of boldine (3.63 mg/L) were irradiated at 300 nm. Studies were conducted using solutions previously bubbled with either nitrogen or oxygen for 6 min. Photoconsumption quantum yield of boldine was evaluated from change in the absorption spectra.

Results and Discussion

Photostability: Boldine was photounstable at the wavelengths irradiated (maximum wavelength 300 nm).

Figure 2 shows the variation in the absorption spectrum of a boldine methanolic solution (10^{-5} M) as a function of irradiation time. It has two major absorption peaks at 282 and 303 nm. It shows the disappearance of the peak wavelength band at 305 nm, and the appearance of new absorption bands in the visible region (480 nm) and in the UVB region (290 nm), which would show photoproduct formation. More work along these lines is in progress.

Table 1 shows the quantum yield of photoconsumption and the kinetics under different conditions (nitrogen, oxygen). The quantum yield of photoconsumption is the ratio between the number of molecules photodecomposed and the light absorbed by these molecules; it is a measure of the efficiency of the reaction. The kinetics under nitrogen and oxygen give information about the mechanism of the photoconsumption (reactive oxygen species may be involved). The results show that photodegradation was oxygen-dependent. The possibility of singlet oxygen generation was proved with deuterium oxide instead of water. The kinetic results, similar to those in aqueous solution, ruled out the presence of singlet oxygen. This agrees with the low value of excited lifetime of boldine ($\tau = 1.44$ ns), and the high fluorescence quantum yield ($\Phi f = 0.54$).

Photoprotection: Before determining the photoprotective capacity of boldine against UVB, it was necessary to evaluate the MED for the guinea pig. Under the employed light conditions, the MED was 5 min observed 4 h after irradiation (n = 10 animals). Then, the SPF value (obtained from Equation 1) was 7.9 ± 0.89 ($p < 0.05$).

Boldine in our tests exhibited a higher SPF value than the homosalate reference[19] with both in vitro and in vivo methods (Table 2). This is an interesting result if we consider the photoinstabilty of the boldine molecule. Nevertheless,

[f] Model T/TE3, No. 2232, Elektrolaboratoriet, Copenhagen, Denmark
[g] Model 8452A, Hewlett-Packard, Guadalajara
[h] Model RF 540, Shimadzu, Tokyo, Japan
[j] Gregg 200, I.S.S., Champaign, Illinois
[k] Rayonet Srinivasan-Griffin, Southern New England Ultraviolet Co.

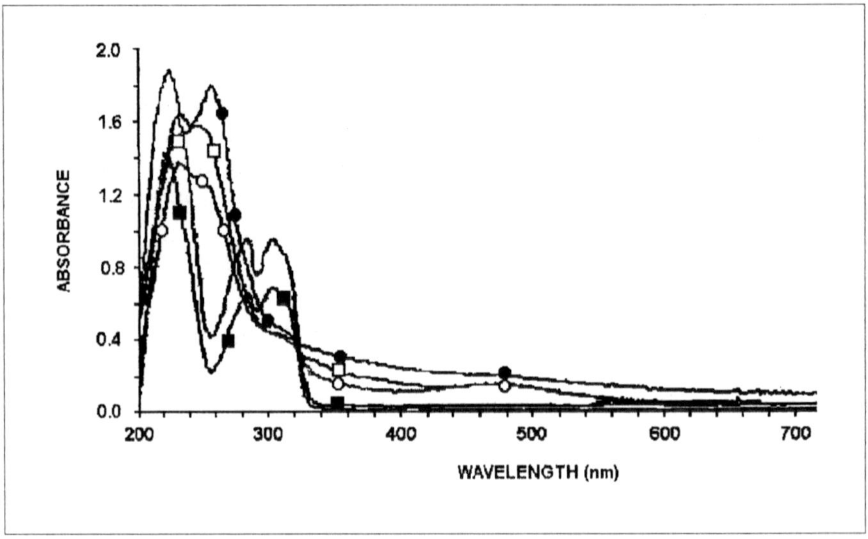

Figure 2. Variation in the absorption spectra of boldine during irradiation with UVB (300 nm) as a function of irradiation time (in seconds)

the photoprotector capacity is probably due to the photoproducts or to a physical block. In fact, as the irradiation occurs, the boldine in the vehicle darkens while the control (vehicle without boldine) remains unchanged.

It is interesting to characterize the boldine photoproducts and to perform photostability studies and SPF determinations in order to establish the roles they have in the photoprotector mechanism. Work is being done on this subject.

The SPF determination in vivo was based on the method evaluated and standardized by Turkoglu et al.[22] This method was chosen because of its high reliability, reproducibility, simplicity and low cost. Turkoglu himself compared the values obtained in animals with those obtained in human beings, and he found a very good correlation (r=0.99).

Skin temperature: Skin temperature is an additional parameter that confirms the photoprotector capacity of boldine. Our statistical analysis (Table 3) illustrates that the skin temperature of the boldine-protected skin is independent of the time of exposure to UVB. The Fischer F test value (1.810) is lower than the critical values F_C=2.58 ($p< 0.05$) and F_C=3.78 ($p <0.01$). On the other hand, we were able to determine that there is a difference between temperatures of skin treated with boldine and untreated skin (Table 3). This result confirms boldine's photoprotector capacity against UVB irradiation.

ROS formation: UV light plays an important role in the production of reactive oxygen species (ROS) that can exert damaging effects on the components of skin or may affect the skin's metabolic processes. Sunscreens, or better, UV reflectants, can be expected to reduce the levels of ROS formation.

Antioxidants are one means for protecting skin against UV-induced ROS damage. It has been reported[5] that in hairless mice, UV irradiation causes reduction in all so-called antioxidant defense systems. The losses of the enzymes (superoxide dismutase, catalase, glutathione peroxidase and glutathione reductase), tocopherol, ubiquinols and vitamin C are significantly higher in the epidermis than in the dermis, suggesting that the primary damage to skin during insulation occurs in the topmost layers. In this context, boldine has been described as antioxidant and cytoprotective.[11]

We showed that the boldine photoprotective capacity, in vitro and in vivo, demonstated SPF values (Table 2) larger than those of homosalate, which was used as reference. The photoprotective action of boldine or its photoproducts is probably due to a combined action as solar filter (UVB absorption) and antioxidant at the topmost skin layer (epidermis). This mechanism has also been described for PABA by Bodaness and Chou, who observed that singlet oxygen can react with PABA to yield spectrally identifiable substances.[23] Thus, PABA and derivatives may protect not only by UV absorption, but also by reaction with ROS formed in irradiated epidermis.[24] In the case of boldine, it is quite unlikely that radical species may be generated in the vehicle used in our tests due to its greasy composition.

In addition, from assumptions based on Rieger's work[25] on the arachidonic acid cascade, it is probable that the photoprotective effect of boldine applied to the guinea pig skin induces an anti-inflammatory action because of a decrease in the prostaglandin production. UVB irradiation steadily increases the skin levels of the prostaglandins PGE_2, $PGF_{2\alpha}$ and PGD_2 and their precursor arachidonic acid until the maximal erythemal level has been reached (about 24 h after irradiation). After 48 h, levels of PGE_2 and $PGF_{2\alpha}$ have returned to normal, but the erythema persists.[4] Using a model similar to our guinea pig in vivo SPF determination, we have undertaken work to evaluate the prostaglandin plasma levels.

The differences observed among the boldine SPF values in vivo and in vitro can be explained from the fact that in the in vivo case, boldine is subject to photoproducts formed by irradiation. In the in vitro case, the boldine solution receives only low-intensity light from the spectrophotometer lamp, so no photodecomposition occurs.

Summary

Boldine was found to be photounstable when irradiated at wavelengths up to 300 nm. The quantum yield of consumption was 5.4×10^{-2} (in oxygen) and 2.0×10^{-2} (in nitrogen). The oxygen-dependent photodegradation mechanism excludes the participation of singlet oxygen species.

The photoprotector capacity of boldine against UVB was demonstrated by in vitro (SPF=5.8 ± 0.6) and in vivo (SPF=7.9 ± 0.89) methods. The photoprotector capacity was confirmed by skin temperature measurements: the temperature of the boldine-protected skin was independent of the time of exposure to UVB. On the other hand, the temperature of the unprotected skin increased significantly during irradiation.

The photoprotection is probably accompanied by an antioxidant action on epidermis.

—**M. Eliana Hidalgo and Iris Gonzalez,**
Faculty of Sciences, University of Valparaíso, Valparaíso, Chile

—**Fernando Toro and Ernesto Fernández**
Chemistry and Pharmacy School, University of Valparaíso, Valparaíso, Chile
—**Hernán Speisky and Inés Jimenez**
Nutrition and Food Technology Institute, University of Chile, Santiago, Chile

References

1. J Scotto and T Fears, The association of solar ultraviolet and skin melanoma incidence among Caucasians in the United States, *Cancer Invest* 5 275-283 (1987)
2. P Plugliese, Concepts in aging and the skin, *Cosmet Toil* 102(4) 19-36 (1987)
3. J Van der Leur, UV Carcinogenesis, *Photochem Photobiol* 39(6) 861-868 (1984)
4. M Rieger, Intrinsic aging, *Cosmet Toil* 110(7) 94-101 (1995)
5. M Rieger, Oxidative reactions in and on skin: Mechanism and prevention, *Cosmet Toil* 108(12) 43-56 (1993)
6. HT Chung, DK Burnham, B Robertson, LK Roberts and RA Daynes, Involvement of prostaglandins in the immune alterations caused by the exposure of mice to UV radiation, *J Immunol* 137 2478-2484 (1986)
7. T Osawa, Plant antioxidants: Protective role against oxygen radical species, *Cosmet Toil* 109(10) 77-81 (1994)
8. ME Hidalgo, E Fernández, W Quilhot and E Lissi, Antioxidant activity of depsides and depsidones, *Phytochemistry* 37 1585-1587 (1994)
9. E Graft, K Empson, J Eaton, Phytic acid: A natural antioxidant, *J Biol Chem* 262 11647 (1987)
10. H Speisky, BK Cassels, Boldo and boldine: An emerging case of natural drug development, *Pharmacol Res* 29 1 (1994)
11. H Speisky, BK Cassels, E Lissi and L Videla, Antioxidant properties of the alkaloid boldine in systems undergoing lipid peroxidation and enzyme inactivation, *Biochem Pharmacol* 41 1575-1581 (1991)
12. MC Lanhers, M Joyeux, R Soulimani, J Fleurentin, M Sayag, F Mortier, C Younos and JM Pelt, Hepatoprotective and antiinflamatory effects of a traditional medicinal plant of Chile Peumus boldus, *Planta Med* 57 110-115 (1991)
13. J Martínez, L Ríos, M Payá, MJ Alcaraz, Inhibition of non-enzymic lipid peroxidation by benzylisoquoline alkaloids, *Free Rad Biol Med* 12 287 (1992)
14. A Cederbaum, E Kukielka, H Speisky, Inhibition of rat liver microsomal lipid peroxidation by boldine, *Biochem Pharmacol* 44 1765-1772 (1992)
15. C Laborde, *Compt Rend Acad Sci* 98 1053 (1984)
16. H Kreitmair, Pharmakologische Wirkung des Alkaloids aus Peumus Boldus Molina, *Die Pharmazie* 7 507-511 (1952)
17. MJ Magistretti, Remarks on the pharmacological examination of plant extracts, *Fitoterapia* 51 67-69 (1980)
18. BK Cassels, M Asencio, P Conget, H Speisky, LA Videla and EA Lissi, Structure antioxidative activity relationships in benzyl-isoquinoline alkaloids, *Pharmacol Res* 31(2) 103-107 (1995)
19. E Fernández, W Quilhot, I González, ME Hidalgo, X Molina and I Meneses, Lichen metabolites as UVB filters, *Cosmet Toil* 111(12) 69-74 (1996)
20. A Meybeck, Objective methods for the evaluation of sunscreens, *Cosmet Toil* 98(3) 51-60 (1993)
21. D Eaton, Luminescence Spectroscopy, in Handbook of Organic Photochemistry, vol 1, JC Scaiano, ed, Boca Raton, Florida: CRC Press (1989) Chap 8
22. M Turkoglu, A Del Sakr and L Lichtin, An in vivo assessment of the sun-protection index, *Cosmet Toil* 104(4) 33-38 (1989)
23. RS Bodaness and PC Chou, Singlet oxygen reacts with inhibitors of ultraviolet mediated damage to skin: para-aminobenzoic acid and its derivatives, *Biochem Biophys Res Com* 87 1116-1123 (1979)
24. M Rieger, Reactions of oxygen affecting skin products, *Cosmet Toil* 104(10) 83-90 (1989)
25. M Rieger, The arachidonic acid cascade, *Cosmet Toil* 100(6) 31-53 (1985)

A New Photostabilizer for Full Spectrum Suncreens

Keywords: sunscreen, photostability, triplet energy, avobenzone, UVA

Following irradiation with full spectrum UV, avobenzone sunscreens that contain diethylhexyl 2,6-naphthalate demonstrate improved photostability compared to those without.

In this article, we, the authors, advocate a non-traditional approach to sunscreen design, one that aims at providing full spectrum sun protection by attenuating at least 90% of all solar UV radiation. We introduce diethylhexyl 2,6-naphthalate (DEHN)[a], a new chemical additive for sunscreens developed in the laboratories of C. P. Hall, and we present experimental data showing the photostabilizing effect of this chemical on sunscreens containing the widely used UV filter, avobenzone. We discuss the photochemistry of sunscreen photostability, the nature of solar irradiance, and the traditional, anti-sunburn approach to sunscreen design. Finally, we present sunscreen formulations that incorporate DEHN and we show the results of both in vitro and in vivo studies of their performance.

Background

At a time when more people are using more sunscreen than ever before, the lifetime risk of developing malignant melanoma is skyrocketing—more than doubling in the past 20 years to 1 in 74 today.[1] One possible explanation is that, protected from burning, people are prolonging their exposure to longer wavelength UV radiation, known as UVA, which most sunscreens block only partially or hardly at all.[2] Although no causal link has been established between exposure to this radiation and melanoma, the evidence is suggestive and the consensus among doctors is growing that sunscreens should block this radiation as well.[3,4]

Of the three parts of solar UV radiation, the shorter wavelength UVB portion, from 290 to 320 nm, is regarded as the most deleterious. Direct links have been made between UVB exposure and acute sunburn, mutation induction, immune suppression, cell mortality and skin cancer.[5-9]

But the evidence is now overwhelming that excessive exposure to any part of the solar UV spectrum, including UVA II (320-340 nm) and UVA I (340-400 nm), is harmful. Apart from its possible role in melanoma, UVA has been shown to cause a

[a] The trade name, HallBrite TQ, is registered to The C. P. Hall Company.

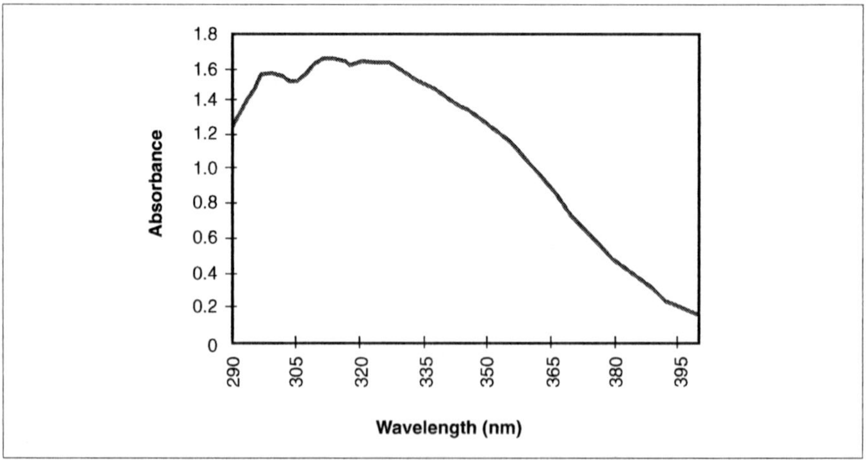

Figure 1. Solar UV absorbance of a traditional SPF 30 sunscreen.

wide variety of chemical and biological effects including generation of reactive oxygen species, DNA damage, lipid peroxidation, increase in elastin fibers, collagen cross-linking, epidermal thickening, and an increase in the number of dermal cysts.[10-16]

Full spectrum UV protection: Traditional sunscreen design focuses on attenuating the sunburn component of UV radiation, principally in the 290 to 320 nm range, sometimes extending to 340 nm (see sidebar on traditional sunscreen design).

In our laboratories, we approach sunscreen design from two simple premises:

- All sun protection derives from UV radiation attenuation.
- The best sun protection derives from radiation attenuation across the entire solar UV spectrum, from 290 nm to 400 nm.

Our objective, therefore, is to facilitate the formulation of sunscreens that reliably attenuate well over 90% of *all* solar UV radiation. As we see it, sunscreens that perform to this level provide their users with full spectrum protection.

We believe the term "full spectrum" could solve the question raised at the American Academy of Dermatology Consensus Conference held February 4, 2000. This conference brought the medical, regulatory and scientific communities together to try and reach a consensus on UVA protection by sunscreens. One conclusion was that SPF must remain as a way to advise consumers on UVB protection. The participants also agreed that UVA protection is required in all sunscreens and that a simple label must be developed to show consumers the level of protection.

Because the term "broad-spectrum" has been used and misused,[22] applying this term to products that protect against both UVB and UVA has become muddled and should be abandoned. We suggest using full spectrum as a permitted claim provided the product attenuates more than 90% of all solar UV radiation. Products that do not offer this protection could not be labeled as either full spectrum or broad-spectrum.

Achieving full spectrum protection: Achieving full spectrum protection in a sunscreen requires the incorporation of UV filters that, alone or in combination,

Traditional Sunscreen Design

To most people, sun protection means protection from sunburn. This painful and potentially serious skin injury in light-skinned people is caused by overexposure to solar UV radiation. Indeed, sunscreens are rated by their ability to increase people's tolerance to UV radiation, as measured by the dose required to provoke a slight reddening or erythema, which is the first sign of sunburn. This rating, of course, is the Sun Protection Factor, or SPF. To this day, SPF remains the only meaningful sun protection claim allowed by regulations in the US.

We all know that SPF is imperfect because consumers do not use sunscreens as directed. For example, they don't apply sunscreens at the level mandated in regulatory tests, and they fail to apply sunscreens 30 minutes before going out into the sun. However, SPF is universally accepted and now would be impossible to change. Although run by slightly different methods in different countries, SPF gives about the same results everywhere.

Human studies have shown that the action spectrum for sunburn, known as the erythemal action spectrum, is concentrated in the range from 290 to 320 nm with a tail extending out to about 340 nm.[17] Not surprisingly, sunscreens today are designed first and foremost to attenuate the radiation that causes sunburn. This is the so-called UVB portion of the spectrum from 290 to 320 nm. It is not much of an oversimplification to say that the greater the attenuation in this range, the higher the SPF. As SPF requirements increase, sunscreen formulators add absorbance to the right of 320 nm to eliminate the effects of radiation out to 340 nm, often referred to as UVA-II. When attenuation approaches 100% of the radiation between 290 and 320 nm and a significant portion of the radiation from 320 to 340 nm, voila, SPFs of 30 and above are the result. No more sunburn!

Figure 1 shows the absorbance profile of a popular commercial sunscreen with a labeled SPF of 30. One can see that the sunscreen absorbs best in the UVB and short-wave UVA region, but its absorbance trails off in the region of long-wave UVA. Calculations based on the area under the curve indicate that attenuation of UVB is 97% as expected, but attenuation of UVA is only 81%.

absorb throughout the solar UV spectrum. It also requires UV filters that provide the needed magnitude of attenuation within the concentration limits as defined by country regulations and/or cosmetic acceptability (see sidebar on organic UV filters).

A major obstacle to delivering full spectrum protection is the dearth of acceptable UV filters that provide significant attenuation across the entire UVA range from 320 nm to 400 nm. Only avobenzone is currently approved for general use in the US.[b] It absorbs broadly enough and with sufficient magnitude to attenuate more than 90% of UVA radiation. Avobenzone needs help to remain photostable; the very radiation it absorbs can cause it to undergo chemical reactions that degrade its absorbance.[23-25]

Solar UV Irradiance

High-frequency electromagnetic radiation beyond the visible includes UV radiation, X-rays, and gamma rays. All this radiation is ionizing and therefore harmful to body tissues, living cells and DNA.[18]

Of the solar radiation striking the upper atmosphere, about 9% is in the UV wavelength range of 200 to 400 nm.[19] Fortunately, the sun's emissions are highly modified by Earth's atmosphere. Radiation of less than 290 nm is, for practical purposes, eliminated, and radiation from 290 to approximately 400 nm is strongly attenuated. To put that into relevant perspective, the body of a sunbather is struck by 10^{21} photons every second, and about 1% of these are photons of ultraviolet radiation.[20]

Within the UV spectrum itself, the shorter, more energetic wavelengths are the ones most strongly attenuated by the atmosphere. Consequently, of the total UV radiation reaching the surface, the portion between 290 and 320 nm accounts for less than 6%.[21] Figure 2 is a graph of solar UV irradiance recorded in Albuquerque, New Mexico at noon on July 3 and illustrates this important fact.

Avobenzone: A Powerful Tool

Absorbance of UVA and UVB: When the FDA approved avobenzone for general use in sunscreens in 1997, formulators in the US joined their counterparts in the rest of the world in having available an extremely powerful tool to attenuate UV radiation.[26] The maximum permitted level in the US is 3.0%, in the EU 5.0% and in Japan 10.0%. Unquestionably, in terms of breadth and magnitude of absorbance and its nearly 20-year history of safe use around the world, avobenzone is the leading candidate to extend sun protection throughout the UVA portion of the spectrum.

An often overlooked property of avobenzone is its significant absorbance of UVB (Figure 3). In fact, our data indicates that at 306 nm, avobenzone absorbs almost twice as well on a molar basis as ethylhexyl salicylate (formerly octyl salicylate). At 320 nm, avobenzone's molar absorptivity is about equal to oxybenzone's molar absorptivity at the same wavelength.[27] Avobenzone maintains a significant magnitude of absorbance throughout the UVB band.

Analyzing sunscreen photostability: In the laboratory, in vitro experiments were conducted to evaluate the absorbance of various avobenzone formulations both before and after irradiation with 10 MED from a solar simulator.[c] For general reference, 10 MED is equivalent to approximately two hours of sunlight.

Absorbance is defined as log (1/T) where T (transmittance) is the ratio of radiation detected after passage through the test vehicle to radiation emitted by a radiation source. Attenuation is defined as 1–T, or, when referred to as a percentage, as 100(1–T). For reference, absorbance of 2 equals 99% attenuation, absorbance of 1.52 equals 97% attenuation, and absorbance of 1 equals 90% attenuation.

---

[b] Avobenzene is available from several suppliers under the trade names Parsol 1789 (Roche Vitamins, Parsippany, NJ), Solarom BMBM (Frutarom, Haifa Bay, Israel), Neo Heliopan Type 357 (Haarman and Reimer, Holzminden, Germany), Eusolex 9020 (E.Merck, Darmstadt, Germany), and Uvinul BMDM (BASF, Ludwigshafen, Germany)

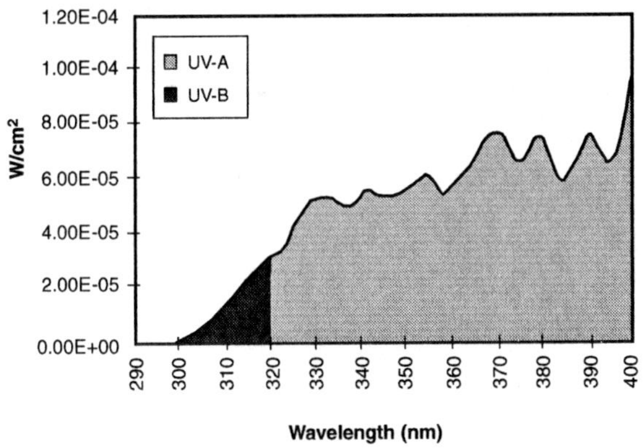

Figure 2. Solar irradiance in Albuquerque, New Mexico, at noon on July 3

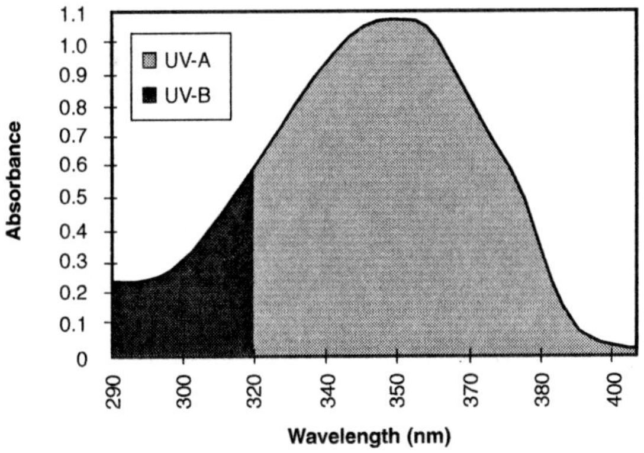

Figure 3. Solar UV absorbance of avobenzone (10 ppm) in cyclohexane

Transmittance data for the sample formulations was obtained by a transmittance analyzer[d]. The analyzer's software[e] integrated the area under the absorbance curve and reported the percentages of radiation attenuation.

Figure 4 illustrates both the promise of avobenzone and its problem. Before irradiation, the 1% avobenzone sunscreen attenuates 75% of the UVB radiation, and 81% of the UVA radiation. Following irradiation, attenuation falls to 57% of UVB and 56% of UVA. Attenuation of UVB is reduced by 24% and attenuation of UVA is reduced by 30% by the 10 MED exposure.

Photostabilizing with DEHN: Figure 5 illustrates what happens when DEHN is substituted for another ester in the formula at 4%. Except for the substitution,

[c] Model 16S Solar Simulator equipped with a WG 320 filter (transmits UV > 290 nm), output monitored by a PMA 2105 UV-B DCS Detector (biologically weighted) and controlled by a PMA 2100 Automatic Dose Controller (Solar Light Co., Philadelphia, Pennsylvania)

Organic UV Filters

Sunscreens contain organic chemicals or metal oxide particles that absorb, reflect, or scatter ultraviolet radiation and prevent it from striking the skin. Most countries regulate these filters by a pre-approval process of creating a positive or permitted list of sunscreen filters. In the C. P. Hall laboratories, research has been limited to organic UV filters; therefore this discussion will be confined to this class of compounds.

Most organic UV filters absorb in a fairly narrow band on either side of their peak absorbance. This band is abbreviated by the symbol λ_{max} (lambda max). The magnitude of a UV filter's absorbance (its ability to attenuate radiation) is a function of its molar absorptivity at its λ_{max}, which is abbreviated as the symbol ε (epsilon). Table 1 lists the most important UV filters approved for general use in the US and in the EU. It indicates their λ_{max}, ε, and maximum permitted concentration by weight in the formulation.

For any solution or sunscreen, it is a fundamental principle of physical chemistry that the magnitude of absorbance is directly proportional to the concentrations of the UV filters present times their molar absorptivity. So theoretically, a sunscreen's total absorbance is the sum of the absorbances of the individual filters. Formulators can choose from several organic filters that absorb between 290 and about 340 nm, so it's a fairly simple matter for them to combine these in various ways to attenuate the shorter wave portion of the solar UV spectrum.

However, their choices are much more limited when they seek to attenuate the longer wavelengths from 320 to 400 nm. As is readily apparent in Table 1, only menthyl anthranilate and avobenzone absorb near 350 nm, and only avobenzone has the molar absorptivity to attenuate significant radiation at or below its legally allowable maximum concentration.

this formulation is identical in every way to the one depicted in Figure 4. Before irradiation, this formulation attenuates 82% of the UVB radiation and 80% of the UVA. After irradiation with 10 MED, attenuation of UVB is 80% and UVA is 77%. For both UVB and UVA, loss of attenuation is less than 4%.

As we shall see, some very good things happen to sunscreen formulations when avobenzone is formulated with DEHN. But first, let's discuss the chemistry and photochemistry of this material.

Diethylhexyl 2,6-Naphthalate

DEHN is the diester of 2,6-naphthalene dicarboxylic acid, and 2-ethylhexanol, a branched C8 primary alcohol. Figure 6 shows the molecular structure, molecular formula, and molecular weight of DEHN.

[d] UV1000S UV Transmittance Analyzer, Labsphere Inc., North Sutton, NH

[e] UV1000S Version 1.21, Labsphere Inc. This software uses 290 nm and 315 nm as the limits of the UVB integral rather than the more commonly used 290 and 320 nm. Similarly, the software uses 315 and 400 nm as the limits of the UVA integral.

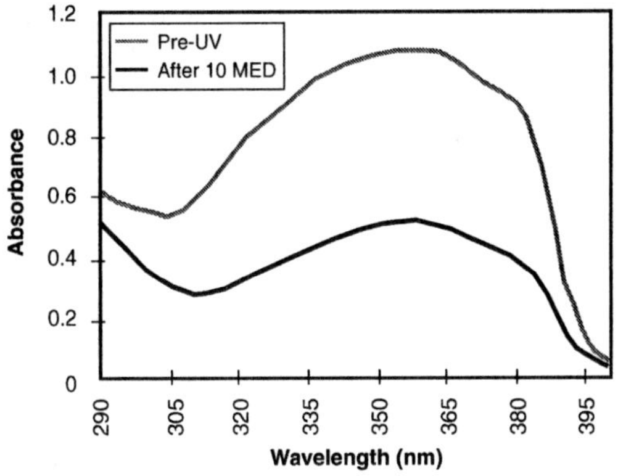

Figure 4. Photostability of a 1% avobenzone formulation with no stabilizer

Table 1. Organic UV filters

	$\lambda_{max.}$	ε	US	EU
UVB Filter				
Methylbenzylidene camphor	300	24,500	NA	4%
Homosalate	306	4,300	15%	10
Ethylhexyl salicylate	307	4,900	5	5
Phenylbenzimidazole sulfonic acid	310	28,250	4	8
Ethylhexyl methoxycinnamate	311	23,300	7.5	10
Padimate O	311	27,300	8	8
Ethylhexyl triazone	313	110,000	NA	5
UVB/UVA Filter				
Oxybenzone	288/325	14,000/9,400	6	10
Octocrylene	303	12,600	10	10
UVA Filter				
Menthyl anthranilate	336	5,600	5	NA
Avobenzone	358	34,720	3	5

NA=Not Approved

The physical properties of this molecule[28] can be inferred to a large extent from its structure. It is a semi-viscous (546 cSt at 25°C by the Kinematic method) liquid at room temperature and has a freeze point of less than 5°C. It has a high refractive index of 1.53. Its specific gravity is 1.02. It is quite lipophilic. It is insoluble in water, propylene glycol and glycerin. It is freely soluble in most oils such as mineral oil, castor oil, and typical cosmetic esters.

Although it may seem unexpected and fortuitous, DEHN is an excellent solvent for lipophilic solids such as the UV filters oxybenzone (benzophenone-3), avobenzone, and ethylhexyl triazone (formerly octyl triazone).

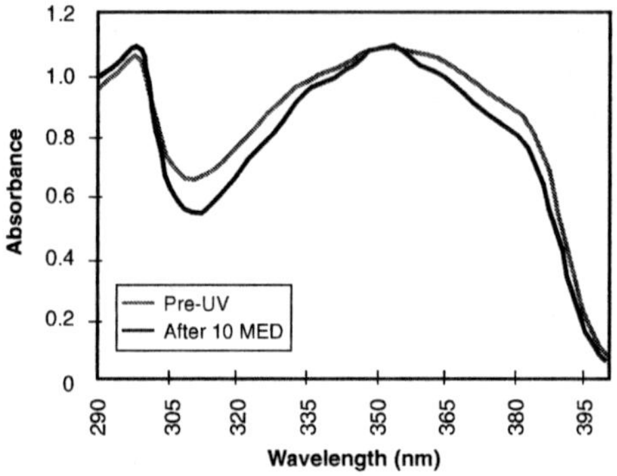

Figure 5. Photostability of a 1% avobenzone formulation with 4% DEHN

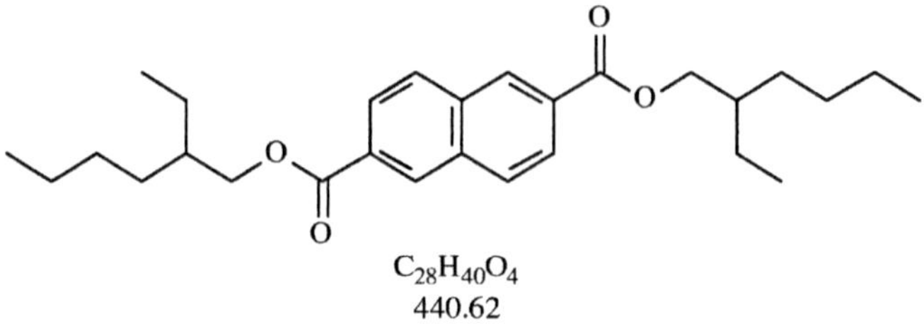

Figure 6. Molecular structure, formula and weight of DEHN. Its CAS Registry Number is 127474-91-3.

Figure 7 shows the UV absorbance of DEHN. In the solar UV range, it has a peak absorbance (λ_{max}) of 294 nm and a molar absorptivity (ε) of about 9,000. It has two small peaks at 332 and 350 nm, and molar absorptivity at those peaks of about 1,000 and 2,000, respectively.

As shown in Figure 7, the solar UV absorbance of DEHN is very weak. By comparison, the absorbance of ethylhexyl methoxycinnamate exceeds 0.8 at 310 nm, and the absorbance of avobenzone is approximately 1.1 at 355 nm.

The ability of DEHN to photostabilize avobenzone is a function of its capacity to act as an acceptor of triplet energy (see sidebar on photochemistry of photostability). Avobenzone has a triplet energy of about 60 kcal/mol.[29] Based on published values for similar compounds, DEHN has a triplet energy of 57-60 kcal/mol and, therefore, may behave as an acceptor of avobenzone's triplet energy. Work is ongoing to provide a more precise characterization of the photophysical properties of DEHN. Its triplet energy and other properties of interest will be published in due course.

Sunscreen Applications

Table 2 shows the formulas for several model sunscreens used in the experiments described below. The first two sunscreens, marked A and B, both contain 3%

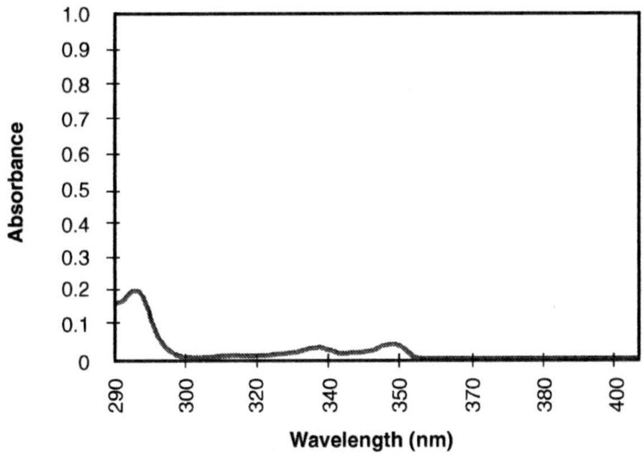

Figure 7. Solar UV absorbance of DEHN (10 ppm) in cyclohexane

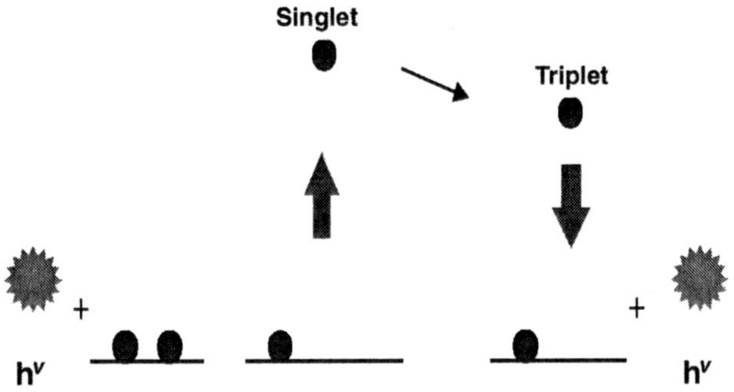

Figure 8. Schematic representation of photon absorption resulting in the excitation of an electron to the singlet state, the decay to the triplet state, and the emission of a photon before returning to the ground state

avobenzone and no other UV filters at all. DEHN has been added to Formula B at 4% and not to Formula A. So basically we have two matched formulas, a negative control and a positive control.

Figures 9 and 10 show the absorbance profiles of these two formulas before and after 10 MED exposures. It is very important to note that 3% avobenzone alone attenuates about 90% of all the radiation between 290 and 400 nm. After 10 MED, unfortunately, UVB and UVA attenuation in Formula A, without DEHN, falls to 77% and 64%, respectively. In stark contrast, Formula B, with 4% DEHN, maintains virtually all of its absorbance across the UV spectrum; after 10 MED its UVB attenuation is 92% and its UVA attenuation is 91%. Most importantly for labelling purposes, the formulation which has been stabilized with DEHN delivers an average in vivo SPF of 12 (Table 2).

Once a photostable foundation of 3% avobenzone has been established, it becomes a simple matter to achieve any desired SPF above 10 by adding UVB

Photochemistry of Photostability

To begin at the beginning: A photon is a quantum or "packet" of electromagnetic energy with an energy equal to Planck's constant (h) times its frequency (ν). The absorption of a photon by an organic molecule causes the excitation of one of a pair of electrons in a low energy orbital to a higher energy unoccupied orbital (Figure 8).[30]

Before absorption, the orbital configuration of the electrons is the "ground" state. Upon absorption, two electronic states are possible. In one, the spins of the two electrons remain paired and, as in the ground state, the net spin of the pair is zero. This is called the "singlet" excited state. In the other, the spins of the two electrons are unpaired, and there is a net spin. This is called a "triplet" excited state because three states can be resolved in a magnetic field.[31]

The energy of both excited states is eventually dissipated as heat (vibration, including both bond stretching and nuclear motion), or heat and light (emission of a photon of lower energy/longer wavelength). Emission of a photon from the singlet state is called "fluorescence." Photon emission from the triplet state is called "phosphorescence." The singlet state may return to the ground state directly, or it may decay to the triplet state.[32]

The singlet state is often short-lived, typically 10^{-9} to 10^{-8} seconds. Therefore, reactions that proceed from it must be quite rapid. Of more importance to the sunscreen formulator are reactions that proceed from the (usually) much longer-lived triplet state, which may last 10^{-4} seconds or longer.[33]

During the triplet state lifetime, the excited molecule looks and behaves as a diradical,[34] from which many chemical reactions are possible. In general, these reactions can be grouped into four categories: photoaddition/substitution, cycloaddition, isomerization, and photofragmentation.[35-38] Of particular importance to the sunscreen formulator are reactions between like or different UV filter molecules, those between UV filter molecules and sunscreen excipients, and isomerizations or fragmentations of the UV filter molecules. Any one of these reactions may alter or destroy the UV absorption capacity of the sunscreen formulation.

The excited molecule may react (to produce isomers or new products), or return to the ground state in its original form. Clearly, the latter is the preferred outcome for sunscreen formulators (and users) because, among other reasons, the UV filter molecule is again available to absorb a photon.

Many factors determine the pathway an excited molecule will take. Among these factors are the triplet energy, the triplet lifetime, the identity and concentration of the reactants, and the rates and activation energies of each competing reaction. Under certain conditions, the excited molecule may return to the ground state (and its original form) by transferring its energy to a nearby molecule. The excited molecule becomes a "donor" and the nearby molecule becomes an "acceptor." Upon the transfer of energy, the donor returns to ground state and the acceptor is elevated to the excited state.[39]

Table 2. Formulas for model sunscreens used in photostability experiments

Ingredient	A	B	C	D	E	F
Oil Phase						
Avobenzone	3.0%	3.0%	3.0%	3.0%	3.0%	3.0%
Ethylhexyl salicylate	-	-	5.0	5.0	5.0	2.0
Oxybenzone	-	-	-	-	4.0	2.8
Ethylhexyl methoxycinnamate	-	-	-	-	-	5.2
Diethylhexyl naphthalate	-	4.0	-	4.0	5.0	7.5
Isopropyl myristate	8.0	4.0	4.0	-	4.0	1.0
Hexyldecyl benzoate (and) butyloctyl benzoate	7.5	7.5	6.5	6.5	4.5	-
Stearyl alcohol	0.3	0.3	0.3	0.3	0.3	0.3
PPG-2 myristyl ether proprionate	1.0	1.0	1.0	1.0	0.5	-
Polyglyceryl-3 methyl glucose distearate	3.0	3.0	3.0	3.0	3.0	3.0
C30-38 Olefin/isopropyl maleate/MA copolymer	1.0	1.0	1.0	1.0	1.0	1.0
Water Phase						
Water (*aqua*)	qs	qs	qs	qs	qs	qs
Disodium EDTA	0.05	0.05	0.05	0.05	0.05	0.05
Glycerin	4.0	4.0	4.0	4.0	4.0	4.0
Butylene glycol	2.0	2.0	2.0	2.0	2.0	2.0
Phenoxyoxythenol (and) methyl-paraben (and) ethylparaben (and) propylparaben (and) butylparaben	0.7	0.7	0.7	0.7	0.7	0.7
Carbomer	0.2	0.2	0.2	0.2	0.2	0.2
Triethanolamine	0.2	0.2	0.2	0.2	0.2	0.2
in vivo static SPF*	6.3(6)	12.1(10)	13.0(10)	17.3(16)	32.8(30)	NA

*Studies conducted on five human test subjects by Consumer Product Testing Company, Fairfield, NJ.
First value given is the average of the five scores. Value in parentheses is the lowest of the five scores.

filters to the formula. Table 2 also shows two more matched formulas, marked C and D, this time adding 5% ethylhexyl salicylate to 3% avobenzone. As you can see in Figure 11, ethylhexyl salicylate increases attenuation of UVB to 94-95% and attenuation of UVA climbs slightly to 93%. After irradiation with 10 MED, however, Formula C without DEHN loses a bit of its UVB attenuation, to 93%, but a lot of its UVA attenuation, falling to 84%. In contrast, Formula D (Figure 12), with 4% DEHN, maintains 94% attenuation of UVB and 92% attenuation of UVA. This level of attenuation adds up to an average in vivo SPF of 17 (Table 2).

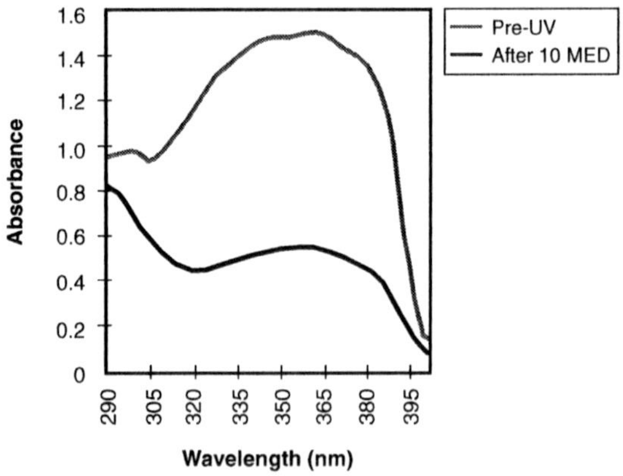

Figure 9. Photostability of Formula A (after 10 MED): 3% avobenzone formulation with no stabilizer

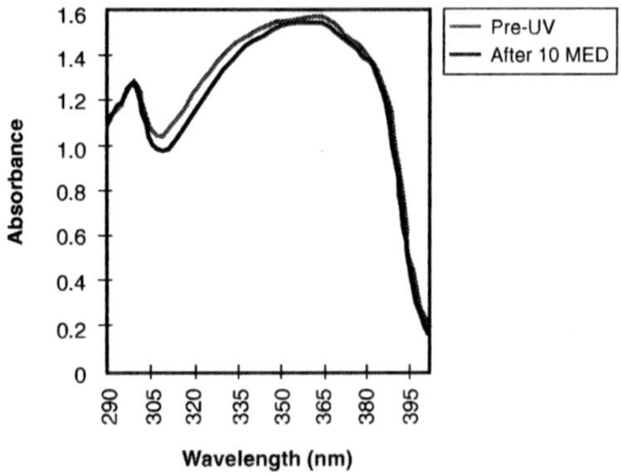

Figure 10. Photostability of Formula B (after 10 MED): 3% avobenzone formulation with 4% DEHN

Formula E in Table 2 shows a formula to which 4% oxybenzone has been added to the 5% ethylhexyl salicylate and 3% avobenzone and the formulation has been stabilized with 5% DEHN. With oxybenzone's contribution to absorbance in both the UVB and short-wave UVA portions of the spectrum, this formula exhibits significantly increased attenuation across the spectrum, to 97% of UVB and 94% of UVA. Figure 13 shows the absorbance profiles of this formulation before and after irradiation with 25 MED, roughly equivalent to 6 hours in the sun. Most important, this photostable formulation delivers an average in vivo SPF of 32 (Table 2).

Ethylhexyl Methoxycinnamate and Avobenzone

Every technology has its limitations, and the limitation of this photostabilization technology is that it will not completely photostabilize the combination of

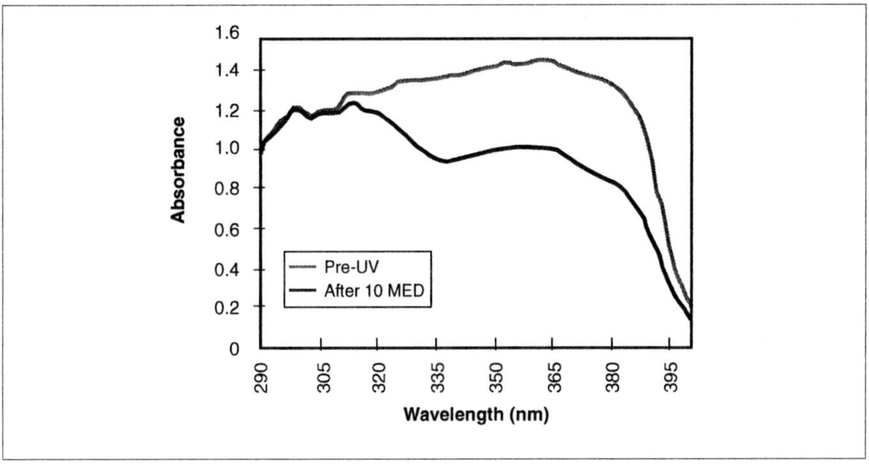

Figure 11. Photostability of Formula C (after 10 MED): 5% ethylhexyl salicylate and 3% avobenzone formulation with no stabilizer

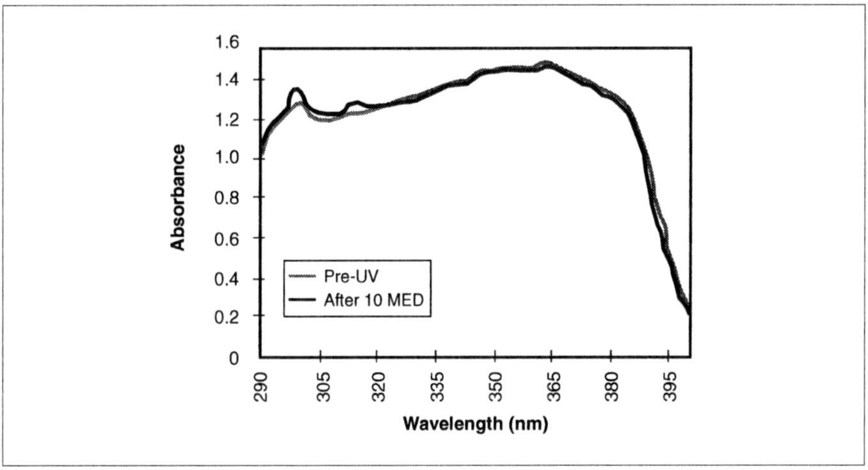

Figure 12. Photostability of Formula D (after 10 MED): 5% ethylhexyl salicylate and 3% avobenzone formulation with 4% DEHN

ethylhexyl methoxycinnamate (formerly octyl methoxycinnamate) and avobenzone. It can, however, help formulations containing this combination, as illustrated in Figures 14 and 15.

In this experiment, a commercial sunscreen and a close equivalent model sunscreen stabilized with 7.5% DEHN were exposed to 5 hours of sunlight side-by-side on a hot summer's day in Chicago. Before exposure, both sunscreens demonstrated an in vitro SPF of 50. Both also exhibited 97% attenuation of UVB and 95% attenuation of UVA. After 5 hours of sunlight, the commercial sunscreen declined in attenuation to 95% UVB and 84% UVA, and to SPF 26 (Figure 14); the stabilized model sunscreen maintained

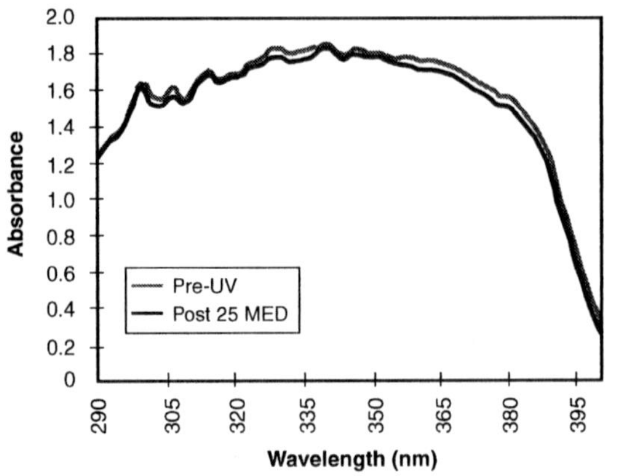

Figure 13. Photostability of Formula E (after 25 MED): 5% ethylhexyl salicylate, 4% oxybenzone, 3% avobenzone with 4% DEHN

attenuation of UVB at 97% and declined in attenuation of UVA to 87%, and to SPF 36 (Figure 15). This is approximately a 40% improvement in SPF stability.

The most stable sunscreens, however, are achieved by adding DEHN to avobenzone, and leaving out the ethylhexyl methoxycinnamate. This is one of our suggestions for formulating with DEHN (see sidebar presenting guidelines for formulating with DEHN).

Conclusion

Unquestionably, people with light skin are best served by sunscreens that attenuate radiation across the entire solar UV spectrum, 290-400 nm. Sunscreens that provide at least 90% attenuation over the entire UV spectrum are deserving, in our opinion, of being called full spectrum sunscreens.

We have presented just a few of the many formulations and allowed UV filter combinations (from the US and Europe) that can be used to achieve full spectrum protection. The key to all of them is the establishment of a photostable foundation on the UVA side of the spectrum. The combination of avobenzone and the photostabilizer diethylhexyl 2,6-napthalate provides a photostable foundation than can, when properly formulated, attenuate more than 90% of both UVB and UVA radiation. Once that's done, UVB attenuation can be staged to give consumers their choice of the level of sun protection they want, as measured by SPF.

We should add that, in our experience to date, the inclusion of diethylhexyl 2,6-naphthalate improves the performance of every sunscreen, regardless of the UV filter combination.

—**Craig Bonda,** *The C. P. Hall Company, Bedford Park, Illinois, USA*
—**David C. Steinberg,** *Steinberg & Associates, Inc., Plainsboro, New Jersey, USA*

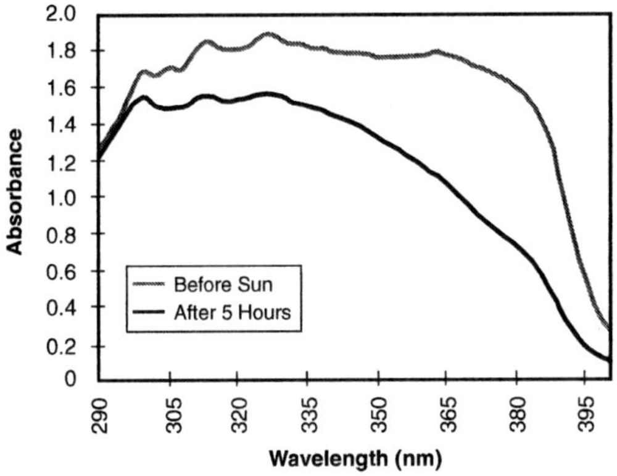

Figure 14. Photostability of an SPF 30+ commercial sunscreen (after 5 hours in the sun) with ethylhexyl methoxycinnamate and avobenzone

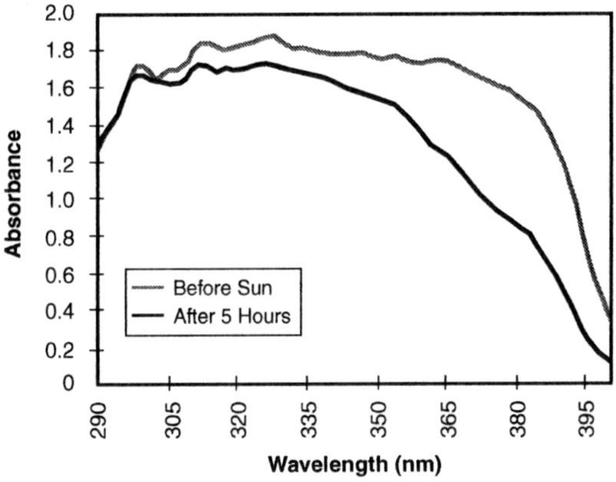

Figure 15. Photostability of an SPF 30+ commercial sunscreen (after 5 hours in the sun) with ethylhexyl methoxycinnamate, avobenzone and DEHN

References

1. HI Hall, DR Miller, JD Rogers and B Brewerse, Update on the incidence and mortality from melanoma in the United States, *J Am Acad Dermatol* 40 35-42 (1999)
2. P Autier et al, Sunscreen use and duration of sun exposure: A double-blind, randomized trial, *J Nat Cancer Inst* 91 1304-1309 (1999)
3. AJ Swerdlow and MA Weinstock, Do tanning lamps cause melanoma? An epidemiological assessment, *J Am Acad Dermatol* 38 89-98 (1998)
4. American Academy of Dermatology Consensus Conference, UVA Protection in Sunscreens, Washington, DC, February 4, 2000
5. J Jagger, *Solar-UV Actions on Living Cells*, New York: Praeger Press (1985)
6. ML Kripke, Immunological mechanisms in UV radiation carcinogenesis, *Adv Cancer Res* 34 69-106 (1981)

7. LA Applegate, D Lautier, E Frenk and RM Tyrrell, Endogenous glutathione levels modulate the frequency of both spontaneous and long wavelength ultraviolet induced mutations in human cells, *Carcinogenesis* 13 1557-1560 (1992)
8. NA Soter, Acute effects of ultraviolet radiation on the skin, Sem Dermatol, 9 11-15 (1990)
9. I Serre et al, Immunosuppression induced by acute solar-simulated ultraviolet exposure in humans: Prevention by a sunscreen with a sun protection factor of 15 and high UVA protection, *J Am Acad Dermatol* 37 187-194 (1997)
10. HS Black, Potential involvement of free radical reactions in ultraviolet-mediated cutaneous damage, *Photochem Photobiol* 46 213-221 (1987)
11. ML Cunningham, JS Johnson, SM Giovanazzi and MJ Peak, Photosensitized production of superoxide anion by monochromatic (290-405 nm) ultraviolet irradiation of NADH and NADPH coenzymes, *Photochem Photobiol* 42 125-128 (1985)
12. MJ Peak and JD Peak, in *The Biological Effects of UVA Radiation*, F Urbach and RW Gange, eds, New York: Praeger Press (1986) pp 42-52
13. MA Pathak and MD Carbonare, in *Biological Responses to Ultraviolet A Radiation*, F Urbach, ed, Overland Park, Kansas: Valdenmar (1992) pp 189-208
14. DL Bissett, DP Hannon and TV Orr, *Photochem Photobiol* 46 367-378 (1987)
15. LH Kligman and P Zheng, The protective effect of a broad-spectrum sunscreen against chronic UVA radiation in hairless mice: A histologic and ultrastructural assessment, *J Soc Cos Chem* 45 21-33 (1994)
16. LH Kligman, PP Agin and RM Sayre, Broad spectrum sunscreens with UVA I and UVA II absorbers provide increased protection against solar-simulating radiation-induced dermal damage in hairless mice, *J Soc Cos Chem* 47 129-155 (1996)
17. AF McKinley and BL Diffey, A reference action spectrum for ultraviolet induced erythema in human skin, *CIE Journal* 6 12-22 (1987)
18. *Encyclopedia Britannica*, http://www.britannica.com/bcom/eb/article/4/0,5716,108504+6,00.html
19. *CRC Handbook of Chemistry and Physics*, 75th ed, Boca Raton, Florida: CRC Press (1994) pp 14-12
20. Ibid, *Encyclopedia Britannica*
21. Labsphere UV-1000S Instruction Manual, Rev 2, p 36
22. Fed Reg 58(90) (May 12, 1993)
23. RM Sayre and JC Dowdy, Photostability testing of avobenzone, *Cosmet Toil* 114(5) 85-91 (1999)
24. N Tarras-Wahlberg et al, Changes in ultraviolet absorption of sunscreens after ultraviolet irradiation, *Soc for Invest Derm* 113 547-553 (1999)
25. CA Bonda and PJ Marinelli, The photochemistry of sunscreen photostability, presentation to the Seventh Florida Sunscreen Symposium, Orlando, Sep 24, 1999
26. Fed Reg 62(83) (Apr 30, 1997)
27. Tech bulletin for Eschol 567, Bellview, NJ: ISP/Van Dyk (Sep 1991)
28. US Pat 5,993,789, CA Bonda, PJ Marinelli, YZ Hessefort, J Trivedi and G Wentworth, assigned to The C. P. Hall Company (Nov 30, 1999)
29. H Gonzenbach, TJ Hill and TG Truscott, The triplet energy levels of UVA and UVB sunscreens, *J Photochem Photobiol* 16 377-379 (1992)
30. NJ Turro, Modern Molecular Photochemistry, Menolo Park, California: Benjamin/Cummings (1978) p 3
31. Ibid, p 23
32. Ibid, pp 4-6
33. Ibid, pp 90, 105, 352
34. Ibid, pp 364-365
35. Ibid, Ch 10
36. Ibid, Ch 11
37. Ibid, Ch 12
38. Ibid, Ch 13
39. Ibid, Ch 9

Stability Analysis of Emulsions Containing UV and IR Filters

Keywords: chemical filters, rheology, viscocity, thermogravimetry, differential scanning calorimetry, UV spectrophotometric analysis, liquid chromatography

A basic o/w emulsion containing UV and IR filters was stable in tests of its rheology and thermal decomposition, and in evaluation by UV spectrophotometry and high performance liquid chromatography.

The stability studies of solar protection emulsions and quantification of chemical filters is of great importance. These emulsions are considered unstable. Very few studies could be found in the literature regarding these types of formulations.

Radiation and Skin

Exposure of human skin to excessive solar radiation can lead to erythema, dermatitis and even skin cancer. The radiation reduces Langerhans cells, responsible for the immunological resistance of the skin. UV filters provide protection against erythema.[1,2]

UVB radiation can provoke tremendous damage to the cutaneous immunological system. UVB is responsible for erythema, damage to the DNA and skin cancer.[3] Kollias and Baqer showed that melanin presents significant protection against UVA radiation and partial protection against UVB radiation.[4]

Solar filters must encompass ample protection against UVA, UVB and infrared (IR) radiation. Physical filters must be mingled rationally with the chemical filters so together they provide efficient protection across a broad area of the UV spectra.[5,6] Cutaneous protection is also required in the near IR radiation region because radiation in this region possesses energy strong enough to modify the rotation and vibration of molecules.[7-9] Kligman has demonstrated that UV and IR radiation acting together can damage skin elastic fibers to a greater extent than either one of them acting alone.[10]

The goal was to evaluate aspects of the stability of a basic, non-ionic o/w emulsion containing UVA, UVB and IR filters. Stability studies were performed, by varying time and temperature, using rheological measurements, thermal analysis, UV spectrophotometry and high performance liquid chromatography.

Formula 1. O/W Solar Protecting Emulsion

A.	Silicon oil	3.00%w/w
	Glyceryl monostearate	2.00
	Cetostearyl alcohol	2.00
	Mineral oil	3.00
	Non-ionic wax	2.50
	Phenoxyethanol	0.80
	Butylated hydroxytoluene	0.05
	Benzophenone-3	2.00
	Octyl methoxycinnamate	4.00
	Titanium dioxide	2.00
	Corallina officinalis (Phycocorail, SECMA)	0.50
B.	Propylene glycol	5.00
	Xanthan gum	0.30
	Imidazolidinyl urea	0.50
	Magnesium aluminum silicate (Veegum Ultra, RT Vanderbilt)	0.50
	Water (*aqua*), distilled	qs 100.00

Procedure: Heat A and B at 75-80°C. After heating, pour B over A while stirring, until mixture reaches ambient temperature.

Measuring Emulsion Stability

Several solar formulations were prepared. For the purposes of this study, the one that presented the best physical stability was used.

Preparation of emulsion: The o/w emulsion (Formula 1) prepared in this work contained benzophenone-3 (UVB 290–320 nm), octyl methoxycinnamate (UVA 320–400 nm) (Figure 1) and an IR radiation blocker. The IR blocker[a] is composed mainly of small oceanic seaweeds (*Corallina officinalis*) (IR 800–2500 nm).

The UV and IR blockers were obtained from ChemyUnion, São Paulo, Brazil. The remaining formula ingredients were supplied by ChemyUnion, or by Galena, Pharmaspecial or Merck, all located in São Paulo.

Rheology: In non-Newtonian liquids, viscosity is dependent on a velocity gradient that gives rise to plastic, pseudoplastic and dilatant curves. Solar emulsions normally present pseudoplastic behavior, exhibiting a reduction in viscosity when the spreading velocity increases. Normally, these emulsions also present thixotropic behavior, which is defined as the capacity of a system to present lowest viscosity when subjected to shear and its capacity to recover its original structure during definitive period of time after the shear is removed.[11-13]

Förster and Herrington performed rheological studies on o/w-type cosmetic emulsions before and after submitting samples to different temperature, agitation and centrifugation conditions.[14]

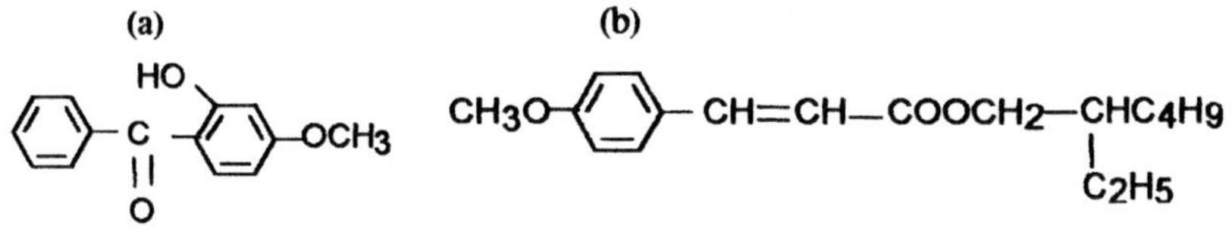

Figure 1. Chemical structures of sunscreens: (a) benzophenone-3 and (b) octyl methoxycinnamate

Primorac et al. studied the temperature influence on the rheological behavior of emulsions.[15] The samples studied presented a decrease in viscosity with gradient increase in temperature. All the samples also presented thixotropy.

In the study, the emulsion's rheology following various periods of storage up to two months at ambient temperature, 10°C, 35°C and 45°C was measured[b]. For each measurement 13 g of the emulsion and an SC4-29 spindle was used.

Heat degradation: Thermogravimetry (TG) reports the variation of mass as a function of time and/or temperature. The curves obtained furnish relative information regarding composition and thermal stability of the sample, the intermediate products and the residue formed.

The derived thermogravimetric curve (DTG) is the first derivative of the TG curve. A DTG curve allows better identification of the thermal decomposition steps. Through TG/DTG curves one can find temperatures corresponding to the beginning and end of a reaction, as well as the corresponding variations of mass. This method can be used to evaluate the thermal stability and the thermodecomposition steps of emulsions containing solar filters.[16,17]

In the study, calcium oxalate (4.8 to 5.3 mg) to calibrate our equipment[c] was used. Both the calibration and the sample thermogravimetric measurements were made under the following experimental conditions: platinum crucible; dynamic air atmosphere of 50 mL/min; heating program from ambient temperature to 900°C at a heating rate of 10°C/min.

Differential scanning calorimetry: Differential scanning calorimetry (DSC) is a technique based on measuring the difference of energy supplied to a substance and a reference material that is thermally inert, while the substance and the reference material are subjected to a controlled temperature program.[18,19]

In the study, equipment was calibrated[d], again using calcium oxalate (1.7 to 2.3 mg). Both the calibration and the sample DSC measurements were made under the following experimental conditions: aluminum crucible (partially closed); dynamic nitrogen atmosphere of 50 mL/min; heating program from ambient temperature to 600°C at a heating rate of 10°C/min.

UV spectrophotometric analysis: UV spectrophotometric analysis using a spectrophotometer[e] with a 1 cm cell, spectral slit width of 2 nm was performed.

These were the experimental conditions: chart speed, 300 nm/min; absorbance limit, from 0 to 1; quartz cell of 1.0 cm; and solvent, 96% ethanol.

To prepare standard solutions, 50.0 mg of benzophenone-3 and 50.0 mg of octyl methoxycinnamate was dissolved in 96% ethanol in a 100 mL volumetric flask. Dilutions were made in order to obtain the following final concentrations: 8.0, 9.0, 10.0, 11.0 and 12.0 μg/mL of benzophenone-3; and 6.0, 7.0, 8.0, 9.0 and 10.0 μg/mL of octyl methoxycinnamate.

Liquid chromatography: High performance liquid chromatography (HPLC) has been used by others for quantitative determination of some sunscreen agents.[20,21]

In this study, liquid chromatographic analyses were performed using a liquid chromatograph[f] equipped with a loop injector (20 μL), a variable UV detector, and a LiChrospher[g] 100 RP-18 (5 μm) with a LiChroCART[g] (4 mm x 125 mm) column. The mobile phase was methanol:water (87:13 v/v). The samples were chromatographed at room temperature, with an injection volume of 20 μL and a flow rate of 1.0 mL/min. Ultraviolet detection was made at 290 nm.

The quantification of chemical filters was done through the HPLC method described, using standards of benzophenone-3 and octyl methoxycinnamate as references.Two standard solutions were prepared. For one, 250 mg of benzophenone-3 was weighed into a 100 mL volumetric flask, methanol as added to make a solution, and then more methanol was added to bring the volume to 100 mL. For the other, 500 mg of octyl methoxycinnamate was used and the same procedure was followed. After dilutions, the final solutions contained 4.5-18 μg/mL of benzophenone-3 and 8-36 μg/mL of octyl methoxycinnamate.

Similarly, an emulsion sample solution was prepared by weighing an amount of the emulsion equivalent to 5 mg of benzophenone-3 and 10 mg of octyl methoxycinnamate into a 100 mL volumetric flask. Methanol (approximately 95 mL) was added and ultrasonicated for 20 min. The volume was completed to 100 mL with methanol. Then, the solution was filtered through cotton. After dilutions, the final solutions contained 10 μg/mL of benzophenone-3 and 20 μg/mL of octyl methoxycinnamate.

HPLC measurements were taken at 7, 15, 30, 45 and 60 days.

Results and Discussion

Rheology: Results of rheological studies can be observed in Figure 2. Samples stored at different temperatures and periods of time presented thixotropy, hysteresis and pseudoplastic characteristics like the majority of cosmetic emulsions. A decrease in thixotropy was observed in samples stored at 10°C when compared with those maintained at ambient temperature. Samples stored at 35°C presented similar behavior of those stored at ambient temperature. Considerable increase in thixotropy was observed in samples stored at 45°C.

As expected, an accentuated decrease in viscosity also was observed with an increase in the storage temperature. One explanation might be that a rising

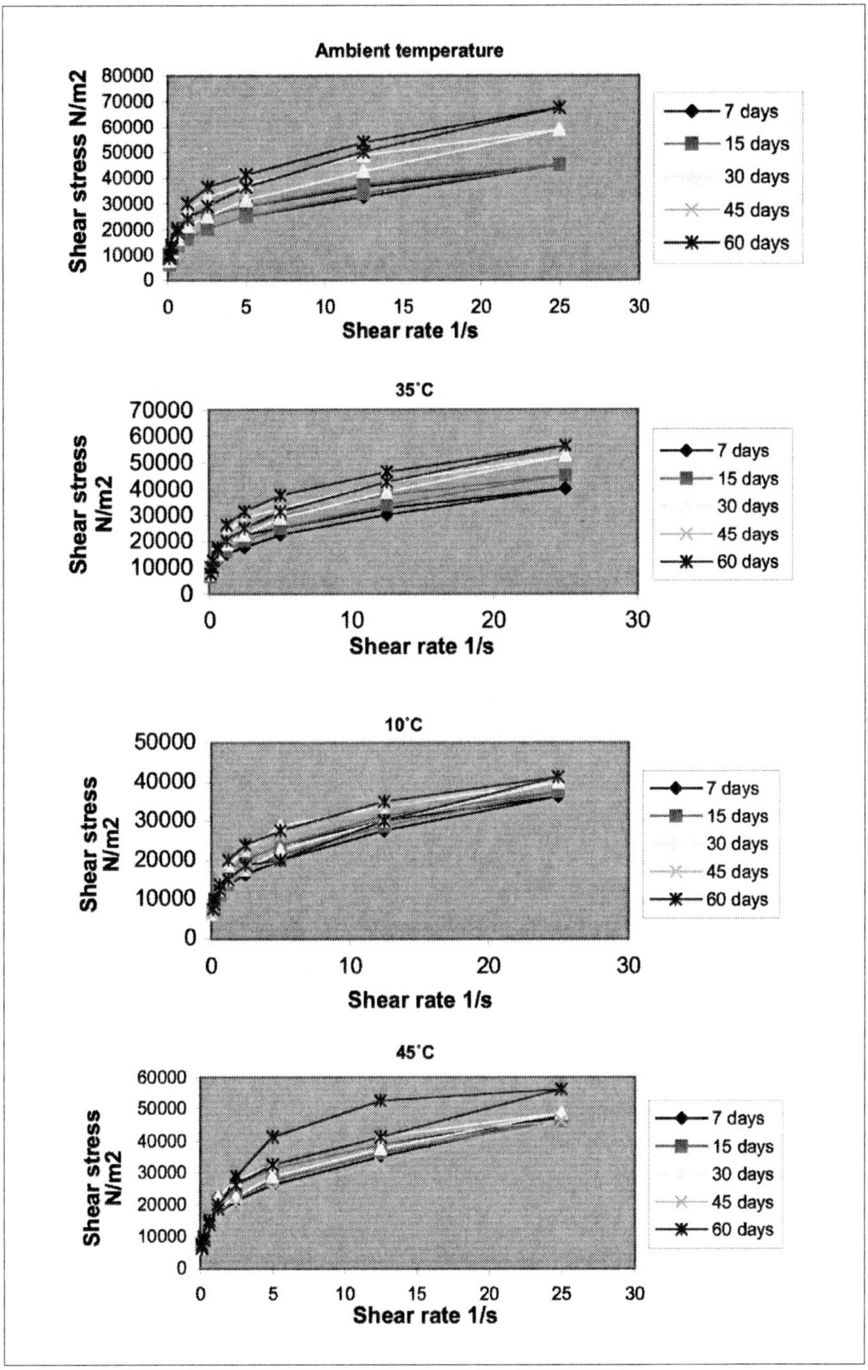

Figure 2. Rheology of sunscreen emulsion samples stored at selected conditions of time and temperature: ambient (top left), 10°C (top right), 35°C (bottom left), 45°C (bottom right)

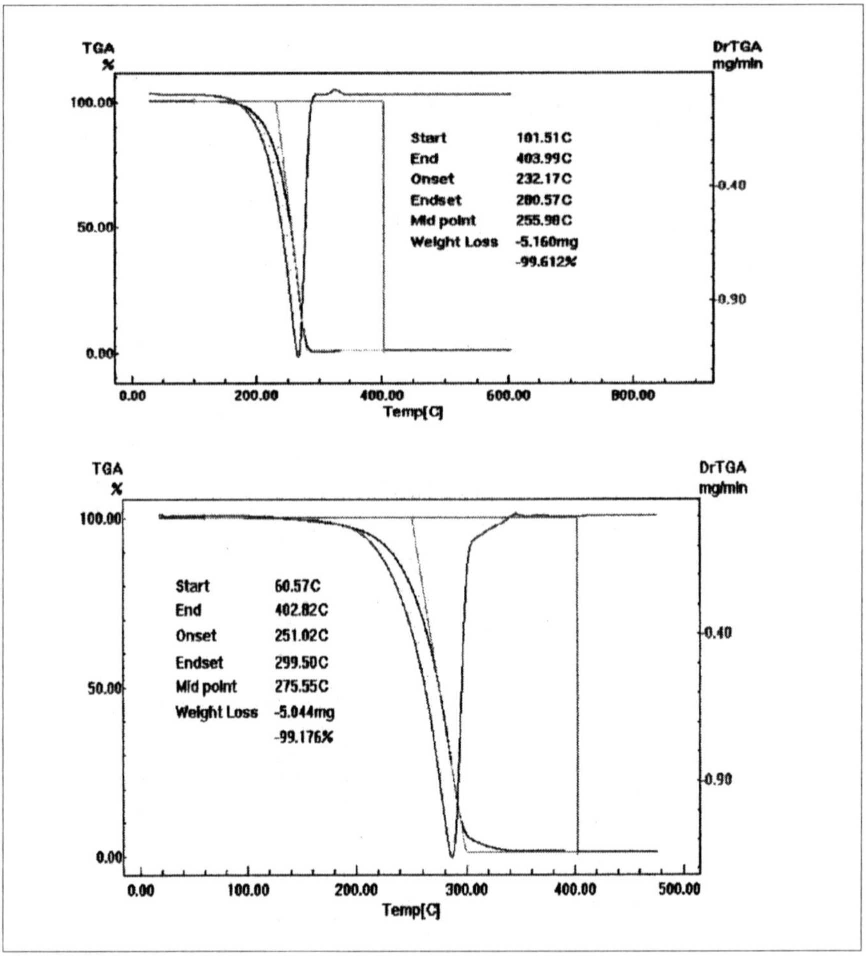

Figure 3. Thermal behavior of UV filters in the tested emulsion. Sample weight was 4.0-6.0 mg for TG/DTG and 1.0-2.0 mg for DSC. Curves are for TG/DTG (left) and DSC (right). Filters are benzophenone-3 (above) and octyl methoxycinnamate (below).

temperature decreases agglomeration between emulsion globules, resulting in decreased emulsion viscosity.

Heat degradation: TG/DTG and DSC curves can be observed in Figure 3. The thermic behavior of each component was related to weight variation at given temperature (TG/DTG) and to phase transition or changing of physical state (DSC).

Understanding the thermal profile of each chemical component of the emulsion could help in determining stability aspects of emulsion. The thermal degradation of emulsion component was directly proportional to a rise in temperature.

UV Spectroscopy: Figure 4 shows UV spectra of benzophenone-3 and octyl methoxycinnamate. UV spectroscopy was used to help identify chemical filters and

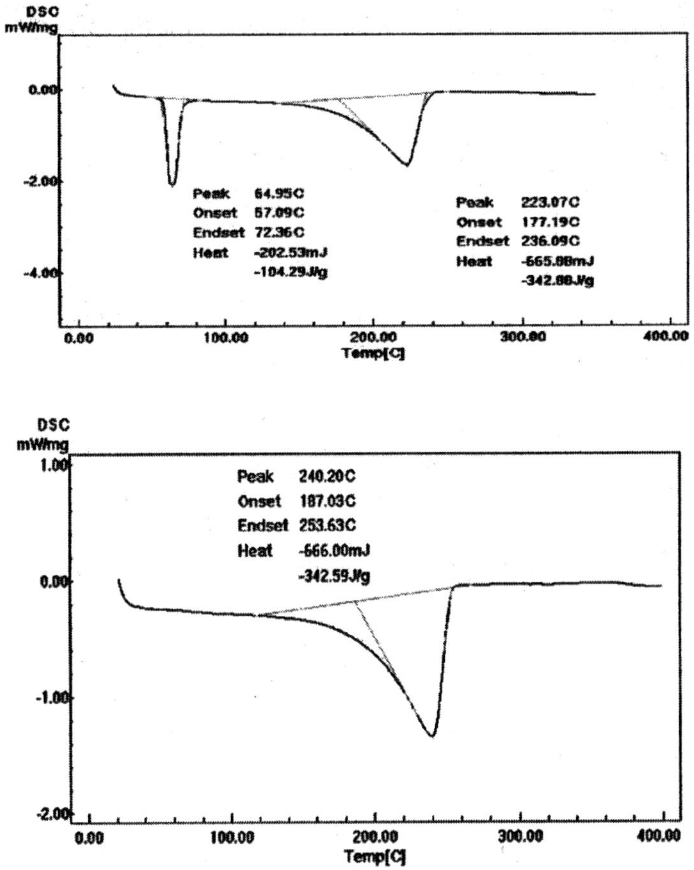

Figure 3. Continued

to determine their UV absorption profile. The determined absorption maxima was used as absorption wavelength in the HPLC method described above.

It was observed that with the degradation of chemical components there was a shift in absorption maxima.

HPLC: HPLC was used for the quantitative determination of both sunscreens in the emulsion and in the standard. Chromatograms are shown in Figure 5.

The accuracy of this method was determined through a recovery test. In a recovery test, known quantities of reference standards are added to the sample solution and analyzed through a proposed method, which was HPLC in this case. The recovery of standard is calculated based on the area obtained by injecting sample solution containing a known quantity of standard, and is expressed as recovered percentage.

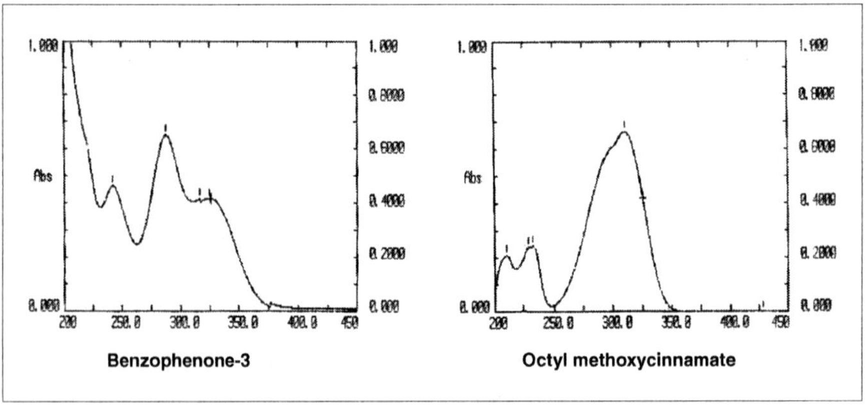

Figure 4. UV spectra of benzophenone-3 (10.0 µg/mL) and octyl methoxycinnamate (8.0 µg/mL) in 96% ethanol

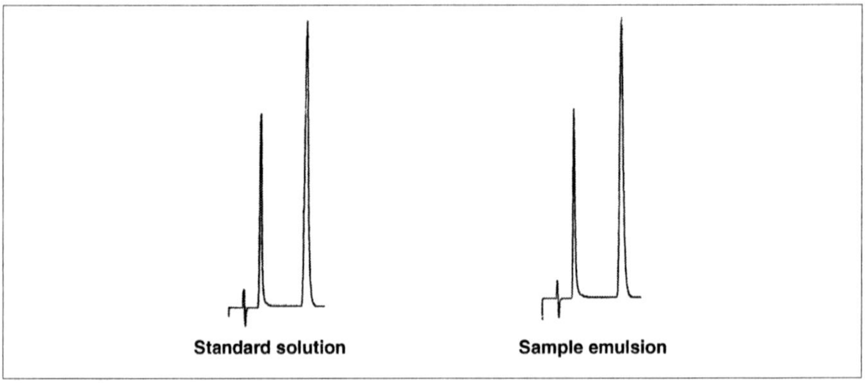

Figure 5. Chromatograms of (1) benzophenone-3 (10.0 µg/mL) and (2) octyl methoxycinnamate (20.0 µg/mL) in a standard solution and in the sample emulsion

The accuracy of a method is defined as the agreement between the results obtained by the proposed method and the reference value. The accuracy is one of the validation parameters recommended by US Pharmacopeia, AOAC International and ICH. One of the easiest ways to determine the accuracy of a method is through a recovery test.

Table 1 shows the results of recovery tests in the study. The percentage of standard recovery was calculated from the obtained areas in the chromatograms.[20] Analytical measurements of the amount of recovered standard confirmed the accuracy of the chromatogram area method. Table 1 shows the results of three tests with each of the sunscreen actives. The average percentage recovery of components was between 98.3 and 100.5%, thus confirming the accuracy of the method.

Table 1. Recovery of UV filter standard solutions added to a sample of the emulsion. Recovery was determined by HPLC.

UV filter standard solution	Standard added (mg/mL)	Standard found (mg/mL)	Recovery (%)
Benzophenone-3	6.0	5.92	98.70
8.0	8.04	100.53	
10.0	9.92	99.15	
Octyl methoxycinnamate	12.0	11.81	98.38
16.0	15.86	99.11	
20.0	19.79	98.93	

Note: HPLC analysis used a Merck LiChrospher 100 RP-18 (5 mm) column; methanol:water (87:13 v/v) as mobile phase; flow rate of 1.0 mL/min; and UV detection at 290 nm.

Summary

The aim of this work was the preparation and analytical evaluation of emulsions containing UVA, UVB and IR filters. Stability studies were performed, by varying time and temperature, using rheological measurements, thermal analysis, UV spectrophotometry and high performance liquid chromatography.

Rheological behavior was similar for emulsion samples stored at different times (up to two months) and temperatures (10°C, 23°C, 35°C and 45°C). Most of the samples showed thixotropic behavior.

Thermal analysis, such as thermogravimetry (TG) and differential scanning calorimetry (DSC), demonstrated the thermal decomposition and purity of the emulsion samples.

High performance liquid chromatography enabled the separation and quantitative determination of benzophenone-3 and octyl methoxycinnamate. The HPLC method we used to determine the degradation products of chemical filters was found to be helpful in accessing stability of formulations. The sample formulations were analyzed at specified temperatures and time intervals to determine the chemical filters concentrations in the formulations.

The techniques described here are useful for obtaining an analytical profile of solar protection emulsions. Such a profile is important for evaluating the physical and chemical stability of those emulsions.

—Maria Inês Rocha Miritello Santoro, Fátima Cristina Franco e Silva and Erika R. M. Kedor-Hackmann, *Department of Pharmacy, Faculty of Pharmaceutical Sciences, University of São Paulo, Brazil*

References

1. LC Junqueira and J Carneiro, *Histologia básica*, 8 ed, Rio de Janeiro: Guanabara Koogan (1995) p 301
2. G Pauly and M Pauly, Immunoprotection in daily-use cosmetics, *Cosmet Toil* 111(12) 47 (1996)
3. CR Taylor, RS Stern, JJ Leyden and BA Gilchrest, Photoaging/photodamage and photoprotection, *J Am Acad Dermatol* 22 1 (1990)
4. N Kollias and H Baqer, The role of human melanin in providing photoprotection from solar mid-ultraviolet radiation (280–320 nm), *J Soc Cosmet Chem* 39 347 (1988)
5. E Khury, TA Nakazawa and AMQR Silva, Fotoprotetores de alta eficiência, *Cosmet Toil* (Ed Port) 7(4) 41 (1995)
6. RM Sayre, N Kollias, RL Roberts, A Baqer and I Sadiq, Physical sunscreens, *J Soc Cosmet Chem* 41 103 (1990)
7. X Briand and NM Secma, A skin care constituent from the Sea, *Drug Cosmet Ind* 154 38 (1994)
8. C Trullas, J Coll, R Del Rio, M Lecha and C Pelejero, Evaluating the IR photoprotection factor of physical sunscreens by bioengineering methods, in Proceedings of the 18th IFSCC Congress in Venice in 1994, vol 1 (1994) p 73
9. L Violin, F Girard, P Girard, JP Meille, and M Petit-Ramel, Infrared photoprotection properties of cosmetics products: correlation between measurement of the anti-erythemic effect in vivo in man and the infrared reflection power in vitro. *Int J Cosmet Sci* 16 113 (1994)
10. LH Kligman, Intensification of ultraviolet-induced dermal damage by infrared radiation, *Arch Dermatol Res* 272 229 (1982)
11. PE Miner, Emulsion rheology: Creams and lotions, in *Rheological Properties of Cosmetics and Toiletries*, D Laba, ed, New York: Marcel Dekker (1993) p 313
12. HN Naé, Introduction to rheology, in *Rheological Properties of Cosmetics and Toiletries*, D Laba, ed, New York: Marcel Dekker (1993) p 9
13. M Nebuloni, Tecniche termoanalitiche per lo studio di prodotti farmaceutici, *Boll Chim Farm* 129 87 (1990)
14. AH Förster and TM Herrington, Rheology of two commercially available cosmetic oil in water emulsions, *Int J Cosmet Sci* 20 317 (1998)
15. M Primorac, LJ Dakovic, M Stupar and D Vasiljevic, The influence of temperature on the rheological behaviour of microemulsions, *Pharmazie* 49 780 (1994)
16. J Canotilho, ATB Sousa and JAMC Pinto, Análise térmica. Interesse em tecnologia farmacêutica, *Rev Port Farm* 42 5 (1992)
17. D Giron, Thermal analysis in pharmaceutical routine analysis, *Acta Pharm Jugosl* 40 95 (1990)
18. M Lonashiro and I Giolito, A nomenclatura em análise térmica - Parte II, *Cerâmica* 34 163 (1988)
19. WW Wendlant, *Thermal Analysis*, 3rd ed, New York: Wiley (1986) p 814
20. R Jiang, CGJ Hayden, RJ Prankerd, MS Roberts and HAE Benson, *J Chromatogr*, B 682 137 (1996)
21. V Vanquerp, C Rodriguez, C Coiffard, LJM Coiffard and Y Holtzhauer, *J Chromatogr*, A 832 273–277 (1999)

Photostability of Menthyl Anthranilate in Different Formulations

Keywords: menthyl anthranilate, UVA absorber, photostability, method, active

Menthyl anthranilate is relatively stable in its pure form and an effective UVA absorber. This article explores the photostability of menthyl anthranilate in different formulations irradiated by artificial UVA light.

Sunscreens are widely used to protect human skin from the harmful effects of ultraviolet radiation. In recent years, organic UV absorbers have seen increased use because of the need for skin protection in preventing phototoxic skin reactions, skin cancer and premature aging of the skin.[1]

In photobiological studies, the UV radiation spectrum is divided into three bands: UVC (200-290 nm), UVB (290-320 nm) and UVA (320-400 nm).[2] When we are outdoors we are exposed to both UVA and UVB radiation. UVA penetrates the skin more efficiently than UVB and it penetrates into the reticular dermis, where the histological changes of solar elastosis are seen.[3] The increasing concern about the deleterious effects of exposure to UVA radiation in human skin has led to studies on topical sunscreens prepared with organic-chemical UVA/UVB absorbers.[4]

Sunscreens can be examined in two main categories: organic or chemical filters against UVA and UVB rays, and inorganic or mineral light-protection filters. Chemical sunscreens are more cosmetically appealing and can selectively absorb UVB and/or UVA. They reduce sunburn and allow longer periods of sun exposure.[5]

We will show that menthyl anthranilate (MA) is an effective UVA absorber. It is oil-soluble and can be used in waterproof sunscreens. When used in combination with UVB sunscreens, it provides broad-spectrum protection.[6] If it is also photostable, it can provide long-lasting protection.[1,7]

Determining the Photostability of MA

The aim of this study was to investigate the photostability of MA and its formulations under artificial UVA by using a TLC-scanner. Additionally, we examined whether different formulations affected the photostability, and if the position of the UV curve changed as a result.

Table 1. Composition of MA formulations

	F1 Cream	F2 Lotion	F3 Gel
Cyclomethicone	10%	-	-
Mineral oil (*paraffinum liquidum*)	-	10%	-
Cetyl alcohol	3	0.5	-
Stearic acid	15	3	-
Lanoline	2	1	-
Glycerine	5.5	2	-
Cetyl trimethyl ammonium chlorur (Carsoquat CT-429, Johnson & Johnson)	3	2	-
Menthyl anthranilate	4	4	4%
Propylene glycol	-	-	39
Carbomer (Carbomer 954, Spectrum)	-	-	2
Methylparaben	0.15	0.15	0.15
Water *(aqua)*, distilled	qs 100.00	qs 100.00	qs 100.00

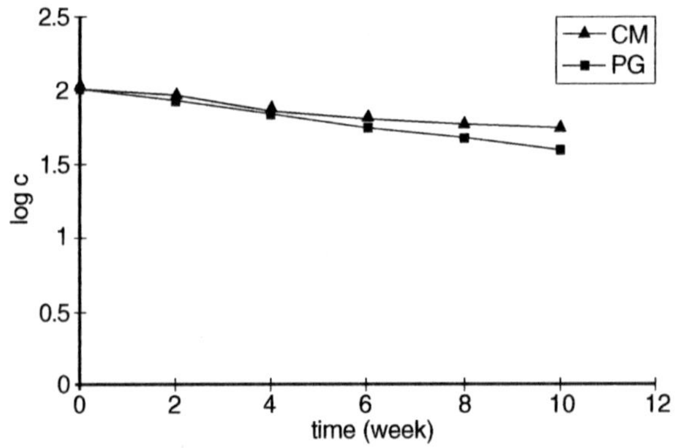

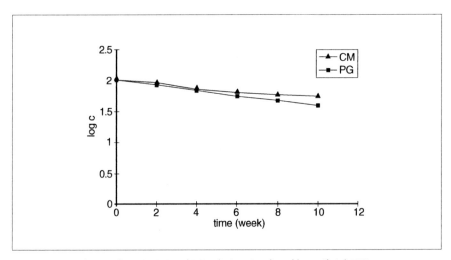

Figure 1. Degradation of MA in CM and PG solutions irradiated by artificial UVA

The proposed method, previously used in our laboratory, was found to be a simple, rapid and accurate method for the assay of active substances.[8] In this work, solutions of MA and three different formulations containing MA were exposed to artificial UVA from a fluorescent lamp, and then the degradation of MA was measured.

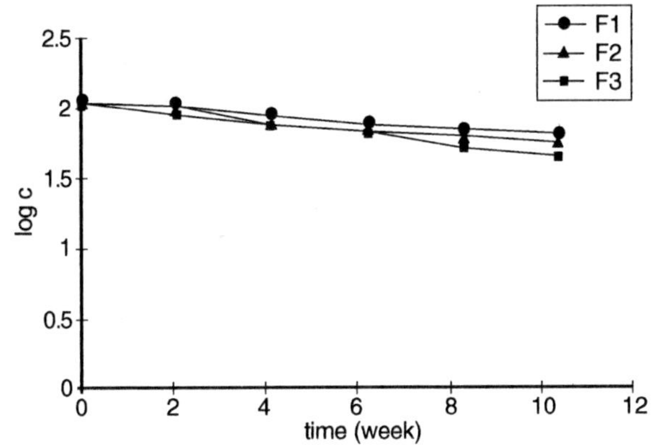

Figure 2. Degradation of MA in cream (F1), lotion (F2) and gel (F3) formulations irradiated by artificial UVA

Materials and Methods

The UV filter we used was MA, a 5-methyl-2-(1-methylethyl) cyclohexanol-2-aminobenzoate.[a] A cetyl trimethyl ammonium chlorur[b] and all other chemicals used in the formulations were purchased from the supplier of the MA.

Diethylether, chloroform and silicagel HF 254 were from another supplier.[c] All other materials were of analytical grade.

We prepared a cream (F1), a lotion (F2) and a gel (F3) according to formulations widely used in commercial cosmetics (Table 1).

TLC Assays

For thin layer chromatography (TLC) silicagel HF 254 was used as an adsorbent. N-hexane:diethylether:ammonia (7:2:0.1) system was used as the developing solvent. Visualization was accomplished by fluoresence quenching under UV light at 254 nm.

The area of the spots was measured at 338 nm on a TLC-scanner[d] (D2-lamp, Lamda=340 nm, Mode=ABS, AZS=on, Linearizer=0).

Photostability Studies

Eleven Philips UVA flourescent bulbs (40 watt) were used as a light source in the light stability test cabinet. Because they irradiate uniformly in the 320-400 nm range, these bulbs are a reasonable substitute for natural sunlight in this wavelength

[a] Purchased from Johnson & Johnson, Istanbul, Turkey
[b] Carsoquat CT-429, purchased from Johnson & Johnson
[c] Purchased from Merck, Darmstadt, Germany
[d] CS-920, Shimadsu, Japan

range. Temperatures in the range of 40°-80°C have been reported in the literature on acceleration tests for chemical changes of an active ingredient. We chose to maintain a temperature of 40°C ± 2 in the test cabinet.

Determining MA in solutions: Cyclomethicone (CM) and propylene glycol (PG) solutions of MA in glass bottles were placed in the cabinet along with the three formulations and a reference solution of MA, also in glass bottles. They were positioned 10 cm from the light source. Then the percent degradation was determined weekly by the TLC technique described earlier.

We ran the photostability study for 10 weeks, which is comparable to the three-month period typically used in the industry for chemical stability tests. At the end of ten weeks, MA and the degradation products were identified using the TLC technique. The Rf values of MA and the degradation products were also determined. Rf is a ratio of two distances: The numerator is the distance from the initial contact with the solvent to the center of the spot; the denominator is the distance that the solvent has traveled. Rf is a characteristic value for each compound.

To examine the effect of the temperature alone, unirradiated CM and PG solutions of MA were kept at 40°C and the results were compared with those of the photostability studies.

Determining MA in formulations: 0.25 g of the formulation sample was extracted with 4 mL methanol and shaken for 1 hr in a horizontal shaker.[e] 0.25 g sodium sulfate was then added and the mixture was left overnight. 10 μL of clear supernatant solution was used for TLC. After elution with the solvent system and location by visualizing under UV light at 254 nm, the determination of MA was made using the TLC technique mentioned above.

The UV absorption spectrums of MA in the formulations before and after irridation were also compared.[f] The spot visualized was physically removed (scraped) and eluted with 10 mL of 96% ethanol for analysis by a spectrophotometer.[f]

Results and Discussion

It has been shown that sunscreens can interact with vehicle components, and those interactions can affect the efficacy of the sunscreens.[9,10] Furthermore, it has been shown that sunscreens will undergo degradation as a result of being exposed to UV light.[10-12] Even though MA is stable as a pure form, it shows degradation of 15.1%, 17.1% and 14.8% in solutions of ethanol-water, isopropyl myristate and mineral oil, respectively.[10]

Degradation of MA in solutions: In this study, we determined the percentage of MA degradation in CM solution and in PG solution. It was found to be 45.24% and 61.74%, respectively, at the end of the three months. Two degradation products were formed in both solutions. The Rf values of MA and the degradation products were found to be 0.31, 0.12 and 0.4, respectively. Although the Rf values indicate that there were two degradation products, we did not do the mass spectrum and NMR analysis to identify these products.

[e] Kühner mini-shaker, B. Braun, Germany
[f] UV-1028 spectrophotometer, Shimadzu, Japan

The plots of concentration of the MA solutions versus time are given in Figure 1. A linear relation was obtained when the equation for the first order reaction was applied.

We found no degradation products in the MA solutions kept at 40°C. This suggests that the degradation was caused by UVA radiation.

Degradation of MA in formulations: The photostability of chemical sunscreens in formulations is of great interest. Therefore we investigated the degradation of MA in three different formulations in the light stability test cabinet at 40°C. We found that MA was relatively stable in these formulations; the MA degradation that occurred was found to be slower than in the PG solution. By the way, the same degradation products found in the solutions were also observed in the formulations.

At the end of three months, the losses of MA were 40.01% in F1, 47.87% in F2, 57.57% in F3 (Figure 2). The degradation was fitted to first order reaction. F1, F2 and F3 were all found to be stable. These results were in good accordance with those reported in the literature.[13]

MA is *ortho*-disubstituted. This provides excellent internal stabilization via hydrogen bonding. As a result, when it is placed into various solvents MA does not undergo a shift in its maximum absorbance, as do *para*-disubstituted sunscreens. Therefore, any significant solvent shifts were seen only after irradiation.[14,15]

UV absorption spectrum of MA in formulations: In this study, we obtained the UV absorption spectrums of MA before irradiation (in ethanol) (Figure 3) and after irradiation (in formulations) (Figures 4, 5 and 6). No significant difference was observed between the spectrums. The wavelength of maximum absorbance was found to be the same. In the spectrum of F3, the shoulder seen in the peak wavelength band at 249 nm suggested degradation due to irradiation.

We also measured the UV absorbances of MA in formulations at 338 nm in order to see if the irradiation-induced changes in absorbance were consistent from formulation to formulation at a different peak in the MA absorption spectrum. At 338 nm, the absorbance of MA in formulations decreased gradually during this period. A larger decrease of absorbance in F3 was noted. These results were in good accordance with the results of the TLC technique (Figure 2). It could be suggested that MA and the degradation products had almost the same chromophoric properties and the concentrations of degradation products were not too high. More work along these lines is in progress.

Conclusion

This study was based on three different formulations of MA. The study had three goals: to investigate the photostability of MA in CM and PG solutions; to investigate the photostability of MA in three formulations under artificial UVA by TLC-scanner technique; and to examine the UV absorbance spectrums of MA in formulations before and after irradiation.

We found that degradation was fitted to the first order reaction in all cases. The loss was mostly found in PG solution and in the gel formulation.

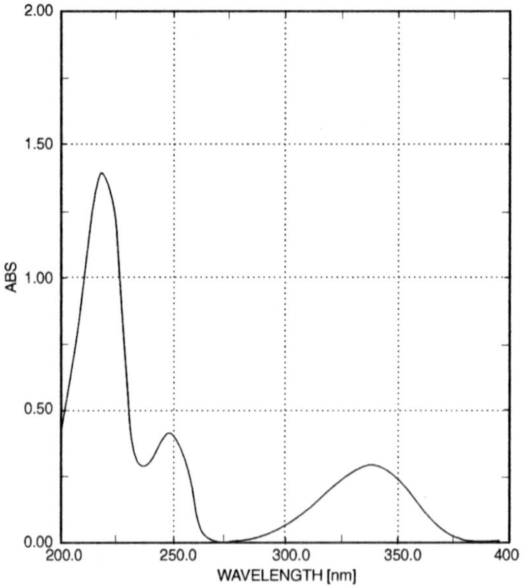

Figure 3. UV absorption spectrum of MA in ethanol before UVA irradiation

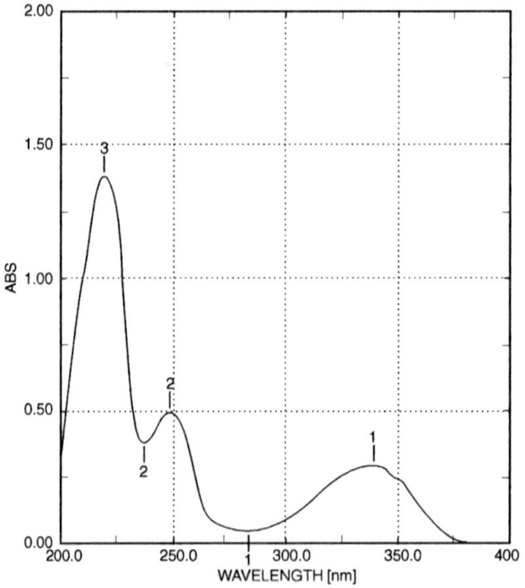

Figure 4. UV absorption spectrum of MA in F1 cream after UVA irradiation

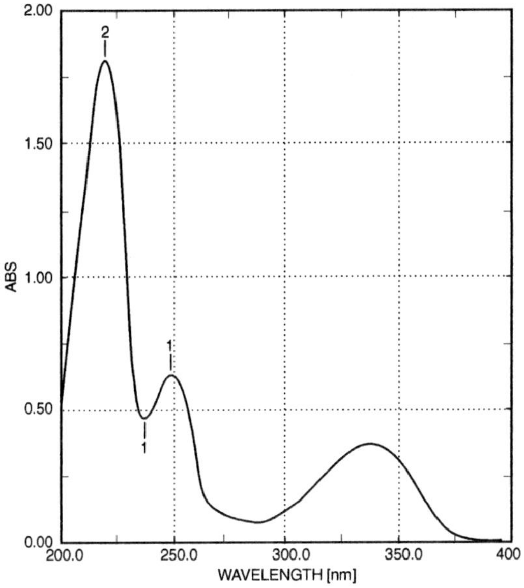

Figure 5. UV absorption spectrum of MA in F2 lotion after UVA irradiation

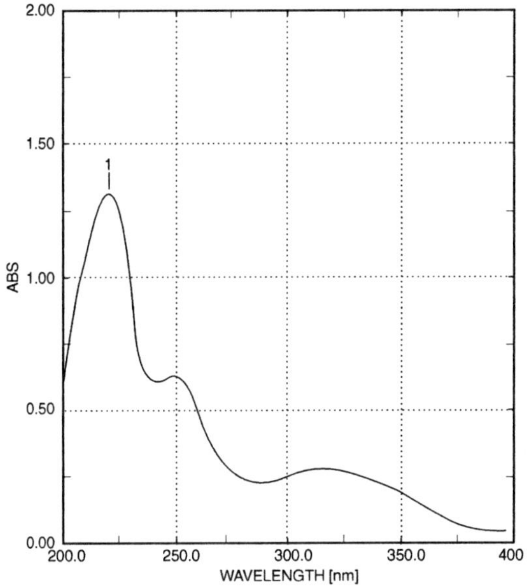

Figure 6. UV absorption spectrum of MA in F3 gel after UVA irradiation

From this study, we learned that the cream form of MA was the most suitable formulation for protecting against UV radiation. We also learned that even if the solutions of MA were stable over three months at 40°C, MA was not photostable when irradiated with UVA.

—Ozgen Ozer, Dilsen Menemenlioglu and Gokhan Ertan,
Pharmacy Faculty, University of Ege, Izmir, Turkey

References

1. ME Hidalgo, I Gonzales, F Toro, E Fernández, H Speisky and I Jimenez, Boldine as a sunscreen, *Cosmet Toil* 113(9) 59-66 (1998)
2. M Turkoglu, AA Sakr, JL Lichtin, EV Buchler and JJ Kruezmann, An in-vivo assessment of the sun protection index, *Cosmet Toil* 104(4) 33-38 (1989)
3. BL Diffey, A method for broad spectrum classification of sunscreens, Int *J Cosm Sci* 16 47-52 (1994)
4. NA Shaath, Encyclopedia of UV absorbers for sunscreen products, *Cosmet Toil* 102(3) 21-26 (1987)
5. TM Macleod and W Frainbell, A study of chemical light screening agents, *J Dermatol* 92 417-425 (1975)
6. DC Steinberg, Sunscreen encyclopedia regulatory update, *Cosmet Toil* 111(12) 77-88 (1996)
7. M Stockdale, UVA Sunscreens-Methods for assessing their efficacy, *Cosmet Toil* 102(3) 111-115 (1987)
8. O Ozer, S Asıcı and G Ertan, The hydrolysis of tenoxicam in acidic and alkaline solutions and its stability in artificial gastric and intestinal media, *Sci Pharm* 64 71-82 (1996)
9. NA Shaath, On the theory of ultraviolet absorption by sunscreen chemicals, *J Soc Cosm Chem* 38 193-207 (1987)
10. NA Shaath, HM Fares and K Klein, Photodegradation of sunscreen chemicals, *Cosmet Toil* 105(12) 41-44 (1990)
11. A Deflandre and G Lang, Photostability assessment of sunscreens, benzylidene camphor and dibenzoyl methane derivatives, *Int J Cosm Sci* 10 53-62 (1988)
12. A Kammeyer, W Westerhof, PA Bolhuis, AJ Ris and H Hische, The spectral stability of several sunscreening agents on stratum corneum sheets, *Int J Cosm Sci* 9 125-136 (1987)
13. M Turkoglu, Design and quality control of sunscreen formulations, *T Clin Cosm* 1 108-111 (1998)
14. K Klein, Menthyl anthranilate, *Cosmet Toil* 105(12) 75-77 (1990)
15. LE Agrapidis-Paloympis and RA Nash, The effect of solvents on the ultraviolet absorbance of sunscreens, *J Soc Cosmet Chem* 38 209-221 (1987)

Photostability of Sun Filters Complexed in Phospholipids or Beta-Cyclodextrin

Keywords: photodegradation, photostability, sun filters, safety factor, skin sensitivity, irritation

Laboratory tests of three commonly used sun filters show that the filters encapsulated in β-cyclodextrin show a photostability significantly higher than that of the same filters uncomplexed or complexed with phospholipids.

The sun produces a great quantity of energy, which is transmitted by radiation of various types (see sidebar). When sun rays reach human skin, they interact with the epidermal cells. They are also able to interact with sun filters, which are substances applied on the skin to prevent sunburn and erythema. UV radiation, in particular, produces a high activation energy and induces a partial degradation or alteration process in the filter so that the filter reduces or loses its capacity for skin protection.[1] This process is called photodegradation.

The photodegradation of the sun filter molecule originates sub-molecules potentially dangerous for the skin because they may induce sensitization and skin irritation. Consequently, the photostability of sun filters is very important in evaluating the protection period, but it also represents a safety factor involving skin sensitivity and irritation.

Several techniques are available to reduce photodegradation:

- Different sun filters can be used in the same product so that a synergistic effect is obtained.
- Specific sun filters that are stable at a specific wavelength can be used.
- The sun filter molecule can be protected by complexing or encapsulating it with specific substances. This process protects the filter molecule from radiation so that the interactions are reduced or eliminated.

Filter complexing or encapsulating is achieved by interaction with substances such as phospholipids or cyclodextrins. These are able to produce closed structures in an aqueous solution. Because of their bipolar chemical structure, phospholipids are able to produce a vesicle structure (liposomes), while cyclodextrins are able to produce channel/cage structures. Inside this structure the sun filter can be situated so that it is protected from the external sun rays.

The aim of the study reported here was to compare complexed versus non-complexed sun filters on the basis of their stability and activity persistence following exposure to UVA and UVB radiation. Three sun filters were tested: octyl methoxycinnamate (OMC), butyl methoxydibenzoylmethane (BMDM), and benzophenone-3 (BEN3). Two types of complexes were used: phospholipids[a] and β-cyclodextrines.[b] They were chosen on the basis of further advantages, such as increased bioavailability for cyclodextrins[2] and better skin adhesiveness and water resistance for phospholipids.[3]

Materials

The sun filters and complexes tested in this study are listed in Table 1, along with their suppliers. The UVA radiation source was a high–emission 8 W lamp type BLB, with emission between 315 and 400 nm, a maximum peak at 352 nm and a

Table 1. Tested products and suppliers

Sun filter or complexed filter	Supplier
Octyl methoxycinnamate (OMC)	Merck
Butyl methoxydibenzoylmethane (BMDM)	Merck
Benzophenone-3 (BEN3)	Merck
Phospholipid complex of OMC	I.R.A.
Phospholipid complex of BMDM	I.R.A.
Phospholipid complex of BEN3	I.R.A.
Cyclodextrin complex of OMC	I.R.A.
Cyclodextrin complex of BMDM	I.R.A.
Cyclodextrin complex of BEN3	I.R.A.

Table 2. Radiation energy to which each sample has been exposed

Radiation	Wave length (nm)	Lamp power (μW/cm² at 1 m)	Produced energy 30 min/m² at 1 m (J/m²)	Produced energy 30 min/m² at 0.5 m (J/m²)
UVA	315-400	300	2,700	5,400
UVB	280-305	12	216	432

[a] Liposystem complex is a registered trademark of IRA, Usmate Velate, Italy.
[b] Cyclosystem complex is a registered trademark of IRA.

radiation power of 300 μWatt/cm^2 at 1 m distance. The UVB radiation source was a 7.9 W lamp with emission between 315 and 280 nm, a maximum peak at 306 nm and a radiation power of 12 μWatt/cm^2 at 1 m distance.

Methods

The test was an attempt to reproduce a realistic situation by applying to glass slides a thin layer of sun filter solution, which was then irradiated. The photostability of sun filters depends on the specific chemical stability of the filter, on the radiation intensity, and also on the filter thickness. Filter thickness is a factor because the total radiation absorbed by a substance results from the absorption index (specific to the substance) and the amount of substance that the UV radiation must cross. The theoretical thickness of the sun filters applied on the slides was around 168 µm, which corresponds realistically to the thickness obtained by a normal sun product application on the skin before sun exposure.

Sample preparation: For the UVA tests, 36 samples were prepared on glass slides—four samples for each of the 9 tested sun filters. Two samples were irradiated with the minimum dose, and two with the maximum dose. For the UVB tests, 36 additional samples were prepared and similarly radiated. 300 mg of test product (corresponding to 100 mg of uncomplexed sun filter) was applied upon each slide, with a surface of 18.75 cm^2, generating a product thickness of 168 µm. To have comparable weights between complexed and uncomplexed filters, the uncomplexed filter was dispersed in an inert substance (maltodextrin).

In order to obtain a homogeneous product distribution on the slide, the sun filters, having been dissolved in ethanol, were applied uniformly upon the slide and then vacuum dried. After irradiation, the filters were dissolved in absolute ethanol, filtrated at 0.45 micron and brought back to an ethanol volume of 10 mL.

Samples were applied in a known quantity in order to have always the same concentration of the different sun filters on the slides.

Irradiation dosages: Each sample was irradiated for 30 min at 1 m distance and for 30 min at 0.5 m distance, so that the irradiation dose was doubled. The minimum dose of radiation energy was calculated as follows for UVA exposures:

$$\begin{aligned} \text{Energy}_{\text{min}} &= \text{Lamp power x exposure time} \\ &= 300\ \mu\text{W x 900 sec} \\ &= 270{,}000\ \mu\text{J/cm}^2 \\ &= 2700\ \text{J/m}^2 \end{aligned}$$

The maximum dose was double the minimum dose. Table 2 shows the calculated energies for UVA and UVB at 1 m and 0.5 m.

Radiation dosage estimation was based upon the amount of radiation required to produce sunburn in an individual exposed to 5.6 h of Mediterranean summer sun.

Sun filter determination: Both before and after UV irradiation, the sun filter concentration (w/w) was determinated by HPLC[c] (High Pressure Liquid Chromatography) and by absorbance and transmittance spectrophotometry. By

Solar Radiation

The sun's energy is transmitted by radiation. Sun rays are a set of different radiations characterized by a specific wavelength, which is measured in nanometers (nm). The spectrum of sun energy is shown in Table 3. The maximum radiation intensity is at approximately 450 nm.

Part of the solar radiation—X-rays, γ-rays, UVC rays and some of the UVB rays—is absorbed by the ozone layer. The other part reaches the Earth with energy strong enough to be able to interact with the cells of human skin. This type of interaction may induce the production of free radicals, which are molecules able to prime hydrolytic chain reactions. Due to their highly energetic status, free radicals can damage cell membranes and DNA. In fact, free radicals are mainly responsible for erythema and abnormal cell proliferation, which may result in skin tumors.

The quantity of radiation energy can be described as follows:

$$E = hv$$

where E is the radiation energy, h is the constant of proportionality and v is the radiation frequency.

Another form of this statement is as follows:

$$E = hc/\lambda$$

where c is light speed and λ is the radiation wavelength.

The radiation energy is inversely proportional to its wavelength. This means that radiation with a short wavelength has a high energy and is able to cause more problems when it reaches the skin.

Sun rays reach the earth with different angles of incidence depending on the season. The incidence goes from 0° at the equator to a maximum of 42° at poles. Part of this radiation is absorbed during passage through the atmosphere and transformed into different kinds of energy. Infrared radiation above 1500 nm is absorbed by the atmospheric moisture and by CO_2, while UV rays shorter than 175 nm are absorbed by oxygen. UVC rays between 175 and 290 nm are absorbed by the ozone layer; 286 nm is the shortest wavelength found in the polar regions. Long UVC and short UVB wavelengths able to reach the earth vary from 287 to 305 nm, due to the seasonal variations of the ozone layer.

The absorbed sun radiation diffuses in the atmosphere, but it maintains its high energy, enabling it to cause erythema even if direct sun rays are not present.

When formulating sun protecting products, the following factors should be considered:

- Quantity of radiation on the earth surface (related to factors such as latitude, hour, cloudiness, humidity);
- Energy level of each radiation type (related to its wavelength);
- Specific sunburn power of each wavelength (see Table 4).

However, it must be also considered that SPF is calculated without taking into account the time factor, but it depends on the stability in time of the sun filter. Therefore, in order to guarantee a real SPF, it is very important that a photostable sun filter is used.

Table 3. Types of solar radiation and their wavelength ranges

Radiation	Wavelength range
γ-rays	500 fm - 50 pm
X-rays	50 pm - 10 nm
UVC	100-280 nm
UVB	280-315 nm
UVA	315-400 nm
Visible rays	400-760 nm
Infrared rays	760 nm - 1 mm
Microwave transmission	10-300 mm
Television broadcast	0.3-10 m
Radio broadcast	10 m - 100 Mm

Table 4. Selected types of solar radiation, the percentage they represent of the total solar radiation, and the energy of each required to produce erythema.[4]

Radiation type	Wavelength (nm)	Energy %	Energy necessary for erythema (μW)
Infrared near and far	>780	44	-
Visible	380-780	51	-
UVA	320-380	4	1,000-10,000
UVB	290-320	0.4	1-100

comparison with a reference product, these analytical techniques allow the identification and quantification of the unaltered sun filter, either in the free or complexed form. Since the only treatment applied to the samples was UV irradiation, a reduction of sun filter amount in irradiated samples is considered to indicate the instability of that filter to UV rays.

A standard solution (an uncomplexed reference product solution from Merck) was prepared by weighing exactly 100 mg of each reference sun filter, pouring it in a 100 ml flask and dissolving it with 100 ml of isopropanol. Then 2 ml of the solution were diluted in 100 ml of isopropanol. This represents a 100% pure sun filter solution.

For the HPLC analysis of the filters, the column was RP18, 25 cm length, 5 mm diameter, 5 µm mesh. The liquid phase was a mixture of methanol and water in an 80:20 ratio. The detector was set for UV at 313 nm.

Technicians injected 20 µL of standard solution, and repeated the injections until a uniform response was obtained. Then they injected 20 µL of sample solution (prepared by dissolving the test product in isopropanol and filtrating it with Millipore at 0.45 µm) until a reproducible response was obtained. At the detector fixed wavelength, the response is visualized as a peak, whose area is proportional to the concentration of the substance in the solution.

After determining the peak area related to the sun filter in the standard solution (As) and in the sample solution (Ac), one can calculate the dosage through the following formula:

$$\text{\% Sun filter in the sample} = \frac{AC \times Ps}{AS \times Pc} \times 100$$

where:

Ps = weight of the standard in the injected solution

Pc = weight of the sample in the injected solution

A decrease of sun filter concentration in a sample after UV irradiation is indicative of photoinstability, because it means that the missing amount was modified by UV rays so that it cannot be identified any more by the analytical instrument.

Results

The obtained data are reported in Table 5 and graphed in Figures 1, 2 and 3. In every case, the cyclodextrin-complexed filter was more stable or at least as photostable as the other filters, and the improvement was most noticeable in cases of high-energy UVB, whereas in the case of low-energy UVA the results showed fluctuation or little increase, difficult to accurately define because of the small percentage difference. This is easily explained by the fact that, of course, the lower energy of UVA has a smaller impact on the chemical structure of any sun filter molecule as confirmed by the two cases (BEN3 at low-dose UVA and BMDM at high-dose UVA), in which complexing appeared to have no effect on filter photostability.

[c] Chromatograph SP 8450, Spectra Physic, San Jose, CA USA.

Table 5. Residual percentage of complexed and uncomplexed sun filters after selected doses of UVA and UVB irradiation

		Residual sun filter after irradiation at these doses (J/cm²)			
Filter	Complexing agent	2,700 UVA	5,400 UVA	216 UVB	432 UVB
	none	98.5%	98.0%	88.0%	70.5%
OMC	phospholipid	97.5	97.0	89.5	70.0
	cyclodextrin	99.5	99.5	95.0	91.5
	none	97.0	97.0	91.0	72.5
BMDM	phospholipid	97.0	97.0	90.0	73.5
	cyclodextrin	98.0	97.0	95.0	87.5
	none	98.0	97.5	86.0	71.5
BEN3	phospholipid	98.0	96.0	86.0	70.0
	cyclodextrin	98.0	97.5	93.0	81.0

Legend

OMC = octyl methoxycinnamate

BMDM = butyl methoxydibenzoylmethane

BEN3 = benzophenone-3

Phospholipid complexing improved filter photostability in only 2 cases out of 12 after irradiation. In 6 of the cases, the phospholipid-complexed filter was less stable than the uncomplexed filter, and in 4 cases phospholipid complexing made no difference.

Octyl methoxycinnamate: When irradiated with UVA, the photostability of this filter (Figure 1) progressively increased from OMC in a phospholipid complex, to uncomplexed OMC, to OMC in a cyclodextrin complex. When irradiated with UVB, the photostability of OMC in a cyclodextrin complex was markedly higher than OMC either uncomplexed or in a phospholipid complex, where the photostability values were comparable.

Butyl methoxydibenzoylmethane: Under UVA irradiation, this filter (Figure 2) appeared to be the most photo-unstable of the three filters; it was also the least affected by complexing. When irradiated by UVB, the uncomplexed BMDM was the most photostable of the three filters.

Benzophenone-3: Under UVA irradiation, this filter (Figure 3) showed no photostability improvement from complexing; in fact, the only change seen was a reduction in photostability when BEN3 was complexed in phospholipids. Under UVB irradiation, even in a complex, BEN3 was the least photostable of the tested filters.

Summary: The data show that complexing the filter with cyclodextrins increases considerably its photostability, while complexing it with phospholipids doesn't

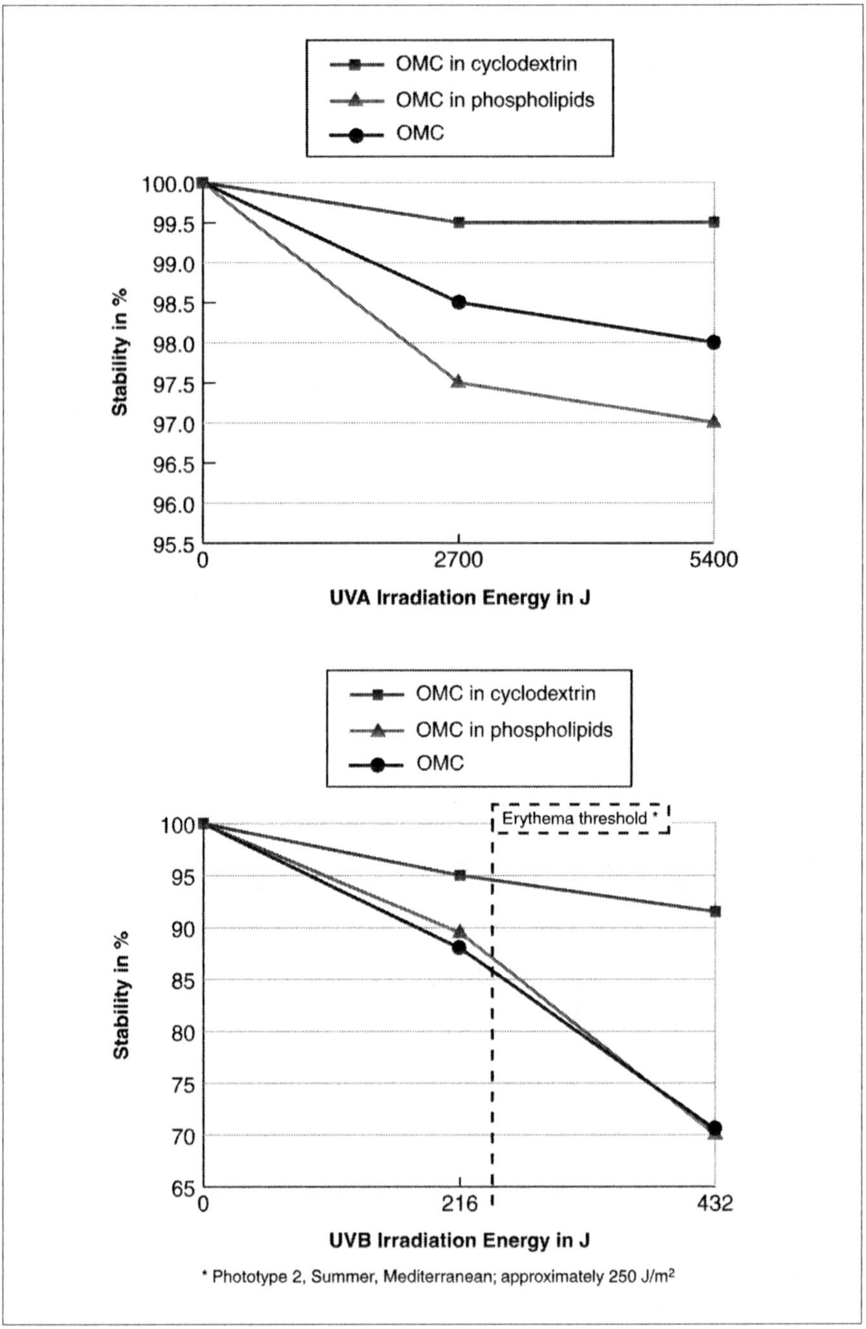

Figure 1. Photostability of octyl methoxycinnamate to UVA (above) and UVB (below)

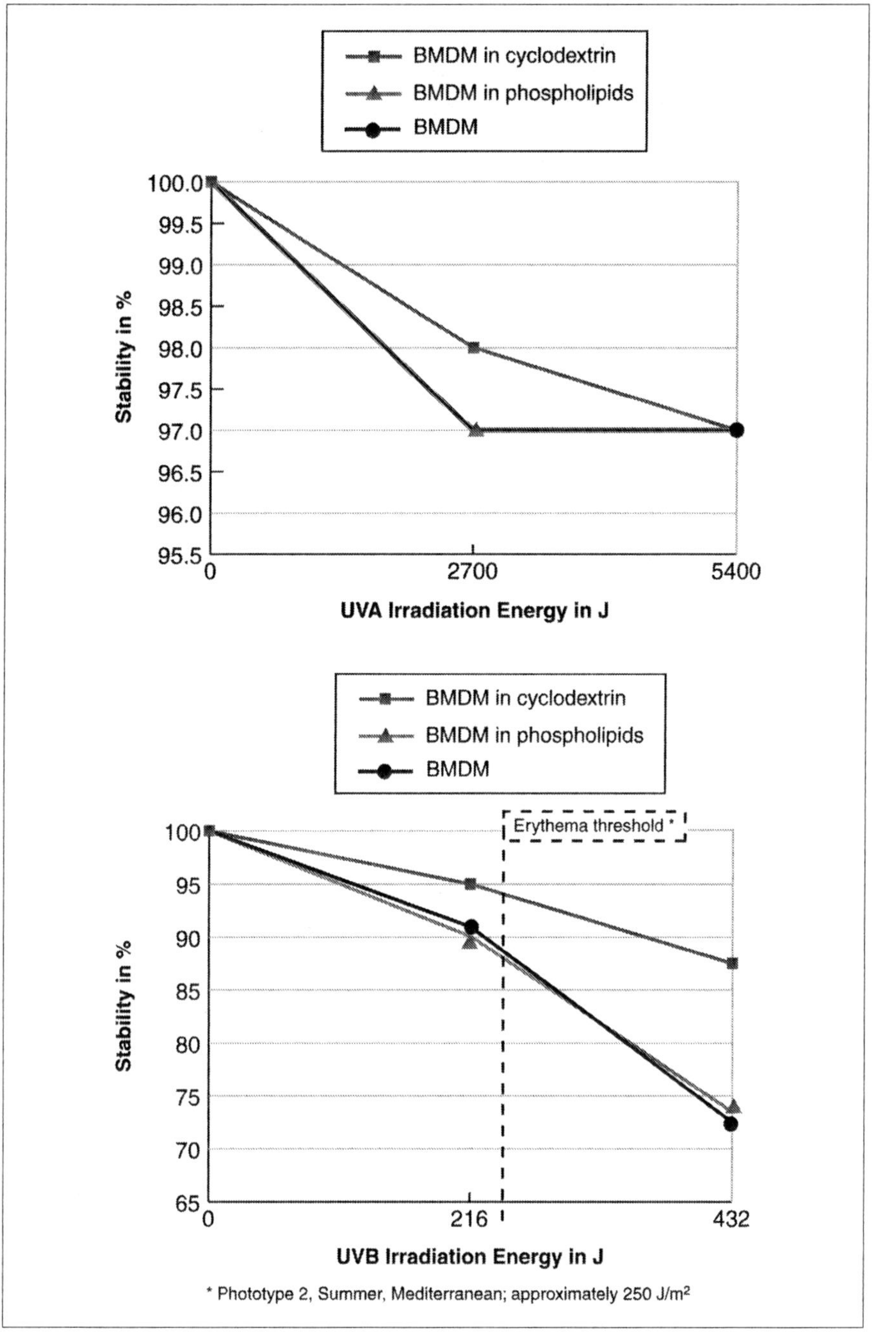

Figure 2. Photostability of butyl methoxydibenzoylmethane to UVA (above) and UVB (below)

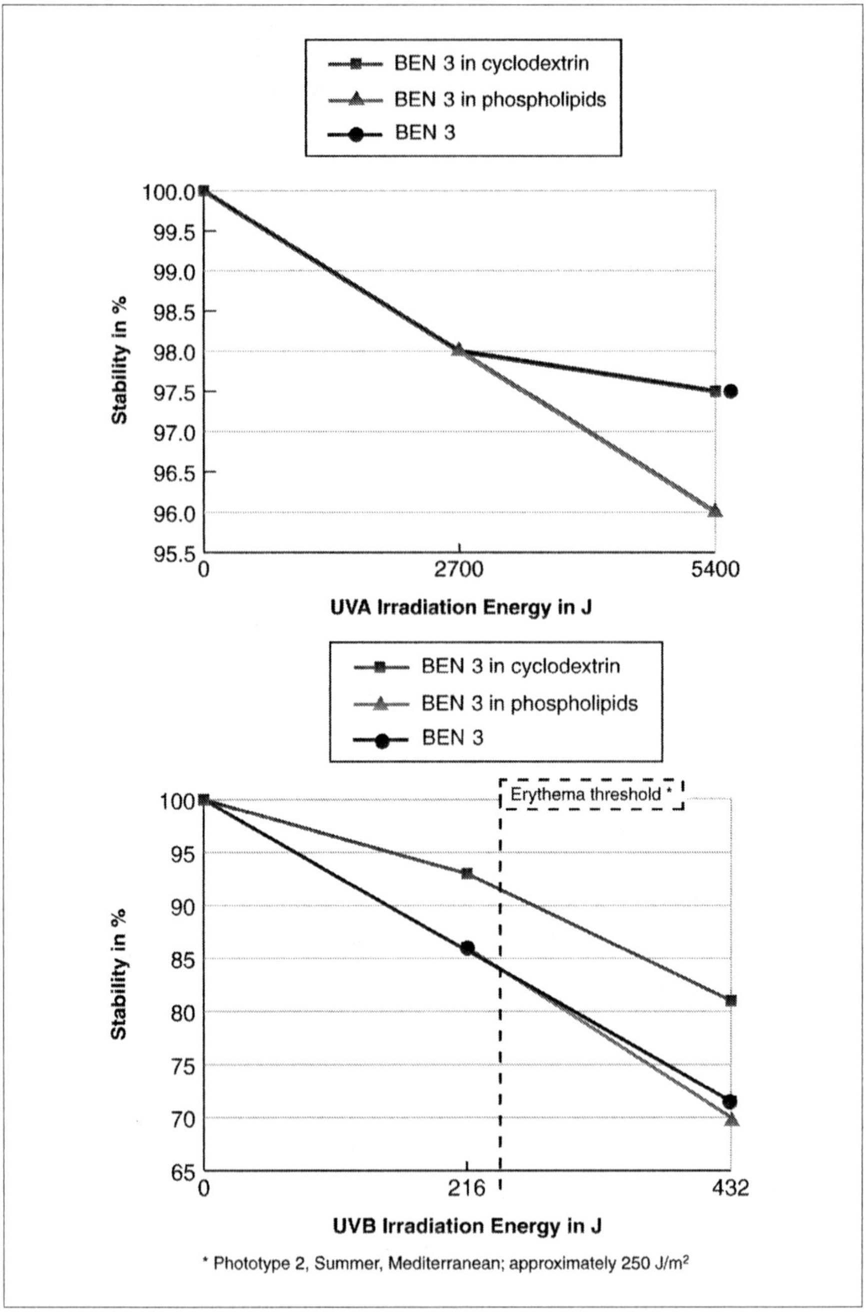

Figure 3. Photostability of benzophenone-3 to UVA (above) and UVB (below)

produce any advantage in terms of stability. The probable explanation of this observation is that phospholipids consist of unsaturated fatty acids, characterized by repeated double bonds. UV radiation probably induces breakage of the double bond with consequences on the stability of both the complexing system and the sun filter itself. Moreover phospholipids have no specific filter capacity.

Conclusion

As can be seen from the results illustrated in the different graphs, at comparable irradiation energy the solar filters encapsulated in β-cyclodextrins present a photostability significantly higher than that of the same filters in the free form or complexed with phospholipids. In the last case, it even seems that the complex with phospholipids increases the photodegradation.

The study was not carried out on a specific finished cosmetic formulation containing the solar filters at issue, but on the pure active ingredients in the free form or complexed with cyclodextrins or phospholipids. This made it possible to individually and directly observe the effect of the different types of UV radiation (UVA, UVB, UVC) on the filter, independent of other factors that could affect UV activity. Moreover, what was evaluated was not the persistence of the sun protection factor (SPF), but rather a direct analytical measurement, by HPLC, of the residual presence of each solar filter, either in the free form or as a complex. This choice was made in order to avoid possible interferences coming from the presence of solar screens, such as mica or titanium dioxide, in finished cosmetic formulations that could alter the effect of the radiation on the solar filter itself. The determination of SPF in fact does not provide sufficient information on the stability of the filter and therefore was not used in this study.

The present study does not clarify the reasons for the stability or instability observed. However, some hypotheses can be formulated.

The stability in the case of cyclodextrin may be explained by the fact that the solar filter, once included in the molecular cavity of the β-cyclodextrin, is linked in this site by quite stable chemical bonds. Thus, it is poorly reactive to the action of possible free radicals such as H^+ and OH^-, forming either as a consequence of the interaction of the UV radiation with water or after the direct action of the radiation on the filter itself. Therefore, a filtering action of the β-cyclodextrin on the UV radiation can be supposed, but it should be ascertained.

The situation of the phospholipids complex is different. In fact, in this case, the complex does not have the goal of increasing the photostability. Instead, its purpose is to make the filter more adherent to the skin so that less of the filter is lost due to the washing action of water or sweat. As a matter of fact, the UV radiation increases the photoinstability of solar filters complexed in phospholipids.

One can, however, formulate a hypothesis regarding stability in a phospholipids complex that is similar but opposite to the hypothesis relative to β–cyclodextrins. In fact, the chemical structure of phospholipids is rich in sequential double bonds that make these molecules highly susceptible to oxidative processes and breakage of the double bonds. The unavoidable consequence is the production of free radicals that determine a further destabilization of the solar filter molecule exposed to the UV radiation. The drawback could be avoided by using hydrogenated phospholipids; this is difficult to put into effect, however, because of the high additional costs that would result.

Recently, cyclodextrins have met with outstanding application success, because they are employed in the formation of complexes of a wide variety of molecules, used either in the cosmetic or the pharmaceutical field. Molecules such as retinol and vitamin C included in cyclodextrins acquire stability to oxidation and increase their bioavailability in percutaneous absorption. Similarly, other molecules such as arbutin and dihydroxyacetone reduce their toxicity and increase both their stability and bioavailability when included in cyclodextrins.

In the pharmaceutical field it is important to mention that only the complexes with cyclodextrins and phospholipids are effectively employed in drug delivery systems, where their action has been scientifically demonstrated.

—**Ugo Citernesi,** *I.R.A. Istituto Ricerche Applicate srl, Usmate Velate, Italy*

References

1. G Leone, Attualità in tema di sicurezza ed efficacia dei prodotti per la protezione solare, *Cosm Dermatol* 64 25-26 (1999)
2. U Citernesi and M Sciacchitano, Cyclodextrins in functional dermo-cosmetics, *Cosmet Toil* 110(3) 53-61 (1995)
3. U Citernesi and M Sciacchitano, Phospholipid/Active ingredient complexes, *Cosmet Toil* 110(11) 57-68 (1995)
4. M Ambrosio, I prodotti solari, *Cosm Dermatol* 41 7-27 (1992)

Green Tea and Skin Photoprotection

Keywords: epicatechins, EGCG, antioxidant, polyphenols, UV radiation

Chronic exposure of solar ultraviolet (UV) light to human skin is responsible for various skin disorders. Treatment with polyphenolic compounds from green tea has been shown to prevent UV light-induced skin photodamage in laboratory animals and human skin models.

Because of its characteristic flavor and pharmacological properties, tea (*Camellia sinensis*) is, after water, the most popular beverage consumed worldwide. Tea is manufactured from the leaf and bud of the plant and is commercially available in two main forms: green and black.[1-4] Of the total tea production, about 80% is consumed as black tea, primarily in Western countries and some Asian countries, while about 20% is consumed as green tea, mostly in Asian countries.[1-4]

The basic steps in manufacturing green and black tea are similar, differing mainly in the development of their distinctive aromas and in the fermentation process. The fermentation process is dependent upon the oxidation states of different polyphenolic compounds present in tea leaves. The term "green tea" refers to the product manufactured from fresh leaves of the tea plant by steaming and drying them at elevated temperatures, being careful to avoid oxidation and polymerization of the polyphenolic components.

Because most of the chemopreventive effects of tea leaves are determined on green tea or its polyphenolic constituents present in in vitro or in vivo animal or human models, in this article we focus our attention mainly on the photoprotective actions of green tea against the adverse effects of solar ultraviolet (UV) light.

Green Tea Polyphenol Chemistry

The polyphenols present in green tea are known as epicatechins or epicatechin derivatives. The major epicatechins found in green tea are epicatechin, epicatechin-3-gallate, epigallocatechin and epigallocatechin-3-gallate (EGCG). Their chemical structures are shown in Figure 1. Laboratory studies indicate that these polyphenolic constituents or epicatechin derivatives are antioxidant in nature and have been shown to function as anti-inflammatory and anti-carcinogenic agents in various biological models in vitro and in vivo.[1-4]

Skin Absorption, Permeability and Stability

Green tea polyphenols are water soluble and hydrophilic in character, but also easily soluble in organic solvents like acetone. Rapid absorption of EGCG was observed in human skin when dissolved in acetone, whereas slower absorption of EGCG was observed following topical application of the hydrophilic ointment-based vehicle.[5]

The stability of EGCG in hydrophilic ointment USP can be increased by mixing it with a small amount of antioxidant such as butylated hydroxytoluene. The stability of EGCG in hydrophilic ointment USP with 0.05% BHT at room temperature is more than three months while with 0.1% BHT, EGCG is stable for about six months.[5] This observation suggests that topical EGCG formulations in hydrophilic ointment USP could be used in clinical trials investigating prevention of skin cancer or other inflammatory skin disorders.

Solar UV Light: An Environmental Hazard

Chronic exposure of human skin to solar UV radiation is the primary cause of cutaneous malignancy[6] and is responsible for more than one million new cases of cutaneous malignancy each year in the USA alone, making it the most hazardous environmental carcinogen known.[6,7]

UV radiation, particularly UVB (290-320 nm), is a complete carcinogen in the skin and, in conjunction with appropriate chemical agents, has been demonstrated to act as both a tumor initiator[8] and tumor promoter.[7] UVB-induced immune suppression is considered a risk factor for skin cancer development.[9,10] In addition, exposure to solar UV radiation is known to induce erythema, edema, sunburn cell formation, a hyperplastic response and photoaging.[11-13]

There is also abundant proof that UV exposure to skin induces reactive oxygen species (ROS) such as singlet oxygen, peroxy radicals, superoxide anion and hydroxyl radical, resulting in oxidative stress and the pathogenesis of UV-induced adverse effects in the skin. Thus, the induction of oxidative stress has been associated with the onset of several disease states including rheumatoid arthritis, Parkinson's disease, reperfusion injury, inflammation, photoaging and cancer.[14] It has also long been recognized that photochemical damage induced by UV light in the chromosomal DNA of skin cells, predominantly in the form of cyclobutane pyrimidine dimers (CPD), play an important role in UV-induced immunosuppression[15,16] and carcinogenesis.[17]

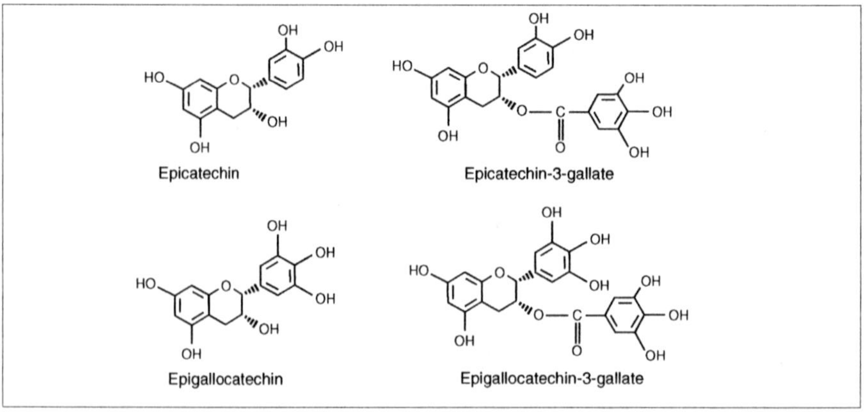

Figure 1. Chemical structures of major green tea polyphenols

Dietary Botanicals as Prevention

In recent years, interest has grown in the use of alternative medicines manufactured from naturally occurring botanicals. This interest has received considerable attention by researchers, the pharmaceutical industry and consumers for the protection of human skin from the damaging effects of environmental stimuli, including solar UV radiation. Botanicals possessing antioxidant properties are the most studied group of compounds. Extensive studies demonstrate the efficacy of naturally occurring antioxidants for protection against UV radiation-induced inflammation and cancer in mouse models. These antioxidants are vitamin E,[18] green tea polyphenols,[2] vitamin C[18] and all-trans retinoic acid.[19] The clinical efficacy of green tea polyphenols in humans has also been tested.[20-23] These antioxidants can be targeted for intervention at the initiation, promotion or progression stages of multistage skin carcinogenesis or other skin disorders.[3,7,24]

Studies have shown that dietary or environmental mutagens and carcinogens, including solar UV light to which humans are constantly exposed, exert their biological effects, at least in part via the generation of ROS and free radicals. Dietary intake of naturally occurring antioxidants, therefore, has been appreciated as an important strategy against the toxic effects of these mutagenic and carcinogenic agents.[2,25,26]

Photoprotective Effects

Anti-inflammatory effects of green tea polyphenols: Inflammatory markers like erythema, edema and hyperplastic epithelial responses are often used as early markers of skin tumor promotion. In our tests, SKH-1 hairless mice were fed green tea polyphenols (hereafter called GTP) as a sole source of drinking water, and then subjected to UV radiation. Significant protection against UV-induced cutaneous edema, erythema, and depletion of the antioxidant-defense enzyme system in the epidermis was found, as well as formation of prostaglandin metabolites via inhibition of cyclooxygenase activity.[27]

Topical treatment of mouse skin with GTP before a single low-dose UV exposure decreased UV-induced hyperplastic response, myeloperoxidase activity (a marker of tissue infiltration), and the number of infiltrating inflammatory leukocytes.[28] We showed that topical application of GTP or EGCG to the backs of human subjects 30 min. before UV irradiation resulted in significantly less erythema development, decreased myeloperoxidase activity and infiltration of leukocytes as compared to UV alone (without GTP or EGCG treatment) exposed skin sites.[20-23]

EGCG was also found to inhibit UVB-induced production of prostaglandin metabolites, which play a critical role in inflammatory disorders and in proliferative skin diseases.[20] These observations suggest a possible protective mechanism involved in the anti-inflammatory effects of green tea.

Anti-carcinogenic effects of green tea polyphenols: Nonmelanoma skin cancers, including basal cell and squamous cell carcinomas, represent the most common malignant neoplasms in humans.[6] Although many environmental and genetic factors contribute to the development of skin cancers, the most important is chronic exposure to solar UV radiation. Chronic oral feeding to mice of GTP in drinking water during the entire period of UVB exposure was found to result in a significantly lower tumor burden as compared to non-GTP-fed control animals.[29]

Topical application of EGCG inhibited photocarcinogenesis in mice with no visible toxicity.[30] The water extract of green tea as a sole source of drinking water to mice afforded protection against UVB radiation-induced tumor initiation and tumor promotion,[31] and it also induced partial regression of established skin papillomas in female mice.[32]

Studies of Photoprotective Potential

Studies were performed to define the mechanism of action of green tea polyphenols mainly in animal models, however some studies are also being made with human subjects.

Antioxidant potential of green tea polyphenols: The addition of epicatechin derivatives to mouse epidermal microsomes resulted in decreased photo-enhanced lipid peroxidation.[33] We showed that topical treatment with EGCG before UV exposure to mouse and human skin significantly reduces UVB-induced development of erythema, myeloperoxidase activity, hydrogen peroxide production and leukocyte infiltration.[23,34]

We showed that green tea, specifically EGCG, has the ability to block UVB-induced leukocyte infiltration in mouse as well as in human skin, and thus may be able to inhibit UVB-induced production of ROS by these infiltrating leukocytes.[20,34] Although ROS help the host destroy invading microorganisms, excessive and uncontrolled production can also damage host tissues and predispose to various diseases. Thus, the application of EGCG may prove helpful in ameliorating the harmful effects caused by UVB exposure through decreased ROS production.

Zhao et al[35] demonstrated that oral administration of green tea extract prior to and during multiple psoralen plus UVA treatments reduced hyperplasia, hyperkeratosis, erythema and edema formation in murine skin. Treatment of green tea extract to EpiDerm[a], a reconstituted human skin equivalent, also inhibited psoralen plus UVA-induced 8-methoxypsoralen-DNA adduct formation and p53 protein accumulation.[35]

Recently, we also observed that EGCG treatment of mouse skin before UV exposure resulted in a decreased number of hydrogen peroxide producing and inducible nitric oxide synthase expressing cells, and reduction in the production of hydrogen peroxide and nitric oxide both in epidermis and dermis of UVB irradiated sites.[28] Similar observations were also found in Caucasian skin where EGCG was topically applied before a single physiologic dose of UVB (4x MED) exposure.[23]

Additionally, EGCG treatment was found to inhibit UV-induced epidermal lipid peroxidation (41-84%), and protect antioxidant enzymes in human skin.[23] Inhibition of UV-induced lipid peroxidation in human skin by EGCG is a characteristic feature, which may protect human skin from solar UV light-induced basal cell and squamous cell carcinogenesis and photoaging as well. Based on the evidence of photoprotective effects of EGCG in animal and human systems, it appears that EGCG induces preventive effects by acting at different active sites of the ROS generating cascade, as shown in Figure 2. These data suggest that green tea may have the potential to reduce the risk of UV-induced, oxidative stress-

[a]EpiDerm, Mat Tek, Ashland, Massachusetts, USA

mediated skin diseases in humans. We also showed that topical application of green tea polyphenols before UV exposure to human skin markedly decreased the formation of cyclobutane pyrimidine dimers induced in DNA by UV radiation.[21] These cyclobutane pyrimidine dimers are critical in initiating the process of UV-induced mutagenesis and carcinogenesis.

To study further the antioxidant potential of EGCG, we conducted in vitro experiments using normal human epidermal keratinocytes (NHEK) to determine the effects of EGCG against UVB-induced oxidative stress-mediated cell signaling events, which play a critical role in the tumor promotion stage of carcinogenesis. In this study, treatment with EGCG to NHEK was found to inhibit UVB-induced intracellular release of hydrogen peroxide. Concomitant inhibition of UVB-induced oxidative stress-mediated phosphorylation of epidermal growth factor receptor and mitogen-activated protein kinases signaling also were observed.[36] These observations indicate that EGCG could play an important role in the attenuation of oxidative stress-mediated cellular signaling responses, which are essentially involved in various skin disorders in humans.

Immunopreventive potential of green tea polyphenols: Several studies suggest an important role for cytokines in regulating the immune response to cutaneous malignancies. These studies also suggest that release of cytokines subsequent to UVB radiation plays a significant role in UVB-induced immunosuppression and, thus, may be an important factor in the growth and development of immunogenic UVB-induced skin tumors.[37-39] It is reported that a number of tumors, including some melanomas and non-melanoma skin cancers, appear to produce IL-10.[37-39]

The immunosuppressive effects of IL-10 might be one possible mechanism by which these tumors escape immunologic control. We found that topical application of EGCG before a single low dose of UV (72 mJ/cm^2) exposure to mice reversed UV-induced immunosuppression to a contact sensitizer, reduced UV-induced production of IL-10 in skin and draining lymph nodes, and significantly increased the production of IL-12 in draining lymph nodes.[34]

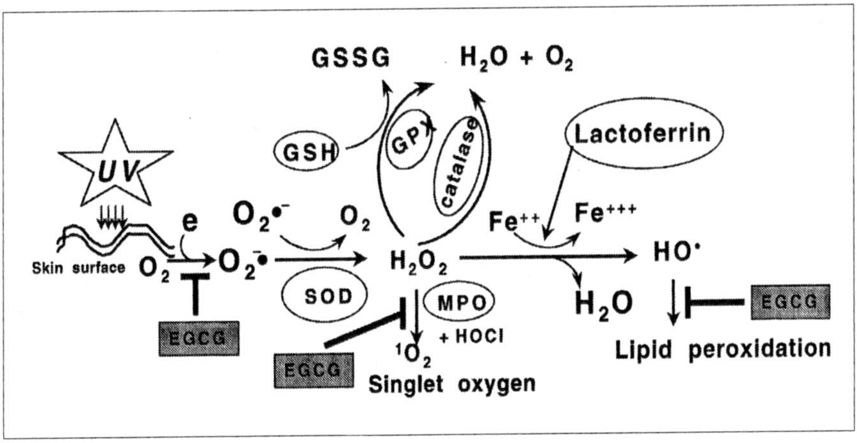

Figure 2. Blocking heads show the probable sites of preventive action of EGCG against UV light-induced oxidative stress in the skin

Such alterations in IL-10 and IL-12 production by treatment with EGCG to mouse skin seem to be mediated through modulation in the functioning of antigen presenting cells and also through blocking of UVB-induced infiltration of leukocytes in the UVB irradiated skin sites. The schematic diagram shown in Figure 3 demonstrates the mechanism of immunopreventive effects of EGCG against UV light-induced adverse effects.

Prospects for Human Use

Green tea polyphenols possess strong anti-oxidant, anti-inflammatory and anti-carcinogenic properties. Therefore, supplementation of skin care and cosmetic products with green tea extract may have a profound impact on various human skin disorders in the years to come. However, extensive clinical trials are still required to provide detailed information about the optimum concentration of green tea polyphenols to be used in skin care and cosmetics products, their half-life period and stability, and the long-term effect of these antioxidants on human skin.

—**Santosh K. Katiyar and Craig A. Elmets,** *Department of Dermatology, School of Medicine, University of Alabama at Birmingham, Birmingham, AL USA*

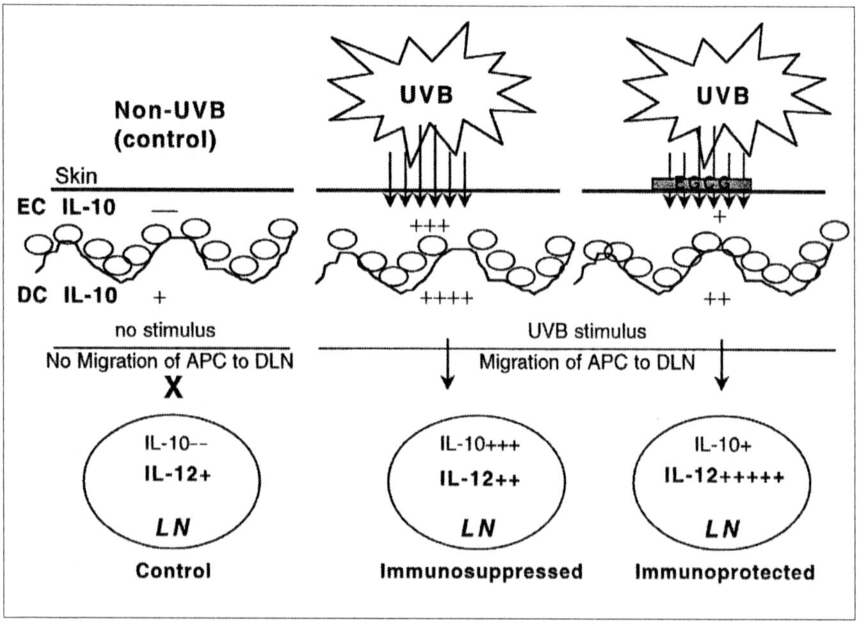

Figure 3. Schematic diagram depicts the preventive effect of EGCG on UVB-induced immune responses through alterations in immunoregulatory cytokines IL-10 and IL-12 production. EGCG treatment of skin increased IL-12 production in lymph nodes while decreasing immunosuppressive cytokine IL-10 production in the skin and lymph nodes as well. EC = epidermal cells; DC = dermal cells; APC = antigen presenting cells; LN = lymph nodes; DLN = draining lymph nodes.

References

1. SK Katiyar, N Ahmad and H Mukhtar, Green tea and skin, *Arch Dermatol* 136 989-994 (2000)
2. SK Katiyar and H Mukhtar, Tea antioxidants in cancer chemoprevention, *J Cell Biochem* (S) 27 59-67 (1997)
3. SK Katiyar and H Mukhtar, Tea in chemoprevention of cancer: Epidemiologic and experimental studies, *Int J Oncol* 8 221-238 (1996)
4. SK Katiyar and CA Elmets, Green tea polyphenolic antioxidants and skin photoprotection, *Int J Oncol* 18 1307-1313 (2001)
5. K Dvorakova, RT Dorr, S Valcic, B Timmermann and DS Alberts, Pharmacokinetics of the green tea derivative, EGCG, by the topical route of administration in mouse and human skin, *Cancer Chemother Pharmacol* 43 331-335 (1999)
6. DL Miller and MA Weinstock, Nonmelanoma skin cancer in the United States: Incidence, *J Am Acad Dermatol* 30 774-778 (1994)
7. H Mukhtar and CA Elmets, Photocarcinogenesis: Mechanisms, models and human health implications, *Photochem Photobiol* 63 355-447 (1996)
8. LH Kligman, FJ Akin, AM Kligman, Sunscreens prevent ultraviolet photocarcinogenesis, *J Am Acad Dermatol* 3 30-35 (1980)
9. T Yoshikawa, V Rae, W Bruins-Slot, JW vand-den-Berg, JR Taylor and JW Streilein, Susceptibility to effects of UVB radiation on induction of contact hypersensitivity as a risk factor for skin cancer in humans, *J Invest Dermatol* 95 530-536 (1990)
10. CK Donawho, HK Muller, CD Bucana and ML Kripke, Enhanced growth of murine melanoma in ultraviolet-irradiated skin is associated with local inhibition of immune effector mechanisms, *J Immunol* 157 781-786 (1996)
11. MF Naylor, Erythema, skin cancer risk, and sunscreens, *Arch Dermatol* 133 373-375 (1997)
12. M Goihman-Yahr, Skin aging and photoaging: An outlook, *Clin Dermatol* 14 153-160 (1996)
13. AR Young, Cumulative effects of ultraviolet radiation on the skin: Cancer and photoaging, *Semin Dermatol* 9 25-31 (1990)
14. JP Kehrer, Free radicals as mediators of tissue injury and disease, *Crit Rev Toxicol* 23 21-48 (1993)
15. LA Applegate, RD Ley, J Alcalay and ML Kripke, Identification of molecular targets for the suppression of contact hypersensitivity by ultraviolet radiation, *J Exp Med* 170 1117-1131 (1989)
16. ML Kripke, PA Cox, LG Alas and DB Yarosh, Pyrimidine dimers in DNA initiate systemic immunosuppression in UV-irradiated mice, *Proc Natl Acad Sci USA* 89 7516-7520 (1992)
17. RW Hart, RB Setlow and AD Woodhead, Evidence that pyrimidine dimers in DNA can give rise to tumors, *Proc Natl Acad Sci USA* 75 5574-5578 (1976)
18. KL Keller and NA Fenske, Uses of vitamins A, C, and E and related compounds in dermatology: A review, *J Amer Acad Dermatol* 39 611-625 (1998)
19. Z Wang, M Boudjelal, S Kang, JJ Voorhees and GJ Fisher, Ultraviolet irradiation of human skin causes functional vitamin A deficiency, preventable by all-trans retinoic acid pre-treatment, *Nat Med* 5 418-422 (1999)
20. SK Katiyar, MS Matsui, CA Elmets and H Mukhtar, Polyphenolic antioxidant (-)-epigallocatechin-3-gallate from green tea reduces UVB-induced inflammatory responses and infiltration of leukocytes in human skin, *Photochem Photobiol* 69 148-153 (1999)
21. SK Katiyar, A Perez and H Mukhtar, Green tea polyphenol treatment to human skin prevents formation of ultraviolet light B-induced pyrimidine dimers in DNA, *Clin Cancer Res* 6 3864-3869 (2000)
22. CA Elmets, D Singh, K Tubesing, MS Matsui, SK Katiyar and H Mukhtar, Green tea polyphenols as chemopreventive agents against cutaneous photodamage, *J Am Acad Dermatol* 44 425-432 (2001)

23. SK Katiyar, F Afaq, A Perez and H Mukhtar, Green tea polyphenol (-)-epigallocatechin-3-gallate treatment to human skin inhibits ultraviolet radiation-induced oxidative stress, *Carcinogenesis* 22 287-294 (2001)
24. SD Hursting, TJ Slaga, SM Fischer, J DiGiovanni and JM Phang, Mechanism-based cancer prevention approaches: Targets, examples, and the use of transgenic mice, *J Natl Cancer Inst* 91 215-225 (1999)
25. BN Ames, Dietary carcinogens and anticarcinogens, *Science* 221 1256-1264 (1983)
26. LW Wattenberg, Inhibition of carcinogenesis by naturally occurring and synthetic compounds. In: *Antimutagenesis and Anticarcinogenesis, Mechanisms II*, Y Uroda, DM Shankel and MD Waters, New York, NY, Plenum Publishing Corp, pp 155-166 (1990)
27. R Agarwal, SK Katiyar, SG Khan and H Mukhtar, Protection against ultraviolet B radiation-induced effects in the skin of SKH-1 hairless mice by a polyphenolic fraction isolated from green tea, *Photochem Photobiol* 58 695-700 (1993)
28. SK Katiyar and H Mukhtar, Green tea polyphenol (-)-epigallocatechin-3-gallate treatment to mouse skin prevents UVB-induced infiltration of leukocytes, depletion of antigen presenting cells and oxidative stress, *J Leukoc Biol* 69 719-726 (2001)
29. ZY Wang, R Agarwal, DR Bickers and H Mukhtar, Protection against ultraviolet B radiation-induced photocarcinogenesis in hairless mice by green tea polyphenols, *Carcinogenesis* 12 1527-1530 (1991)
30. HL Gensler, BN Timmermann, S Valcic, GA Wachter, R Dorr, K Dvorakova and DS Alberts, Prevention of photocarcinogenesis by topical administration of pure epigallocatechin gallate isolated from green tea, *Nutr Cancer* 26 325-335 (1996)
31. ZY Wang, MT Huang, T Ferraro, CQ Wong, YR Lou, M Iatropoulos, CS Yang and AH Conney, Inhibitory effect of green tea in the drinking water on tumorigenesis by ultraviolet light and 12-O-tetradecanoylphorbol-13-acetate in the skin of SKH-1 mice, *Cancer Res* 52 1162-1170 (1992)
32. ZY Wang, MT Huang, CT Ho, R Chang, W Ma, T Ferraro, KR Reuhl, CS Yang and AH Conney, Inhibitory effect of green tea on the growth of established skin papillomas in mice, *Cancer Res* 52 6657-6665 (1992)
33. SK Katiyar, R Agarwal and H Mukhtar, Inhibition of spontaneous and photo-enhanced lipid peroxidation in mouse epidermal microsomes by epicatechin derivatives from green tea, *Cancer Lett* 79 61-66 (1994)
34. SK Katiyar, A Challa, TS McCormick, KD Cooper and H Mukhtar, Protection of UVB-induced immunosuppression in mice by the green tea polyphenol (-)-epigallocatechin-3-gallate may be associated with alterations in IL-10 and IL-12 production, *Carcinogenesis* 20 2117-2124 (1999)
35. JF Zhao, YJ Zhang, XH Jin, M Athar, RM Santella, DR Bickers and ZY Wang, Green tea protects against psoralen plus ultraviolet A-induced photochemical damage to skin, *J Invest Dermatol* 113 1070-1075 (1999)
36. SK Katiyar, F Afaq, K Azizuddin and H Mukhtar, Inhibition of UVB-induced oxidative stress-mediated phosphorylation of mitogen-activated protein kinase signaling pathways in cultured human epidermal keratinocytes by green tea polyphenol (-)-epigallocatechin-3-gallate, Submitted for publication in *Toxicol Appl Pharmacol* (2001).
37. W Dummer, JC Becker, A Schwaaf, M Leverkus, T Moll and EB Brocker, Elevated serum levels of interleukin-10 in patients with metastatic malignant melanoma, *Melanoma Res* 5 67-68 (1995)
38. RD Granstein, Cytokines and photocarcino-genesis, *Photochem Photobiol* 63 390-394 (1996)
39. J Kim, RL Modlin, RL Moy, SM Dubinett, T McHugh, BJ Nickoloff and K Uyemura, IL-10 production in cutaneous basal and squamous cell carcinomas. A mechanism for evading the local T cell immune response, *J Immunol* 155 2240-2247 (1995)

Artemia Extract: Toward More Extensive Sun Protection

Keywords: photodamage, artemia extract, protective, Immunoblotting studies, ELISA method

An extract from the plankton* Artemia salina *protects DNA from UV damage and decreases UV-induced inflammatory cytokines, suggesting a way to help sunscreens protect skin from UV damage.

In recent years, people have become increasingly aware that UV exposure can cause skin damage, such as photoaging and carcinogenesis. Therefore, people are exhibiting a growing interest in protecting their skin from the harmful effects of UV radiation. It is well-known that UVB induces a variety of cellular damage, including DNA photodamage with formation of cyclobutane pyrimidine dimers,[1] and consequent mutation.[2] Moreover, there is increasing evidence that UVA contributes to skin carcinogenesis and premature aging.

Sunscreens are now widely used in an attempt to block and/or decrease UV radiation's undesirable effects, such as photoaging and skin cancer. Also widely used is SPF (sun protection factor) to designate the protective properties of sunscreens. In fact, SPF indicates only the sunscreen's ability to protect skin from UV-induced erythema. Studies have suggested that because erythema and sunburn are absent in sunscreen-protected skin, sunscreens may lead to a more intensive exposure to sun irradiation, which in turn increases the risk of DNA damage, immunosuppression and skin cancer.[3,4]

Recent studies have found the rate of tumor formation in populations that use sunscreens to be higher than tumor formation in populations that have no sun exposure, thus raising questions about the failure of sunscreens.[4] Concerned with this problem, we were interested in finding a new, natural way of helping protect human skin from UV damage. In this article, we describe the UV-protective properties of Artemia extract and, consequently, the potential of Artemia extract use in sun-protective products.

Artemia Extract

Artemia extract is prepared by marine biotechnology from a specific plankton, the *Artemia salina,* that lives in hyper-mineralized lakes. Artemia is characterized by its great capability of withstanding and surviving various sudden environmental aggressions. Artemia extract is rich with phosphorylated nucleotides and, in particular, with diguanosine tetraphosphate (GP4G), which our recent studies have demonstrated to have a great protective effect on DNA.[5]

Materials and Methods

Viability of cultured cells: Human fibroblasts were treated with Artemia extract 3% for 24 h, then exposed to UVB irradiation. The MTT (3-(4,5-cimethylthiazol-2-2,5-diphenyl tetrazolium bromide) colorimetric assay[6] was carried out 24 h later.

Skin organ culture and staining: We used the established skin organ culture method.[7,8] Test substances (a cream containing Artemia extract and a placebo cream) were applied topically on the skin samples. Designated skin samples were irradiated and all skin samples were submitted to standard Hematoxylin and Eosin (H&E) or to immunostaining 24 h later.

Immunostaining studies: Cultured skin samples were studied by routine direct immunofluorescence (DIF) microscopy following standard methods.[9,10] For heat shock protein studies, rabbit polyclonal antibody anti-Hsp70 was used. Mouse anti-CD1a antibody was used for Langerhans cell staining.

Immunoblotting studies: Immunoblotting studies[11] were performed using NuPAGE 12% gel and polyclonal rabbit anti-human Hsp70 antibody, monoclonal mouse anti-human p21, or anti-mutant and wild-type human p53 antibody.

Comet assay: Comet assay[12,13] was conducted on cultured fibroblasts. In brief, after irradiation, cells were harvested and cell lysate was subjected to electrophoresis, then stained with propedium iodide, after which the individual cell comet aspect "head and tail" was examined.

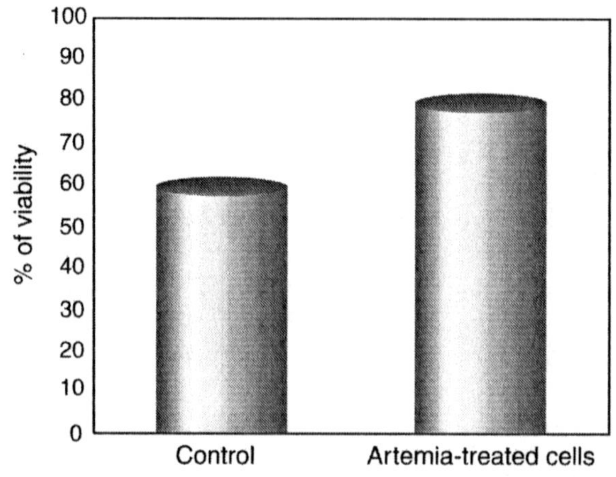

Figure 1. Cell viability after irradiation with 100 mJ/cm^2 UVB

Tunel assay: Cultured human keratinocytes received 1% of Artemia extract 24 h before and after their irradiation with 100 mJ/cm² of UVB. 24 h later, the Tunel assay[14] was performed.

Enzyme-linked immunosorbent assay: Standard ELISA method[15,16] was carried out on human HaCat cells. Cells received Artemia extract 3% for 24 h, followed by UVB irradiation. Total cytokine level was assessed for IL-1 alpha, IL-8 and TNF-alpha. The level of the cytokines was expressed in pg/mg protein.

Results and Discussion

Artemia extract and UV protection of human cells: MTT viability test demonstrated that cell preparation with Artemia extract prior to stress protected the cells from UVB (100 mJ/cm²) aggression and increased their viability by 20% (Figure 1). Likewise, observation of the morphology and aspect of the cells showed that Artemia extract also preserved their morphology from UV-induced stress signs.

Artemia extract and protection of DNA: The principal constituent of Artemia extract is a molecule called GP4G. Previous studies had demonstrated that the GP4G molecule has a remarkable effect on cellular DNA.[5] This strongly suggests that the Artemia effect we see on DNA is due mainly to the GP4G molecule in the extract.

With the Tunel method we studied the effect of GP4G on DNA protection from UVB-induced DNA damage and consequent programmed cell death or apoptosis.

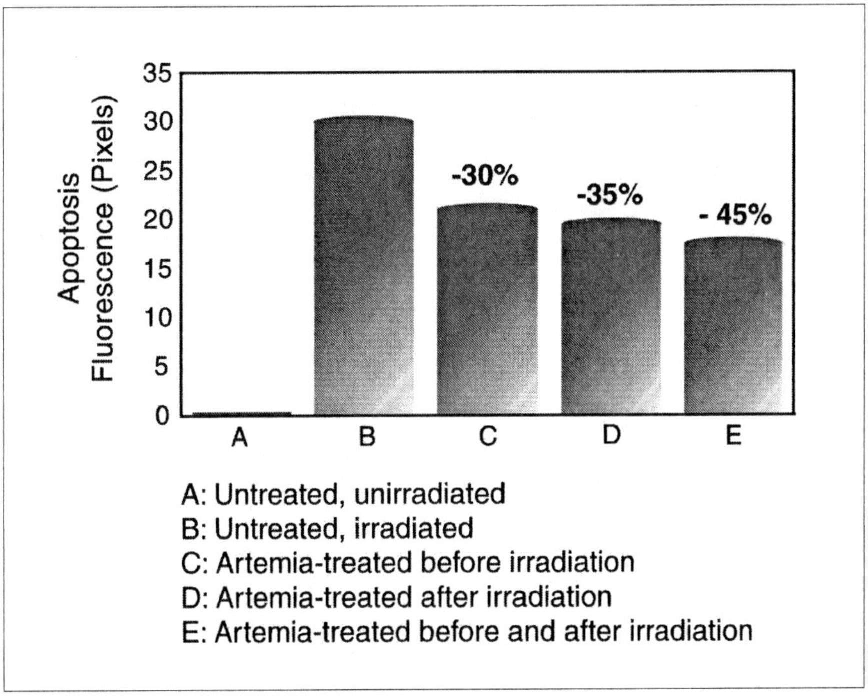

Figure 2. Tunel assay. GP4G-Artemia extract effect on UVB-induced apoptosis.

Our studies demonstrated that adding Artemia extract into the cells after UVB exposure (100 mJ/cm^2) decreased UVB-induced apoptosis by 35%. Interestingly, GP4G protection reached 45% when the extract was added before and after UVB irradiation (Figure 2), which suggests its role in both DNA protection and repair.

Comet assay: It is known that there is a linear dose response curve for thymine dimers' formation in UV-irradiated cells,[17] and that depending on the UV dose and DNA damage, the cell uses its repair mechanism or goes into apoptosis. Therefore, we performed comet assay using UVB doses of 30, 60, and 100 mJ/cm^2 on human fibroblasts previously treated or not treated with Artemia extract. Our results showed that, at the low UVB dose (30 mJ/cm^2), there was very little UVB-induced DNA degradation in the cells treated with the extract, compared to the untreated cells (Figure 3). Similarly, with higher UVB doses of 60 and 100 mJ/cm^2, Artemia extract decreased DNA degradation, as shown in Figures 4 and 5.

p53 and p21 expression: Other researchers have reported that after UV irradiation, tumor suppressor gene p53 wild-type is involved in two mechanisms: a) both G_1 cell cycle arrest and DNA repair, and b) apoptosis.[18-20] The mechanism in play depends on the degree of UV damage. Moreover, it has also been demonstrated that p21 helps protect the cell from UV-induced apoptosis.[21-23]

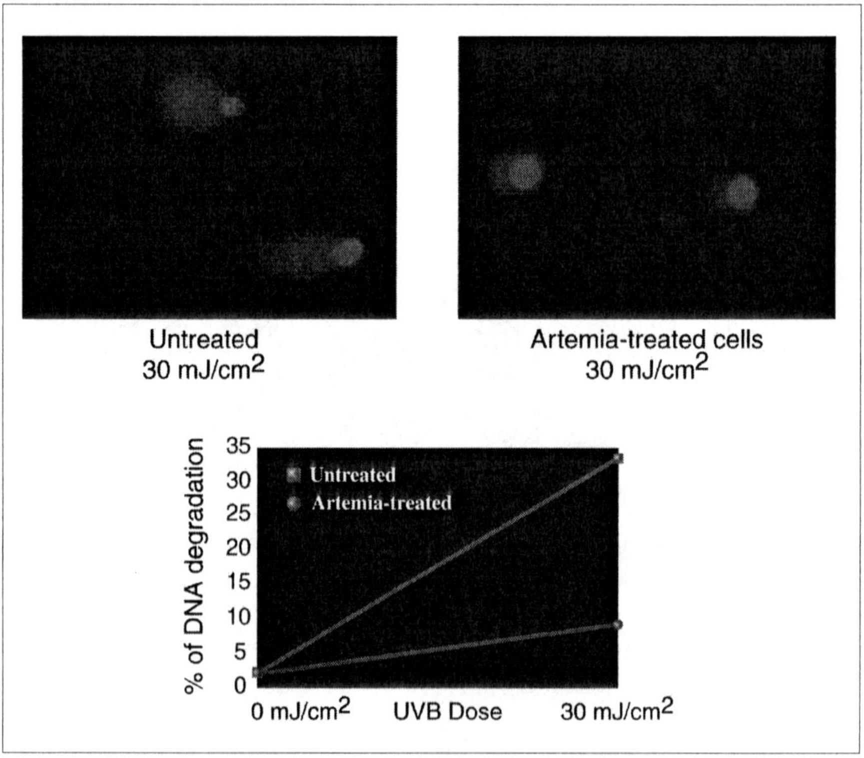

Figure 3. DNA degradation by Comet assay, in cells irradiated with 30 mJ/cm^2 of UVB

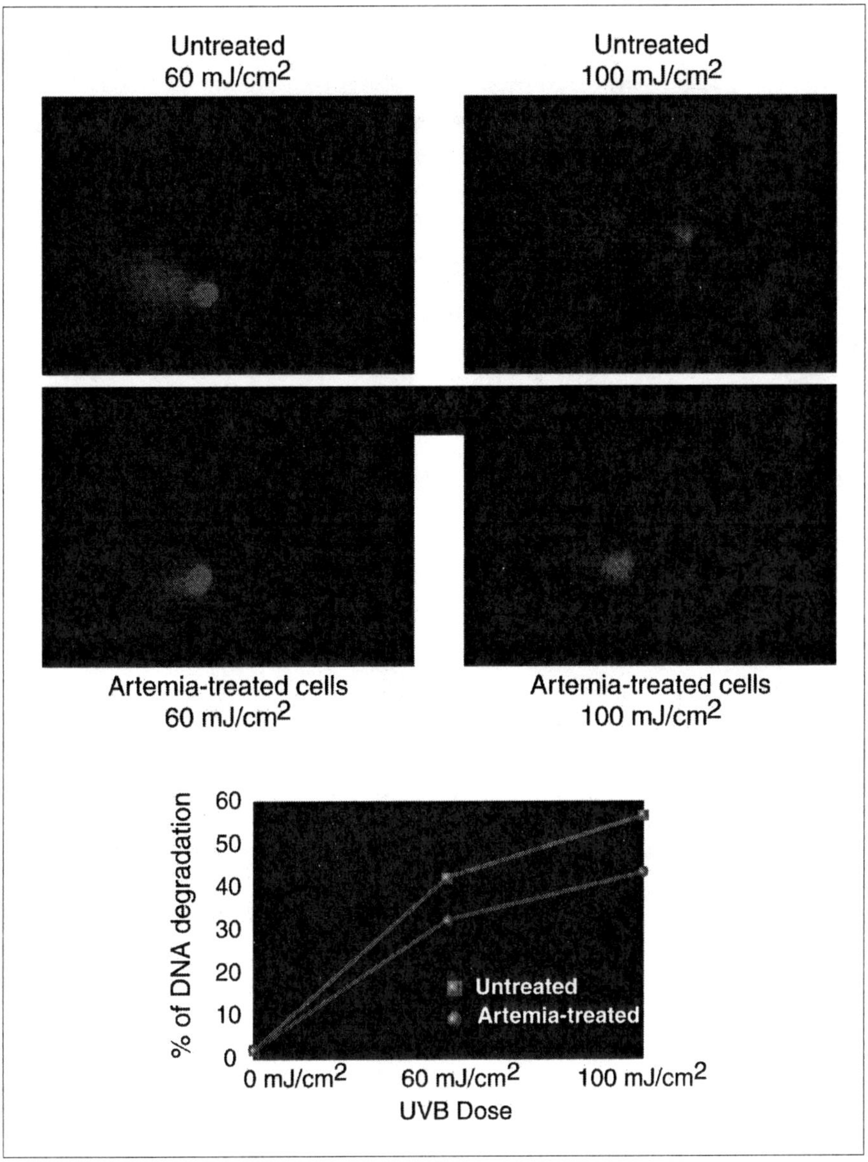

Figure 4. DNA degradation by Comet assay, in cells irradiated with 60 and 100 mJ/cm^2 of UVB

Our results showed that at low UVB doses, both p53 and p21 are upregulated in Artemia-treated cells more than in control (Artemia-untreated) cells, which align the activation of DNA repair pathway by transient cell cycle arrest through p53 and p21. Interestingly, with higher UVB doses, our results were similar and showed that in irradiated cells treated with Artemia extract, both p53 and p21 were higher than in control cells that failed to repair their damaged DNA (Figure 6). These results,

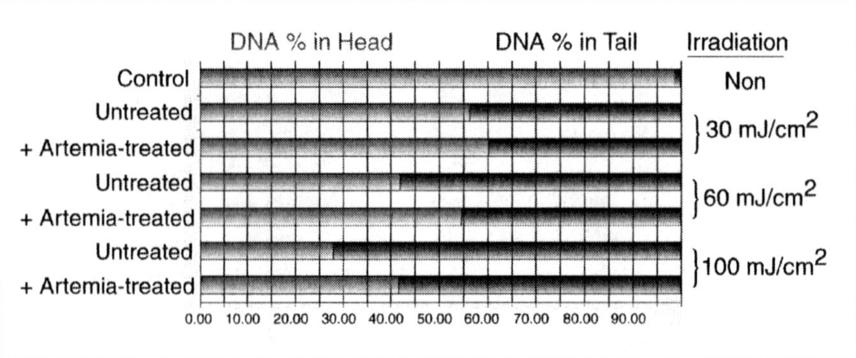

Figure 5. Analysis of DNA degradation percentage in "head & tail" of the Comet

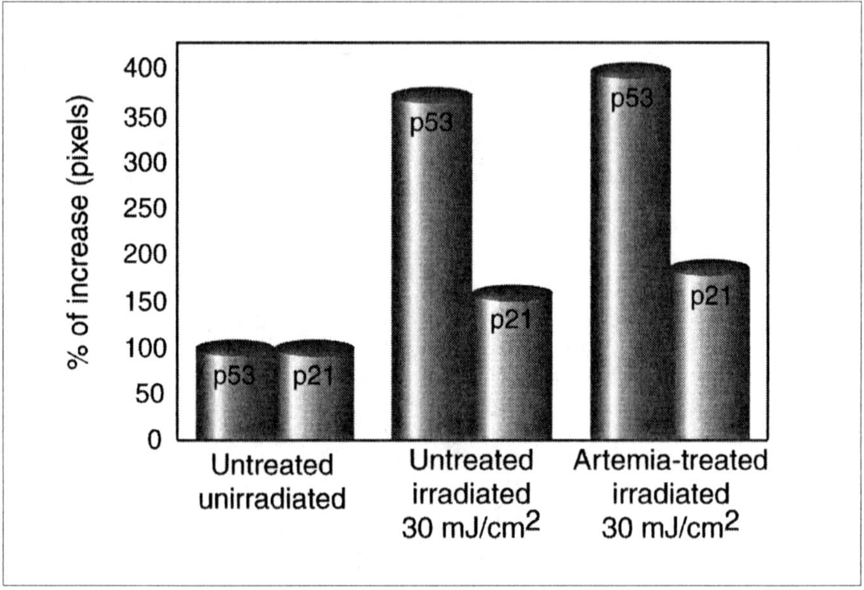

Figure 6. p53 and p21 expression after irradiation with 30 mJ/cm² of UVB

together with the data above, suggest a consistent activation of the repair pathway rather than the apoptosis pathway, in cells treated with Artemia extract.

Skin protection from UVB by Artemia extract: In order to confirm the above results on human skin, we treated skin samples either with Artemia extract 3% in a cream or with placebo, prior to exposure to UVB at selected doses in the range 100-200 mJ/cm². Skin structure and morphology, evaluated after H&E staining, revealed that skin samples treated with Artemia extract exhibited great skin structure preservation with rare occurrence of sunburn cells,[24,25] in contrast to the control skin which exhibited extended signs of damage with much sunburn cell formation, as shown in Figure 7.

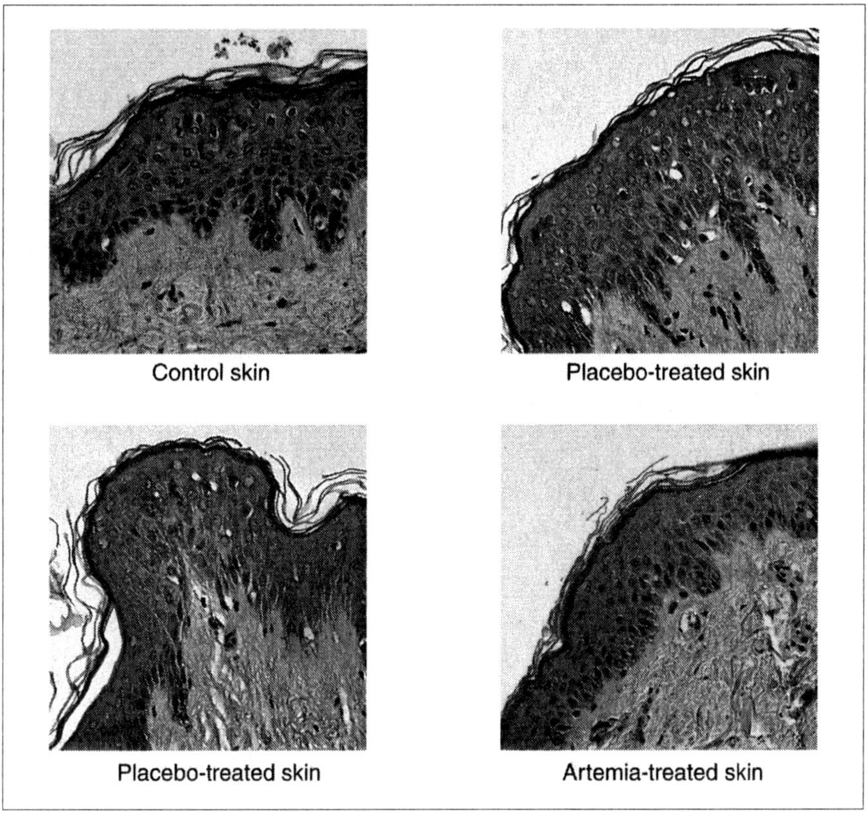

Figure 7. H&E staining of human skin irradiated with 200 mJ/cm^2 of UVB

Moreover, CD1a immunostaining of Langerhans cells revealed that Artemia-treated skin samples were similar to unirradiated control skin, and exhibited no decrease in the number or distribution of Langerhans cells in the skin. On the other hand, UV-irradiation of placebo-treated skin induced Langerhans cell depletion from the epidermis (Figure 8), as is usually seen after UV exposure.[25,26]

Hsp70 induction by Artemia extract in cultured human cells and skin: It is well known that heat shock proteins (Hsp) or molecular chaperons, and in particular Hsp70, play an important role in protecting the cell from different types of stress.[27-29] Because Artemia is able to withstand various environmental stresses, we investigated whether the induction of Hsp70 was one of Artemia extract's essential mechanisms of action.

Different studies, conducted on cultured human fibroblasts, showed that, within 3-6 hours, Artemia extract 3% significantly induced Hsp70 synthesis in these cells. These results were confirmed by Hsp70 mRNA studies (data not shown).

Moreover, our recent studies[30] have shown that Artemia extract induction of Hsp70 is stress-free. Similar studies on human skin showed that the application of Artemia extract 3% in a cream formula significantly induced and increased Hsp70

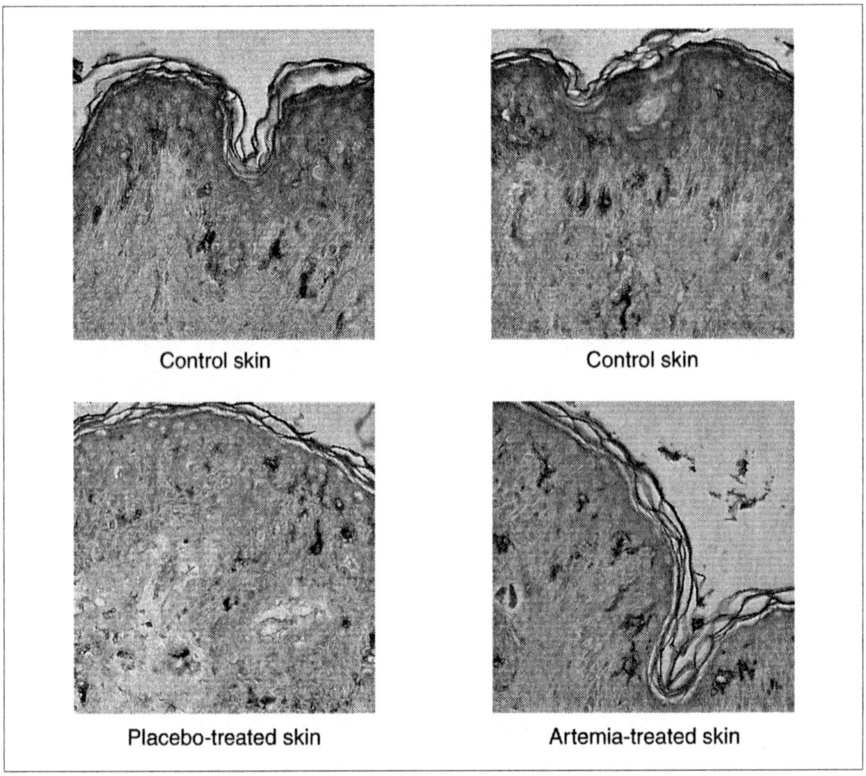

Figure 8. CD1a staining of Langerhans cells in human skin irradiated with 200 mJ/cm^2 of UVB

protein expression in the skin, compared to the placebo-treated control skin where Hsp70 was found at the basic level (Figure 9).

Furthermore, we completed the studies above using 1 and 2 J/cm^2 of UVA alone or followed by UVB irradiation (50-150 mJ/cm^2) on cultured keratinocytes and fibroblasts in order to study the influence of UVA. These results confirmed the above findings concerning the protection Artemia extract offers to the cells. Our results also showed the induction of Hsp70 in the UV-stressed cells, and that this induction was higher when the cells were treated with Artemia extract, prior to their irradiation.

Artemia extract's effect on inflammatory cytokines: Prior data have shown that the induction of Hsp70 inhibits the biosynthesis of IL-1 and other cytokines in human cells and, thereby, serves a protective role by suppressing the proinflammatory response.[31] Because our studies have demonstrated that Artemia extract induces Hsp70 in human cells, we looked at Artemia-induced Hsp70 and irradiated HaCat cells, and used ELISA to investigate the anti-inflammatory effect of Artemia extract on the following cytokines:

- IL-1 alpha and IL-8 in dose course studies. Total IL-1 alpha and IL-8 levels were determined by ELISA after exposure of the cells to UVB

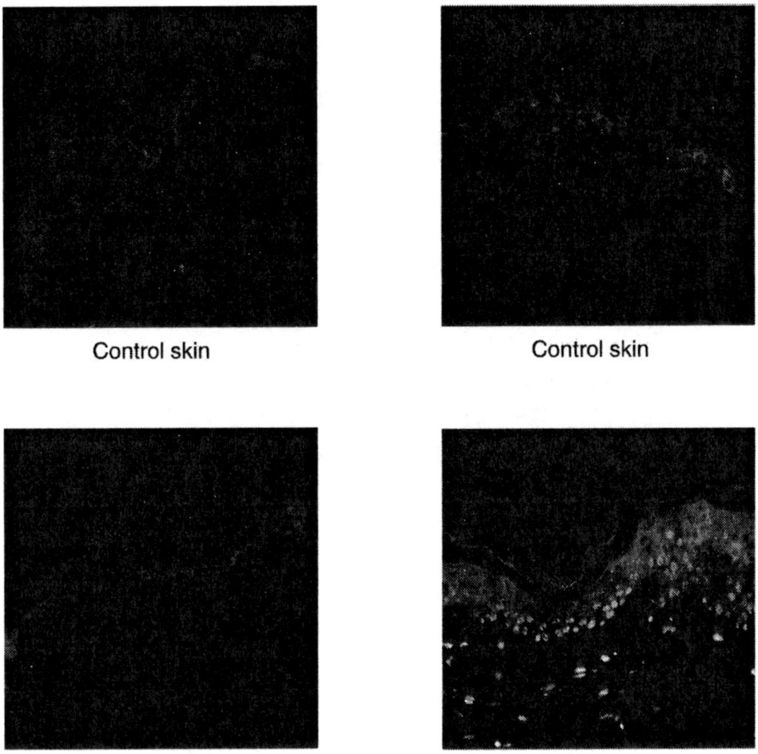

Figure 9. Hsp70 induction in human skin by Artemia extract. Immunofluorescence staining

doses of 0, 10, 20, 30, 40 mJ/cm^2. Our studies demonstrated that IL-1 alpha and IL-8 level of synthesis decreased considerably in the Artemia-Hsp70-induced cells by 20-30% for IL-1 alpha and 25-40% for IL-8, compared to the control cells (Figures 10 and 11).

- IL-1 alpha and IL-8 in time course studies. Total levels of IL-1 alpha and IL-8 were evaluated at the following time points: 0, 6, 8, 18 and 24 h. Interestingly, a significant decrease in UV-induced IL-1 alpha level was found in the Artemia-Hsp70-induced cells at different time points, compared to the control cells (Figure 12). A maximum decrease in IL-1 alpha was observed at 18-24 h after UV irradiation. Similarly, a decrease in IL-8 level in irradiated Artemia-Hsp70-induced cells was found over all time points. Interestingly, the maximum decrease of UV-induced IL-8 was observed within 6-8 h after UV irradiation (Figure 13), which is of interest for 'early' control of the inflammatory process.
- TNF-alpha studies. TNF-alpha is another potent multifunctional mediator of inflammation and immune responses. As shown in Figure 14, our results

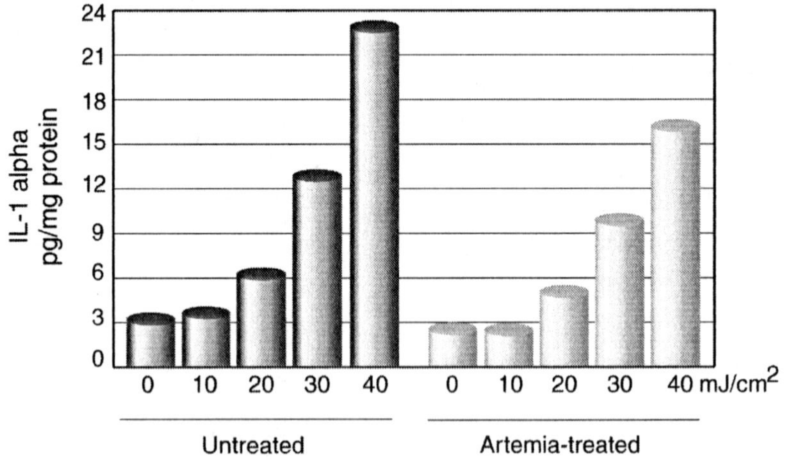

Figure 10. Artemia extract downregulation of total IL-1 alpha synthesis in cells irradiated with different UVB doses

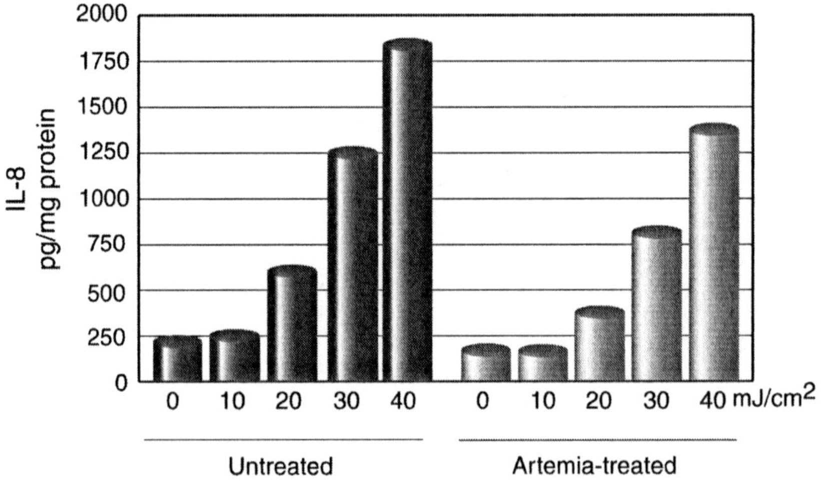

Figure 11. Artemia extract downregulation of total IL-8 synthesis in cells irradiated with different UVB doses

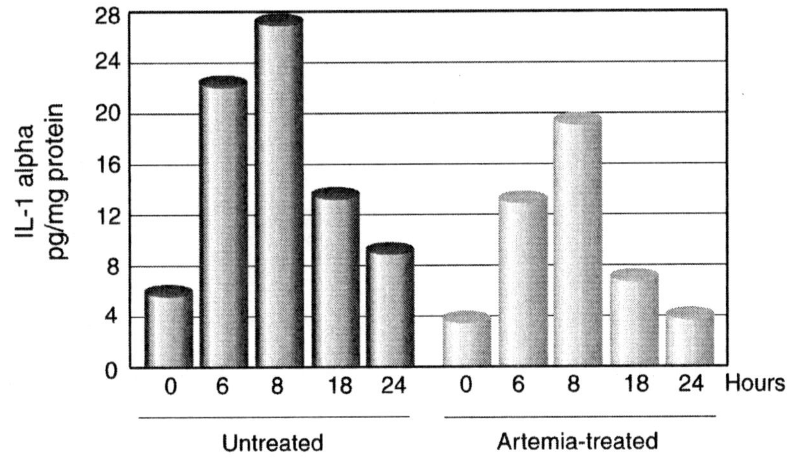

Figure 12. Time course studies of Artemia extract effect on total IL-1 alpha synthesis in cells irradiated with 30 mJ/cm^2 of UVB

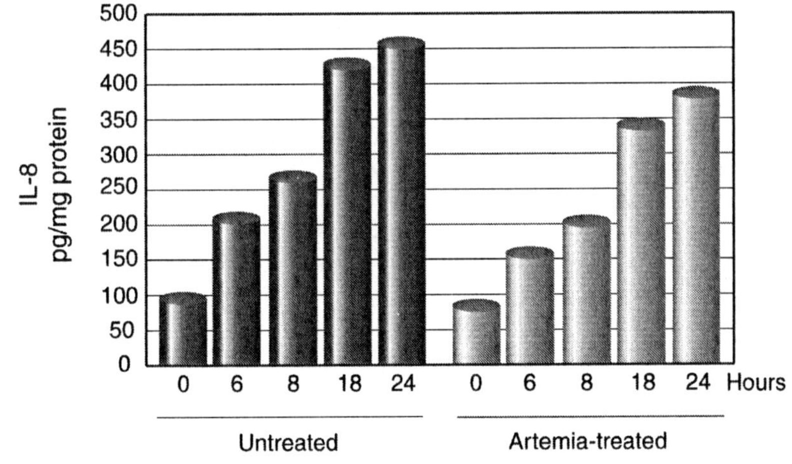

Figure 13. Time course studies of Artemia extract effect on total IL-8 synthesis in cells irradiated with 30 mJ/cm^2 of UVB

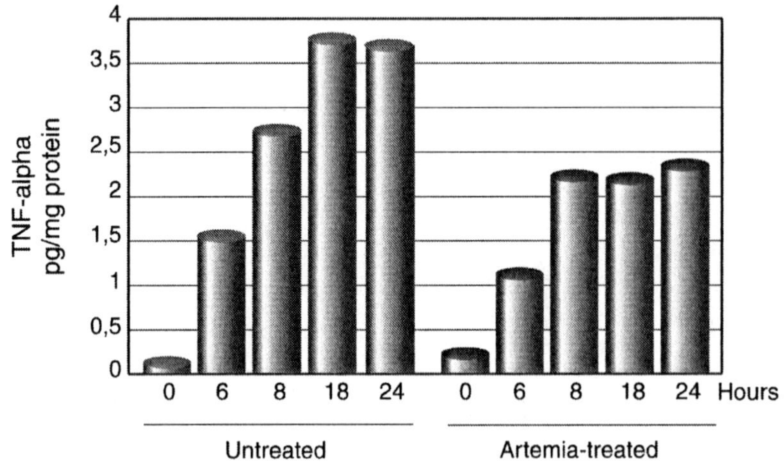

Figure 14. Time course studies of Artemia extract effect on total TNF-alpha synthesis in cells irradiated with 30 mJ/cm^2 of UVB

demonstrate that TNF-alpha level is decreased in Artemia-Hsp70-induced cells at all the time points, and this decrease is statistically significant at the 24-h time point, after irradiation. This result is very interesting and corresponds to the decrease in the maximum inflammatory effect of TNF-alpha.

Conclusion

This data demonstrates that Artemia extract displays two unusual characteristics. It has a high content of GP4G, which protects DNA and decreases DNA damage. It induces Hsp70 in a stress-free setting, while also having an anti-inflammatory effect on inflammatory cytokines.[32]

These studies reveal that Artemia extract offers an interesting contribution in the search to meet the present desire for new "agents and natural mechanisms" that offer more comprehensive skin protection from sun damage, and that reinforce the protection of sunscreens. Therefore, Artemia extract has potential applications in sun care products.

—N. Domloge, E. Bauza, K. Cucumel, D. Peyronel and C. Dal Farra
Vincience Research Center, Sophia Antipolis, France

References

1. ML Kripke, PA Cox, LG Alas and DB Yarosh, Pyrimidine dimers in DNA initiate systemic immunosuppression in UV-irradiated mice, *Proc Natl Acad Sci USA* 89(16) 7516-7520 (1992)

2. A Ziegler et al, Sunburn and p53 in the onset of skin cancer, *Nature* 372(6508) 773-776 (1994)
3. J Westerdahl, C Ingvar, A Masback and H Olsson, Sunscreen use and malignant melanoma, *Int J Cancer* 87(1) 145-150 (2000)
4. MF Naylor, HW Lim, JK Robinson, MA Weinstock, DS Rigel and MR Verschoore, Photoprotection, the *AAD 2001*, Proceedings of the 59th Annual Meeting of the American Academy of Dermatology, Washington DC, March 2-7, 2001
5. C Dal Farra., K Cucumel, D Peyronel and N Domloge, DNA protective effect of diguanosine tetraphosphate, *J Invest Dermatol* 114(4) 808 (2000)
6. E Borenfreund, H Habich and N Martin-Alguacil, Comparison of two *in vitro* cytotoxicity assays – the neutral red and tetrazolium MTT tests, *Toxic in vitro* 2(1) 1-6 (1988)
7. J van de Sandt, J van Shoonhoven, W Maas and A Ruten, Skin organ culture as an alternative to *in vivo* dermatotoxicity testing, *ATLA* 21 443-449 (1993)
8. S Boisnic, MC Branchet-Gumila, L Benslama, Y Le Charpentier and J Arnaud-Battandier, *Eur J Dermatol* 7 271-273 (1997)
9. N Domloge-Hultsch, P Bisalbutra, R Gammon and KB Yancey, Direct immunofluorescence microscopy of 1 mol/L sodium chloride-treated patient skin, *J Am Acad Dermatol* 24(6) 946-951 (1991)
10. N Domloge-Hultsch, P Benson, R Gammon and KB Yancey, A bullous skin disease patient with autoantibodies against separate epitopes in 1 mol/L sodium chloride split skin, *Arch Dermatol* 128 1096-1101 (1992)
11. S muller, V Klaus-Kovtum and JR Stanley, A 230-kd basic protein is the major bullous pemphigoid antigen, *J Invest Dermatol* 92 33-39 (1989)
12. EJ Morris, JC Dreixler, KY Cheng, PM Wilson, RM Gin and HM Geller, Optimizing of single-cell gel electrophoresis (SCGE) for quantitative analysis of neuronal DNA damage, *BioTechniques* 26(2) 282-289 (1999)
13. O Ostling and KJ Johanson, Micro-electrophoresis study of radiation-induced DNA damages in individual mammalian cells, *Biochem-Biophys Res Commun* 123 291-298
14. F Washio, M Ueda, A Ito and M Ichihashi, Higher susceptibility to apoptosis following ultraviolet B irradiation of xeroderma pigmentosum fibroblasts is accompanied by upregulation of p53 and downregulation of bcl-2, *Br J Dermatol* 140(6) 1031-1037 (1999)
15. WG Phillips, M Feldmann, SM Breathnach and FM Brennan, Modulation of the IL-1 cytokine network in keratinocytes by intracellular IL-1 alpha and IL-1 antagonist, *Clin Exp Immunol* 101 177-182 (1995)
16. A Guéniche, J Viac, G Lizard, M Charveron and D Schmitt, Effect of nickel on the activation state of normal human keratinocytes through interleukin 1 and intercellular adhesion molecule 1 expression, *Br J Dermatol* 131 250-256 (1994)
17. AR Young, CS Potten, O Nikaido, AG Parsons, A Boenders, JM Ramsden and CA Chadwick, Human melanocytes and keratinocytes exposed to UVB or UVA in vivo show comparable levels of thymine dimers, *J Invest Dermatol* 111 936-940 (1998)
18. G Li and VC Ho, p53-dependent DNA repair and apoptosis respond differently to high- and low-ultraviolet radiation, *Br J Dermatol* 139 3-10 (1998)
19. J Cotton and DF Spandau, Ultraviolet B-radiation dose influences the induction of apoptosis and p53 in human keratinocytes, *Radiat Res* 147 148-155 (1997)
20. ML Smith, IT Chen, Q Zhan, PM O'Connor and AJ Fornace, Involvement of the p53 tumor suppressor in repair of UV-type DNA damage, *Oncogene* 10 1053-1059 (1995)
21. N Bissonnette and DJ Hunting, p21-induced cycle arrest in G_1 protects cells from apoptosis induced by UV-irradiation or RNA polymerase II blockage, *Oncogene* 16 3461-3469 (1998)
22. S Inohara, K Kitagawa and Y Kitano, Coexpression of $p21^{waf1/cip1}$ and p53 in sun-exposed normal epidermis, but not in neoplastic epidermis, *Br J Dermatol* 135 717-721 (1996)
23. M Loignon, R Fetni, AJE Gordon and EA Drobetsky, A p53-independent pathway for induction of $p21^{waf1/cip1}$ and concomitant G_1 arrest in UV-irradiated human skin fibroblasts, *Cancer Res* 53 3390-3394 (1997)

24. F Daniel, L Brophy and W Lobitz, Histochemical response of human skin following ultraviolet irradiation, *J Invest Dermatol* 37 351-357 (1961)
25. BA Gilchrest, NA Soter, JS Stoff and MC Mihm, The human sunburn reaction: Histologic and biochemical studies, *J Am Acad Dermatol* 5 411-422 (1981)
26. V Rae, T Yoshikawa, W Bruins-Slot, JW Streilein and JR Taylor, An ultraviolet B radiation protocol for complete depletion of human epidermal Langerhans cells, *J Dermatol Surg Oncol* 15 1199-1202 (1989)
27. M Jäättelä, Heat shock proteins as cellular lifeguards, *Ann Med* 31 261-271 (1999)
28. TKC Leung, MY Rajendran, C Monferies, C Hall and L Lim, The human heat-shock protein family, *Biochem* J 267 125-132 (1990)
29. MG Santoro, Heat shock factors and the control of the stress response, *Biochem Pharmacol* 59 55-63 (2000)
30. TJ Hall, Role of Hsp70 in cytokine production, *Experientia* 50 1048-1053 (1994)
31. K Cucumel, JM Botto, E Bauza, C Dal Farra, R Roetto and N Domloge, Artemia extract induces Hsp70 in human cells and enhances cell protection from stress, the *SID 2001*, Proceedings of the 62nd Annual Meeting of the Society for Investigative Dermatology, Washington DC, May 9-12, 2001
32. French Pat 0007606, Préparations cosmétiques ou dermo-pharmaceutiques renfermant un extrait de zooplancton qui contient et induit des HSP, N Domloge, D Peyronel and C Dal Farra (Jun 15, 2000)

Heat- and UV-Stable Cosmetic Enzymes from Deep Sea Bacteria

Keywords: biotechnology, BSE, manufacturing, molecular modeling, microorganisms, bioinformatics, encapsulation sphere technologies

Here, the author describes some of his recent successes with identifying extreme enzyme systems from hydrothermal-residing marine bacteria. These extraordinarily adaptive microorganisms may provide innovative reparative approaches to common skin exposures (i.e., air pollution, smoking and resulting changes in skin conditions).

Editor's note: Regarding biotechnology and raw active ingredients, the cosmetic industry globally has provided a plethora of phytochemicals, skin delivery systems, tissue-engineered derived activities, hydroxy acids and aqueous/marine compounds. These materials have a winning combination: demonstrable performance in finished formulations and—even more importantly—safety during repeated topical application.

One remaining area of concern is cosmetic actives derived from cattle. The real or perceived fear of viral contamination, specifically from bovine spongiform encephalopathy (BSE), has forced the industry to evaluate new sources of novel actives, specifically plants and microorganisms. New manufacturing extraction processes, bioinformatics and molecular modeling are the new millennium tools of active cosmetic ingredient discovery and efficacy. The manufacturing production of these new molecules can be moderated through genetic cloning, cell culture or even through a bioreactor to ensure a very precise performance profile in a finished skin or personal care product.

Molecular modeling, bioinformatics and encapsulation sphere technologies are some of the new tools for the raw material suppliers. Innovation and consumer performance needs will no doubt reinforce the solid marriage of biotechnology and the cosmetic industry in the search for novel cosmetic actives.

The use of protective enzymes in cosmetics has been resisted because of their well-known instability. However, we report here on heat- and UV-stable enzymes obtained from hydrothermal vents deep in the ocean. In vitro and cell culture studies were carried out to test the potential for cosmetic applications. Catalase-like, SOD-like and glutathione peroxydase-like activities, all heat activated, were found to protect human skin fibroblasts, cell membrane lipids and collagen contraction capacity.

The Origin of Life: An Extreme Event?

According to an increasing number of scientists, life began to appear on Earth some 4 billion years ago, not in a "primordial soup" on the surface, but in the depths of the ocean close to the hydrothermal vents ("mini-volcanoes"), which were only discovered a few decades ago.

The combination of complex chemistry, richness in metals (catalysts) and the high temperature and pressure at those sites was, according to the researchers, propitious to the advent of the first complex structures, those which now remain to testify to that distant epoch, the *Archae*. Today, these and other "extremophiles" live in the immediate vicinity of these "black smokers" at –2,000 meters, –4,000 meters or even deeper. They are astonishing microorganisms that multiply in environments that were too hostile for life: high temperatures (>70°C); pressures 200- to 400-fold greater than atmospheric pressure; high local concentrations of sulfur and heavy metals; low quantities of carbon-based organic materials; and low oxygen levels.

Those extreme bacteria, by their very existence, show life's great adaptive capability and have led scientists to dream of physiological characteristics that could be of benefit to humans. Obviously, in order to survive under those extreme environmental conditions, these organisms must have equipped themselves with particularly effective and specific defense mechanisms.

The Skin and Its Defenses

Every day, our skin is subject to aggressions of variable violence that threaten its equilibrium, function and visible beauty: chemicals, pollution, stress, electromagnetic radiation (infrared, ultraviolet), abrasion and even mechanical injury. The consequences of those aggressions may be multiple, irrespective of whether or not they are visible: inflammation, burning, edema, actinic lesions, premature aging and diseases in the long term.

The mechanisms that induce lesions of the skin are also very varied. Nonetheless, we now recognize that a major part of the harmful effects of daily aggression is mediated by molecular species known as "free radicals" or radical oxygen species (ROS). Free radicals are preferentially generated by inflammatory processes and UVA radiation, then potentiated by heavy metals (Fenton's reaction).

The disequilibrium giving rise to accumulation of ROS in the body may have two origins:

- Overproduction due to exogenous stimulation by UV radiation (particularly UVA), heat, bacterial invasion, shock or local hypoxia;

Figure 1. Deep sea hydrothermal vent (courtesy of IFREMER, the French Research Institute for Exploitation of the Sea)

- Accumulation through a deficit in the enzymatic capability for conversion of those reactive species into inactive species.

The body, including the skin, possesses defense systems, which may be innate (enzymes) or exogenous (vitamins, particularly vitamin E), to inactivate toxic ROS.

In eukaryotes, the key detoxification enzymes are superoxide dismutase (SOD), catalase, glutathione peroxidase (GSH peroxidase or GPO). The enzymes intervene to ensure transformation of the initial superoxide ion into water via the following reaction route:

$$O_2^{\circ}- + O_2^{\circ}- + 2H^+ \longrightarrow H_2O_2 + O_2$$
$$H_2O_2 \longrightarrow H_2O + \tfrac{1}{2}\,O_2$$
$$H_2O_2 + 2GSH \longrightarrow 2H_2O + GSSG$$

where GSH is reduced glutathione, and GSSG is oxidized glutathione.

In short, among other factors, skin aging is due to a vicious circle where UVA radiation creates free radicals that lead to lesions in the genes (DNA) which, in turn, lead to reduced efficacy of the protective enzymes and a further increase in free radical presence.

This dangerous process is exacerbated by the fact that the protective enzymatic activity at the surface of the skin also seems to be directly reduced, or even destroyed, under the impact of aggression by radiation or chemicals. Certain findings suggest that the catalases, SOD and GPO, are less present and active in summer than in winter. Could this be due to their inactivation by radiation? It looks like it. Yet, it is precisely under those circumstances that the skin needs protection most.

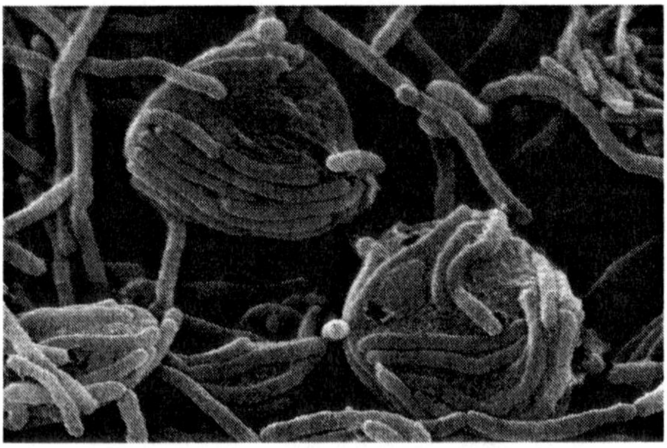

Figure 2. Scanning electron micrograph of *Thermus thermophilus* (X 10,000)

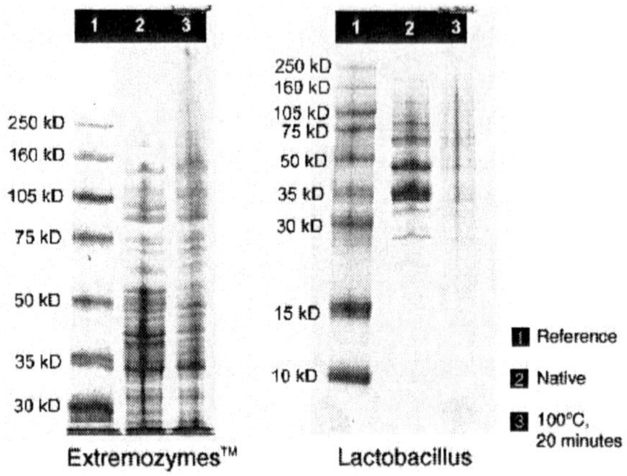

Figure 3. Protein profiles for EEC (left) and Lactobacillus (right) by denaturating electrophoresis

In partnership with the French research organization CNRS (Centre National pour la Recherche Scientifique), Sederma harnessed certain strains of microorganisms from the depths of the ocean collected in the mid-oceanic Great Rift and around the submarine trenches and volcanoes with marked hydrothermal activity.[1]

Experimental

The thermophile strain known as *Thermus thermophilus* GY1211 (Figure 2), isolated at –2,000 meters on the flanks of a "black smoker" in the Guayamas Bay of California, lives under 200 bars of pressure at an optimal temperature of 72°C but

Why was* Thermus thermophilus *selected?

The genus *Thermus* is characterized by its ability to live at high temperatures in an oxygen-poor medium (anaerobic respiration based on reduction of nitrates takes over in the absence of oxygen) and in acidic (sulfuric acid) or basic (bicarbonate/carbonate, silica) environments. *Thermus* may express very different phenotypes as a function of the environmental context of the habitat.

This great plasticity of the strains is doubtless all the more marked when the bacteria on the flanks of the volcanoes of the ocean floor are confronted with eruptions of hot water loaded with metals, inorganic salts and sulfides while remaining able to withstand temperature variations and marked pressure. *Thermus thermophilus* thus shows an especially adaptive physiology enabling its survival under a great variety of conditions.

readily withstands 85°C. The organism also withstands high saline concentrations (halophilic at 2% sodium chloride concentrations). The organism has been subjected to comprehensive characterization, growth-optimization and proteomic studies.

The remarkably adaptive physiology of *Thermus thermophilus* enables it to survive under a variety of conditions (see sidebar). This suggests that the cells' components – proteins, enzymes, organelles, DNA – are probably more resistant than their equivalents, particularly in mesophilic bacteria (bacteria living at room temperature). *Thermus thermophilus* GY1211 thus promised all the wealth of the deep-sea bacteria, a new source of active substances for cosmetic applications.

For industrial use, fermentation is conducted at high temperature in a medium appropriate to the strain in order to achieve an optimal quantity of biomass for extraction of the active substances of interest.

Once the fermentation process has been arrested and sufficient biomass has been obtained, tangential microfiltration separates the biomass from the culture medium. Using a specifically chosen detergent, it is possible to extract the cell content containing the active enzymes. Tangential ultrafiltration is then used for concentration and purification of the extract. The resulting concentrated ferment is available as a commercial product.[a] In this article we will call it EEC (Extremophile Enzymes Complex) in reference to the term "Extremozymes" coined by Bernan et al. in 1997[2] to describe enzymes stable in extreme conditions.

Electrophoretic analysis (Figure 3) shows the protein composition of *Thermus* differs completely from that of Lactobacillus, taken as an example of a "normal" bacterium. It also shows the extreme temperature stability of EEC; treatment at 100°C for 20 min totally destroys the proteins of *Lactobacillus*, but the proteins in EEC remain intact.

Enzymatic Activities Measured In Vitro

SOD-like activity: The hypoxanthine/xanthine oxidase system is used to generate the superoxide anion. To detect the presence of the superoxide anion, one can add an indicator, NBT (tetrazolium blue), to the reaction mix. NBT in the presence of $O_2°$- changes color from yellow to blue. This may be monitored using a spectrometer adjusted to a wavelength of λ 570 nm. The intensity of the coloration is proportional to the quantity of $O_2°$-. When the superoxide anion is being converted in the presence of an enzyme, such as SOD, the intensity of the coloration is reduced, compared to that of the control.

In our lab, tests were conducted in triplicate. They showed that the EEC solution has a dose-dependent detoxifying activity against the superoxide anion. The inhibition of NBT conversion reaches 43% and 64% for EEC solution at 1% and 3%, respectively.

Peroxide convertase activity: Hydrogen peroxide (H_2O_2), in the presence of an enzyme endowed with catalase-type activity, is converted to water and oxygen. The oxygen released is quantified by a specific electrode (Clark's electrode). This test is widely known as the oxygraphic test.

Various concentrations of the EEC were transferred to the oxygraph. The oxygen conversion rate (incubation time of a few minutes) was monitored, as a function of time, and converted into units/ml using a standard curve.

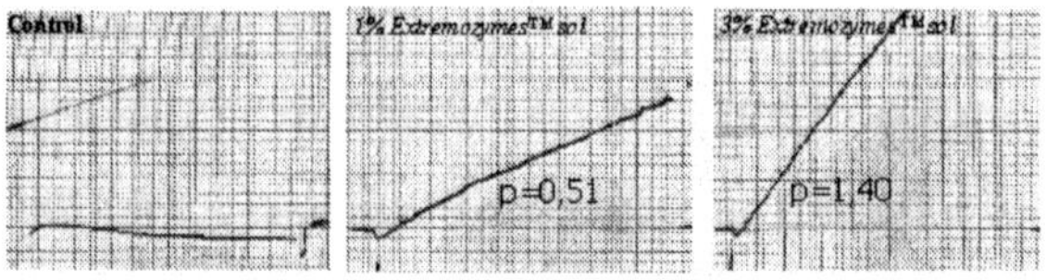

Figure 4. Oxygen conversion (p = slope) in an H_2O_2 medium as a function of time for selected concentrations of EEC

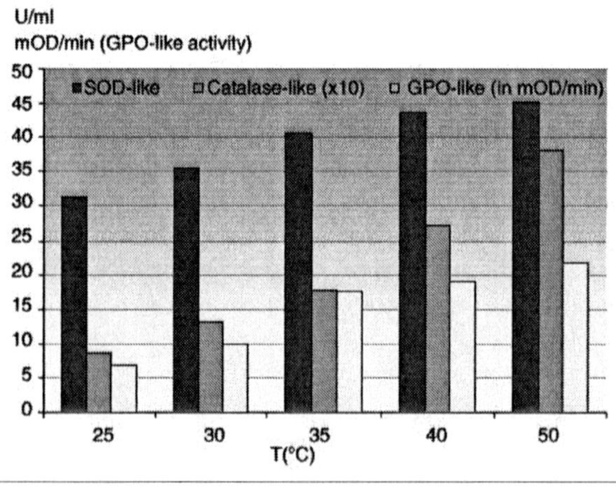

Figure 5. Enzymatic activities of EEC as a function of temperature

As Figure 4 shows, the EEC is endowed with potent H_2O_2 detoxification activity and that activity increases with the concentration used. We believe the conversion of H_2O_2 by EEC can be explained by an enzyme that we will identify as a convertase in the absence of structural characterization.

GPO-like activity: As indicated earlier, the skin possesses several defensive enzymatic systems against free radicals or ROS. After having detected SOD-like and catalase-like activities in the EEC, we investigated the presence of glutathione peroxidase (GPO) activity. The importance of that detoxification activity cannot be overestimated.

It was interesting to observe that even ancestral bacteria such as those found in the depths of the ocean already possess complex systems, which have survived to the present time, as shown by the SIGMA test.

The SIGMA test enables evaluation of glutathione peroxidase activity by spectrophotometric monitoring of the quantity of NADPH consumed in the following coupled reactions:

$$2GSH + H_2O_2 \longrightarrow GSSG + 2H_2O$$
$$GSSG + NADPH,H^+ \longrightarrow 2GSH + NADP^+$$

where GSH is reduced glutathione, the substrate of GPO, and GSSG is oxidized glutathione, NADP is the Nicotine-Adenine-Dinucleotide-Phosphate, NADPH is the reduced, uncharged form of NADP, $NADP^+$ is the ionically charged form of NADP.

Thus, when hydrogen peroxide is converted by glutathione peroxidase, two moles of reduced glutathione and one mole of NADPH are consumed. Glutathione reductase is added to the reaction medium and the NADPH consumed is proportional to the oxidized glutathione formed by glutathione peroxidase. The NADPH disappearance rate at $\lambda = 340$ nm enables calculation of the GPO activity in absorbance units/min.

Once again, the EEC showed marked detoxifying activity (data not shown), simulating the activity of GPO in higher organisms, with respect to hydrogen peroxide.

Enzymatic activities as a function of temperature: The above experiments were conducted to detect the basic activities of EEC at normal physiological temperatures. However, given that the *Thermus* bacteria usually live at 70-80°C, we decided to study the behavior of the enzymatic activities of EEC as a function of temperature.

We, therefore, analyzed the three discovered activities (SOD-like, catalase-like and GPO-like) in protocols similar to those previously described, but at test temperatures of up to 50°C. The increase in enzyme activity would probably peak at about 70°C (physiological for the strain), but this is devoid of interest in the cosmetic context.

The results of the three experiments are shown in Figure 5. There was a marked increase in enzymatic activity, which was proportional to the temperature increase. The increase in activity was 45% for SOD, 330% for H_2O_2 convertase and 220% for GPO when compared to the activities observed at room temperature (25°C).

Temperature stability compared to conventional enzymes: We conducted tests to compare the temperature stability of enzymatic activity of EEC to that of conventional enzymes. The tests consisted in maintaining commercially available reference enzymes and the EEC in a usual buffer medium at two temperatures—25°C and 50°C — for a few hours. Samples were taken at time T0, then at 30 min, 1 h, 2 h and 4 h, and enzymatic activity was determined at each point. The stability of the enzymatic activity was expressed relative to the values determined at T0.

The results shown in Figure 6 clearly illustrate the difference between the behavior of the EEC and the reference enzymes.

The instability of the catalase and GSH peroxidase activities of the reference enzymes was clearly demonstrated at 25°C and even more dramatically at 50°C. In contrast, the EEC showed stable activity in an aqueous buffer medium over several hours for all types of activity.

Conclusions from the in vitro tests: The results from this set of tests show that the EEC have detoxifying activities for the reactive species of oxygen:

- H_2O_2 convertase;
- O_2°- convertase;
- GPO-like.

The activities were stable over time and under the influence of heat in contrast to what was observed with the reference enzymes: SOD, catalase and GPO. In addition, the activities were heat-inducible with an increase from 1.5- to 4-fold of the baseline activity at 50°C.

We concluded that the stability and heat-activation of the EEC affords the skin a remarkable protection that adapts to the environment. Increased stress,

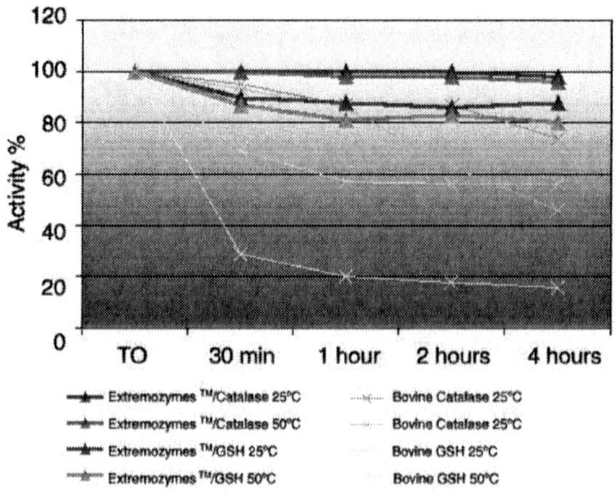

Figure 6. Stability of two types of EEC activity at 25°C and 50°C, shown as a percent of activity at time t=0, compared to activity of bovine enzymes

due to hot climates or exposure to the sun, is countered by a proportional increase in activity.

Efficacy Assessment in Fibroblast Cell Culture

Protection of cell lipid membranes: Cultured fibroblasts exposed to UVA radiation generate radical oxygen species that induce lesions in the cell membranes. The membrane lesions can be measured in terms of the degree of UVA-induced lipid peroxidation and by observation of the cell layers under the microscope.

For our experiment, we cultured fibroblasts to confluence, then incubated them in the presence or absence of EEC for 24 h. The cells were then subjected to UVA radiation at a dose of 15 J/cm^2 in phosphate buffer, then re-cultured with or without the EEC for 5 h. Lipid peroxidation was measured by determining 4-HNE, one of the main end by-products of lipid peroxidation.

As shown in Table 1, 1% EEC showed a protective effect against UVA-induced stress with a 46% mean decrease in membrane peroxidation.

The results were corroborated visually under the microscope. A markedly impaired cell layer was observed with the irradiated control, compared to the control cells not exposed to UVA. In the presence of the EEC, effective protection enabled the maintenance of a cell layer that was very similar to the control.

Protection of collagen layer contraction: In human skin, fibroblasts are embedded in an extra cellular matrix mainly consisting of proteoglycans and collagen. The interactions between the fibroblasts and collagen are very important because they determine the elasticity and firmness of the skin and are involved in wound and tissue repair.

The ability of fibroblasts to adhere to the collagen network and maintain the three-dimensional structure of the dermis is reflected, in vitro, by traction on the fibers of a collagen layer inducing clearly visible "contraction" of the layer. Any impairment in the functional adherence capability of fibroblasts (from membrane lesions, in particular) due to UV radiation results in a loss of wound healing potential (ability to contract the collagen).

To test for protection of the collagen layer's contraction capability, we subjected cultured normal human dermal fibroblasts to UVA radiation at a dose of 15 J/cm^2 in phosphorus buffer. Following irradiation, the fibroblasts were inoculated into collagen layers and incubated for 48 h in the presence or absence of EEC at various concentrations. Control fibroblasts were not exposed to radiation in order to obtain maximum contraction. The collagen gel was observed after 48 h.

Photographs were taken to document collagen layer contraction under the various conditions. While the non-irradiated control (Figure 7a) showed a very contracted collagen layer 48 h after cell inoculation, the UVA-irradiated fibroblasts were deprived of their ability to contract the collagen gel, so it covered the entire surface of the culture dish (Figure 7b).

The repair effect (collagen gel contraction) of the EEC is concentration dependent. Contraction increased with EEC concentration (Figures 7c and 7d) and reached a maximum at 3%, which was identical to the contraction of the control

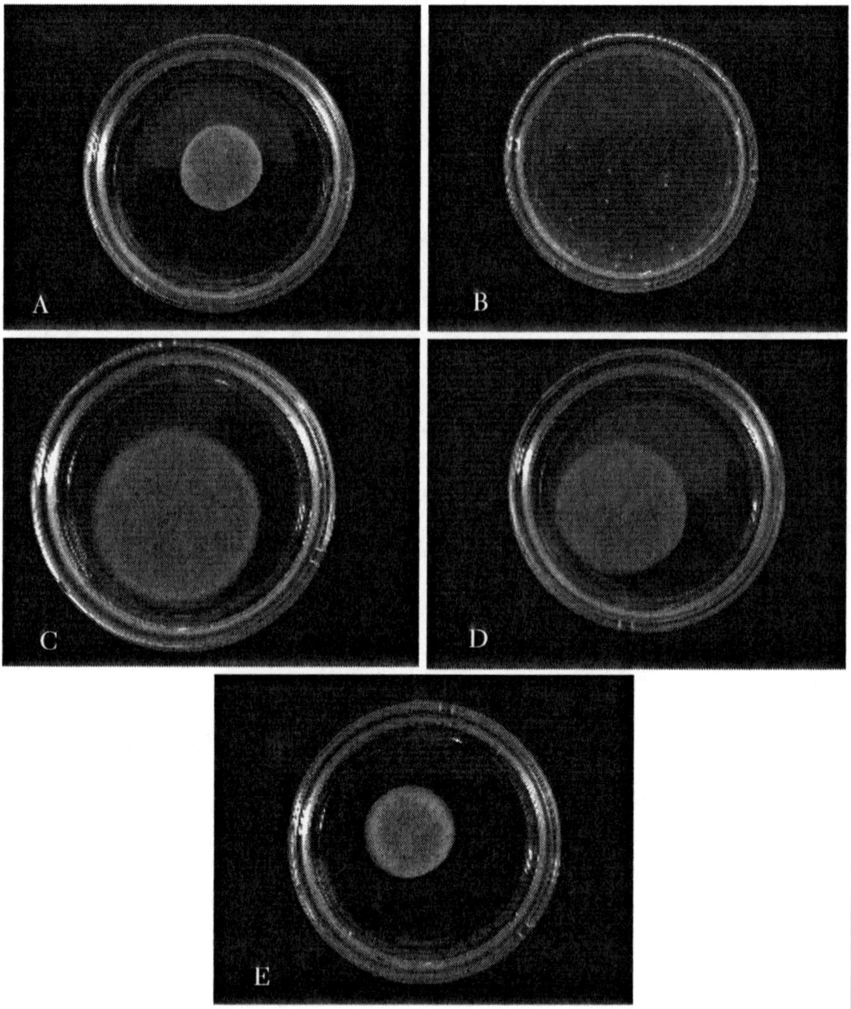

Figure 7. EEC-based protection from UV-induced loss of collagen contraction 48 h after inoculation with irradiated fibroblasts

a) Control collagen (inoculated with non-irradiated fibroblasts)
b) Unprotected collagen
c) Collagen protected with EEC solution at 0.5%
d) Collagen protected with EEC solution at 1.0%
e) Collagen protected with EEC solution at 3.0%

(Figure 7e). These experiments show that the EEC protects the main function of the cell and its architecture.

Epidermal Protection Measured Ex Vivo

The lesions induced by very strong UV irradiation (especially UVA, against which sunscreen protection is frequently inadequate) is reflected by the peroxidation of intercellular and membrane lipids. In vivo, UVA rays cross the epidermis before reaching the dermis. In vitro reconstructed epidermis models are recognized to be good systems for evaluation of the expected effect in vivo. The development of new specific probes, particularly fluorescence probes, enables in situ quantification on cell or tissue cultures.

In our lab to test for epidermal protection, epidermal equivalents[b] in the presence of 3% EEC solution were irradiated with a high dose of UVA (60 J/cm^2), then incubated for 18 h. At the end of that period, the epidermal specimens were removed and lysed by sonication. The fluorescence probe was added to the medium for assay.

Table 2 shows the results, expressed as arbitrary fluorescence units, normalized relative to the quantity of protein extracted.

The UVA-induced oxidation markedly affected the exposed epidermis with an increase in cell oxidation of 650%. In the presence of the EEC, cell oxidation (as evidenced by the fluorescence measured) was reduced by 60% (–7,852 units) relative to the unprotected epidermis. In other words, the EEC effectively protects the epidermis and renders it 2.5-fold less susceptible to the harmful effects of UVA radiation.

Conclusion

Enzymes, functional proteins essential for the survival of living organisms, generally play a very specific role. As catalysts, they facilitate chemical reactions, which could not otherwise take place or would occur too slowly to be of value.

As the term "catalyst" suggests, enzymes are unchanged by the reaction to which they contribute. For a given task, enzymes are practically inexhaustible. However, their protein structure (large fragile molecule) makes enzymes unstable when exposed to exogenous aggression. In most of the industrial applications of enzymes and in living organisms, the life of those biochemical motors is limited: enzymes are quickly inactivated by temperature, radiation or attack by proteases or microorganisms.

Although protective enzymes could be used in cosmetics to protect the skin from the aggressions that accelerate aging, they have not been used because of their well-known instability. Meanwhile, skin exposed to daily aggressions needs increased protection – more than can be provided by the conventional free radical scavengers (such as polyphenols and vitamins E and C), which are quickly exhausted. As was recently shown, the scavengers may even form toxic products, despite their being used for their protective action.[3]

Table 1. EEC-based protection from UV-induced lipid peroxidation

	Test No. 1	Test No. 2
Irradiated control (nmol 4-HNE/10^6 cells)	1.060	0.975
1% EEC sol (nmol 4-HNE/10^6 cells)	0.671	0.442

Table 2. EEC-based protection from UV-induced cell oxidation in epidermal equivalent

Cell Oxidation	Variation from control (Fluo U/mg protein)	(%)
Non-irradiated epidermis	1,715	-
Irradiated epidermis	12,910	+653
Epidermis + 3% EEC sol	5,058	+194

The multifunctional EEC system we have described here is the forerunner of a new generation of protective and anti-aging cosmetic active substances that afford the prospect of cosmetic products whose effects are modulated by the external conditions imposed on the skin. The EEC, obtained by culturing extremophile microorganisms, enables protection of the skin against pro-oxidizing attacks. The protection increases with increasing temperature or increasing intensity of UVA radiation and is stable in cosmetic formulations.

—Claire Mas-Chamberlin, François Lamy, Philippe Mondon, Sebastien Scocci, Laurent de Givry, François Vissac and Karl Lintner,
Sederma, Le Perray en Yvelines Cedex, France

References

1. VT Marteinsson et al, Isolation and characterisation of Thermus thermophilus Gy 1211 from a deep-sea hydrothermal vents, *Extremophiles* 3 247 (1999)
2. VS Bernan et al, Marine micro organisms as a source of new natural products, *Adv Appli Micro* 43 57 (1997)
3. F Zülli et al, UV-A induced formation of toxic compounds in vitamin E, Proceedings 2001 IFSCC Conference, Stockholm, May 7-9, 2001, IFSCC, SCANCOS (2001) pp 239-247

A Photoprotection Polymer for Hair Care

Keywords: ultraviolet radiation, photodegradation, melanin, color fading, tensile strength, spectroscopy, substantivity, hair morphology, photostability

UV protection hair care products in the current marketplace are mainly monomeric and adopted from skin care, therefore lacking substantivity and water solubility. This chapter reviews the detrimental effects of each range of UV light to hair and presents experimental results on the sun protection performance of polyamide-2 for both hair morphology and artificial hair color fading.

Many researchers have implicated UV light as causing hair damage.[1,2] This damage to the hair can be manifested in a number of ways, including loss of color and tensile strength, the degradation of disulfide bonds and an increase in hair surface roughness. As consumer awareness about photodamage to hair has increased, sun protection products are expected more than ever to perform in a more visually perceptible way. However, the UV protection products for hair care in the current marketplace, mainly monomeric and adopted from skin care, lack substantivity and water solubility. Thus, they do not possess conceivable UV protection efficacy.

Polyamide-2[a], a photoprotective polymer, was designed specifically for UVB and UVA protection of hair. This chapter reviews the detrimental effects of each range of UV light to hair and then presents the experimental results showing the sun protection performance of polyamide-2 for both hair morphology and artificial hair color fading. Finally, photostability, an important subject for all sunscreen products, will be discussed in detail.

Effects of UVA and UVB on Hair

Ultraviolet radiation is divided into three primary regions: UVC (< 290 nm), UVB (290–320 nm) and UVA, including UVA II (320–340 nm) and UVA I (340–400 nm). Of the three ranges of UV radiation, the shortest wavelength, UVC, is filtered out by the stratosphere and is not likely to contribute to the photodamage reactions of hair proteins. UVB radiation is the primary cause of natural photodamage since the most significant chromophores in hair proteins absorb UVB.[3] Photodegradation of hair is

---

[a] Solamer is a product of Nalco Company, Naperville, Illinois, USA

believed to follow the photolysis mechanism proposed by Togyesi.[4] Cystine, tyrosine, phenylalanine and tryptophan absorb UVB radiation, resulting in the formation of free radicals. Homolytic scission of disulfide bonds occurs. Dryness, reduced tensile strength, rough surface texture, brittleness and hair color bleaching are the direct consequences from UVB induced hair photodegradation. Although UVA is not the primary cause of UV damage to hair, scientists have found that visible and UVA light are largely responsible for artificial hair color fading.[5] In summary, both UVB and UVA radiation play a role in the photodamage of hair, the former associated with deteriorated hair integrity and the latter leading to dye fading.

Polyamide-2

Hair fibers contain a natural melanin coloring component that provides partial sun protection to the hair. It acts in a sacrificial manner and its degradation results in the lightening of hair color. However, melanin in hair cannot fully protect against UV damage when hair is repeatedly exposed to sunlight. Synthetic UV protection molecules are therefore necessary to protect hair against the adverse effects of solar radiation. Polyamide-2 was specifically designed to absorb a broad range of UVA II and UVB. The development of polyamide-2 represents an advanced technology in hair protection from UV rays.

As shown in Figure 1, both chromophore groups of methyl salicylamide and 1,3-diimine (or its tautomer), the former used as a capping agent and the latter incorporated in the polymer chain, provide a wide range of UVB and UVA II absorption.

Polyamide-2 is cationic when pH is less than nine due to the protonated nature of amine functionality. This cationic characteristic gives the polymer good substantivity on hair, which was proven in the following evaluation. Polyamide-2 is also water soluble, so it can be easily formulated in water-based hair care formulations. The importance of combining the substantivity and UV protection properties of polymeric materials with typical properties of water solubility, low toxicity and lasting protection efficacy is self-evident.

Hair Treatment and Test Formulations

Hair samples and treatment: Virgin blond and medium bleached brown hair was purchased[b]. Hair was washed by applying 1 g sodium laureth sulfate (SLES) on a 2 g hair tress from top to bottom and massaged for 1 min. The tresses were rinsed under 40° +/- 2°C tap water for 1 min and soaked in deionized water overnight.

Three treatments were applied on separate hair tresses: surfactant solution, shampoo and conditioner formulations. There are three variants of each formulation: no polymer (control), polyamide-2 and benchmark (polyquaternium-59 with butylene glycol). For the treatment, 1 g of the respective test formulation was applied to each washed tress for 1 min, rinsed under DI water for 30 seconds and air-dried before the next treatment. The treatment was performed a total of three times.

[b] International Hair Importers & Products, Inc. Glendale, New York, USA
[c] L'Oréal Feria, Multi-Faceted, Shimmering Color 67 is a product of L'Oréal, Paris, France.

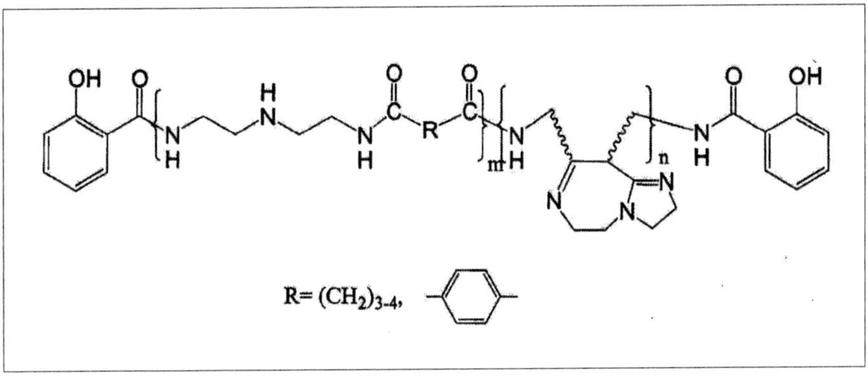

Figure 1. Chemical structure of polyamide-2

For the artificial color fading study, pre-washed, medium bleached brown hair was dyed[c] following the manufacturer's instructions. The dyed hair then received the same 1 g of the respective test formulation, as described above.

Test formulation: Surfactant solution used 10% sodium lauryl sulfate (SLS) with 2% (solid basis) test polymer. Clear shampoo and conditioner formulations are listed in Formulas 1 and 2.

Sunlight simulation: An accelerated weathering tester[d] with a 340 UVA bulb has a spectrum similar to sunlight in the range of 290-400 nm. The irradiation of the fluorescent bulb is at 340 nm maximum with an energy dose of 450 J/cm^2. Hair tresses of 1.5 g each were arranged in a single layer and were placed 10 cm from the UV lamps. Each side of the hair tresses was equally exposed and total exposure time was 400 h.

Evaluation

Work on updated methods for evaluating the performance of polyamide-2 was completed. These evaluations encompass a range of performance characteristics including hair color, hair morphology, hair integrity, substantivity and photostability.

Color fading: The total color change of $\Delta E = [9^{*}\Delta L^2 + \Delta a^2 + \Delta b^2]^{1\text{-}2}$ was used to calculate color fading.[6] It was determined by measuring[e] the L, a, b Tristimulus color values of hair tresses, where "L" stands for lightness, "a" for color of green to red, and "b" for blue to yellow. For the sun protection study, virgin blond hair was selected as the test substrate due to its susceptibility to sun damage. The same three treatments described above (i.e., surfactant solution, shampoo and conditioner with three variants – no polymer, polyamide-2 and benchmark) were applied to hair tresses. The determined changes in–Δ E values are presented in Figure 2.

The total color change (Δ E) represents the hair color difference of hair tresses before and after UV exposure. Since the bleaching effect of brown melanin in hair is an indication of hair photodamage, unquestionably, a lower Δ E value represents

[d] QUV Accelerated Weathering Tester is a product of Q-Panel Lab Products, Cleveland, Ohio, USA

[e] Hunter Colorimeter LabScan XE, Hunter Associates, Reston, Virginia, USA

Formula 1. Clear shampoo formulation

Water (*aqua*)	qs to 100% w/w
Ammonium laureth sulfate (Standapol EA-1, Cognis)	43.0
Cocamidopropyl betaine	9.3
Sodium lauroyl sarcosinate (Hamposyl L-30, Hampshire Chemical Corp.)	5.0
Testing polymer	2.0 (solid)
Preservative	0.2
Citric acid	qs to pH 5.8 - 6.10
Sodium chloride	1.0

Formula 2. Clear conditioner formulation

Water (*aqua*)	qs to 100% w/w
Hydroxyethylcellulose (Natrosol 250 HHR, Hercules)	0.8
Testing polymer	2.0 (solid)
Panthenol, 50%	0.7
Cetrimonium chloride (Varisoft 300, Degussa)	2.0
Glycerin	1.0
Polysorbate 20 (Tween 20, ISP Technologies)	0.5
Preservative	0.2
Sodium hydroxide (or) citric acid	qs to pH 5.5

better sun protection. One can see from Figure 2 that hair tresses treated with formulations containing polyamide-2 show the lowest Δ E values in all three treatments. Similarly, total color change also was used to evaluate artificial hair color fading. In addition, color changes in lightness index (Δ L), in red-green index (Δ a), and in blue-yellow index (Δ b) also was included in the evaluation. As one would expect, the smallest color change represents the least artificial hair color fading after UV exposure.

Manually dyed and dried hair tresses were treated with conditioner formulations (control, polyamide-2 and benchmark). Polyamide-2 treated tresses show the lowest values for all indices. The color differences among these tresses were so apparent that they could be easily identified by visual inspection. The polyamide-2 treated hair remained red while the other two largely faded to brown.

Tensile strength: The strength of hair directly reflects its health. Hair strength comes from the helical chains of keratin polypeptides that are stabilized by hydrogen bonding, cystine cross-linkages, coulombic interactions and hydrophobic

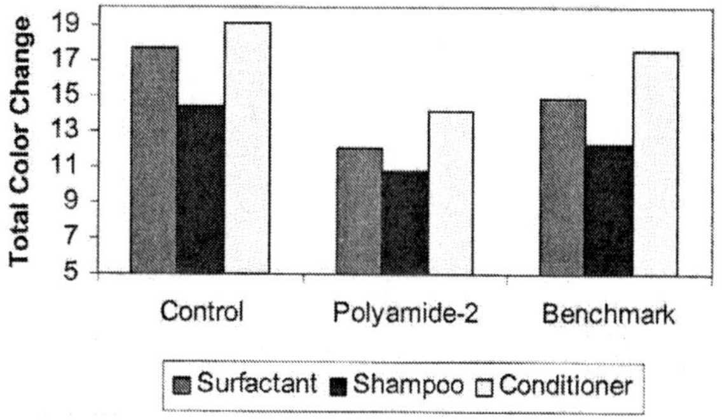

Figure 2. Total color change evaluation for hair protection

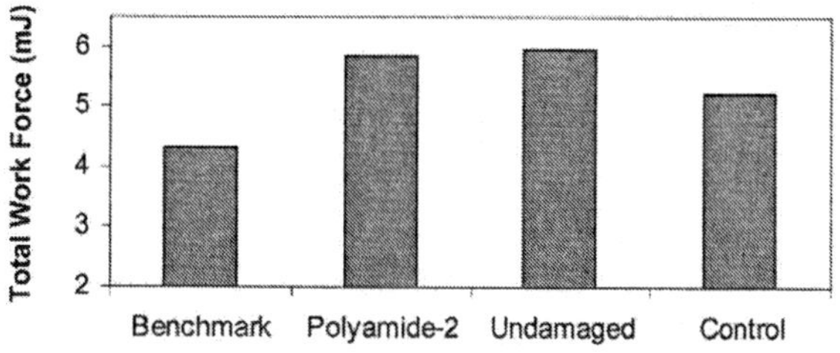

Figure 3. Tensile strength

interactions.[7] Scission of cystine linkages due to excessive UV exposure will disrupt the helical structure and weaken the hair. Thus tensile strength, the total work required to break a hair strand, is another means of evaluating the photodamage of hair fiber.

Hair tresses that were treated with a conditioner formulation containing either water (control), polyamide-2, or the benchmark were exposed to UV for 400 h. An unexposed hair tress was also included in the test to provide an example of undamaged hair. Sixty-five hair strands were randomly selected from each tress and hair diameter was measured using a fiber dimensional analysis system[f].

The hair samples were then placed in a miniature tensile tester[g]. To eliminate the effect of hydrogen bonding, the test was conducted in a wet condition. Each individual hair strand was pulled at a rate of 20 mm/min until breakage occurred. Load and extension were automatically recorded and converted to stress and strain using application software[g]. The mean values were then calculated and a Tukey HSD

[f] Fiber Dimensional Analysis System, Model LSM 5000, is a product of Mitutoyo, DiaStron Limited, UK.
[g] DiaStron Miniature Tensile Tester Model 170/670 and model MTTWIN Version 5.0 are products of DiaStron Limited, UK.
[h] JMP statistical software is a product of SAS Institute, Cary, North Carolina, USA

statistical analysis was performed for all testing pairs (ANOVA one-way analysis of variance from JMP statistical software[h]) (Figure 3).

It was found that the tress that received no UV protection from the control showed an obvious reduction in tensile strength compared to undamaged hair proving hair undergoes photodegradation.

The overall decrease in tensile strength upon UV exposure is believed to be related to the scission of cystine disulfide bonds in the hair proteins. This will be explained in more detail in the next section. Hair strands protected by polyamide-2 required a significantly higher breaking force than those that were unprotected or treated by the benchmark. Hair protected by polyamide-2 showed no statistical difference from undamaged hair. Although the cause or mechanism is unknown, the benchmark-treated hair was weaker than the control. The statistical analysis proves a significant difference at the 95% confidence interval. The test results indicate that polyamide-2 provides effective protection of the hair from UV damage.

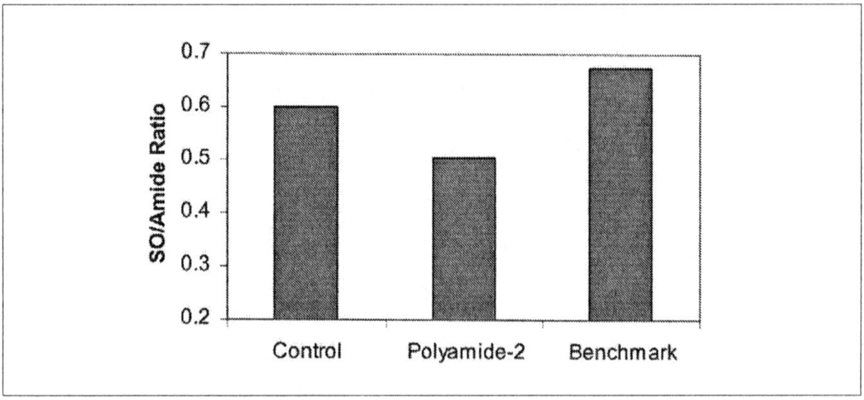

Figure 4. FTIR analysis of SO/Amide-I band

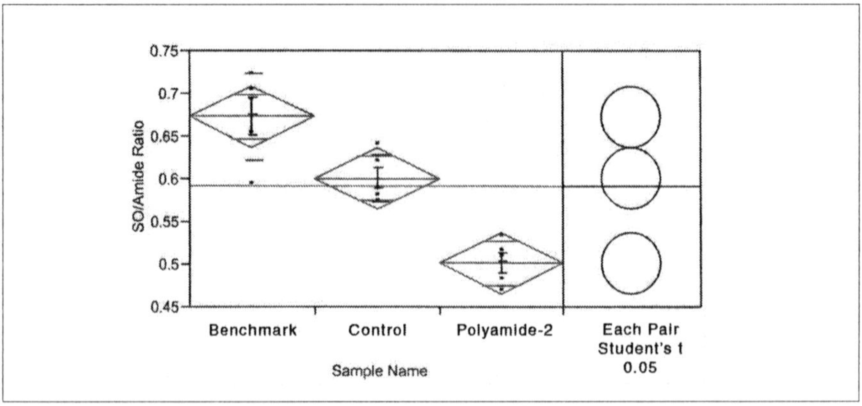

Figure 5. Statistical analysis of FTIR results

Hair integrity via FTIR spectroscopy: It is generally accepted in the industry that the photodamage of hair protein proceeds via oxidation of cystine following the S-S scission pathway, whereby oxidative scission yields S-sulfonic acid that is finally degraded to cysteic acid. While the S-S bonds are susceptible to this degradation, the Amide I band is not affected by UV radiation. The ratio of the S-O bond/Amide I band measured by Fourier Transform Infrared Spectroscopy (FTIR spectroscopy) is, therefore, an indication of the degree of photo-degradation of hair fibers. In other words, the higher the S-O/Amide I ratio, the more severely degraded the hair.

A hair tress treated with conditioner formulation and UV exposed was pressed on a single reflectance cell with the hair aligned along the axis of the beam path of an FTIR spectroscope[i]. One hundred twenty eight scans were signal averaged. The detector employed was deuterated triglycine sulphate (DTGS). The Amide I peak at 1633 cm^{-1} was used as an internal standard. Peak height of the SO band at 1041 cm^{-1} was measured and normalized to the Amide I band. Each type of hair tress was measured five times at ambient temperature and humidity.

It is clear that the polyamide-2 treated hair tress showed a lower amount of oxidative products than both the benchmark and control (Figure 4). Unexpectedly, the benchmark showed a higher mean value of SO/Amide I ratio than the control, which is consistent with the prior tensile strength measurement; that is, benchmark-treated hair seems to have more deterioration than the control. For the statistical analysis, the center lines of the means diamonds are the group means (Figure 5).

The top and bottom of the diamonds form the 95% confidence intervals for the means. It can be said that the probability is 95% that this confidence interval contains the true group mean. If the confidence intervals shown by the means diamonds do not overlap, the groups are significantly different at 95% confidence. None of the confidence intervals displayed overlap (Figure 5); therefore, the results are statistically valid.

Substantivity: As mentioned earlier, sun protection products commonly used in hair care today have been adopted from those used for skin care. Octylmethoxy-cinnamate (OMC) is a typical example. However, these oil-soluble UV filters do not normally have affinity to the hair and, in many cases, they only deposit in minimum amounts from rinse-off formulations.[8] Substantivity, therefore, is of much importance since it is a key factor to determine the UV protection efficacy for hair care.

The authors developed a non-traditional weighing method to measure substantivity using a fine micro-balance in a humidity- and temperature-controlled environment. Pre-tabbed hair tresses were washed and dried and then the initial mass (W_0) was recorded using a 5 decimal-place micro-balance[j].

The hair was then treated in a 5% polymer solution (solid basis) for five minutes and rinsed under deionized water for 30 sec and dried. The treatment was performed three times in total and then the treated hair was weighed again (W_t). The hair tab was carefully removed and weighed as W_b. The substantivity is expressed as:

$$\textit{Substantivity}\ \textbf{(mg Polymer/g Hair)} = [(W_t - W_0)] / (W_0 - W_b)$$

[i] Nicolet Avatar 360 FTIR is a product of Varian Instruments, Walnut Creek, CA, USA
[j] Mettler AE 163 micro-balance is a product of Mettler Toledo, Columbus, Ohio, USA

Table 1. Means and standard deviations of substantivity

Test Name	Mean value of substantivity (mg Polymer/g Hair)	No. Of replication	Standard Deviation	Std Err Mean	Lower 95%	Upper 95%
Benchmark	26.2101	5	6.26019	2.7996	20.110	32.310
Control	4.0566	5	1.29621	0.5797	2.794	5.320
Polyamide-2	34.2807	5	1.90899	0.8537	32.421	36.141

The absolute values of substantivity are largely related to the type of test method used. Some traditional methods such as ^{14}C radio labeling and gel permeation chromatography can detect the polymer deposition on hair at low levels when a low concentration of treatment solution is used.[9] The typical substantivity measured in these methods is around 7-10 µg polymer/mg of hair, or 7-10 mg of polymer/g of hair.[10] The substantivity listed (Table 1) is expected to be higher than these conventional methods due to the higher concentration treatment solution used. Nevertheless, the data provides a relative comparison for the amount of deposition of polyamide-2 compared with the benchmark (Table 1). Obviously, polyamide-2 shows better deposition, which results in better sun protection in rinse off formulations.

Hair morphology via AFM: Atomic Force Microscopy (AFM) is a technique that allows for direct interaction between a sharp probe and a sample surface as the probe is brought into contact with this surface. Using piezo electronics and laser deflection, these interactions can be precisely measured. A common measurement provided by this technique is to compute height variations in a surface profile as recorded within the scanned area.

The specific method used here was to incorporate a low force contact probe to directly measure[k] points within the scanned surface of each hair fiber. The method would easily scan over the surface while maintaining constant tip contact via an electronic feedback system within the AFM. This method generally would correspond to the technique of surface profilometry.

The collected data points are then plotted along a Z-scale defined by a specified and matching color palette for each image. This palette was then projected as a 3D image by using a defined image pitch (30°) and image rotation (45°) to qualitatively exhibit the distinct surface profiles.

Three hair strands were tested using the AFM method: undamaged (virgin), control, and polyamide-2 treated. The latter two had been UV exposed. Typical 3-D projected images from these separate samples are shown (Figures 6–8). The undamaged image displays a smooth cuticle edge indicating no evidence of damage.

The polyamide-2 treated hair has an image similar to the undamaged image even after UV exposure. The control image displays clear evidence for cuticle

[k] VEECO Methrology, Digital Instruments Div. Dimension 3100 is a product of VEECO Metrology, LLC, Santa Barbara, California, USA

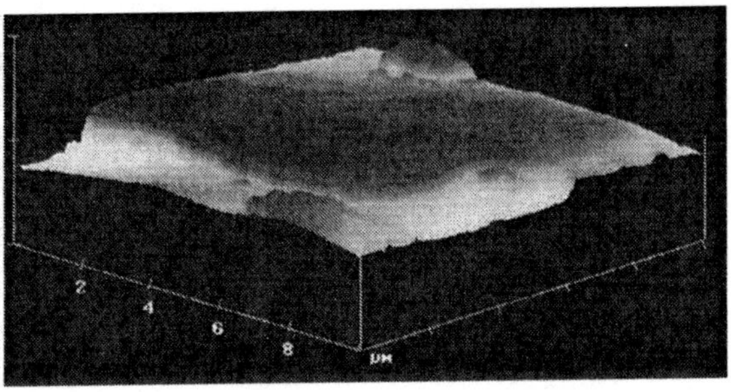

Figure 6. AFM image for undamaged hair

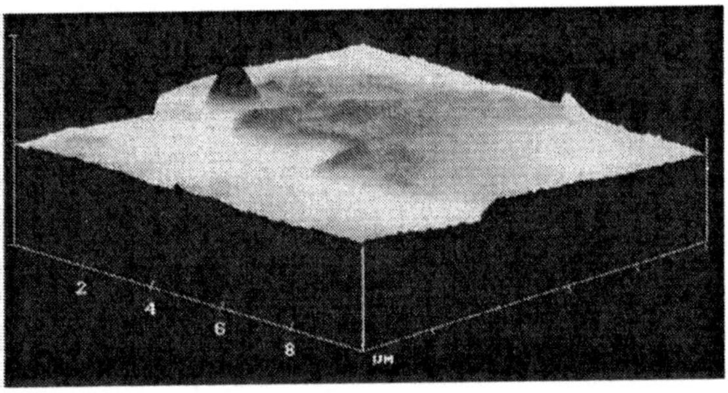

Figure 7. AFM image for UV protected hair

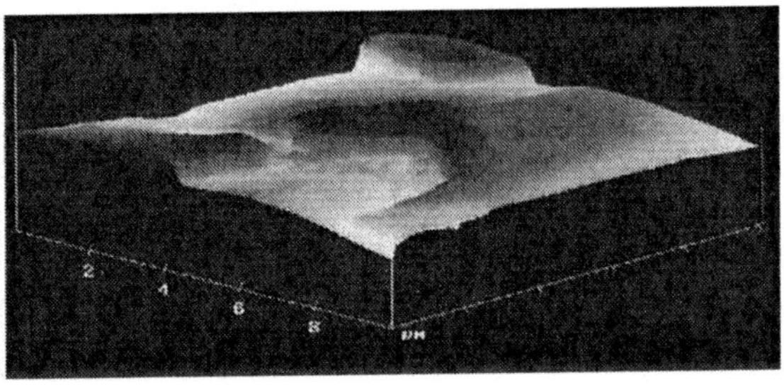

Figure 8. AFM image for UV unprotected hair

edge damage where a portion of this edge has been broken off. The similar findings were also observed from an SEM study,[11] where the damaged cuticle cells were found to be thinning or fusion due to photodegradation. The AFM method provides a different perspective to identify the photodamage and prevention. This work was also supported by the frictional analysis of the cuticle, which proves that

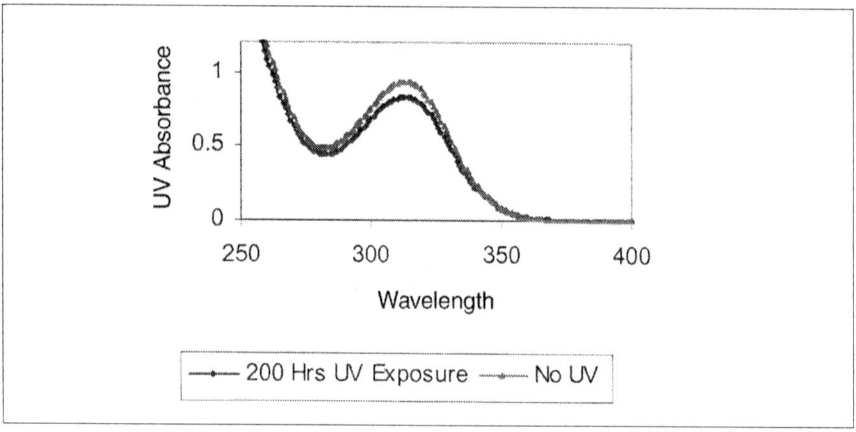

Figure 9. UV absorbance before and after UV exposure for polyamide-2

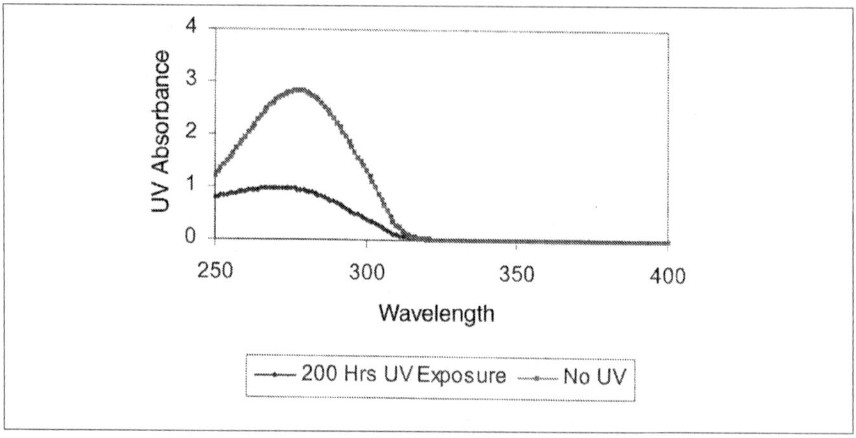

Figure 10. UV absorbance before and after UV exposure for polyquaternium-59 with butylene glycol

the polyamide-2 treated hair tresses have a significantly lesser friction coefficient than the control.[12]

Photostability: If a UV filter loses its absorbance over time, it may be called photounstable. The cause of the photolabile or photoinstability can be explained by the energy dissipation mechanism. The absorption of UV radiation by some organic molecules can cause the excitation of one of a pair of electrons in a low energy orbital to a higher energy unoccupied orbital.[13] An excited molecule can return to its ground state by a combination of several mechanistic steps. Two of these steps, fluorescence and phosphorescence, involve the release of a photon of radiation.[14] The other deactivation steps are radiationless processes. The favored route to the ground state is the one that minimizes the lifetime of the excited state. Thus, if deactivation by fluorescence is rapid with respect to the radiationless processes, such emission is observed. Fluorescence is, therefore, considered one of the pathways that dissipates the energy in a non-destructive pathway. On the other hand, if a

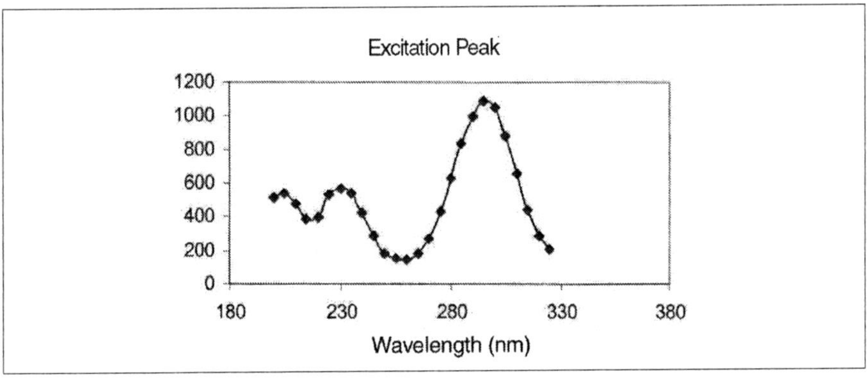

Figure 11. Fluorescence excitation peak for polyamide-2

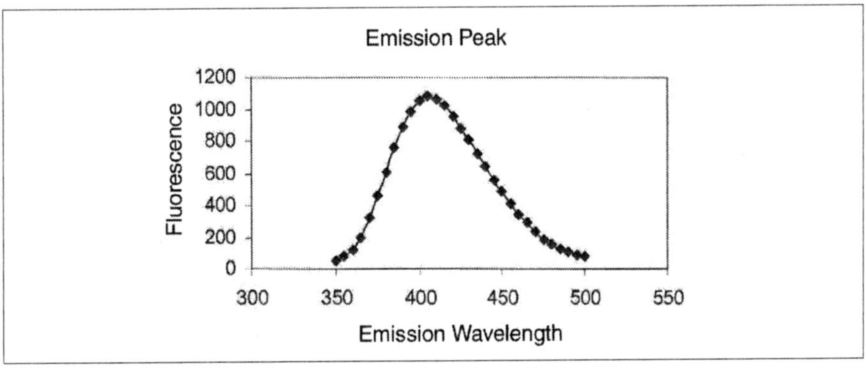

Figure 12. Fluorescence emission peak for polyamide-2

radiationless path has a more favorable rate constant, fluorescence is either absent or less intense. Energy dissipation can also follow the destructive pathways, which possibly belong to radiationless steps where the fragmentation, or certain types of isomerization and bimolecular reactions likely happen.[13] If the destructive pathway predominates, the molecule will be photounstable.

To study the photostability, both polyamide-2 and the benchmark were prepared at 10% in water. The solutions were scanned for UV absorption before and after UV exposure (200 hours) using a UV-visible spectrophotometer[m] to detect the change in absorbance (Figures 9–10).

The absorbance of solutions was recorded against the wavelength of 400 to 250 nm. Polyamide-2 shows almost no reduction in UV absorbance compared to the benchmark. Fluorescence microscopy[n] was used to further understand the energy dissipation mechanism (Figures 11–12).

Polyamide-2 solution was prepared at 1 ppm concentration in water and scanned at 30000 nm/min speed with sampling interval 10 nm. Both figures clearly indicated that polyamide-2 has fluorescence property. It absorbs the radiation near 300 nm

[m] Cary 3 Bio UV Visible Spectrophotometer is a product of Varian Instruments, Walnut Creek, California, USA

and emits near 400 nm. The longer wavelength emitted is due to the phenomenon called "Stokes shift." The same fluorescence microscopy was also tested for benchmark, but no fluorescence was observed. This investigation not only provides a substantial understanding of the photostable nature of polyamide-2, it also explains why polyamide-2 exhibits high UV protection efficacy on hair.

Summary

Polyamide-2 is a wide-range, UV-absorbing polymer that was specifically designed to incorporate a unique chromophore group and methyl salicylate endcaps. Further, additional functionality was added to ensure water solubility. These novel structural attributes provide a wide-range of UV absorption capabilities and ease-of-formulation. Polyamide-2 has been proven to be an effective UV absorbing polymer that offers numerous benefits when formulated into hair treatment products. It helps maintain hair tensile strength, lessens the degradation of cystine disulfide bonds and minimizes the fading of hair color (both natural and artificial). Further, this innovative UV-absorbing polymer is photo-stable, and it has been demonstrated to be substantive and improve hair morphology. Polyamide-2 is an important tool that can be used to protect hair from the harmful effects of UV radiation.

—Yin Z. Hessefort, Wayne M. Carlson and Mingli Wei,
Nalco Company, Naperville, Illinois, USA

[n] F-4500HITACHI Fluorescence Microscopy is a product of Hitachi High-Technology, UK.

References

1. S Nacht, Sunscreens and hair, *Cosm Toil* 105 (December 1990)
2. SB Ruetsch, Y Binhua and YK Kamath, Role of melanin and artificial hair color in preventing photo-oxidative damage to hair, *IFSCC* magazine 7, No. 2 (April/June 2004)
3. CM Pande and J Jachowicz, Hair photodamage - Measurement and prevention, *J Soc Cosmet Che* 44 109–122 (March/April 1993)
4. E Tolgyesi, Weathering of hair, *Cosmet Toil* 98 29–33 (1983)
5. B Locke and J Jachowicx, Protection of artificial hair color, *J of Cosmetic Science* 54(2) p. 212, (March/April 2003)
6. J Jachowicz, M Helioff, and C Rocafort, Photodegradation of hair and its photoprotection by a substantive photofilter, *Drug and Cosmetics Industry* (December 1995)
7. C Zviak, *The Science of Hair Care*, New York: Marcel Dekker (1986)
8. C Pande and J Jachowicz, Hair photodamage: Measurement and prevention, *J Cosmet Sci* 44 109-122 (1993)
9. AR Sykes and PA Hammes, The use of merquat polymers in cosmetics, *Drug and Cosmetics Industry* (Feb 1980)
10. B Blanco, BA Durost and RR Myers, Gel permeation chromatography: An effective method of quantifying the absorption of cationic polymers by bleached hair, presented at the Annual Scientific Seminar of the Society of Cosmetic Chemists, Nashville, TN (May 1997)
11. SB Ruetsch, Y Kamath and HD Weigmann, Photodegradation of human hair: An SEM study, *J Cosmet Sci* 51 103–125 (Mar/Apr 2000)
12. RW Cloud and YZ Hessefort, Characterization of Friction Properties for Photo-Degraded Hair Fibers Using AFM, to be presented at the 23rd IFSCC (October, 2004)
13. NJ Turro, *Modern Molecular Photochemistry*, Menlo Park, CA: Benjamin/Cummings: (1978) p 3

Protecting Against UV-Induced Degradation and Enhancing Shine

Keywords: shine, fading, UV radiation, Polysilicone-15, UV protection, photostable

The data presented shows polysilicone-15 demonstrates the ability to perform as a UV filter for hair protection and decreases the combing force necessary after irradiation to sample hair tresses in addition to enhancing the shine attribute of a formulated product.

Healthy, shiny hair starts with proper protection and care. Unfortunately, hair is exposed to daily stress, and without proper treatment it will become weakened and may appear unhealthy and dull. In a recent nationwide survey conducted in the United States, consumers demonstrated a high awareness to what UV radiation can do to hair; 61% of 1,002 respondents indicated they knew that sun visibly damages their hair.[1]

Artificially colored hair and natural uncolored hair of Asian and Caucasian origin are susceptible to fading upon prolonged exposure to UV radiation. Polysilicone-15[a] has been shown to reduce such fading.[2] The integrity of hair was shown to be affected by exposure to UV radiation as demonstrated in one study in which the content of amino acids, such as cysteine and tryptophane, present in hair were protected against UV-induced degradation.[3]

In this paper, we will explore two additional effects of polysilicone-15 on hair: hair combability and hair thermal stability via differential scanning calorimetry (DSC). These measurements demonstrate the structural damage of hair caused by UV radiation and how polysilicone-15 can help to protect the hair against this damage.

Polysilicone-15

Polysilicone-15 has the benefits of a silicone plus the added benefit of UV protection because the molecule has UV filter moieties attached to the silicone backbone. The silicone provides shine, conditioning and smoothening to the hair while the UV filters provide protection against UV degradation and color change of the hair.

[a]Parsol SLX is a product of DSM Nutritional Products AG

Chromophores

Organic molecules in sunscreens contain a chromophore that absorbs light at certain wavelengths. Chromophores are groups of atoms bonded together that share electrons. When wavelengths of light, whether visible or UV, match the resonance of these electrons, they will be absorbed and cause an excitation. This energy can then be released in many different ways.

The organic molecules in sunscreens usually emit a thermal release of energy through vibrational relaxation or nonradiative decay. The molecule goes from the excited state back to the ground state where it can continue to absorb more UV light. Today, most sunscreens contain molecules than have chromophores that absorb both UVB and UVA. The three most common protective molecules found in most daily moisturizers were oxybenzone, octisalycate and oxtinoxate.

Source: University of Pennsylvania School of Arts and Sciences Web site. Available at: http://www.sas.upenn.edu/~hasty/work.html. Accessed: January 17, 2005.

Table 1. Specifications of polysilicone-15

Specifications	
Appearance:	Clear, pale yellow, liquid
Specific extinction E(1%, 1cm):	160 – 190
γ max:	310 – 314 nm in ethanol
Refractive index:	1.44
Specific gravity:	1.015 – 1.045

Polysilicone-15 is comprised of polysiloxane chains (Figure 1) containing benzyl malonate chromophores (see sidebar on Chromophores). It is photostable and has a good safety profile due to its large molecular size. Polysilicone-15 is approved in the European Union as a UVB filter with an allowed dosage of up to 10% in sun care products. It is not an approved sunscreen for skin protection in the United States, but it may be used for hair care. The specifications of polysilicone-15 are given in Table 1.

Protecting Hair Combability

The outermost layer of keratin fibers consists of approximately three-quarters protein and one quarter lipid. UV degradation of amino acids will result in a fragmented keratin structure and significant structural changes of the surface cuticle cells (e.g. lifting, thinning). UV-induced lipid degradation will also contribute to altered surface properties. The state of the cuticle governs the frictional properties of hair fibers

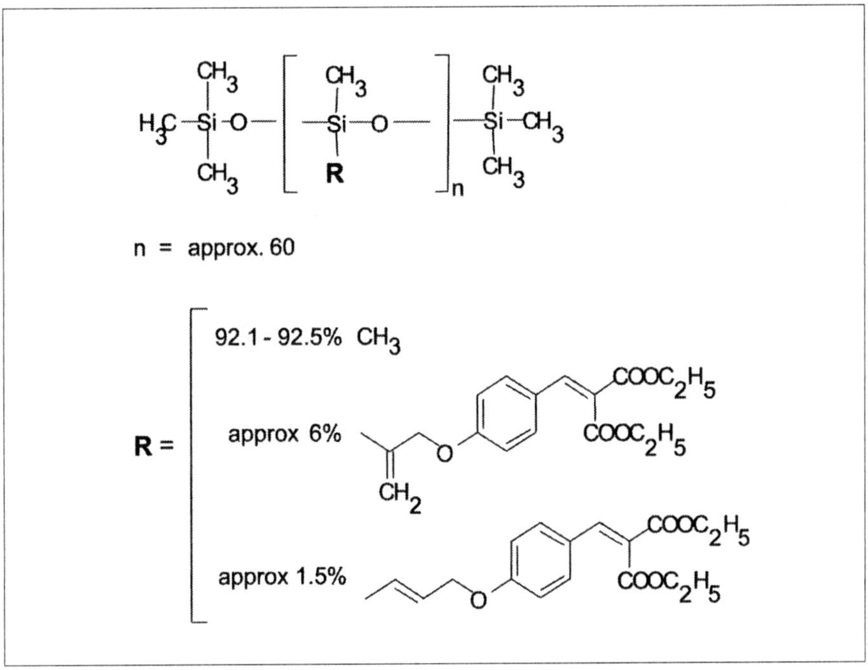

Figure 1. Structure of polysilicone-15

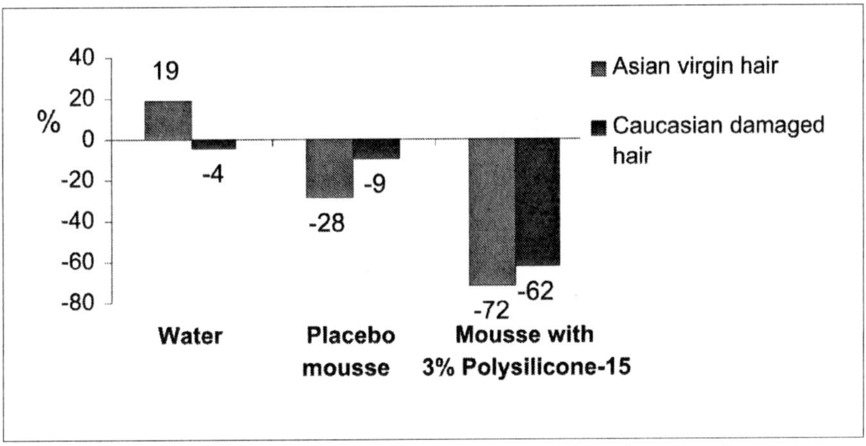

Figure 2. Relative change in combing force (after application of formulation, irradiation and washing)

and determines how the hair feels to the touch and how easily it combs out. These physical changes lead to a rough hair surface and can directly relate to the stronger forces required to comb UV damaged hair. Combability measurements are taken to assess any alteration of the hair cuticle due to external stress and have been proven sensitive tools to detect UV-induced hair damage.[4,5]

Methods: The wet combing forces on hair tresses–Caucasian (bleached and permanently colored) and virgin Asian samples–were measured[b] before and after irradiation [Irradiance (300-400 nm): 6.5mW/cm^2 @ 70% r.h and 23°C, Atlas ES25 Weather-Ometer; Dose: (over 140 h): 3.2kJ/cm^2] under environmentally controlled conditions (20°C, 65% r.h.). Before irradiation, tresses were washed, dried, acclimated, wetted and initial combability was measured. The tresses were then treated with the leave-on formulations (Formula 1) and irradiated. Prior to the final combability measurements, the hair tresses were again washed, dried, acclimated, wetted and final combability was measured.

Each tress was massaged with a generous amount of mousse for 2 min. The tresses were then combed to remove excess mousse and dried under climatized conditions. The tresses were reacclimated back to 60% humidity (wet combability) and combed at 100mm/min on a length of 10 cm and the tensile strength was measured[c].

Results: Figure 2 illustrates the level of protection attained when treating the hair with a leave-on conditioner containing a UV filter prior to irradiation.

Both the placebo and test mousses led to decreased combing forces after irradiation and washing. The treatment with pure water resulted in an increase of combing force after UV irradiation. This confirmed that higher forces are needed to comb UV damaged hair. The placebo mousse provided some protection due to the conditioning effect of cocamidopropyl oxide. However, under the same conditions, the swatch treated with 3% polysilicone-15 (formula B) showed a much stronger effect on reducing the combing force for both hair types.

The difference between damaged Caucasian hair (bleached and permanently colored) and virgin Asian hair is limited. This indicates both hair types are affected by the treatment. Although the dark melanin pigments (eumelanin) present in virgin Asian hair are known to act as natural UV absorbers, they are present in the cortex of the hair and not in the cuticle. This means that the proteins and lipids in the cuticle of both hair types are equally sensitive to UV irradiation and the resulting degradative processes. Since the state of the cuticle governs the frictional properties of hair fibers, it is not surprising to see that both Asian and Caucasian hair were similarly affected in terms of UV-induced loss of combability.

Protecting Hair Structure

Background on differential scanning calorimetry: As illustrated in the previous studies, the condition of the hair cuticle determines important attributes for the consumer, such as ease of combability and shine. However, other attri-

[b] Atlas ES25 Weather-Ometer, Atlas Material Testing Technology, LLC.
[c] INSTRON 1122, Instron

Formula 1. Leave-on mousse formulations in hair combability study

	A	B
Water *(aqua)*	ad 100% w/w	ad 100% w/w
Cocamidopropyl oxide	0.40	0.40
Polysilicone-15	-	3.00
Preservative	0.60	0.60
Propane/butane	10.00	10.00

Formula 2. Leave-on mousse formulas in hair structure study

	A	B	C
Water *(aqua)*	ad 100% w/w	ad 100% w/w	ad 100% w/w
Cocamidopropyl oxide	0.40	0.40	0.40
Polysilicone-15	-	1.00	3.00
Preservative	0.60	0.60	0.60
Propane/butane	10.00	10.00	10.00

butes including hair strength are determined by the condition of the hair cortex. The complex structure of hair can be simplified using the matrix/filament model proposed by Feughelman.[6]

The main part of the fiber is composed of an amorphous matrix into which crystalline α-helical filaments are embedded. The amorphous matrix contains proteins with a high sulphur content and is responsible for structural fiber integrity. The crystalline α-helical filaments contain proteins with a low sulphur content and are responsible for fiber elasticity.

DSC is a method of choice in assessing the state of both amorphous matrix and crystalline α-helical filaments of keratin fibers.[7-9] Upon heating, the structure of the hair samples becomes denatured and this denaturation can be quantified by measuring the two following endpoints:

- Denaturation Temperature (T_D) which indicates the structural integrity of amorphous matrix
  T_D (virgin hair) > T_D (damaged hair)
- Denaturation Enthalpy (δH_D) which indicates the content of native α-keratin in the crystalline filaments.
  δH_D (virgin hair) > δH_D (damaged hair)

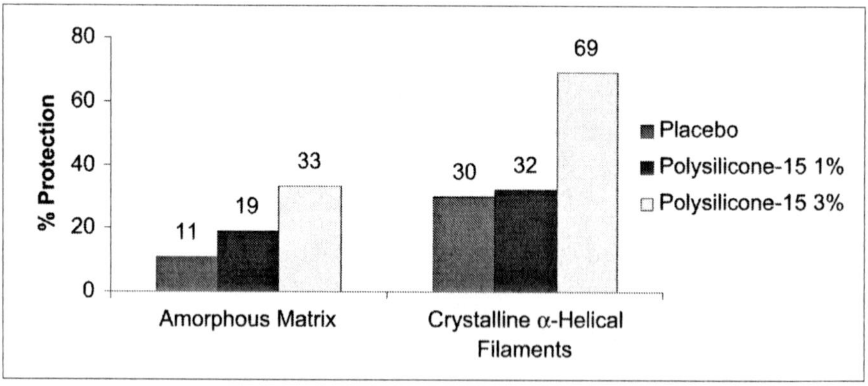

Figure 3. UV protection of hair structure using DSC measurements (bleached Caucasian hair)

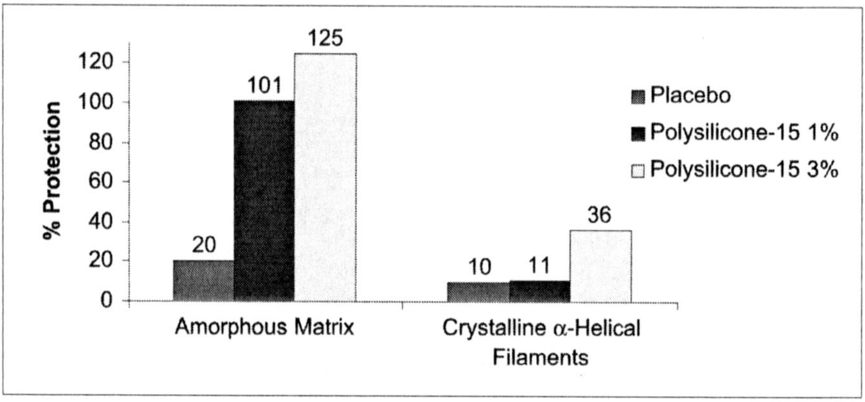

Figure 4. UV protection of hair structure using DSC measurements (virgin Asian hair)

Methods: We measured the denaturation temperature (T_D) and denaturation enthalpy (δH_D) of hair tresses – Caucasian (bleached and permanently colored) and virgin Asian samples – before and after irradiation. Before irradiation (under the same parameters in the first study, at a dose over 150 h: 3.4 kJ/cm^2), the tresses were washed and towel dried. DSC measurements were made on untreated hair. The tresses were then treated with the formulations in Formula 2 and irradiated. Final DSC measurements were made.

Each tress was massaged with a generous amount of mousse for 2 min. The tresses were combed to remove excess mousse and dried under environmentally controlled conditions. DSC measurements were taken: wet state [(50 µl water + 10 mg hair), 80-180°C, 10°C/min, n=2].

Results: Figures 3 and 4 illustrate the level of protection achieved when treating the hair with a leave-on conditioner containing a UV filter prior to irradiation.

Protection (matrix, %)

$$\frac{(\delta TD_{Untreated} - \delta TD_{Formulation})}{\delta TD_{Untreated}} \times 100$$

Protection (α-helical, %)

$$\frac{(\delta H_{Untreated} - \delta H_{Formulation})}{\delta H_{Untreated}} \times 100$$

The protection of the hair follicle's matrix and α-helical components can be expressed using the following equations:

Using this equation, calculations reveal the hair will be protected if the use of the test formulation versus no treatment leads to no change in denaturation temperature/enthalpy. When compared to no treatment, no protection is obtained when the change in denaturation temperature/enthalpy for the test formulation equals the values obtained for the untreated hair.

For both hair types, the amorphous matrix and the crystalline α-helical filaments are significantly protected when polysilicone-15 is present in the formulation (in both B and C samples). The amount of protection is clearly concentration dependent. Interestingly, the best protection of the amorphous matrix was obtained for virgin Asian hair while the best protection of the crystalline α-helical filaments was obtained with bleached Caucasian hair.

Enhancement of Hair Shine

Background on hair shine: Consumers perceive shiny hair as healthy hair. Shine is "in" and it is the number one claim for most hair care products. The surface of a hair fiber is such that perceived hair shine not only depends on the intensity of visible light reflected at a 90-degree angle (specular reflection) but also on the light reflected at other angles (diffuse reflection). Polysilicone-15, being a liquid, does not recrystallize when used in leave-on applications and spreads uniformly on the hair fiber. Its relatively high refraction index enhances hair shine as indicated by the results below.

Formula 3. Leave-on formulas for shine enhancement

	A	B	C
Isododecane	ad 100% w/w	ad 100% w/w	ad 100% w/w
Polysilicone-15	-	2.00	4.00
Phytantriol	0.25	0.25	0.25
Vitamin E acetate	0.25	0.25	0.25
Fragrance *(parfum)*	0.05	0.05	0.05
Methyldibromo glutaronitrile (and) phenoxyethanol	0.20	0.20	0.20

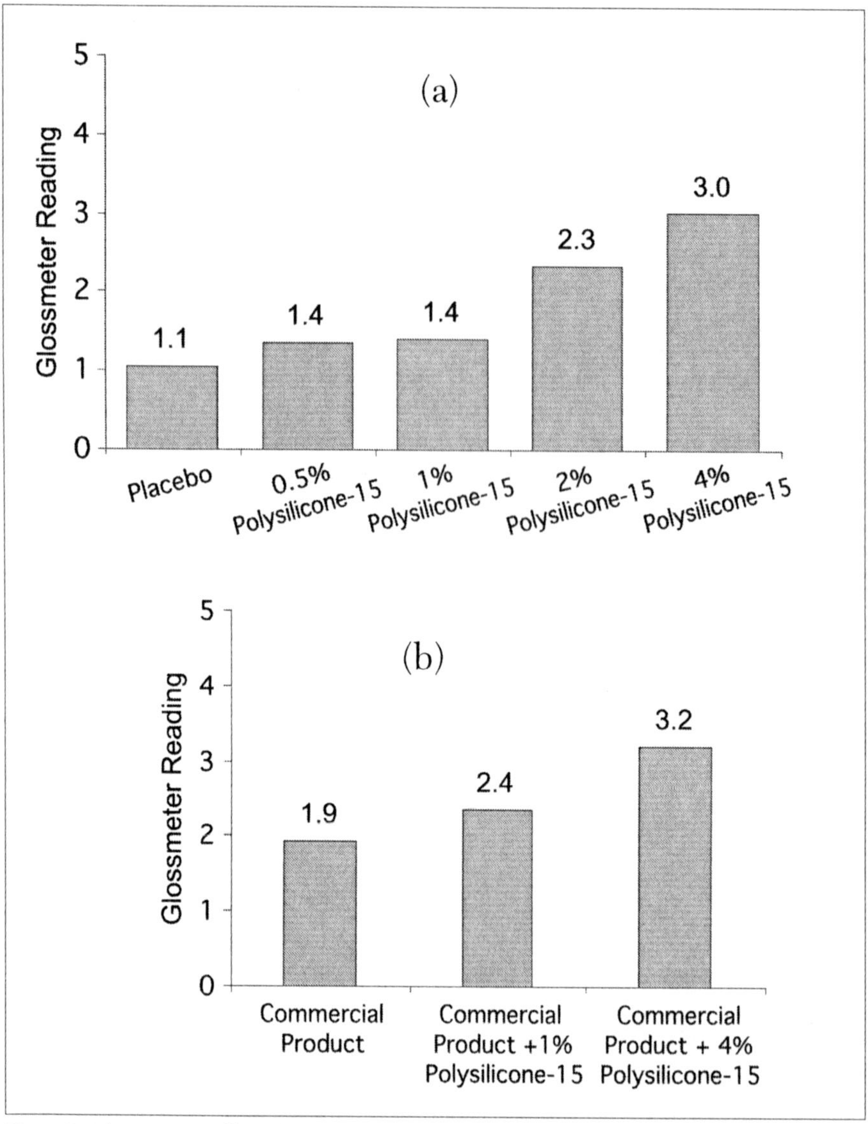

Figure 5. Enhancement of hair shine (Glossmeter measurements) using polysilicone-15 in aqueous solution (a) and commercial hair protection spray product (b)

Method: Varying amounts of polysilicone-15 in a UV protection spray (Formula 3) was applied by spraying the formulations on hair tresses at a 10-cm distance using six pumps from a pump bottle. The product was allowed to dry for 30 min before measurements were taken using a gloss meter.

Figures 5 and 6 illustrate the level of shine enhancement obtained by adding polysilicone-15 to the formulations.

Discussion: The chemical binding of chromophores attached to a silicone polymer, such as for polysilicone-15, is a new concept in the development of effective types of sun filters. The experimental data presented in this paper indicate that the performance of a UV filter for hair protection cannot be predicted by solely considering its capacity in absorbing UV radiation [E (1%, 1cm) value]. Whenever compared to the filter ethylhexyl methoxycinnamate, polysilicone-15 always performed better in terms of hair protection, despite its relatively low absorption value (about one fourth compared to EHMC). The high performance of polysilicone-15 is linked to its structure; internal results have shown that the distribution and the loading of the UV chromophores on the siloxane chains are optimized for highest performance in terms of UV protection. The silicone moiety of polysilicone-15 favors uniform spreading on the hair fibers while the polymeric structure offers substantivity in rinse-off applications.

In the combability study, the very significant decrease of combing forces observed after irradiation is the result of the properties of the two moieties of polysilicone-15 (UV chromophore and silicone polymer) working in synergy.

Ingredients that enhance hair shine usually have high refraction indexes and spread uniformly on the hair fiber. Polysilicone-15, with a refractive index of 1.44 and its silicone backbone, increases perceived hair shine when added in cosmetic formulations as shown in the shine study.

Additionally, panel studies have shown that no build-up effect is observed upon consecutive use of a shampoo containing polysilicone-15.[10] This is probably also due to the structure of the molecule. The large size of the chromophores attached to the silicone backbone sterically disfavor excessive buildup.

Polysilicone-15 also delivers benefits such as improving fragrance retention on hair tresses using leave-on formulations.[11]

Conclusion

Our studies indication polysilicone-15 protects the structure of hair against UV radiation as measured by DSC and enhances combability of hair after exposure to UV radiation. Furthermore, it enhances the shine attribute of a formulated product.

—Philippe Maillan, Anna Gripp, Fintan Sit, Roland Jermann and Hörst Westenfelder,
DSM Nutritional Products AG, Basel, Switzerland

References

1. USA Caravan Consumer Survey (2002)
2. P Maillan, UV protection of artificially coloured hair using a leave-on formulation, *Int J Cosmet Sci* 24 1-7 (2002)
3. H Gonzenbach, W Johncock and K-F De Polo, UV damage on human hair, *Cosm Toil* 113 43-49 (1998)
4. J Jachowicz, M Helioff and C Rocafort, Photodegradation of hair and its photoprotection by a substantive photofilter, *DCI* (Dec 1995)

5. P Maillan, Protecting hair combability from UV irradiation using a leave-on formulation, *Cosm and Toilet Mfr Worldwide* 22-26 (2003)
6. M Feughelman, Physical properties of hair, Hair and hair care, Editor: DH Johnson, *Cosm Sci and Techn* 17, M Dekker 13-32 (1997)
7. F-J Wortmann an H Deutz, Characterizing keratins using high pressure differential scanning calorimetry, *J Appli Polymer Sci* 48 137-150 (1993)
8. F-J Wortmann, C Springob and G Sendelbach, Investigations of cosmetically treated human hair by differential scanning calorimetry in water, *J Cosmet Sci* 53 219-228 (2002)
9. H Schmidt and F-J Wortmann, High pressure differential scanning calorimetry and wet bundle tensile strength of weathered wool, *Textile Res J* 64 (11) 690-695 (1994)
10. Study report, Institute Dr Schrader, Holzminden, Germany (June 2004)
11. WO 2003035022 A1, DSM Nutritional Products.

Photoprotection from Ingested Carotenoids

Keywords: ingested, UV-induced skin damage, β-carotene, carotenoids, Dunaliella salina

Dunaliella salina *alga derivative increases skin's pigmentation and reflectivity*

Recently the photoprotective and antioxidative properties of β-carotene and other carotenoids have added functionality to this traditional source of provitamin A (see sidebar). For example, it can be ingested to provide additional sun protection for persons who are already using a high-SPF topical sunscreen. Raab cites its photoprotective action and good compatibility and recommends prescribing it for photodermatosis, ingestion of phototoxic drugs or prevention of UV-induced skin damage.[17] An already existent phototoxic dermatitis caused by, for example, sulfonamides or antibiotics, reacts well to a secondary administration of β-carotene.

In a comprehensive study, we measured the changes that occur in various parameters of the skin after oral administration of β-carotene and other carotenoids derived from a natural source, the alga *Dunaliella salina*. We focused our attention on the question of whether ingestion of natural mixed carotenoids could be expected to reduce the sun sensitivity in volunteers.

Methods

Design of the study: Using 20 test subjects with healthy skin, we studied the effect of ingested natural mixed carotenoids on skin color, photosensitivity and reflection spectra in different skin locations:

- Forehead
- Inside of the forearm
- Back of the hand
- Palm of the hand
- Back
- Back with erythema

We took initial measurements before the study began to document the baseline values for each test subject. Then, for six weeks, we gave each subject a daily dosage of 50 mg (two 25 mg capsules) of natural mixed carotenoids derived from the alga *Dunaliella salina* dissolved in vegetable oil. To determine the effects on the skin color, photosensitivity and reflection spectra, we took measurements at four weeks and again at six weeks.

We also completed a lifestyle questionnaire for each test subject, documenting factors such as the subject's dietary habits, alcohol consumption, smoking habits and use of medication.

Determining skin color: Skin color depends largely on melanin pigmentation and hemoglobin oxygenation. Pigmentation is due to the presence of melanin pigment in the epidermis, and involves the interaction of melanocytes and keratinocytes. The skin's pigmentation is subject to a regulatory process, which can be influenced by external factors. Such factors include UV-irradiation, oxidation, inflammation and hormonal processes as well as coloration after ingestion of certain foods (carotenoids).

In this study, we measured the skin color at each of the selected sites with an instrument[a] that converts skin color to a three-dimensional coordinate system in which absolute numerical values of shade, saturation and luminosity reveal even slight changes in color. Our procedures to obtain these values conformed with standards (L-, a-, b-systems) of the Commission Internationale de l'Éclairage (CIE).[5]

Determining photosensitivity: We assessed the photosensitivity of each test subject by irradiating[b] the subject's back with UV light and then using different levels of light to determine the minimum erythema threshold dosage (MED). We measured the redness exactly and expressed it as the a-value (redness value) in the erythema zone 24 h after irradiation in comparison with the color value immediately before irradiation.[6]

Reflection-spectroscopy: Reflection-spectroscopy is a non-invasive method for quantitative substance analysis in vivo. We used it to obtain detailed information about the mode of action of carotenoids in the body.

Reflection-spectroscopy involves irradiating the tissue with either white light[19,20] or monochromatic light.[3] Some of this light reflects from the surface, and some penetrates the tissue before it is reflected. The reflected light is recorded and evaluated with the help of suitable sensors. The light that penetrates the tissue is more or less attenuated by the substances in its path. The absorption at different wavelengths is characteristic for individual substances. The reflected light therefore contains information about the type and amount of substances with which it has interacted on its path through the tissue.

During the past 20 years, completely new mathematical methods of signal analysis have been developed. These include strictly deterministic as well as statistical methods for quantitative analysis of reflection spectra.[7,8]

The reflection spectrometer[c] used in our study consists of a 50 Watt halogen lamp that emits continuous light in the 300-1500 nm range. An optical system guides the light into a bundle of glass fibers, in which 1000 individual fibers, each with a diameter of 50 microns, transmit the light to the surface under examination. Inside the light guide, 1000 additional glass fibers of the same size transmit the reflected light to a holographic grating that separates the wavelengths between 400

[a] Minolta Chromameter CR200, Minolta, Hamburg, Germany
[b] SOL 3, Fa. Hönle, Munich, Germany

and 1100 nm into a spectrum. The grating has 1200 lines/mm, giving it a resolution of 1 nm. A photosensor converts the optically separated light into an analog signal. A 16-bit analog-to-digital converter prepares the signal for further processing in a computer system.

The electronic system can record up to 400 spectra each second. The high speed of signal registration matches the speed of the physiological processes under examination. For instance, in vivo quantitative determination of hemoglobin is necessary for the understanding of a large number of physiological processes. The hemoglobin concentration in the capillaries increases rapidly during a systole and falls again more slowly during the diastole. At a heart rate of 120 beats/min, the duration of the systole is 80-100 msec. That means the system can record 32-40 spectra during each systole. Therefore the concentration of the hemoglobin should not change significantly while any single spectrum is being recorded.

Evaluating reflection spectra: In order to evaluate the reflection spectra, which are the measured spectra of the skin, one must record the so-called basic

β-Carotene: Vitamin, Photoprotectant, Antioxidant

There are many risks associated with acute and chronic irradiation of the skin with ultraviolet light. In particular, these include premature aging of the skin, UV-induced hyperkeratosis and UV-induced atrophy. A further consequence is the provocation of skin diseases, precancerosis and carcinoma, including examples such as epithelioma, basalioma and possibly malignant melanoma.

A suntan is a symbol of attractiveness and social success, and as long as this is the case it is essential to provide clear information about the use of internal and external protective agents. Some people with very sensitive skin already apply topical sunscreen products with a high sun-protection factor. For these people, additional protection can be especially worthwhile.

UV irradiation releases free radicals, primarily in the epidermis, so there is a need for substances that can suppress the harmful action of these free radicals, or combine with or eliminate them after they have been formed. β-Carotene (provitamin-A) is such a substance. Interest was originally focused mainly on its provitamin-A character, but in recent years the photoprotective and antioxidative properties of not only β-carotene but also other carotenoids have gained in importance.

β-Carotene converts to retinol in the intestine, but this metabolization only occurs until retinol reaches a normal level in the plasma. Therefore, β-carotene cannot cause vitamin-A intoxication. This is confirmed by the fact that it is well tolerated at typical dosages of 180-200 mg/day, for example by patients with protoporphyria, for which β-carotene is now the preferred remedy. Depending on individual absorption, however, oral dosages of more than 50 mg/day can cause the skin to acquire an orange coloration that has no negative effects and is totally reversible when the dosage is reduced.

c Multi-Scan OS10, program version BCV1.5, MBR, Herdecke, Germany spectra. Basic spectra for a substance are spectra measured with the substance in solution, usually in a cuvette, under conditions of no scattering. To do this, we removed the carotenoid mixture from a capsule and dissolved it in ethanol at a ratio of 1:80. Then we introduced this solution into a 10 mm quartz cuvette positioned in front of a flat, reflecting white surface.

For the evaluation of the skin spectra of the test subjects, the amount of measured β-carotene in the skin (as the main component) can be determined as a function of the hemoglobin concentration in the wavelength zone from 425-500 nm, because the concentration of the hemoglobin in the wavelength zone above 520 nm is only insignificantly affected by the presence of β-carotene. For the mathematical calculation of the concentration of β-carotene in the skin, this means that it is sufficient to carry out a multivariant multicomponent analysis.[13] We used the spectra of the oxygenated and deoxygenated hemoglobin and of the β-carotene as the basis of the vector space. Furthermore, we recorded the skin spectrum of the relevant test subject each time a test-subject measurement was carried out. In order to do this, we interrupted the supply of blood to the forearm tissue by exerting pressure on the artery of the upper arm. We used the spectrum of the blood-free tissue for calibration and corrective measurements, especially of β-carotene.

Results from the Questionnaire

Data obtained from the questionnaire concerning dietary habits, alcohol consumption, smoking habits and use of medication revealed no peculiarities that could have impaired the course of the study.

Results of the Coloration Measurements

Coloration measurements during the course of the study revealed only a slight increase in the a-values (skin redness), and that occurred only in the area of the forehead. By contrast, the b-values (yellow component of the skin color) exhibited

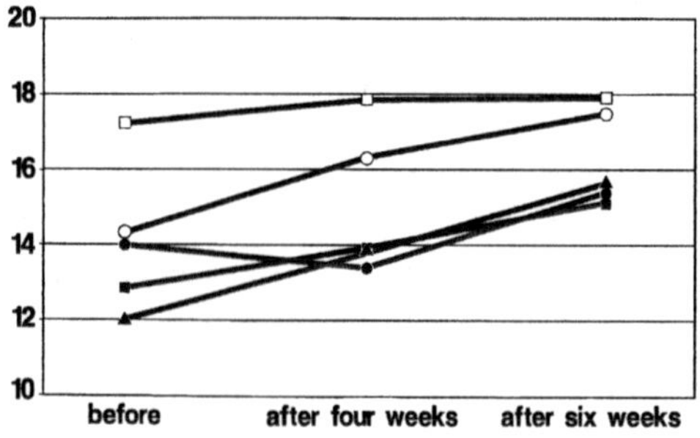

Legend	Six-week difference
palm of the hand	3.61
forehead	2.95
back of the hand	2.17
arm	1.46
back	0.66

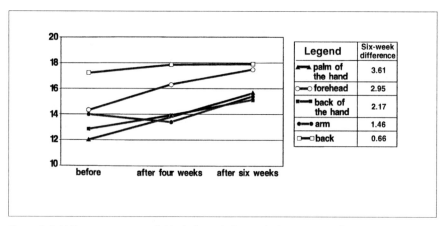

Figure 1-1. Yellow component of skin before, during and after carotenoid ingestion in 20 subjects

more marked changes, as shown in Figure 1-1, which also shows the differences in the average b-values before and after the six-week study.

One would expect a lowering of the L-value (skin luminosity) to accompany darkening of the skin. However, the L-values we obtained from the individual measuring zones revealed no significant changes that could indicate a slightly darker skin shade.

Influence of Photosensitivity

To ascertain the photosensitivity of the test subjects, we determined the MED at various times before, during and after ingestion of the natural mixed carotenoids. In a designated erythema zone on the backs of our test subjects, we measured the a-values 24 h after each of the three irradiations to determine the redness of the skin associated with each erythema. These measurements revealed that the skin redness decreased as the study progressed, from a-values averaging 12.06 before the carotenoid capsules were consumed, to 10.92 after four weeks (second irradiation) and 10.52 after six weeks (third irradiation).

For comparison, we plotted the yellow component (Figure 1-2) of the skin color of the test subjects' backs and found rather good correlation between the increase in the yellow component and the decrease in photosensitivity of the same area of skin.

Results of Reflection-Spectroscopic Studies

β-Carotene concentration: Table 1-1 shows the β-carotene concentration at various skin sites during the study. In general, we found that the β-carotene was relatively uniformly distributed into the different areas of each test subject's body. However, at four weeks we did find higher β-carotene concentrations in the back region and in the palm of the hand (0.264 and 0.243 nmol/g tissue, respectively). In addition, we found slightly lower β-carotene concentrations at sites such as the

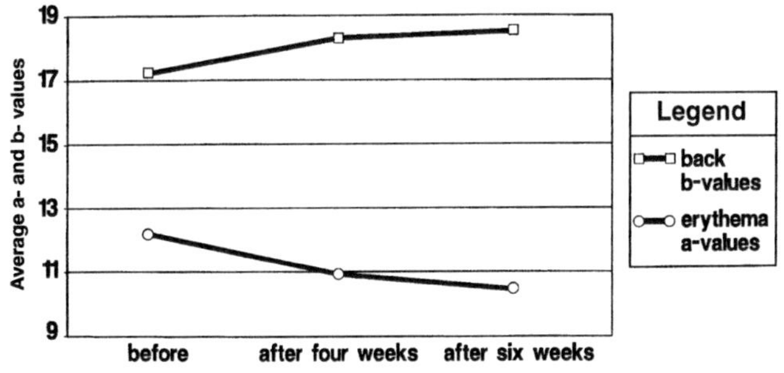

Figure 1-2. Yellow and erythema red components of skin before, during and after carotenoid ingestion in 20 subjects

Table 1-1. β-Carotene concentrations (nmol/g tissue) at various skin sites before, during and after ingestion of carotenoids (averages for 20 subjects)

	Before	At 4 weeks	At 6 weeks
Forehead	0.095	0.194	0.226
Palm of the hand	0.145	0.243	0.286
Back of the hand	0.077	0.229	0.252
Back	0.109	0.264	0.289
Inside of forearm	0.157	0.256	0.303

forehead and the back of the hand (0.194 and 0.229 nmol/g tissue, respectively) where there was more exposure to sunlight and, perhaps, more consumption of β-carotene to perform the much discussed cell-protection function. A higher tissue concentration after six weeks indicates that the maximum concentration in the tissue had not been reached after four weeks of taking carotenoids.

We also found that the hemoglobin content remained steady during the study. This is an important indication that no systemic cardiovascular changes are to be expected while the preparation is being taken.

Hemoglobin concentration: We obtained a further measurement parameter by means of spectroscopic examination of the erythemas generated by UV irradiation. When examining an erythema spectroscopically, one must distinguish between two different optical components of the erythema. These components are caused by two different physiological processes. On the one hand there is an increase in the amount of blood in the irradiated area of the skin when an erythema is generated. This increase corresponds to an increase in the hemoglobin concentration. On the other hand, if the erythema is generated by UV light, as it is in this case, the pigmentation of the skin also undergoes a change. Because both the change in hemoglobin concentration and the change in skin pigmentation take place in the same wavelength zone from 450 to 650 nm, one must distinguish between them, otherwise large errors can occur in the erythema determination. Figure 1-3 shows the correlation between pigmentation and hemoglobin concentration change associated with an erythema. The changed spectra after carotenoid supplementation shows a lower absorption, leading to a smaller increase of erythema, and consequently a smaller hemoglobin increase.

Before the first irradiation, the hemoglobin concentration in the tissue at the location on the back where the erythema was induced was 0.2 mg/ml. An average value of 0.2 mg/ml was also measured at a reference point on the lower arm. After the first irradiation (prior to carotenoid ingestion), hemoglobin concentration on the back increased by 0.13 mg/ml. After the second irradiation (at four weeks), hemoglobin concentration again increased by an average of 0.13 mg/ml. By contrast, after the third irradiation (at six weeks), hemoglobin concentration increased by only 0.07 mg/ml. This means that the third irradiation, after six weeks of administration of β-carotene, resulted in only half the previous increase in hemoglobin. In contrast to the first two irradiations, the third irradiation produced an effect that was much less pronounced in terms of microcirculation.

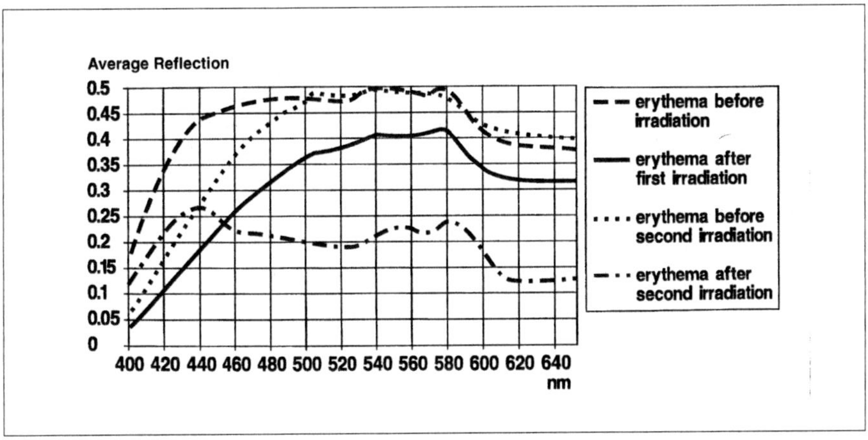

Figure 1-3. Reflection spectra for back skin from 20 subjects before and after two incidents of irradiation, showing the relation between pigmentation and erythema

Pigmentation: A change in pigmentation causes an increase or decrease in scattered radiation.[9] An increase in pigmentation is accompanied by increased reflection of light from the skin. In spectroscopic terms, this means that less light penetrates the skin and is absorbed there.

The scatter in the observed wavelength range from 400 to 1100 nm is largely independent of wavelength and is expressed in a uniform shift of the spectrum parallel to the x-axis towards smaller optical density values. The amount of scatter can be calculated with the help of the theory of Kubelka and Munk, which states that the multiplicative constant becomes smaller.[11,12] The quotient for the extinction at 650 nm is a direct measure for the pigment change. A higher value indicates more pigmentation.

As the β-carotene content of the skin increases, so does the pigmentation, as indicated by the skin's reflection capability. The increase in pigmentation caused by the β-carotene concentration depends on the initial pigmentation. Therefore the increase in pigmentation caused by β-carotene is an additive effect.

Overall, we found that the reflection capacity of the skin was increased by a factor of 2.3 as a result of taking carotinoid capsules. In other words, the amount of incident light that was reflected increased by this factor, irrespective of the wavelength. Therefore, carotenoid supplementation caused a beneficial change in the absorption characteristics of the skin, improving its protective function.

Accompanying the increase in reflection capacity was a parallel shift in the spectra, pointing toward less absorption and therefore more reflection of the incident light. One could view this, in simplified terms, as an additional protective function against incident light, if the carotenoids are stored in the tissue as titanium dioxide is stored—in the form of finely distributed particles. This would produce a positive change in the absorption conditions by increasing the natural pigment action.

Relation Between Fat Intake and β-Carotene Absorption

Of the more than 600 different carotenoids identified so far, only a few break down in the human body into vitamin A in amounts that are of physiological and nutritional relevance. Quantitatively, β-carotene is the most important vitamin A precursor. As a fat-soluble substance, it is absorbed in the small intestine. Absorption from vegetable nutritional components is in the 30-60% range, depending on the proportion of fat in the diet. 15,15-Deoxygenase, an enzyme present in the intestinal cells and the liver, breaks the central double bond of β-carotene, forming two molecules of vitamin A (retinol). The enzymatic conversion depends on the organism's supply situation, the level of β-carotene and protein intake, the vitamin E supply and the simultaneous fat absorption. The complete reaction mechanism has yet to be explained. The β-carotene is stored mainly in the fatty tissue and the liver. There are numerous indications that the total mass of fat in the body has an influence on the liberation of β-carotene and therefore on plasma concentration and availability for the target tissue.[1,2,10,15,18]

Studies carried out by Prince and Frisoli on β-carotene accumulation in serum and skin showed that the amount of β-carotene in serum increased by a factor of 2.5 when the test subjects were fed a diet rich in fat.[16] A reference group under a low-fat diet exhibited no significant increase in serum concentration.

In our study, analysis of the individual results of the test subjects revealed marked differences in the incorporation of β-carotene in the tissue. Although the lifestyle data obtained from a questionnaire concerning smoking and alcohol consumption, diet and drug intake gave no indications that any appreciable effects on the metabolism were to be expected, a comparison of the individual body weights showed that people with a high weight apparently take up especially large amounts of β-carotene in the tissue. It is at least conceivable that these people have a higher fat intake than others, therefore indirectly favoring the absorption of β-carotene into the tissue.

Relation Between β-Carotene Content and Skin Photosensitivity

Besides influencing various physiological parameters of the skin by administering capsules of carotenoids derived from the alga *Dunaliella salina*, this study aimed above all to answer the question of whether ingestion of natural mixed carotenoids can be expected to lower the sensitivity of voluntary test subjects to the sun. We found that administration of 50 mg carotenoids each day produced a continual increase in β-carotene concentration even after six weeks.

In general the distribution of β-carotene in the five skin areas we examined was relatively uniform. However, we observed slightly lower β-carotene values from areas (such as the forehead and the back of the hand) that were more exposed to the sun. This observation agrees with studies made by other authors, who also detected a decrease in the concentration of β-carotene after exposure to sunlight.[4,14] Notice that this study was carried out during the winter months; this effect would possibly have been much more pronounced during the summer, when the light would have been stronger. It is possible that consumption of β-carotene increases under UV irradiation. In view of the release of free radicals, above all in the epidermis, by UV radiation, this could be taken as an indication that in this situation β-carotene acts directly as a radical scavenger.

Gollnick et al examined the skin-protective action of β-carotene against UV damage in the Berlin-Eilat study.[4] In a double-blind, placebo-controlled study involving the participation of 20 female student volunteers who were exposed to the sun under controlled conditions in Eilat (Israel), the authors found that β-carotene clearly increases the erythema threshold and therefore decreases the risk of sunburn. Although sunburn could not be completely prevented under intensive sunlight, the authors found that regular intake of β-carotene provides a protective effect. In comparison to the placebo group, the photoprotection was clearly observed. Simultaneously they found that the level of β-carotene in the serum decreases sharply under the influence of intensive exposure to the sun's rays, and in some circumstances can drop below a critical value. For this reason, Gollnick et al suggested that people with skin Type I (that is, the more sensitive Celtic type) protect themselves from sun-induced skin damage by taking regular doses of β-carotene. We agree.

During our own study, we used reflection spectroscopic analyses to obtain more information about the action mechanism and the effect on the metabolism of administering carotenoids. We found that the skin redness in an erythema zone on the backs of our 20 test subjects decreased following three irradiations as the study progressed. This observation correlates with the reflection spectroscopy analyses, also of the erythema zone, which showed that the hemoglobin concentration increased by only 0.07 mg/ml after the third irradiation, in comparison with 0.13 mg/ml at the start of the study. In contrast to the first irradiation, therefore, the effect on the microcirculation was much smaller.

It was also clear that the pigmentation of the skin as a result of taking carotenoids performed an additional protective function. Carotenoid ingestion changed the reflection capacity of the skin by a factor of 2.3. Moreover, the reflection is not wavelength-dependent; irrespective of the particular wavelength, the amount of reflected incident light increases by a factor of 2.3. During the course of the study we observed that the increase in the β-carotene concentration in the tissue (Figure 6) was accompanied by a parallel shift in the spectra. The light-protection effect is attributable to the fact that less light is absorbed in the tissue and more light is reflected from the surface of the skin. The protective factor of 2.3 indicates that people with sensitive skin and correspondingly weak natural protective systems can be provided with double their usual amount of protection. This is of importance especially in combination with topically applied screening agents, since the action of chemical or physical light-protection filters can be doubled indirectly. As is the case with titanium dioxide, the β-carotene in the skin takes the form of finely dispersed protective particles, which have a positive effect on the absorption conditions by increasing the natural pigment action.

The product tested was well tolerated and none of the subjects showed any side effects whatsoever.

The study of natural mixed carotenoids derived from the alga *Dunaliella salina*, with regard to the photoprotective action of the skin, revealed numerous positive effects. However, many new questions were raised concerning the exact action mechanism, especially under the influence of incident light. These questions could be answered in appropriately designed follow-up studies.

—Ulrike Heinrich, Mathilde Wiebusch and Hagen Tronnier
Institute for Experimental Dermatology, University of Witten/Herdecke, Witten, Germany
—Holger Jungmann
Institute for Bioscientific Research, University of Witten/Herdecke, Witten, Germany

References

1. Beta-Carotin, *VitaMinSpur* 6 175-178 (1991)
2. KH Biesalski, Wirksamkeit von Beta-Carotin bei der Prävention von Krebs, *VitaMinSpur* 5 Suppl I 3-32 (1990)
3. B Chance, Spectrophotometry of intracellular respiratory pigments, *Science* 120 767-775 (1954)
4. H Gollnick, SC Chun, C Hemmes, K Sundermeier, W Hopfenmüller and HK Biesalski, Hautschutz durch Betacarotin gegen UV-Schäden, *Haut* 7 1-2 (1993)
5. U Heinrich and H Tronnier, Experimentelle Untersuchungen über Wirkungseintritt und -verlauf lokal aufgetragener Corticoide am Beispiel von Betametasondipropionat, *Akt Dermatol* 17 327-331 (1991)
6. U Heinrich and H Tronnier, Vergleichende Studie zur antiphlogistischen Wirkung von Ciclopiroxolamin , *Dt Dermatologe* 42 855-861 (1994)
7. WR Hruschka, Data analysis: Wavelength selection method, in *Near Infrared Reflectance Spectroscopy* , PC Williams, ed, St Paul, Minnesota: American Cereal Association (1987) pp 35-55
8. T Isaakson and T Naes, The effect of multiplicative scatter correction (MSC) and linearity transform in *NIR spectroscopy, Applied Spectroscopy* 42 1273-1284 (1988)
9. G Kortüm, *Reflexionsspektroskopie*, Berlin-Heidelberg-New York: Springer Verlag
10. NJ Krinsky, The biological properties of carotenoids, *Pure & Appl Chem* 66 1003-1010 (1994)
11. P Kubelka, New contributions to the optics of intensely light scattering materials, *J Opt Soc* 44 330 (1990)
12. P Kubelka and F Munk, Ein Beitrag zur Optik der Farbanstriche, *Z Techn Physik* 12 593 (1931)
13. H Martens and T Naes, *Multivariate Calibration*, Chichester: J Wiley (1993)
14. M Mathews-Roth, MA Pathak, J Parrish, F Fitzpatrick, A clinical trial of the effects of oral beta-carotene on the responses of human skin to solar radiation, *J Invest Derm* 59 349-383 (1972)
15. JA Olson, *Carotenoids and Vitamin A: An Overview of Lipid-Soluble Antioxidants Biochemistry and Clinical Applications*, ASH Ong and L Packer, eds, Basel: Birkhäuser Verlag (1992)
16. MR Prince and JK Frisoli, Beta-carotene accumulation in serum and skin, *Am J Clin Nutr* 57 175-181 (1993)
17. W Raab, Carotin-Wirksamer Schutz vor Schäden durch UV-Licht, VII. Stuttgarter Mineralstoff-Symposium (1995)
18. SS Shapiro, DJ Mott and LJ Machlin, Kinetic characteristics of β-carotene uptake and depletion in rat tissues, *J Nutr* 114 1924-1933 (1991)
19. R Wodick and DW Lübbers, Ein neues Verfahren zur Bestimmung des Oxygenierungsgrades von Hämoglobin-Spektren bei inhomogenen Lichtwegen, erläutert aus der Analyse von Spektren der menschlichen Haut, *Pflügers Arch* 342 29-40 (1973)
20. R Wodick and DW Lübbers, Quantitative evaluation of reflection spectra of living tissues, Hoppe-Seyler's *Z Physiol Chem* 335 583-594 (1974)

Sunscreen Formulation and Testing

Keywords: emulsion systems, efficacy, sunscreen actives, sunscreen, stratum corneum

The author reviews SPF testing methods and sunscreen components (actives, active solvents, water-resistance agents and emulsifiers) that assist the formulator in the art and science of sunscreen formulation.

Sunscreen formulation is both an art and a science. Sunscreens appear to be simply an emulsion system with sunscreen actives included in the formulation. However, this is far from reality. Even a formulator experienced in the development of emulsion systems will experience a myriad of difficulties during sunscreen formulation. This review will be successful if the reader understands that sunscreen formulation is complex and is best left to the experienced artist.

Formulation greatly influences the efficacy of the sunscreen actives. Efficacy of a sunscreen is defined as its ability to protect the skin against ultraviolet-induced burning, or Sun Protection Factor. The SPF is influenced by the type of sunscreen active(s), by the emulsion's oil phase, by the emulsion's water phase, by the emulsification process, and perhaps by other factors. Because sunscreen efficacy is defined by its SPF, the task of the formulator is to create a formulation with the highest SPF, at the lowest cost, and with highly favorable cosmetic properties.

This article reviews many of the tools available to the sunscreen formulator: sunscreen actives, formulation types, formulation characteristics and in vivo testing methodology.

Interaction Between Skin and UV Radiation

Skin: Skin is composed of two bands of defined tissue separated by a thin membrane. The dermis is the inner band. The epidermis, the outer band, is composed of four layers: stratum basale, stratum spinosum, stratum granulosum and stratum corneum.

The stratum corneum is the outer-most layer of the epidermis and is evident to the naked eye. It is about 20 cell layers thick, with no viable cells and no blood supply. The stratum corneum provides a protective barrier, called the acid mantle or horny layer, that keeps moisture inside the body and the environment outside. The stratum corneum provides some protection against UV radiation and can alter itself to provide additional protection.

UV radiation: Ultraviolet radiation is composed of UVC (200-290 nm), UVB (290-320 nm) and UVA (320-400 nm).

Our atmosphere blocks UVC, so only UVB and UVA reach the surface of the Earth. Generally, UVB penetrates only into the epidermis, while UVA penetrates all the way to the dermis. While both UVA and UVB cause skin changes, UVB is about 50 to 100 times more energetic than UVA at inducing these changes. UVB radiation is more energetic than UVA radiation at inducing erythema, melanogenesis, DNA damage and squamous cell carcinoma.

Interaction between UV and skin: UV radiation that impinges on the surface of the skin can be reflected back toward the environment, can be absorbed by substances on the surface of the stratum corneum or can continue deeper into the skin. The physics of ultraviolet radiation is reviewed elsewhere.[1] The stratum corneum thickens as it adapts to UV exposure, so it can absorb more UV.

Interaction between UV and sunscreen actives: Theoretically, sunscreen actives could perform their task of reducing or preventing UV-induced burning in one of three ways. One way (chemical) would be to absorb the UV radiation to prevent the radiation from damaging viable tissue. A second way (physical) would be to reflect the UV radiation. The third way (biological) would be to reduce inflammation either by blocking the biological inflammatory response or by enhancing biological repair. A few sunscreen actives may operate in more than one way.

While a chemical or physical sunscreen active must stay on the stratum corneum to be effective, a biological sunscreen active would have to reach the viable tissues to be effective.

A chemical sunscreen active is believed to protect the viable tissues by absorbing the UV radiation and transforming it into less damaging radiation such as heat or light. Chemical sunscreen actives might also generate free radicals in response to UV radiation. Regardless of the mechanism, chemical sunscreen actives should absorb the UV radiation before it can reach the viable tissues. For this to occur, the sunscreen active must maintain a high concentration in the stratum corneum for several hours.

SPF Testing

SPF is not well understood by the public, or even within the sunscreen profession. SPF is a ratio of the ability of a person to burn with the sunscreen relative to his

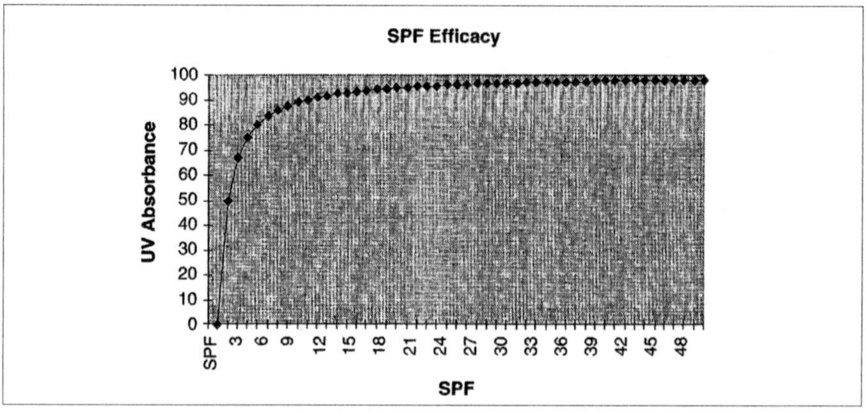

Figure 1. SPF efficacy versus UV absorbance

ability to burn without the sunscreen. Thus, if a person burns in 10 min without the sunscreen, but does not burn until 150 min with the sunscreen, then the SPF of the sunscreen is 150/10 = 15. Consequently, the better a sunscreen protects the user's skin against sunburn, the higher the SPF. The goal of the formulator is to develop the highest SPF possible using the least amount of sunscreen actives, because the sunscreen actives are expensive and may be irritating.

Requirements for higher SPF increase the difficulty for the formulator to generate a safe, efficacious and pleasant product. Generally, higher SPF values protect better because SPF is a measure of sunscreen's efficacy. As shown in Figure 1, a SPF 2 sunscreen will block 50% of the sunburn response.

The role of very high SPF values (SPF > 30) may be unclear to the consumer. An SPF 25 blocks 96% of the sunburn response, while an SPF 50 only blocks 2% more, or 98% of the total sunburn response. Very high SPF products are more expensive, may be more irritating to the consumer's skin and eyes and may offer little extra protection for the average consumer. Very sun-sensitive individuals may benefit from the extra protection, however, suggesting that a market niche exists.

In the recent Final Monograph on sunscreens, the US Food and Drug Administration (FDA) established the upper limit for SPF at 30+.[2] Any product offering a SPF greater than 30 can only exhibit 30+ on the label unless FDA approval is obtained.

SPF test: The SPF test approved by the FDA in its Final Monograph defines the only method to determine the SPF of a sunscreen.[2] The procedure is simply to determine the Minimal Erythema Dose (MED) on at least 20 but not more than 25 qualified subjects. The MED is that amount of ultraviolet radiation required to produce the first perceptible redness reaction with clearly defined borders at 22 to 24 hours after irradiation.

To determine the MED, a series of 5 exposures of increasing energy is administered to the subject's unprotected skin. Each exposure is 25% greater than the previous exposure. At 22 to 24 hours after exposure, a trained grader other than the person who conducted the irradiation or who applied the sunscreen evaluates the redness of each exposure site. The MED on unprotected skin, or MED_{US}, is used to calculate the radiation exposures for the sunscreen-protected site.

To determine the MED in the presence of the sunscreen, MED_{PS}, the first step is to apply the sunscreen to the subject's skin, usually the back. After a wait of at least 15 min for the sunscreen to dry, the treated area is exposed to seven geometrically increasing doses of radiation. The geometric progression is dependent on the predicted SPF of the sunscreen, as shown in Table 1, where X is the product of the expected SPF and the subject's MED.

The MED_{PS} is the lowest dose of radiation that produces the first perceptible redness reaction with clearly defined borders of the exposure site at 22 to 24 hours after exposure. The SPF value is the ratio of the energy required to produce the MED_{PS} to that required to produce the MED_{US}. Based on its SPF, the sunscreen is placed in one of three product category designations, or PCD, as shown in Table 2. Any product with a SPF below 2 should not be labeled as a sunscreen drug product.[2]

Water-resistant or very-water-resistant test: To determine the water resistance of a sunscreen formulation, you must first subject the skin with the sunscreen test material to repeated exposures to fresh water. Fresh water in an indoor pool, whirlpool, or Jacuzzi[a] is maintained at 23° to 32°C for the test. The sunscreen is applied to the skin and allowed to dry.

The subject enters the water and engages in moderate activity for 20 min. The subject exits the water to rest for 20 min, being careful to avoid rubbing off the sunscreen. For a water-resistant claim, the 20 min in the fresh water while engaging in moderate activity is repeated once more, for a total of 40 min in the water. For a very water-resistant claim, the 20 min in the fresh water while engaging in moderate activity is repeated three more times, for a total of 80 min in the water. Each 20-min segment in the water is followed by a 20-min rest period out of the water.

Following the water immersion procedure, the sunscreen sites are allowed to air dry. Then the SPF of the sunscreen is determined in a manner identical to that described above under SPF test. The SPF of the sunscreen determined after 40 total min of water exposure is the SPF placed on the label of a water-resistant sunscreen. The SPF of the sunscreen determined after 80 total min of water exposure is the SPF placed on the label of a very water-resistant sunscreen.

One controversy in water-resistant sunscreens is whether water resistant is the same as sweat resistant. In essence, the debate is whether water coming from outside the skin acts identically to water coming from inside the skin in disrupting sunscreen efficacy. In the 1993 Tentative Final Monograph,[3] the FDA ruled that because the water-resistant test was more stringent than the sweat-resistant test, any product that passed the water-resistant or very-water-resistant test could also make the claim of sweat resistant.

Recommended approach for testing: The target for the novice sunscreen formulator is to develop a sunscreen that will meet the customer's SPF requirements. General principles to meet those requirements will be discussed later. After a few years of formulating sunscreens to meet SPF values, the experienced sunscreen formulator can then begin to formulate sunscreens with improved cosmetic properties. The experienced formulator remembers that SPF formulations must meet the requirements of the tests first and meet consumer requirements second.

[a] Jacuzzi is a registered trademark of Jacuzzi Inc., Walnut Creek, California

Table 1. Geometric progression of irradiation exposures

SPF	Site 1	Site 2	Site 3	Site 4	Site 5	Site 6	Site 7
<8	0.64X	0.80X	0.90X	1.00X	1.10X	1.25X	1.56X
8 - 15	0.69X	0.83X	0.91X	1.00X	1.09X	1.20X	1.44X
>15	0.76X	0.87X	0.93X	1.00X	1.07X	1.15X	1.32X

Table 2. Product category designation of sunscreens

SPF	PCD
30+	high
12-30	moderate
2-12	minimal

Methods exist for clinical testing labs to obtain high SPF values and stay within the FDA testing guidelines. For example, COLIPA has greater defined values for the emission spectrum of solar simulators than the FDA's Final Monograph. When testing a sunscreen for the US market, the solar simulator must only emit between 290 nm and 400 nm with a spectrum similar to sunlight at sea level from the sun at a zenith angle of 10°. A solar simulator can have an increased amount of UVB that then generates a higher SPF relative to the COLIPA-certified solar simulator. Using solar simulators with higher amounts of UVB relative to UVA will increase the SPF of sunscreen products containing UVB sunscreen actives.

UVA protection testing: Currently, there is no FDA-accepted method to test for UVA protection. Developing an accepted method is difficult because of the inability to easily detect any acute changes in skin resulting from non-erythemal UVA radiation (340-400 nm). Sunburn satisfies the requirements for UVB radiation and for erythemal UVA radiation. However, for non-erythemal UVA radiation, no easy method exists to detect changes. It seems that we may be attempting to provide protection from UV radiation that causes only minimal changes in skin biochemistry.

Sunscreen Actives

The FDA has published a Final Monograph that details the sunscreen actives that can be used in OTC sunscreens in the US.[2] This Final Monograph incorporates the two additional actives that were added after the Tentative Final Monograph was published.[4,5] A list of these sunscreen actives appears in Table 3 with additional pertinent information. Many suppliers of these sunscreen actives can be found in different source books, such as the *International Cosmetic Ingredient Dictionary and Handbook*.[6] To obtain additional information, you need only ask one or more of the suppliers.

The Final Monograph includes only maximum concentrations of sunscreen actives when used alone and in combination with other sunscreen actives. The sunscreen active concentrations allowed by the Final Monograph are shown in Table 3.

Sunscreens are safe and effective for the prevention of sunburn as determined by the SPF test. Ultraviolet radiation triggers many biological events, including acute, delayed and chronic skin effects in addition to sunburn. Animal and human studies indicated that sunscreens prevent these biological events. The risk of allergic contact dermatitis or of photoallergic contact dermatitis due to sunscreen actives is low.[7] Gasparro, Mitchnick and Nash published a thorough review on sunscreen safety in 1998.[8]

To formulate a high SPF sunscreen, one must first block much of the UVB, so in selecting sunscreen actives, you must first choose a UVB blocking active. For many years, Padimate O was the UVB sunscreen active of choice because it was very effective, inexpensive and relatively easy to formulate. Recently, octyl methoxycinnamate has become more popular with experienced formulators bowing to concern about increased photosensitization with Padimate O relative to octyl methoxycinnamate and to increased stinging with Padimate O. Although the former concern appears to be unfounded,[7] the public now looks for sunscreens that are "PABA free."

Titanium dioxide and zinc oxide are two new important sunscreen actives for the formulator. Each of these metal oxides displays absorbing properties throughout much of the UVB and UVA spectrum. These actives can impart high SPF values at relatively low concentrations, provide for broad-spectrum protection, and are relatively inexpensive. The difficulty in formulating with titanium dioxide or zinc oxide is creating a product with acceptable consumer qualities. Recent research by Kobo Products and by Sunsmart has advanced the formulation possibilities of these two sunscreen actives.

Sunscreen Active Solvents

Once the formulator has decided which sunscreen active(s) to use in the formulation, the next step is to choose a solvent or solvents to solubilize those sunscreen actives. The solvents allow the sunscreen actives to emulsify and may have additional effects on the orientation of the sunscreen actives after the actives are applied to the skin. Some of these solvents, such as butyloctyl salicylate,[b] may actually stabilize some sunscreen actives against photodegradation.[9-11] Rassat and colleagues published a list of solvents for the newest sunscreen active, avobenzone.[12]

For the inexperienced sunscreen formulator, benzoate esters are perhaps the best starting point. Formulations using C^{12-15} alkyl benzoates are shown in Formulas 2 and 8. A common solvent for the sunscreen active avobenzone is butyloctyl salicylate. Some selected sunscreen solvents are shown in Table 4.

Table 3. FDA-accepted sunscreen actives

INCI Name	Max. Conc.	Protection Range	Trade Name(s)	Selected Supplier(s)
PABA	15%	UVB	PABA	Rona
Avobenzone	3%	UVA	Parsol 1789	Roche
Cinoxate	3%		None	None
Dioxybenzone mid	3%	UVA	Benzophenone-8	American Cyana-
Homosalate	15%	UVB	HMS	Rona
Menthyl anthranilate Reimer	5%	UVA	Neo Heliopan MA	Haarmann &
Octocrylene	10%	UVB/UVA	Excalol 597	ISP Van Dyke
Octyl methoxycinnamate	7.50%	UVB	Parsol MCX Escalol 557	Roche ISP Van Dyk
Octyl salicylate	5%	UVB	Escalol 587	ISP Van Dyk
Oxybenzone Reimer	6%	UVA	Neo Heliopan BB	Haarmann &
Padimate O	8%	UVB	Escalol 507	ISP Van Dyk
Phenylbenzimidazole sulfonic acid	4%	UVB	Eusolex 232 Parsol HS	Rona Roche
Sulisobenzone	10%		UMS 40	BASF
Titanium dioxide	25%	UVB/UVA		Kobo
Trolamine salicylate	12%	UVB	None	None
Zinc oxide	25%	UVB/UVA		Kobo Sunsmart

The more experienced formulator can examine the use of unique co-solvent systems such as the combinations of the alkyl benzoates with isopropanol. Suppliers are now making new sunscreen solvent blends available, such as hexyldecyl benzoate with butyloctyl benzoate and $C^{12\text{-}15}$ alkyl benzoate, stearyl ether benzoate and dipropylene glycol dibenzoate. The subject of sunscreen solvents is a very active area of research within the R&D programs of raw material suppliers as well as of sunscreen formulators. Much useful information on the issue of solvency should be available from experienced sun care formulators during the next several years.

Inorganic Sunscreen Active Dispersions

Inorganic sunscreen actives such as titanium dioxide and zinc oxide present formulation difficulties unlike those of organic sunscreen actives. Inorganic sunscreen actives must have a small particle size to avoid the "whitening" effect, a particle size large enough to absorb UV radiation,[13] and proper wetting and dispersion to avoid agglomeration. The optimum particle size for titanium dioxide seems to be about

Table 4. Selected solvents for organic sunscreen actives

Alkyl salicylate	Isostearyl benzoate
Butyloctyl salicylate	Isotridecyl isononanoate
$C^{12\text{-}15}$ Alkyl benzoate	Isotridecyl isononanoate
$C^{12\text{-}15}$ Triethoxy alkyl benzoate	Methyl gluceth-20 benzoate
Cocoglycerides	Octyldodecyl benzoate
Dipropylene glycol benzoate	Poloxamer 105 benzoate
Isocetyl salicylate	Poloxamer 182 dibenzoate
Isodecyl isononanoate	PPG-15 stearyl ether benzoate
Isodecyl salicylate	Tridecyl salicylate
Isononyl isononanoate	

Table 5. Selected agents imparting water resistance

Film formers
PVP hexadecene copolymer
PVP eicosene copolymer
Tricontanyl PVP
Acrylates/C10-30 alkyl acrylate crosspolymer
Acrylates/t-octylpropenamide copolymer
Hydrophobic barrier formers
Cetyl dimethicone
Maleated soybean oil

15-20 nm, while that for zinc oxide appears to be slightly larger, 15-35 nm. These sizes are best for avoiding the "whitening effect" while maintaining strong UVB and UVA absorption.

Proper wetting and dispersion of the inorganic sunscreen will prevent agglomeration and precipitation in the formula. Some advances have been made by treating or coating the particles to improve dispersion and stability. For example, Schlossman patented the treatment of titanium dioxide with isopropyl titanium triisostearate.[14] This treated material has greater dispersability and minimizes aggregation.

Agents Imparting Water Resistance

Agents imparting water resistance are materials that protect the sunscreen active from being removed easily with water. This is an important characteristic for sunscreens that will be used at the beach or swimming pools, or during times of high physical activity. Under these conditions, the sunscreen should be able to avoid any loss of efficacy from the aqueous environment. Sweating or swimming will

[b] Butyloctyl salicylate is marketed as HallBrite BHB, which is a trademark owned by The C.P. Hall Company, Chicago, Illinois

Formula 1. Sunscreen Lotion (SPF 30) (ISP Van Dyk)

A	Water *(aqua)*	50.40% wt
	Xanthan gum	1.00
B	Glyceryl stearate (and) laureth-23	6.00
	PEG-20 stearate	3.00
	Cetyl lactate	3.00
	$C^{12\text{-}15}$ alkyl lactate	1.00
	Myristyl myristate	4.00
	Octyl methoxycinnamate	7.50
	Benzophenone-3	3.00
	Octyl salicylate	3.00
	Propylene glycol	6.00
C	Titanium dioxide, ultra fine	5.00
	Isocetyl stearoyl stearate	3.00
	Maleated soybean oil	3.00
D	Preservative	1.00
E	Fragrance *(parfum)*	0.10
		100.00

Procedure: Mix C with a roller mill. Disperse A with high speed mixing. Heat to 75°C. Heat B to 80°C and add C. Add BC to A. Mix with homogenizer for 15 min. Mix while cooling to 40°C with a sweep blade. Add D and then E. Mix while cooling to 25°C.

cause a loss of activity from a sunscreen product without water resistance. With these agents, the sunscreen actives are not lost to an aqueous environment.

The first waterproof sunscreen was developed by Johnson & Johnson in 1977. Coppertone quickly followed with a water-resistant product using a polyanhydride resin, PA-18, as the agent imparting water resistance. PA-18 has several characteristics that make it ideal for imparting water resistance in sunscreens. First, it imparts water resistance to the sunscreen—it's effective. Second, it is very inexpensive. Third, it is safe topically as evidenced by years of successful use. Unfortunately, PA-18 is under patent until 2002.[15]

Several other agents can be used to impart water resistance to sunscreen formulations. These are generally based on film forming characteristics or on hydrophobic barrier characteristics. A listing of selected water-resistant agents appears in Table 5.

One important note for the formulator using a film former is to use the lowest possible concentration of emulsifier. The emulsification system dries on the skin

Formula 2. Sunscreen Lotion with Avobenzone (SPF 26) (C.P. Hall Company)

A	Octyl methoxycinnamate	7.50% wt
	Oxybenzone	3.50
	Avobenzone	3.00
	Octyl salicylate	5.00
	Butyloctyl salicylate	5.00
	Hexyldecyl benzoate and butyloctyl benzoate	5.00
B	Tocopheryl acetate	0.20
	Sorbitan oleate	0.40
	PVP/eicosene copolymer	0.75
	Dimethicone copolyol	0.20
	Silica	0.40
	Acrylates/$C^{10\text{-}30}$ alkyl acrylates crosspolymer	0.30
C	Water *(aqua)*	qs
	Disodium EDTA	0.10
	Carbomer	0.20
D	Butylene glycol	2.00
	Preservative	qs
	Panthenol and propylene glycol	0.50
	Hydroxypropyl methylcellulose	0.20
E	Triethanolamine	0.50
		100.00

Procedure: Mix A until dissolved. Add B to A and heat to 50-55°C. Mix C. Add D to C and heat to 50-55°C. With vortex stirring, add AB to CD and stir for 30 min. Cool while stirring. Adjust viscosity with disodium EDTA.

with the waterproofing agents. If too much emulsifier exists, then the addition of water can cause the waterproofing agent to re-emulsify and wash off. This would allow the sunscreen active to wash off as well.

Sunscreen actives cause irritation to the eyes. When a waterproof sunscreen enters the eye, the waterproofing agents adhere to the mucus membrane of the eye. This holds the sunscreen actives in place, causing severe and prolonged irritation. Therefore, this author suggests that waterproof sunscreen products never be tested for eye irritation. Alternatively, the product without the waterproofing agents in the formula might be tested in an eye-stinging assay.

Emulsifiers for Sunscreen Products

The choice of an emulsifier for a sunscreen product is dependent on many variables including the influence of emollients on sunscreen performance (spreadability, water resistance and penetration), your existing internal technology and knowledge base and manufacturing capabilities. The choice of emulsifier affects the absorption spectrum of the sunscreen actives, penetration of the sunscreen actives, the spreading of the actives and the adherence of the actives to the skin.

A relationship appears to exist between the thickness of the film, the spreadability of the product and the efficacy of a sunscreen. This should not be surprising because sunscreen actives are, for the most part, planar molecules; orientation is important. If

Formula 3. Waterproof Sunscreen Formula (SPF 14)
(Kobo Products)

A	Cetyl dimethicone	3.00% wt
	Cyclomethicone	7.50
	Isononyl isononanoate	6.00
	Methyl glucose sesquistearate	0.50
	Dioctyl malate	2.00
	Polyglyceryl-4 isostearate (and) cetyl dimethicone copolyol (and) hexyl laurate	5.00
B	Micronized zinc oxide (and) isononyl isononanoate (and) hexaglyceryl polyricinolate (and) isopropyl titanium triisostearate	21.33
C	Water *(aqua)*	51.07
	Sodium chloride	0.50
	C10 polycarbamyl polyglycol ester	2.50
	Preservative	0.60
		100.00

Procedure: Heat A to 75°C and cool to 65°C with propeller. At 65°C, add B to A under homogenizer. With propeller mixing, add premixed C to AB. Cool to 30°C.

the sunscreen active is positioned properly on the skin, then its ability to absorb UV radiation is maximized. A thicker layer of sunscreen may result in better orientation.

The spreadability and surface tension are important and have been important for sunscreen formulators. Usually, creams have a higher SPF than corresponding lotions. However, with research, lotions can match creams for SPF values. Any sunscreen emulsification system should not be able to oxidize on the skin. Dahms suggests that PEG emulsification systems can undergo auto-oxidation, resulting in an incompatibility between the sunscreen product and the skin.[16]

If spreadability is too great, then the sunscreen product will spread over the skin into the eyes of the user. Sport type sunscreen products are designed to hold the sunscreen actives in place so they do not spread. These formulations are designed for customers with an active lifestyle. The first product of this class was marketed by

Formula 4. SPF 15 Lip Balm (Protameen Chemicals)

A	Isostearyl linoleate	10.00% wt
	Caprylic/capric triglyceride	60.20
	Octyl methoxycinnamate	7.00
	Benzophenone-3	3.00
	Propylparaben	0.10
	$C^{30\text{-}40}$ alkyl methicone	4.00
	Ozokerite wax, white	5.00
	Petrolatum	10.00
B	Fragrance *(parfum)*	0.70
		100.00

Procedure: Heat A to 80°C, mixing until uniform. Cool to 65°C. Add B. Pour into suitable container.

Formula 5. Sunscreen Spray

A	$C^{12\text{-}15}$ alkyl benzoate	10.00% wt
	Octyl methoxycinnamate	7.50
	Octyl salicylate	4.00
B	Isododecane	73.50
	Isohexadecane	5.00
C	Fragrance (parfum)	qs
		100.00

Procedure: Mix A. Add B and continue mixing. Add C.

Coppertone under the name Coppertone Sport. Since its creation, several similar "sport" type formulations have been developed.

Because the majority of raw material costs for a sunscreen are for the sunscreen actives, the formulator must develop products with high SPF values and low concentrations of sunscreen actives. Recently, several companies have developed materials referred to as SPF boosters or enhancers. These raw materials attempt to increase the SPF of a formulation without increasing sunscreen actives. These SPF boosters or enhancers should be evaluated carefully because they are not universally effective. They may enhance the SPF of one particular type of formulation, but not of another.

Sunscreen products are available in many different types of formulations to appeal to a wide variety of customers. Formulas 1 through 5 illustrate various types of creams, lotions, gels, sprays and sticks, but these are not inclusive. The talented formulator will branch from accepted formulation types to create a unique product.

—**Michael Caswell, Ph.D.,** *C.B. Fleet and Central Virginia Community College, Lynchburg, VA USA*

References

1. M Caswell, The theory of sunscreens and tanning, Chap 3 in vol 2 of *The Chemistry and Manufacture of Cosmetics*, ML Schlossman, ed, 3rd edn, Carol Stream, IL: Allured Publishing (2000)
2. *Fed Reg* 64(98) 27666-27693 (May 21, 1999)
3. *Fed Reg* 58(90) 28194-28302 (May 12, 1993)
4. *Fed Reg* 61(180) 48645-48655 (Sep 16, 1996)
5. *Fed Reg* 63(204) 56584-56589 (Oct 22, 1998)
6. *International Cosmetic Ingredient Dictionary and Handbook*, 7th ed, JA Wenninger and GN McEwen, Jr, eds, Washington, DC: Cosmetic, Toiletry, and Fragrance Association (1997)
7. S Schauder and H Ippen, Contact and photocontact sensitivity to sunscreens. Review of a 15-year experience and of the literature, *Cont Derm* 37 221-232 (1997)
8. FP Gasparro, M Mitchnick and JF Nash, A review of sunscreen safety and efficacy, *Photochem Photobiol* 68(3) 243-256 (1998)
9. US Pat 5,788,954, Hydrating skincare and sunscreen composition containing dibenzoylmethane derivative, eg, Parsol 1789, and C12, C16, C18 branched chain hydroxy-benzoate and/or C12, C16 branched chain benzoate stabilizers/solubilizers, CA Bonda and SP Hopper (Aug 4, 1998)
10. US Pat 5,783,173, Stable sunscreen composition containing dibenzoylmethane derivative, eg, Parsol 1789, and C12, C16, and C18 branched chain hydroxybenzoate and/or C12, C16 branched chain benzoate stabilizers, CA Bonda and SP Hopper (Jul 21, 1998)
11. US Pat 5,849,273, Skin care and sunscreen composition containing dibenzoylmethane derivative, eg, Parsol 1789, and C12, C16, and C18 branched chain hydroxybenzoate and/or C12, C16 branched chain benzoate stabilizers/solubilizers, CA Bonda and SP Hopper (Dec 15, 1998)
12. F Rassat, H Gonzenbach and GH Pittet, Use of sunscreens and vitamins in the daily-use cosmetic, DCI 12 16-30 (1997)
13. R Sayre et al, Physician sunscreens, *J Soc Cos Chem* 41 103-109 (1990)
14. US Pat 4,877,604, Method of incorporating cosmetic pigments and bases into products containing oil and water phases, ML Schlossman (Oct 31, 1989)
15. US Pat 4,522,807, Substantive topical compositions, C Kaplan (Jun 11, 1985)
16. GH Dahms, Choosing emollients and emulsifiers for sunscreen products, *Cosmet Toil* 109(11) 45-52 (1994)
17. J Guth, G Martino, D Patel and J Pasapane, Versatility in sun protection, *Cosmet Toil* 105(12) 87-90 (1990)

Formulating Water-Resistant Sunscreen Emulsions

Keywords: SPF, washoff, emulsions, stability

The sunscreen market has been ever-changing and has become more segmented since the publication of the proposed sunscreen monograph more than 20 years ago. The maximum SPF (Sun Protection Factor) ever targeted was 15, and most products were not required to be waterproof. Note that today we aren't permitted to use the term "waterproof" since the FDA believes that this term implies an absolute–it never washes off–and feels that consumers might be inclined not to reapply sunscreen after swimming or perspiring. So, we must now use the term "very water-resistant" in its place.

It is now quite commonplace to see marketed products that claim an SPF of 45 or even greater. I've been told that there are products with claimed SPFs of 100+. I would estimate that a product that provides an SPF of 100+ should be used by fair-skinned people sunbathing at noon on the planet Mercury on a cloudless day, assuming they have cloudless days on Mercury! One could argue whether there is a consumer need/benefit for these very high SPFs, but that is a topic for another column.

Minimizing Hydrophilic Emulsifiers

The most popular sunscreen products used today fall into the category of SPF 15-45 and are designed for beach use. These products are also designed to be very resistant to water. A cosmetic formulator has several strategies to develop sunscreen products that are very water-resistant. We will explore a few of these here. Probably the single most important factor to consider rests with the choice of sunscreen vehicle.

We must first understand the mechanism by which sunscreen products "wash off." If we place a drop of water (very polar) on our skin, the drop doesn't spread; it remains a drop. If we perform the same "experiment" using a drop of mineral oil or IPM (isopropyl myristate), the drop spreads out. We see them spread out. Because most sunscreen actives are lipophilic (esters) in nature, they inherently have an affinity for skin (which is also lipophilic). They don't want to wash off. Something in formulations encourages their wash off. This "something" is a hydrophilic emulsifier. Typical of these types of emulsifiers are soaps (TEA stearate) and nonionic ethoxylates (PEG-100 stearate). Probably the single most important factor to be considered in the design of these products relates this. In order to develop very water-resistant products, you *must minimize* the level of hydrophilic emulsifier.

W/o Emulsions

While emulsions are by far the most popular sunscreen vehicle, they can be very difficult to formulate and can present stability and performance problems.

Without doubt, w/o (water-in-oil) emulsions are more efficient sunscreen vehicles than their o/w (oil-in-water) counterparts. This is due to several factors. Oil-soluble sunscreens are soluble in the external (continuous) phase of w/o emulsions where they are uniformly distributed. We will not consider solubility parameter mismatches here that might result in agglomeration on the micro level. When the emulsion is applied, the sunscreen spreads quickly and uniformly on the lipophilic skin. Thus we can achieve a uniform sunscreen film and an inherently higher SPF.

These products, such as Formula 1, contain hydrophobic emulsifiers (alkyldimethicone copolyols, dimethicone copolyols, polyglyceryl esters, ethoxylated di-fatty esters, and other) that promote the formation of w/o emulsions. Typically they do not contain any hydrophilic emulsifiers and therefore are inherently very water-resistant. In fact, it would be most difficult to formulate a w/o emulsion that did wash off.

It is not necessary to incorporate a film former in these systems to achieve water resistancy. However, you often see film formers utilized to thicken the film on the skin and thus boost the SPF. This SPF boost is caused by increasing the optical path length. The photons (with an energy corresponding to the UVA/UVB wavelengths 290-400 nm) travel a greater distance before reaching the skin, where they do their dastardly deed, and thus have a greater likelihood of bumping into a molecule (sunscreen) that can absorb their energy.

Formula 1. Very Water-Resistant Sunscreen Emulsion w/o (Particulate/Organic) with Expected SPF=30

Cetyl dimethicone copolyol	Emulsifier	5.0%
Octyl palmitate	Emollient	11.0
Cetyl dimethicone	Emollient, SPF booster	2.5
Cyclomethicone	Emollient	7.5
Ceresin wax	Thickener	1.0
Octyl methoxycinnamate	Sunscreen	7.5
Octocrylene	Sunscreen	6.0
Hydrogenated castor oil	Thickener	0.5
Titanium dioxide (microfine)	Sunscreen	5.0
Zinc oxide (microfine)	Sunscreen	5.0
Water (*aqua*), deionized	Bulk-Internal phase	qs
Quaternium-15	Preservative	0.15
Magnesium sulfate	Stabilizer	0.75

O/w Emulsions

Minimize hydrophilic emulsifiers: Formulating o/w emulsions that are very water-resistant is a much more difficult task. In order to prepare classical emulsions, significant levels of hydrophilic emulsifiers are required. These emulsifiers will remain behind on the skin after the water and other volatiles have gone, and will emulsify the sunscreen when water is added (swimming or perspiring). This product will not be very water-resistant. Often the formulator will then add film formers to try to overcome this effect. It is, however, an effort that is doomed to fail.

If hydrophilic emulsifiers must be used, their concentration should be kept as low as possible. Titrate them down, very carefully balancing stability with emulsifier concentration. Water-phase thickeners (such as carbomer, acrylates/steareth-20 methacrylate copolymer, xanthan gum) can assist in this process. Some companies have found that adding a small amount of lipophilic emulsifer further inhibits the re-emulsification of the sunscreen on the skin. This technique can be quite successful.

Form liquid crystal structures: Another approach for formulating o/w emulsions uses emulsifiers that stabilize emulsions via the formation of liquid crystal structures. Most often, the development of a lamellar gel network thickens the external (water) phase and stabilizes the emulsion.

Formula 2. Water-Resistant Lotion with Expected SPF=15+

Water (*aqua*), deionized	Diluent	qs
Tetrasodium EDTA	Chelating agent	0.1%
Glycerin, 96%	Humectant	3.0
Carbomer	Thickener	0.25
Laureth-23	Emulsifier	0.25
Sorbitan sesquioleate	Hydrophobic emulsifier	0.5
Glyceryl stearate	Bodying agent/stabilizer	2.0
PEG-100 stearate	Emulsifier	1.0
PVP/Eicosene copolymer	Film former, SPF booster	2.5
Octocrylene	Sunscreen	10.0
Octyl methoxycinnamate	Sunscreen	7.5
Oxybenzone	Sunscreen	4.0
Octyl palmitate	Emollient, sunscreen solubilizer	5.0
Cyclomethicone	Emollient	5.0
Vitamin E acetate	Label copy	0.25
Alpha bisabolol	Anti-irritant	0.25
Triethanolamine, 99%	Neutralizer	0.25
Propylene glycol (and) diazolidinyl urea (and) methylparaben (and) propylparaben	Preservative	1.0

These emulsions can thin out significantly at elevated temperatures (45°C) and exhibit instability; using a water phase thickener that maintains some viscosity at 45°C (such as xanthan gum or carbomer) is recommended.

Emulsion stabilization can also occur if the liquid crystals form in the immediate vicinity of the oil droplets, structuring themselves into an "onion skin" type layer. This layer acts as a barrier to coalescence and thus improves stability. Most liquid crystal emulsifiers (polyglyceryl esters, alkyl lactylates, cetearyl glucosides, lecithin, phosphate esters, among others) have rather limited water solubility and do not promote wash off of the sunscreen film.

While these emulsions, such as Formula 2, have many advantages over conventional emulsions, including low temperature emulsification and mildness, they can be quite tricky to manufacture. You should avoid high shear after the liquid crystal structures have formed or instability will surely result.

Use polymeric thickeners/emulsifiers: A final approach for developing very water-resistant o/w emulsions rests with the use of hydrophobically modified polymeric "thickeners/emulsifiers." The most popular of this category is C10-30 alkyl acrylate crosspolymer, as shown in Formula 3. One can say that it allows the formulation of "emulsifier-free" products. When neutralized, it forms a mini gel that surrounds the oil/sunscreen droplets and immobilizes them. Thus the product exhibits good storage stability.

When the product is applied to the skin, the salt from the skin coagulates the polymer and the emulsion breaks, permitting the sunscreen to spread onto the skin. Since there is no hydrophilic emulsifier left behind, the emulsion is inherently very water-resistant.

The use of fatty alcohols in this product, along with high shear agitation, will reduce the particle size and improve application qualities. Use of emulsifiers at any significant use levels should be avoided because these emulsifiers will compete for space at the interface and the polymer will not be able to partition itself properly. An unstable product will be the result. This phenomenon is known as flocculation depletion.

—**Ken Klein,** *Cosmetech Laboratories, Inc., Lincoln, Calif., USA*

Formula 3. O/W Polymeric ("Emulsifier-Free") Emulsion Sunscreen with Expected SPF=12+

Water (*aqua*), deionized	Diluent	qs
Acrylates/$C_{10\text{-}30}$ alkyl acrylates crosspolymer	Polymeric thickener/stabilizer	0.2%
Propylene glycol	Humectant, improves freeze/thaw stability	2.5
Xanthan gum	Improves product application	0.05
Octyl methoxycinnamate	Sunscreen	7.5
Triethanolamine	Neutralizer	0.18
DMDH hydantoin	Preservative	0.2
Trisodium EDTA	Chelating agent, viscosity control	0.05
Oxybenzone	Sunscreen	3.5
Cetearyl alcohol	Thickener, improves product application	0.25
Octyl palmitate	Emollient	7.5

Hydrogenated Polydecenes and High-SPF Physical Sunscreens

Keywords: sunscreen, hydrogenated polydecenes, ultrafine titanium dioxide, photostability, physical filters, high SPF

New methods of micronization and coating of physical filters combined with vehicles containing high amounts of hydrogenated polydecenes provide high-SPF sunscreens that show improvements in photostability, effect-to-dose ratio, and skin substantivity.

The requirements of an effective sunscreen include high solubility of organic filters in the vehicles, efficient dispersion of physical filters, absence of crystallization after application onto the skin surface and lack of aggregation phenomena of micronized powders in the product. Moreover, an effective sunscreen requires an appropriate load of inorganic filters as well as their optimum wetting and the formation of a transparent limit layer (i.e., the residual layer on the skin that cannot be further thinned by massage) over the skin surface.

Photostability is another essential requisite for high-SPF products. Indeed, a homogeneous self-adjusting limit layer on the skin is an effective means of preventing a reduced protective effect that may result from uneven or excessive spreading ability of the product. In other words, sunscreens must be formulated in order to have a limited, "consumer-proof" massageability over the skin, so that the residual layer is thick enough to provide an adequate amount of sunscreens into it. Much care has been taken to research on high load of filters and on their chemical and physical nature, but so far little evidence exists on the influence of the oily phase in determining sunscreen performances in actual conditions of use. Also, no exhaustive studies have been carried out concerning the performances of different types of coating on the ease of formulation.

Nowadays, new synthesis and advanced purification processes provide the formulator with pure, chemically defined hydrocarbon molecules having interesting potential for use in sunscreen products with high photostability and the ability to stay on the skin surface without diffusing. This article discusses one example, the hydrogenated polydecenes.

New methods of micronization and coating of physical filters provide a wide range of particle dimensions, different degrees of lipophilic or hydrophilic behavior and enhanced transparency. This article describes two new ultra-fine titanium dioxide-coated particles.

When associated with branched chain hydrocarbons these new micronized physical filters yield an improvement in the effect-to-dose ratio and form a very transparent film over the skin surface. Some practical formulation examples in this article demonstrate the wide range of combinations made possible by such high-technology ingredients.

New Aspects of Sun Protection

Skin is constantly exposed to visible and UV irradiation (see sidebar) in an oxygen-rich environment. This leads to oxidative injury capable of modifying the cell membranes, their lipids and proteins by a mechanism of peroxidation.

UVA I rays are mainly responsible for such action, demonstrated by the observation that UVA-mediated tanning and erythema are oxygen-dependent.[1] This damage is cumulative with erythema and sunburn, loss of physical and barrier properties of the skin and the structural breakdown of the skin's biopolymers (collagen, elastic fibers).[2, 9]

Under the action of UVB and UVB II, DNA (which mainly absorbs at 300 nm) is damaged, dimers are formed and successive inhibition of correct synthesis takes place, affecting cell repair, apoptosis or, even worse, the reproduction of irreversibly damaged cells. In the long term, repeated damage may lead to the development of visible signs and altered characteristics of photo-aged skin. Repeated damage is also thought to increase the risks of inducing skin cancer.[3]

Chronic sub-erythematous exposure to UVA and UVB rays is reported to lead to immuno-suppression, that, as with the above-described damage, can be prevented by appropriate filters.[4] In the short-term, changes in skin moisture content occur after irradiation. Moreover, in the case of individuals who were irradiated in the presence of cosmetics applied over the skin, UVA rays may induce phototoxic and photoallergic reactions.[5]

Hydrogenated Polydecenes

While skin lipids are being oxidized, double bonds are broken by hydroxyl radicals

Factors Affecting UV Irradiance

- Ozone layer thickness
- Scattering from atmospheric gases and pollution
- Season
- Altitude and latitude
- Time of the day
- Clouds
- Reflection from water, rocks, sand
- Transmission through UV-transparent materials
- Absorption by cosmetic deposits
- Transmission through skin layers

and lipid peroxyradicals are formed. Lipid peroxy radicals may react with other lipids and form hydro-peroxides first, followed by breakdown products like aldehydes.

Therefore, it is essential that saturated, UV- and oxidation-fast lipids are employed in sunscreen formulas in order to possibly inhibit the propagation of free radicals into the superficial skin lipids and to keep their structure stable.

The choice of cosmetic lipids having stable structures while maintaining a pleasant and silky feel is quite limited. One option is hydrogenated polydecenes, which are new blends of oligomers derived from the catalytic conversion of extra pure α-decene. Hydrogenated polydecenes exhibit unique branched chain structures. In spite of their saturated non-polar chains, they show good solvent power towards many organic filter structures. Moreover, hydrogenated polydecenes are completely miscible with most cosmetic oils; they require a co-solvent (an emulsifier or a coupling oil) only when combined with non-volatile silicones.

Surprisingly, hydrogenated polydecenes act as efficient wetting agents for micronized inorganic filters, in spite of their low polarity. In some cases up to 80% pigment load in stable suspension can be obtained. This property may be attributed more to their lubricating properties and globular shape than to polar interactions with pigment surfaces. Furthermore, rounding effect (making the surface shapes more spherical) is likely to result in higher evenness of the pigment's surface as well as decreased friction among solid particles.[6]

Being completely saturated, hydrogenated polydecenes exhibit high stability to photo-oxidation, thus providing products with long shelf life and in-use stability.[7]

Last but not least, the unusual lack of skin absorption and particle lubrication enhance the surface efficiency of physical filters.[8]

Trends in Sun Protection Products

Three recent trends can be seen in cosmetic products designed to counteract[10] the negative effects of sun rays.

- Reducing all possible interactions between the skin and the product ingredients, especially when the skin covered by the product is exposed to sun rays.
- Ensuring optimized protection by employing highly efficient filtering systems.
- Protecting the chemical stability of the skin lipids by means of appropriate synergism with the applied substances.

One approach to meeting the objectives of all these trends is to select photostable, low-transdermal delivery vehicles, such as those based on the polydecenes, already mentioned. Their advantage is that they are highly spreadable and do not leave oily residues. They also have good dispersing power for physical filters and are highly miscible with most cosmetic substances. Their photostability and resistance to free radical formation provide a physicochemical quench that interrupts oxidative stress mediated by free radicals.

Another approach is to use filtering substances with low chemical activity, such as inorganic (mineral) sunscreens. Their wide employment in the most significant products on the market is a proof of their extraordinary characteristics: chemical inertia when properly coated, capability to adhere to the stratum corneum without diffusing, transparency in visible light and powerful scattering of UV rays. Moreover, they may provide the correct synergistic effect of simultaneous UVB and UVA protection, which is currently required.

Micronized, Unsuspended TiO_2 for Sunscreens

The presence of a high percentage of solids in a cosmetic formula is a typical problem for the formulator, who has to adjust performance for both the long-term stability and the production process itself. Indeed, the polar thickeners tend to form very viscous films in contact with the stabilizing ions contained in aqueous suspensions or may even react by precipitating over the solid surface of the pigment.

Moreover, the pH of the suspension must allow for the Zeta potential of the solid interface in order to avoid long-term clustering of particles.

Furthermore, the suspending vehicle selected (water or oil) modifies the final feel and also the stability[11] and the partition of the solid in between the two phases of the emulsion in an unpredictable way. Indeed, suspensions are difficult to use in the production stage, because they always pose the risk of non-homogeneity due to settling.

Finally, the batch-to-batch reproducibility of the raw materials may be faulty. It has been reported that inappropriate homogenization after the addition of physical filters may lead to reduced filtering effect.[12]

Improved technology and skill may help the modern formulator overcome the most common problems. However, the main drawback of titanium dioxide suspensions is related to the high amounts employed when preparing the high SPF products currently required by dermatological science. The interaction with the other ingredients becomes very important; rheology changes remarkably during shelf life and spreadability decreases. Sometimes, the overlay of pigments onto the skin acquires a blue shade, mainly after immersion in sea water, due to coarser particles.

For these reasons, new types of micronized, unsuspended titanium dioxides are being introduced in the market. Their main advantages include very narrow particle size ranges, absence of a suspending agent, and several types of coating materials and technology that ensure homogeneous and complete coating. In other words, new micronized titanium dioxides are not anymore a simple category of raw materials whose selection was confined to the type of suspending agent. Instead, they have become a complete category of ingredients to select from according to the requirements of the formulation strategy.

Indeed, from the new types of solid ultra-fine titanium dioxide, the formulator can choose the ingredient that is more suitable to match the following requirements: transparency, skin feel, compatibility with other ingredients in the formula, covering power, load of solids, surface characteristics, level of hydrophobic or hydrophilic character, range of UV spectrum absorbed and even the thickness of the coating layer. Dispersing vehicles can be selected in the formulation phase, by just examining with a microscope the quality of the suspension obtained with the different types of wetting agents available when mixed with the new micronized filters.

This solution was obtained thanks to new production processes that combine a narrow size distribution with excellent dispersibility. The result is an improved skin feel and cosmetic acceptability of the products.

Formulation Strategy

We examined two new types of ultra-fine titanium dioxide. One type (M170) is a very hydrophobic, rutile lattice particle of low particle size (14 nm) and high specific surface area (80 m^2/g), which was heavily treated with alumina and

Formula 1. w/o Sun-Protective Spray Lotion

A. Lauryl dimethicone copolyol	5.00% w/w
Hydrogenated polydecene (Nexbase 2004 FG, Fortum)	30.00
Hydrogenated polydecene (Nexbase 2006 FG, Fortum)	6.00
Ethylhexyl ethylhexanoate	10.00
Caprylic/capric triglyceride	5.00
Methylparaben	0.20
Propylparaben	0.20
Prunus Dulcis (sweet almond) oil	12.00
Simmondsia Chinensis (jojoba) oil	8.00
Tocopheryl acetate	0.50
B. Titanium dioxide (and) alumina (and) dimethicone (UV Titan M170, Kemira)	4.00
C. Water (*aqua*)	10.00
Sodium chloride	0.50
Allantoin	0.10
Imidazolidinyl urea	0.20
Betaine	0.50
Glycerin	5.00
D. *Chamomilla recutita* (matricaria) extract (and) caprylic/capric triglyceride	2.00
Beta-carotene	0.50
E. Fragrance (*parfum*)	0.30
	100.00

Procedure: Heat A to 70°C, add B and continue homogenization for 10 min, then heat C to 70°C and slowly add C to AB with turbine mixing. Cool to 35°C while mixing, then add D and E. Cool to RT. Shake before using.

A W/O pump-spray emulsion, with high massageability, self-regulating limit-layer and enhanced water repellence.

silicone. The small and controlled crystal size allows the achievement of very high SPFs, while the after-coating process imparts superior light stability and excellent dispersability both in emulsions and in premix of oils. Skin feel is also enhanced.

The other type (M262), with slightly larger particle size (20 nm), is very slightly hydrophobic, with a smaller specific surface area (60 m^2/g) and a maximum content of titanium dioxide (85%). The special coating was made with alumina and dimethicone. This coating makes the M262 type compatible with all organic phases as dispersing agents. Moreover, it can even be dispersed in water by adding some wetting agent. The main advantage of the M262 type is the broad spectrum of UVB and UVA protection and the UVB/UVA ratio provided (0.67), and the dispersability in emulsions of different polarity. Light stability of both grades was also excellent.[13]

Coupling of these stable versatile inorganic filters with the skin feel and stability property of polydecenes provided two new tools for creating a wide range of innovative formulations. They allow the formulator to reverse the strategy of pigments use,

Formula 2. Sunscreen Lipstick

A. Hydrogenated polydecene (Nexbase 2008 FG, Fortum)	26.10% w/w
Limnanthes Alba (meadowfoam) seed oil (and) *Butyrospermum Parkii* (shea butter) extract (Fancol VB, Fanning)	11.70
Isopropyl myristate	11.30
Ethylhexyl ethylhexanoate	4.90
B. Titanium dioxide (and) alumina (and) dimethicone (UV Titan M262, Kemira)	12.20
C. Carnauba	9.00
Cera alba (beeswax)	6.30
Ozokerite	5.90
PEG-2 hydrogenated castor oil (and) ozokerite (and) hydrogenated castor oil (Arlacel 582, Uniqema)	4.80
Polyethylene	2.80
Tocopheryl acetate	0.20
BHT	0.30
D. Homosalate	4.30
E. Fragrance (*parfum*)	0.20
	100.00

Procedure: Disperse B into A, previously heated at 70°C. Heat C at 80°C and add into the mix. Add in D and E. Pour into molds.

In this protective lipstick, M262 sunscreen can be substituted partially or totally by the M160 grade, according to the required UVA/UVB protection ratio. The presence of a w/o emulsifier contributes to improving skin substantivity and compatibility with skin moisture content.

Contemporary Challenges for Future Sun Products

- New immunologically significant end-points in SPF measures
- Fast but safe tanning
- Sun-mediated protection
- Better understanding of sub-erythematous protection
- Stimulation of melanogenesis
- Sun exposure without skin aging
- Long-lasting tanning
- Safer filtering action
- World-wide harmonization of SPF numbers
- Immediate visible advantages of the use of sunscreens
- Easier understanding of products communication
- Improved appearance from use of sunscreens

where formulations were adapted to a one-dimension ingredient. Now the formulator has flexible substances able to fulfill the requirements of the more sophisticated 'recipes.' Sunscreens, lip protection formulas and skin care products have already been prepared with these ingredients. Examples are shown in Formulas 1 and 2.

Conclusions

Sunscreen formulations aim at minimizing all possible damage to exposed skin in a consumer-compliant way. Their new role is to provide the user with the rheologic properties necessary to obtain an efficient limit layer of product that does not allow to be thinned excessively while it is being spread over the skin.

Vehicles must provide spreadability and lubricity, low transdermal delivery, good filter dispersion and water resistance. Because of their unique set of properties, extra pure hydrogenated polydecenes seem to be an intelligent choice. They provide an adjustable limit layer and efficient filter trapping. They reduce transdermal delivery of vehicles, and exert a layering effect on filters, thus ensuring a constant grade of dispersion for the formula ingredients.

Filters should exhibit minimum skin penetration, high efficiency, low visual impact, photostability and good skin substantivity. Technologically advanced tailor-made physical sunscreens can provide a narrow range of dimensions and optimized compatibility with vehicles in all phases of product application.

Nowadays, consumer friendly cosmetics should not only protect the skin effectively but should also be comfortable and easy to use. Sun care products of the future will face new challenges (see sidebar).

—**L. Rigano and A. Mezzanotte,** *Rigano Laboratories, Milan, Italy*
—**M. Lohman,** *Fortum Oil & Gas Oy, Fortum, Finland*
—**L. Kujansivu,** *Kemira Pigments Oy, Helsinki, Finland*

References

1. PJ Matts, Influence and biological relevance of UVA radiation in northern Europe, *Proceedings from Conference UVA Protection June 2001*, London: The Royal Society (2001)
2. T Herrling, L Zastrow, K Golz and N Groth, Dangerous free radicals in skin generated by UVA irradiation, *Proceedings Cosmetic Science Conference, Düsseldorf, 2001*, H. Ziolkowsky: Augsburg (2001)
3. S Seité, S Richard, J-L Lèvêque and A Fourtanier, Effect of repeated suberythemal doses of UVA in human skin, *Eur J Dermatol* 7 204-209 (1997)
4. D Moyal, C Courbière, Y Le Corre, O de La Charrière and C Hourseau, Immunosuppression induced by chronic solar-simulated irradiation in humans and its prevention by sunscreens, *Eur J Dermatol* 7 223-225 (1997)
5. S Schauder and H Ippen, Contact and photocontact sensitivity to sunscreens, *Contact Derm* 37 221-232 (1997)
6. L Rigano, M Lohman, Polideceni idrogenati, *Cosm Technol* 4 (4) 30-34 (2001)
7. H Kunttu, Certificate of analysis of free radicals, University of Jyäskylä, Dept. of Chemistry (1.12.99)
8. B.A. John et al, Huntigton Life Science Ltd., Absorption studies on ^{3}H-Nexbase 2006FG (February-May 1999)
9. L Mei-Lane, L Shan-Yang and L Run-Chu, Changes in the skin moisture contents, skin color, and skin protein conformation structures of Sprague Dawley rats after UVB irradiation, *Skin Pharma Appl Skin Physiol* 12 336-343 (1997)
10. L Ferrero, M Pissavini, C Perichaud and I Zastrow, Experimental design application to sunscreen products: demonstration of a synergistic effect between UVB and UVA absorbers, *Proceedings XXI IFSCC International Congress 2000, Berlin*, Augsburg: H. Ziolkowsky (2000)
11. M Kobayashi and W Karlriess, Photocatalytic activity of titanium dioxide and zinc oxide, *Cosmet Toil* 112(6) 83-86 (1997)
12. W Johncock, Sunscreen interactions in formulation, *Cosmet Toil* 114(9) 75-82 (1999)
13. K Heilkkilä, Kemira Pigments datasheets, Kemira, Helsinki 2001.

Controlling the Spreading of Sunscreen Products

Keywords: sunscreen, spreading, leveling, spreading control agents, Beer's Law, Rayleigh's scattering equations, dosage, effectiveness

Spreading control agents used in the formulation of topically applied drugs, especially sunscreens, provide a method to control dosage and improve effectiveness. These materials have recently begun to solve the problem of insufficient product application in actual use.

Most commercial sunscreens fail to deliver their labeled potency in actual usage. SPF value is determined using a 2 mg/cm^2 thick sunscreen film on the subjects' skin, however we have shown that most formulations will spread to a much thinner layer in normal usage. This accounts for results from prior studies which show that consumers fail to apply the correct amount of sunscreen needed in order to produce the labeled level of sun protection. Our investigation examined the spreadability of current commercial sunscreens and the effects of several potential "leveling" agents for their ability to limit the spreading of three standard sunscreen formulas. These "leveled" formulas, when applied in actual usage, provided the full SPF level as labeled.

How Much Sunscreen Do We Really Need?

According to Dr. Brian Diffey, a number of factors affect the amount of UV protection we need. He cites the latitude, the season, the altitude, the skin's sensitivity and the properties of the formulation, among others.[1] Of these variables, we believe that the formulation properties produce a greater level of unpredictability in the potency of sunscreen products than any of Diffey's factors.

It is clear that there are many formulation variables. These variations make it impossible to determine how much sunscreen one might need, especially when the actual use level can vary from 30% to 120% of the FDA standard thickness. However if consumers could rely on receiving the labeled potency from their sunscreen product, their estimate of their own required protection would then depend on more obvious cues, such as weather, skin sensitivity and exposure time.

To most people the term SPF (Sun Protection Factor) is confusing. Many people we surveyed on Miami's South Beach[2] understood that higher SPF numbers indicate more protection. However, none of the more than 250 surveyed beachgoers could tell us very precisely how long their sun product would protect them.

Exposed body surface area is yet another variable that may contribute to consumer confusion about how much sunscreen to use. We measured the skin surface of four typical sunbathers, from large to small. From these measurements we calculated the sunscreen amount needed to provide the FDA's "SPF delivering" standard (2 mg/cm^2) layer. This amount varied from 0.75 oz for a small woman to 1.5 oz for a large man. We believe it is impossible for consumers to judge whether they have dispensed 0.75 oz or 1.5 oz when they are applying their sunscreen. Previous studies suggest that consumers consistently under-apply sunscreen. Even unit dosage, such as packets or towelettes, would have to be "sized" to match the user to avoid misdosage by up to 50%.

Improper dosage of sunscreen product is an endemic and persistent problem, with possibly serious consequences. It is no secret that skin cancer rates are skyrocketing. Various scientists have blamed everything from triclosan to the ozone hole, but the general consensus is a suspicion that inadequate or improper use of sunscreens is to blame. Recent controlled studies of daily application of sunscreen resulted in a significant reduction in skin cancers.

The ability to deliver a product that spreads to its proper thickness would be a major advance in the field. We hope this article contains enough specific information to make that possible throughout the field. Only a few responsible sunscreen producers are currently using this technique. It is a new concept, so new in fact that no patents have yet been issued on the subject.

Strength vs. Potency

Major formula variations are becoming obvious to label readers, now that sunscreens must declare their percentage of "active ingredients" on the label. We surveyed 10 commercial SPF 30 products sold in Florida this year and discovered that they range from 7% to 27% total active ingredient by weight, yet they provide the same level of UVB protection. These products have differing "strengths" but the same "potency." Strength is a chemical term, but potency is biological.

Consumers may be misled to believe that greater percentages of active ingredients (strength) provide greater protection (potency). In fact, just the opposite is often true. Some products with high levels of UV absorbers can thin out on application. Then they provide far less real protection than an efficient formula with low UV absorber percentages but controlled film application. In most drug categories, strength and potency are related. But unlike other drug classes, potency and strength of sunscreens are not related, except in batch-to-batch comparisons of the same product formula.

In sunscreens, the *inactive* ingredients often have a strong effect on the performance of the UV absorbers. Hewitt has shown that sometimes a more than 2× increase in potency can result from changes in refraction of UV by inactive ingredients or from emulsion recovery time.[3] Likewise, our study of leveling agents shows that a sunscreen product's strength provided by the film thickness does not proportionally match potency, as delivered in the SPF we receive in actual usage.

The Effect of Too Thin Application of Sunscreen

Many researchers have addressed the shortcomings of inadequate film formation by sunscreen products, beginning with Stenberg and Larko,[4] who compared SPF produced by 2 mg/cm^2 sunscreen films with film half as thick (1 mg/cm^2), and reported a 50% loss in protection.

These results are generally in agreement with theoretical projections based on Beer's Law of light absorption; yet opposing results have been reported by Gottleib et al.,[5] who recorded no significant loss of protection by films of half thickness. We believe such a divergence of results may be explained by the two competing theories of light blocking, which will be explained in a moment.

To exacerbate the film-thinning effect, O'Neill[6] calculated that "unevenness" in film application can result in drastically reduced levels of protection. Additionally, Stansfield[7] has recently described negative deviations from Beer's Law from photo-instability of UV absorber, projecting how a film of half the initial sunscreen content (strength) can provide less than half the expected protection (potency) (Figure 1).

Two Theories of Light Blocking by Sunscreen Actives

Beer's Law: UV absorbers in sunscreens generally follow the Bouguer-Beer-Lambert Law[8] if they are soluble in their formula vehicle and passing through a film of UV absorber:

$$\%T = I_t/I_o \times 100\%$$

where %T is the percent of transmitted light, I_o is the intensity of the original light beam, and I_t is the intensity of the light beam after transmission.

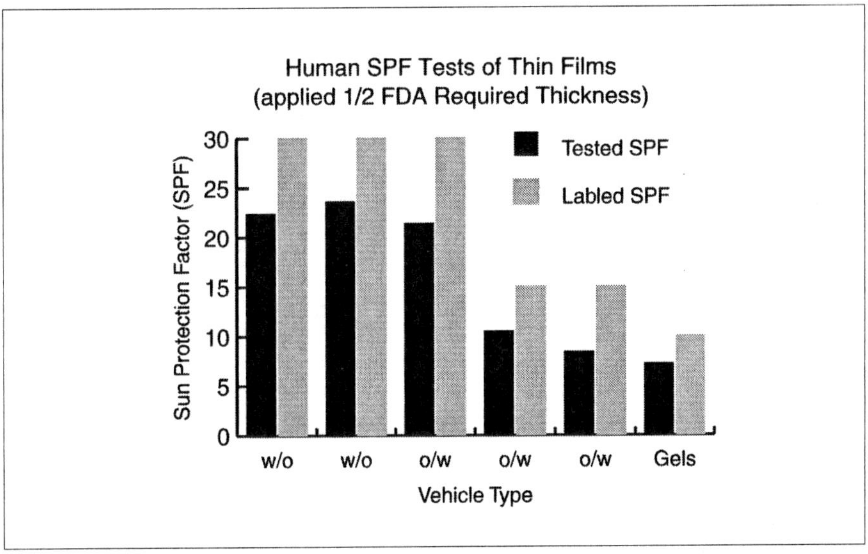

Figure 1. What happens when sunscreen is too thin

Or, to complicate matters, the nontransmission of light is often reported using a nonlinear log scale because it is proportional to concentration of the absorber:

$$A = -\log\{I/I_o\} = \log\{I_o/I\}$$

where A, called absorbance, is the negative log of the decimal value of the transmittance, where an absorbance of 1 signifies 90% light absorbed; an absorbance of 2 signifies 99% light absorbed.

Simplified, this law provides a uniform method to predict the effect of UV absorbers on UV light. What it tells us is that adding some UV absorber to a transparent material blocks a percentage of any UV light that might shine through it. The more you add, the less it appears to do. If you were to add enough absorber to block 25% of the UV, the mixture would then let 75% of the incident UV through. But, if you were to add the same amount of absorber again, the second dose would only stop an added 18.75% of the original beam. This is because the second addition would block 25% of the *remaining* UV light, not 25% of the original UV.

Because a second dose of absorber will block less, and the third even less, we have an example of the law of diminishing returns. As you add more and more UV absorber, and it does less and less, eventually it does almost nothing (Figure 2).

From this we can see that we will never be able block all the UV no matter how much absorber we use. Thus, sunscreen chemists know well that you cannot get much more than an SPF of 8 with octinoxate in mineral oil no matter how much you use. Higher SPF values require tricks including smoke and mirrors (light reflection and scattering)!

This law is more commonly attributed to August Beer (1825-1863), however, it was Pierre Bouguer[9] in 1729 who first determined that the thickness of a sample is inversely related to the passage of light. Around 1760, J.H. Lambert[10] applied the following differential to equate absorbance to sample thickness:

$$dI/I = -a \times dx$$

where "a" is a constant Lambert called opacity and dx is an infinitesimal distance through the sample.

Two and a half centuries later, most studies of sunscreen application have found that the effect of soluble UV absorber is proportional to the thickness of the film. The thicker the film, the better the protection. Such observations are in agreement with Bouguer's hypothesis and Lambert's equation for sunscreens, as well as our human SPF test results on films of varying thicknesses.

Rayleigh's scattering equation: Particulate UV blockers such as titanium dioxide or zinc oxide are the "smoke and mirrors" of the sunscreen science. They do not obey Dr. Beer when used in concentrations above 0.1%. In suspension, pigments tend to reflect or scatter incident UV light waves and the amount of light passing through a film of particles – whether it be the rings of Saturn or an oxide sunscreen layer – diminishes with the concentration of particles. This blocking effect differs from the diminishing returns of Beer's Law.

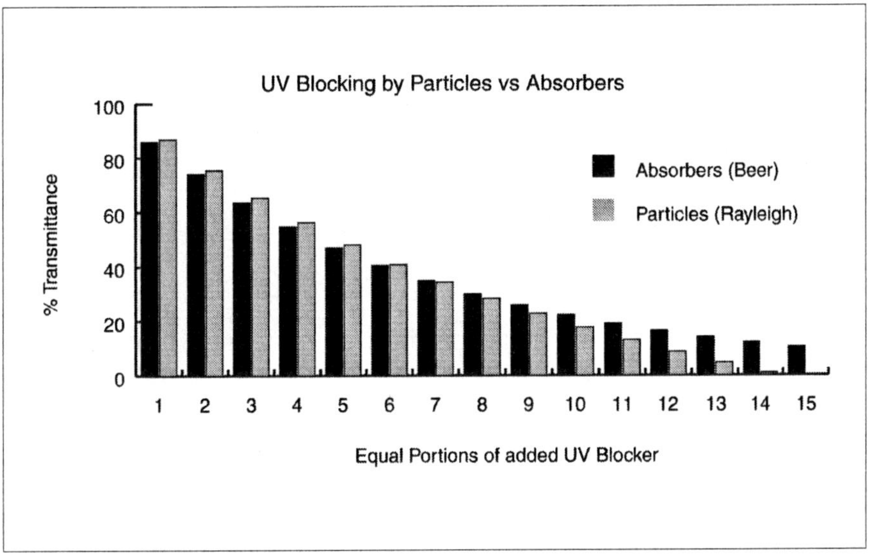

Figure 2. Relative effects of concentration on passage of UV light (Beer's Law vs. Rayleigh's Equation)

For sunscreens, light that is not reflected and that passes through a film is generally defined by what's left after we subtract reflection; this is the transmission (%T) equation.[11] The reflection equation was developed by John Strutt in 1871, before he became Lord Rayleigh. Our equation is a variation of the famous Rayleigh Equation:

$$\%T = (I_o - I_\pi)/(I_o + I_\pi) \times 100$$

where %T is the percent of transmitted light, I_o is the intensity of the original light beam and I_π is the intensity of the light reflected or scattered.[a]

The resulting relationship can be envisioned as what happens when sand is scattered on the floor. The first portion covers some measurable percent of the floor. The second portion fills in the holes and covers almost an additional equal percentage. Soon the floor is covered and the last holes are filled in. If the particles can pack well together, insoluble pigments have the ability to block much higher percentages of light than do soluble absorbers.

In real practice, tight packing by a pigment layer is limited by the vehicle. The other ingredients (oils, emulsifiers, preservatives and thickeners) tend to separate the particles and let UV light pass through. Still, the first SPF 50 sunblocks were achieved using micronized titanium dioxide pigment. If excess pigment beyond the amount required for maximum coverage is used, it could theoretically prevent significant reduction of SPF by thinning the applied film. It is unlikely that most manufacturers would be anxious to add extra micronized pigment that costs 10 to 20 dollars per pound, with no immediate SPF benefit. All of the chosen products were SPF tested at half thickness (1 mg/cm^2) and showed significant loss of protection.

Table 1. Spreading of 16 Commercial Sunscreens

Sample	Brand	Label SPF	Film Thickness	Drying Time (sec)	Brookfield Viscosity (Spindle 7/100 rpm)
1	A	70	1.40	51	1500
2	A	50	1.20	65	1260
3	B*	48	1.92	35	2960
4	C*	45	1.18	45	240
5	D	45	1.60	130	2240
6	E*	36	2.08	55	3400
7	C*	30	1.71	59	6920
8	F	30	1.37	58	5400
9	C	30	1.41	40	2560
10	F	30	1.24	55	4120
11	G	30	1.39	45	2360
12	H	30	1.26	80	2600
13	J	30	1.51	75	3000
14	K	30	1.37	100	1440
15	L	15	1.46	50	6640
16	M	15	1.58	16	3400

* product contains a leveling agent

Spreading Test of Commercial Sunscreen Products

Protocol: Numerous pharmaceutical researchers have addressed the nature of spreading of topical ointments and lotions.[12] We, however, needed to evaluate product spreadability with a quick, easy and reproducible method using a panel of volunteers. In our test protocol, a sample of test product was dispensed by syringe onto the center of the volar (inner) forearm surface. The volunteer was then asked to spread the product "as far as it would go." The area covered and the spreading or drying time were then measured and recorded. We then calculated the applied film's resulting film density (in units of mg/cm^2) as follows:

Film Density = Sample Weight / Covered Area

We found it necessary to control the temperature and humidity to achieve reproducibility and we also discovered that sweating further reduces the film thickness by 30-50% in conventional sunscreen. We also observed that drying time was proportional to the spreading of sunscreen products.

Spreading results: When we measured the spreadability of our commercial survey products we found that individual application thicknesses of these products ranged from 30% to 140% of the U.S. Food and Drug Administration (FDA)[12] standard (2 mg/cm^2). Some products varied up to ±50% within the panel of 5-8 volunteer subjects, while other products showed as tight a variation as 10%. The tightest variations were observed in products containing leveling agents.

In Table 1 we report the average (mean) film densities of films spread by our subjects. These range from 59% to 93% of the FDA standard. Spreadability is a property that can easily compromise the performance of many other topically applied drug products beside sunscreens. We also noted that the application of those few commercial products containing leveling agents (to control spreading) and our test samples were not affected much by temperature and humidity.

SPF results: We performed SPF determinations to verify our presumption that the innate spreadability of a sunscreen results in the delivery of a wide variation of protection from brand to brand.

Our commercial survey product group was chosen to represent a large percentage of the market. It consisted of 16 commercial sunscreen products, including the major mass-marketed brands. Some products were expected to provide as little as 40% of the labeled SPF when used as directed, due to thin application. From our sampling, we projected that more than 80% of the currently marketed sun protection products fail to provide the SPF indicated on their label. Moreover, we determined that the 80% of commercial products that fail to provide the labeled SPF fail because the products can be easily spread out to a less effective thickness.

The 16 commercial sun protection products we purchased ranged in SPF from 10 to 70. These products were evaluated for their average application thickness by 5-8 human volunteers. Volunteers applied 100 mg on their inner forearms, spreading the material until it dried.

External variables in this test series were monitored, including temperature, humidity, sweatiness of the subject, skin topology and sample drying time. Internal variables (which we controlled) affecting the results, included size of sample applied, viscosity and leveling agent (in the formulation tests). In the end, we settled on a 100 mg sample size because it produced results close to several whole body sunscreen application experiments we conducted on the beach in Ft. Lauderdale, Florida, and Gottlieb's comparison[4] of product application on different body parts.

We calculate that our 16 leading sunscreen products probably represent between 30% and 70% of the commercial market.

Materials Affecting Film Thickness

"Leveling agent" is the term used by the technologists in the paint and coatings industry to describe materials that help deliver a level and uniform film. Leveling agents have been widely researched in the printing ink industry. Uneven layering of an applied ink film results in blotchy color. We found several cosmetic materials that acted similar to the ink additives, but were much safer for application to the skin. These materials produced different effects in the various formulas (Figures 3-5).

The rheology of the best leveling agents exhibited dilatent flow under increasing shear. This means that they tended to resist flow when rapid spreading forces were applied. Some combinations of the tested

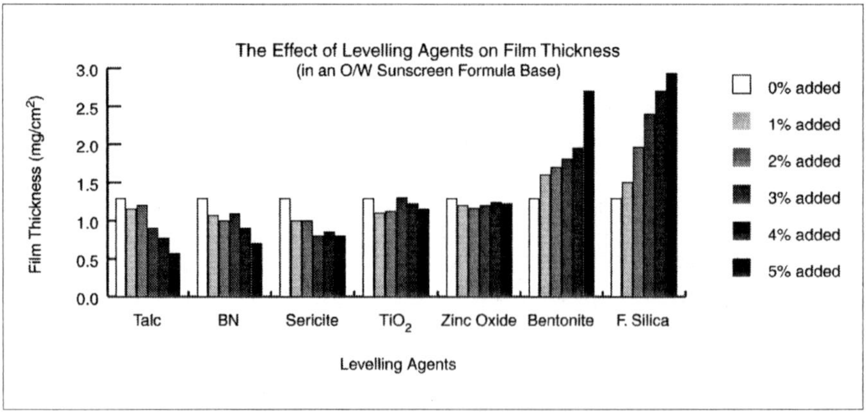

Figure 3. The effect of leveling agents on an o/w vehicle

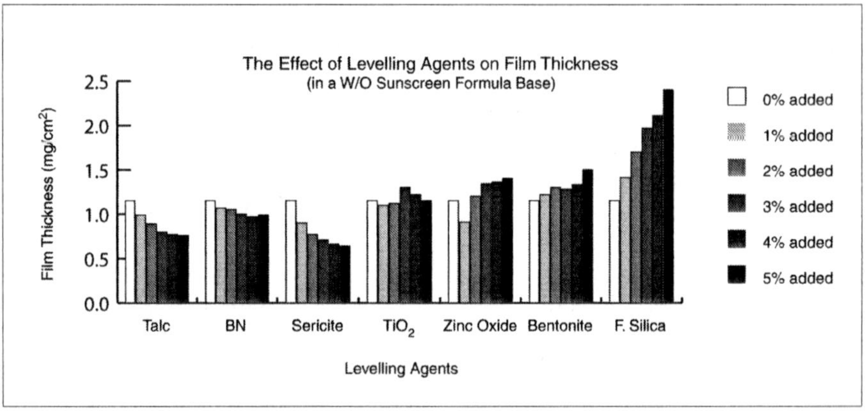

Figure 4. The effect of leveling agents on an o/w vehicle

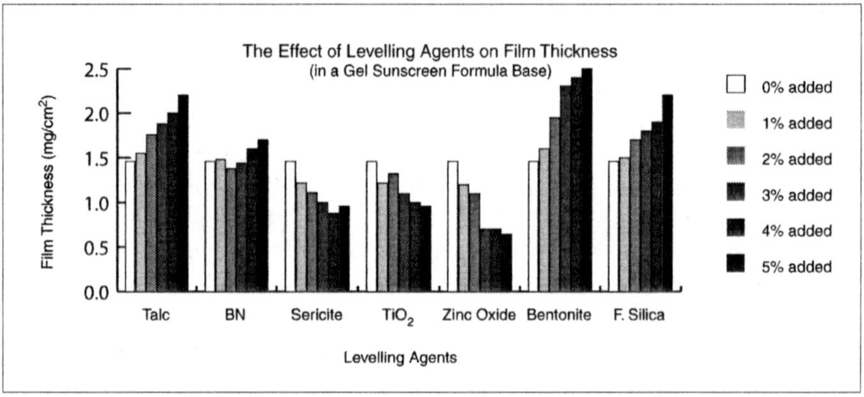

Figure 5. The effect of leveling agents on a gel vehicle

ingredients showed synergism. Ingredients tested for leveling ability included bentonite[a], boron nitride[b], sericite[c], silica[d], talc[e], titanium dioxide[f] and zinc oxide[g].

Formulation Studies

We were surprised to find that many of the materials we screened for their effect on spreading, such as talc and sericite, increased the formula spread instead of controlling it. Of course, the sunscreen base formulas we were applying were much thicker than the ink films we expected to emulate. Nevertheless, we discovered several potent and effective candidates that worked well in our formulas.

Our test formula vehicles were of three varieties; oil-in-water (o/w), water-in-oil (w/o) and gel. Each formula attributes its efficiency to a different principle. The o/w formula was a product with a synergistic UV absorber combination. The w/o formula exhibited exaggerated reflection of UV due to refraction, and the gel formula exhibited strong molecular attraction to keratin.

We discovered that good leveling agents have differing effects in different systems. Silica was the most effective agent in o/w and w/o, but bentonite was more effective in gels. Both emulsion formulas used a combination of absorbers, while the w/o formula included a reflector (titanium dioxide), and the gel used only an absorber (Ensulizole) which is water-soluble.

In our initial in vivo SPF tests, our test formula vehicles – o/w, w/o and gel – all provided approximately 75% of their labeled SPF when applied at half the rated thickness. Because most of the test products spread at 65-75% of their rated thickness, we estimate a mean deficiency of protection of 12-15% across the industry. The w/o and o/w emulsions contained reflectors (scattering UV) and therefore did not show as much loss of protection on thinning as was shown by the gel. Also our results were in agreement with theory which predicts less potential thinning by Rayleigh response (particle reflections) compared to Beer's Law (transparent absorbers). Finally, we observed that formulas with leveling agents show less variability within the test subject panel.

Since we began our investigation into leveling agents, one major marketer and several smaller brands have introduced these additives into their newest formulas. Their products were easily the best performers in our study.

Leveling sunscreen formulations is a very new technology. We expect this new technology to be rapidly adopted throughout the industry because, beyond the health impact, it has a market impact that is hard to ignore. Leveling agents will increase the usage rate and can therefore be expected to increase sales.

[a]Albagel 4444 is a product of Whittaker, Clark & Daniels, South Plainfield, New Jersey USA
[b]BN is a product of Advanced Ceramics, Strongsville, Ohio USA
[c]SL-012 is a product of Presperse, Somerset, New Jersey USA
[d]Cabosil M5 is a product of Cabot, Tuscola, Illinois USA
[e]Melody is a product of Ultra, Red Bank, New Jersey USA
[f]M262 is a product of Presperse, Somerset, New Jersey USA
[g]Z-Cote is a product of BASF, Mount Olive, New Jersey USA

Conclusions

Our in vivo SPF test results support the proposition that thinner sunscreen films indeed deliver diminished UVB protection.

The widespread under-performance of commercial sunscreens is a problem being addressed by new leveling technology. Silica and bentonite were the most effective leveling agents in our test formulations. The test formulations incorporating leveling agents achieved full 2 mg/cm^2 films, or more on uncontrolled application.

In vivo product application testing confirmed the effectiveness of incorporating leveling agents in sunscreens to deliver to the marketplace the full labeled SPF of sunscreen products.

—Christopher D. Vaughan, Susan M. Porter and Sherine Bichara,
SPF Consulting Labs, Inc., Pompano Beach, Florida, USA

References

1. B Diffey, How much sunscreen do we really need, keynote speech at International Cosmetic Exposition, Miami (2003)
2. C Vaughan et al, The Southbeach sunscreen survey, *Cosmet Toil* 38(16) 221-229 (2002)
3. E Leukenbach, U.S. Patent 6540986 (4/1/2003)
4. C Stenberg and O Larko, Sunscreen application and its importance for the sun protection factor, *Arch Dermatol* 121 1400-1402 (1985)
5. A Gottlieb et al, Effects of amounts of application of sun protection factors, Ch 28 in *Sunscreens*, Lowe, Shaath and Pathak, eds, 2nd edn, New York: Marcel Dekker (1997)
6. JJ O'Neill, Effect of film irregularities on sunscreen efficacy, *J Pharm Sci* 73 888-891 (1984)
7. JW Stansfield, Sunscreen photostability and UVA protection, *Proceedings of Society of Cosmetic Chemists Annual Scientific Seminar* (2001) pp 40-41
8. DW Ball, (Bouguer-Lambert) Beer's Law, *Spectroscopy* 14(5) 16-17 (1999)
9. P Bouguer, *Essai d'Optique sur la Gradation de la Lumiere*, Paris: (1729)
10. JH Lambert, *Photometria, Klassiker der Exxakten Wissenshaften*, Leipzig: (1892)
11. Lord Rayleigh, On the transmission of light through an atmosphere containing small particles in suspension, and on the origin of the blue of the sky, *Philos Mag* 47 375-384 (1899)
12. G Alka et al, Spreading of semisolid formulations: an update, *Pharmaceutical Technology* 84-105 (Sept 2002)
13. FDA Sunscreen Drug Products – Final Monograph, *Federal Register* 64(98) 27667-27693 (5/51/99)

Sunscreen Formulas with Multilayer Lamellar Structure

Keywords: sunscreen, multilayer lamellar structure, phosphate emulsifiers, emulsion rheology, Power Law, bound water, water wash resistance

The multilayer lamellar structure in sunscreens containing phosphate emulsifiers plays key roles in the emulsion rheology and in enhancing deposition of sunscreen oil on skin surface, thereby improving the SPF water wash resistance.

A sunscreen formula's ability to protect the skin from UV damage depends on a wide variety of factors. These factors include chemical structures of UV filters, concentrations of active ingredients in the formula, and importantly, what concentration can be achieved on application to the skin and how much remains during the water wash process. The emulsifier(s) and fatty components in an applied sunscreen emulsion are the principal influences on the product's effectiveness.

In a previous study,[1] we concluded that a new phosphate emulsifier[a] improved the SPF water wash resistance of sunscreen formulations, and the structural characteristics of the emulsion were also very important. The INCI name of this ingredient is cetearyl alcohol (and) dicetyl phosphate (and) ceteth-10 phosphate. Figure 1 shows a drawing of the two phosphate esters.

In general, phosphate esters demonstrate the following chemical properties:

- The ionic phosphate group constitutes a powerful o/w emulsifying agent.
- The ionic phosphate group is shielded by the alkyl chains, hence the "crypto" in the term "cryptoanionic."
- Unlike carboxylic acid esters, the phosphate ester link is very stable at high and low pH.
- The emulsification characteristics of the ester depend on the degree of neutralization of the free-acid groups (ratio of free-acid/mono ester/di-ester).
- The lipophilic character is critically dependent upon the alkyl chain length, chain-length distribution, and number of EO (ethylene oxide) groups in the molecule.

---

[a] Crodafos CES, Croda Inc., Edison, New Jersey USA

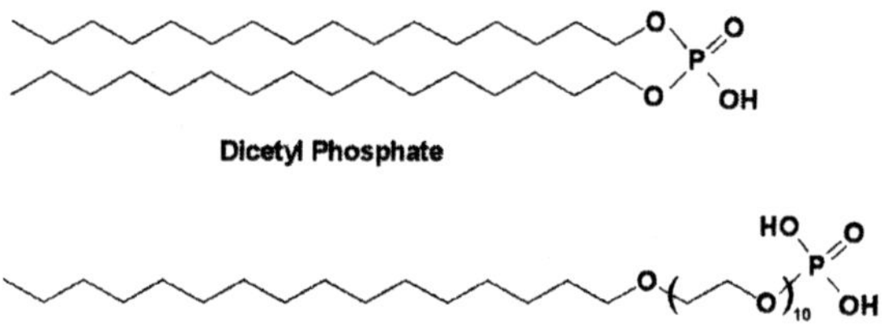

Figure 1. The phosphate esters in a new phosphate emulsifier

Formula 1. Waterproof SPF 30 Sunscreens

	Weight %		
	SPF30PA	SPF30P	SPF30W
Cetearyl alcohol (and) dicetyl phosphate (and) ceteth-10 phosphate (Crodafos CES, Croda)	6.50	6.50	-
Emulsifying Wax NF (Polawax, Croda)	-	-	6.50
Benzophenone-3 (Rona)	5.00	5.00	5.00
Octyl methoxycinnamate (H&R Group)	7.50	7.50	7.50
Octyl salicylate (H&R Group)	5.00	5.00	5.00
Menthyl anthranilate (H&R Group)	5.00	5.00	5.00
Octyl stearate (Crodamol OS, Croda)	5.00	5.00	5.00
Avocado oil unsaponifiables (Crodarom Avocadin, Croda)	2.50	-	-
Sodium hydroxide, 10%	1.54	1.54	1.54
BHT	0.10	0.10	0.10
Carbomer (Carbopol 981, BFGoodrich)	0.13	0.13	0.13
Propylene glycol (and) diazolidinyl urea (and) methylparaben (and) propylparaben (Germaben II, ISP)	1.00	1.00	1.00
Water (*aqua*), deionized	60.73	63.23	63.23
Static SPF	**31.66**	**31.66**	N/D
Waterproof SPF	**30.31**	**30.31**	N/D

- Fatty alcohols work well in combination with phosphate esters made from these alcohols.

In this article, we present our recent studies on sunscreen emulsions with and without the phosphate emulsifier. We observed the emulsion structure, and we measured emulsion rheology, the content of structurally bound water in the

Formula 2. Waterproof SPF 15 Sunscreens

	Weight %	
	SPF15P	SPF15W
Cetearyl alcohol (and) dicetyl phosphate (and) ceteth-10 phosphate (Crodafos CES, Croda)	6.00	-
Emulsifying Wax NF (Polawax, Croda)	-	5.20
Mineral oil	8.00	8.00
Ethylhexyl methoxycinnamate (Escalol 557, ISP)	7.50	7.50
Petrolatum	4.50	4.50
Steareth-10 (Volpo S-10, Croda)	1.00	1.00
Steareth-2 (Volpo S-2, Croda)	0.50	0.50
Cetyl alcohol (Crodacol C-70, Croda)	0.50	0.50
TEA, 99%	0.30	-
Propylene glycol (and) diazolidinyl urea (and) methylparaben (and) propylparaben (Germaben II, ISP)	1.00	1.00
Water (*aqua*), deionized	70.70	71.80
Static SPF	**16.20**	**14.3**
Waterproof SPF	**13.60**	**11.2**

sunscreen formula, and the in vivo fluorescence intensity of the tested sunscreen emulsions applied on human skin before and after the water wash process.

Materials

We prepared sunscreen formulas that provided static SPFs of approximately 30 and approximately 15. In each case we made one version with the phosphate emulsifier and another version in which the phosphate emulsifier was replaced with an emulsifying wax. For the SPF 30 formula, we also made a version containing the phosphate emulsifier and 2.5% avocado oil unsaponifiables to enhance the skin moisturization. In this article we'll identify these five sunscreen formulas as follows:

- SPF30P = SPF 30, with 6.5% phosphate emulsifier;
- SPF30PA = SPF 30, with 6.5% phosphate emulsifier and 2.5% avocado oil unsaponifiables;
- SPF30W = SPF ~30, with 6.5% emulsifying wax NF;
- SPF15P = SPF 16, with 6.0% phosphate emulsifier;
- SPF15W = SPF 14, with 5.2% emulsifying wax NF.

The exact SPF of SPF30W was not determined. (It was made only to study the emulsion structure. Because its formula was similar to those of SPF15P and SPF15W, we believe that the SPF value should be approximately 30.) The

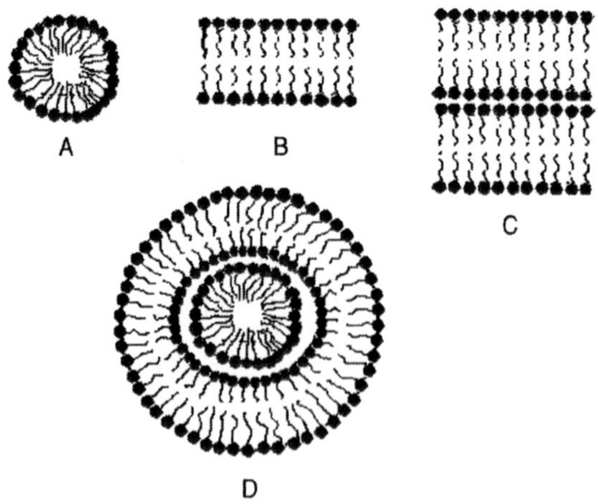

Figure 2. Schematic representations of emulsion particles
A = spherical micelle
B = lamellar micelle
C = multilayer lamellar micelle
D = multilayer lamellar vesicle

complete formulations are given in Formulas 1 and 2, where the static SPF and the waterproof SPF (after 80 minutes of water immersion) are also given.

Methods

Emulsion structure: The phase structure of emulsions depends on various factors, such as the emulsion type, concentrations of surfactants and other ingredients, the molecular structures of emulsifiers and the preparation process (shearing speed and temperature). Generally speaking, micelles exist in two-component systems and their shape can be spherical, lamellar, and multilayer lamellar (Figure 2).

At higher concentrations, surfactant or mixed surfactant solutions can form liquid crystals – lamellar and hexagonal phases. These liquid crystals can be observed using an optical microscope with crossed polarizers.[3-4] It is our interest to look at two unique multilayer lamellar phases: unilamellar and multilamellar phases.

Particles containing only one bilayer have been termed "unilamellar vesicles" (ULV), among which small unilamellar vesicles (SUV) and large unilamellar vesicles (LUV) can be differentiated. Particles containing two and more bilayers have been termed "multilayer lamellar vesicles" (MLV). The MLV have been widely used in pharmaceutical as well as in cosmetic applications as a delivery system for active ingredients. The major benefits obtained from a MLV delivery system are[2]:

- Improved dispersion of difficult-to-solubilize compounds
- Microencapsulation in a vehicle that may enhance penetration into skin
- Improved adhesion on the skin surface and sustained release
- Reduced skin toxicity/irritation from the carrier/solubilizer system

In our experiments, sunscreen emulsion samples were placed on glass slides with cover glasses on a thermal stage with a constant temperature of 25°C. The image of the emulsion was captured using a polarizing microscope[b] with crossed polarizers and later subjected to computerized image analysis[c].

Emulsion rheology: Rheology of an emulsion is of fundamental importance, because flow is always involved in both making and using emulsions. Rheological measurements provide very useful information about the effect of temperature on the viscosity of the emulsion, the influences of ingredients on the emulsion stability and shelf life, and the response of the finished products to different shear rates (i.e., dispensing and application to the skin). These are factors used by the consumer to create a perception of quality and value.

The correlation between shear rate, shear stress and viscosity of liquids can be expressed as the power law:[5]

$$\tau = KD^{n} \qquad [1]$$

$$\eta = \tau/D = KD^{n-1} \qquad [2]$$

where D is the shear rate (1/s), τ is the shear stress (Pascal), η is the calculated viscosity (mPa.s), K is the consistency index, and n is the flow index. K and n are characteristic constants of the material. Liquids with a value of n greater than 1 are shear thickening (viscosity increases with an increase in shear rate) while those with n less than 1 are shear thinning (viscosity decreases with an increase in shear rate). The smaller the n value, the quicker the viscosity decreases with increasing shear rate. When $n = 1$ and $\eta = K$, then η is the Newtonian viscosity, and the viscosity remains constant at various shear rates.

It has been reported[6,7] that the multilamellar vesicles can be formed from regular lamellar phase by shear-induced processes. Three main orientations have been described: At very low shear rate (below 1 s^{-1}) the lamellar layers are mainly parallel to the flow. At very high shear rates, the orientation of lamellae is similar. For intermediate shear rates, the lamellar phase organizes itself into multilamellar vesicles, which are close-packed and fill up space.

The rheological profiles of our sunscreen samples were determined using a rheometer[d] at constant temperature of 25°C. The data analysis was performed using computer software[e].

Determination of structurally bound water in an emulsion: Water, as the continuous phase, is a major component in an o/w emulsion. The water molecules in an emulsion are of two types: "structurally bound" and "free." The "structurally bound" water molecules are defined as chemically associated with other molecules in the emulsion. The "free" water molecules, in fact, are strongly bound themselves together through inter-molecule hydrogen bonding.

It has been reported[8] that differential scanning calorimetry (DSC) can be used to determine the temperature and the enthalpy of a phase transition (freezing/melting) of water during a heating or cooling process. In fact, water molecules contained in emulsions have relatively higher free energy than those in pure wa-

[b]Nikon Optiphot-Pol Polarizing Microscope, Nippon Kogaku KK, Tokyo, Japan
[c] Image-Pro Plus 4.5 image analysis software, Media Cybernetics, Silver Spring, Maryland USA

ter. This means that structurally bound water molecules in emulsions need less activation energy to escape from the emulsion (evaporation), and therefore, they should have lower melting temperature and smaller enthalpy of fusion than those in pure water. The melting point is 0°C and the enthalpy of fusion of pure water is 333.6 J/g,[9] which reflects the highest energy required to break the hydrogen bonding between associated water molecules.

When a portion of water molecules is bound to other molecules of ingredients in an emulsion, the determined enthalpy of fusion of water should be always smaller than the standard value. Increasing the amount of bound water will mean fewer free water molecules and a lower melting/freezing enthalpy value. Therefore, it is possible to use the determined value of freezing enthalpy of water in an emulsion to calculate the percent of bound water (PBW) and free water (PFW) in the emulsion:

$$\text{PBW}(\%) = [1-(H_1/H_0)] \times 100\% \qquad [3]$$
$$\text{FBW}(\%) = (H_1/H_0) \times 100\% \qquad [4]$$

where H_0 is the standard water freezing enthalpy, 333.6 J/g, and H_1 is the determined freezing enthalpy of water in the emulsion sample (Joule/gram). The calculated PBW/FBW value can be used to estimate the emulsion stability. The higher the content of structurally bound water, the more stable the emulsion.

In this paper, the PBW of our emulsion sample was determined instrumentally[f]. About 2 mg of emulsion sample was placed in a hermetic pan and cooled to -20°C. The sample then was heated at 2°C/min of heating to 25°C and the melting point and the enthalpy of fusion of water were calculated.

Quantitative assessment of sunscreen application by in vivo fluorescence spectroscopy: A group of scientists from the UK has conducted extended studies on evaluation of sunscreen performance using fluorescence spectroscopy.[10-13] They have found that the feasibility of using fluorescence spectroscopy for in vivo quantitative assessments of sunscreen substantivity depends on selecting a suitable fluorescent agent, which should be nontoxic, mix readily with sunscreens and be excited in the UV/visible region of the spectrum. They have observed that the fluorescence from Neutrogena Sunblock was sufficiently intense for use as a tool for in vivo measurements of substantivity. This product achieves SPF 15. Its active ingredients are octyl methoxycinnamate, octyl salicylate and menthyl anthranilate.

Based upon the same technique, we determined the fluorescence intensity of our test SPF 30 sunscreens (SPF30P, SPF30PA and SPF30W) and found that the intensities were strong enough for in vivo measurements. The fluorescence intensity of sunscreen on human skin was calibrated and determined using a spectrofluorometer[g] coupled to a bifurcated optical fiber cable. Radiation from the exit slit of the excitation monochromator was conducted via quartz optical fiber to the skin. The optical fiber collected the fluorescent radiation from the skin and conducted this to the entrance slit of the emission monochromator. The signals were processed and the data displayed using appropriate software[h].

[d] DV-III Rheometer, from Brookfield Engineering Lab, Inc., Middleboro, Massachusetts USA
[e] Rheocalc 2.3, from Brookfield Engineering Lab, Inc., Middleboro, Massachusetts USA
[f] DSC (differential scanning calorimeter) Q-100, TA Instruments, New Castle, Delaware USA

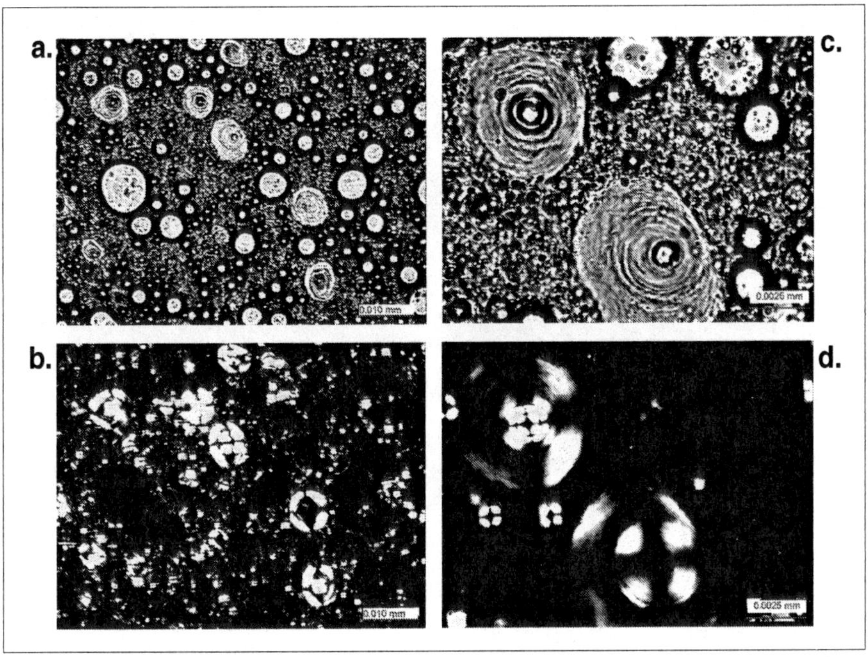

Figure 3. Micrographs of phosphate-emulsified SPF 30 sunscreen formula (SPF30PA) with avocado oil unsaponifiables
a = X100, no polarizer
b = X100, under crossed polarizer
c = X400, no polarizer
d = X400, under crossed polarizer

For the water wash test of sunscreen formulas, the main experimental procedures include:

1. Finding the optimum excitation and emission monochromators and respective bandwidths for the applied sunscreen formula on the skin surface.
2. Establishing the calibration curve, which is the relationship between sunscreen density and fluorescence intensity for the applied sunscreen.
3. Applying a certain amount of sunscreen on the determined skin area and allowing it to dry for 30 minutes.
4. Water wash step: set the running tap water at 30°C with a flow rate of 2.5 L/min; water wash time: 30 seconds; dry the skin for 15 minutes. Repeat the washing and drying steps for two more cycles.
5. Determining the fluorescence intensity after the water wash step and calculating the remained sunscreen density on the skin.
6. Comparing the original sunscreen density to the remained density after water wash and calculating the remained sunscreen percentage.

[g] Fluorolog-3 spectrofluorometer, ISA Instruments SA, Inc, Edison, New Jersey USA

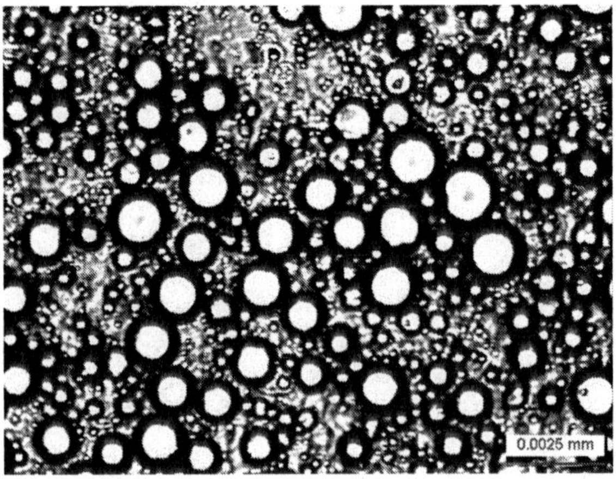

Figure 4. Micrograph of wax-emulsified SPF 30 sunscreen formula (SPF30W) at X400, no polarizer

Results on Emulsion Structure

Micrographs of sunscreen formulation SPF30PA (with the phosphate emulsifier and avocado oil unsaponifiables) are presented in Figure 3. From Figures 3a and 3c (no polarizer; X100 and X400, respectively) it is clearly observed that SPF30PA contains particles having multilayer structure. Figures 3b and 3d (cross polarizer; X100 and X400, respectively) show typical patterns of liquid crystals of these particles and clearly demonstrate the multilayer lamellar phase in the formula. The observed sizes of large multilayer lamellar droplets are in the range 0.5-5.0 μm

Sunscreen formula SPF30P (with the phosphate emulsifier) showed the same patterns of micrographs as SPF30PA and was similarly composed of multilayer lamellar vesicles.

A micrograph of the corresponding formula SPF30W (using Emulsifying Wax NF as the emulsifier) is presented in Figure 4. It can be seen that the sizes of droplets in the formula are smaller and more uniform, but no liquid crystal was observed under crossed polarizers.

A similar difference in the emulsion structure was observed for the SPF15 sunscreen formulas: SPF15P showed multilamellar vesicles and SPF15W did not.

These experimental results strongly indicate that the phosphate emulsifier plays a key role in the formation of the multilayer lamella liquid crystals of the formula. In fact, phospholipids have been widely used to form multilayer lamellar vesicles in both pharmaceutical and cosmetic applications (see sidebar).

Results on Emulsion Rheology

Change in viscosity with shear rate: Plots of viscosity versus shear rate of

[h] DataMax software from ISA Instruments SA Inc, Edison, New Jersey USA

Effect of Emulsifier Molecular Structure on Emulsion Structure

It is interesting to study the effect of modification in molecular structure of a phosphate emulsifier on the emulsion structure of sunscreen formulas. A sprayable sunscreen lotion (Formula 3) with a phosphate emulsifier has been developed. The lotion has low viscosity and the micrograph of the emulsion is presented in Figure 5.

It can be seen that the average droplet size is small and the density of the droplets is low in the lotion. The low viscosity of the formula can be attributed to the increased number of ethylene oxide groups in the molecule. The larger the EO group number in the molecule, the more flexible the molecular configuration, and the more difficult the formation of lamellar structure.

Formula 3. Sprayable Sunscreen Lotion

Cetearyl alcohol (and) dicetyl phosphate (and) ceteth-20 phosphate (Crodafos CS 20 Acid, Croda)	4.00% wt
Xanthan gum	0.20
Glycerin	5.00
TEA 98%	0.10
C12-C15 Alkyl benzoate	3.00
Di-PPG-3 Myreth-10 adipate	5.00
Ethylhexyl methoxycinnamate (Octinoate, ISP)	7.50
Ethylhexyl salicylate (Octisalate, H&R Group)	5.00
Benzophenone-3 (Oxybenzone, H&R Group)	5.00
Propylene glycol (and) diazolidinyl urea (and) methylparaben (and) propylparaben (Germaben II, ISP)	1.00
Water (*aqua*), deionized	64.20
Static SPF = 16.35	
Waterproof SPF = 13.25	

sunscreen formulas SPF30P and SPF30W are presented in Figure 6. It can be seen that these two formulas demonstrated similar rheological characteristics but with different parameters:

- Both formulas were viscoplastic materials and demonstrated thixotropic behavior: Their viscosity decreased with an increase in shear rate.
- The phosphate-emulsified formula showed more pronounced thixotropic characteristics: It had higher initial viscosity at low shear rate and lower viscosity at high shear rates, compared to the wax-emulsified formula.

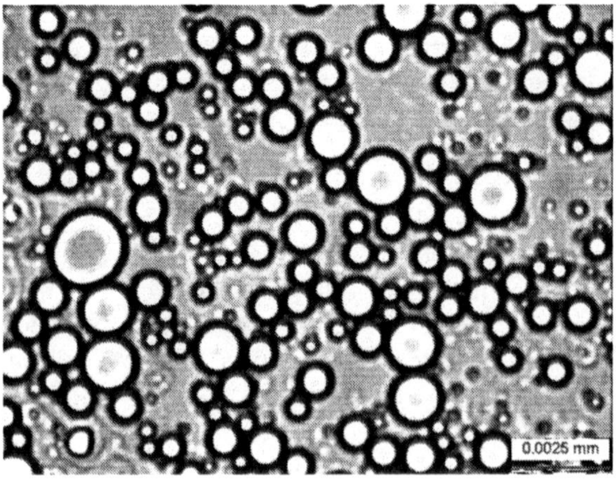

Figure 5. Micrograph of phosphate-emulsified SPF 15 sprayable sunscreen lotion (Formula 3) at X400, no polarizer

The differences in their rheological profiles can be attributed to their different emulsion structures. Indeed, viscosity is a measure of the hydrodynamic radius of particles of liquid under shear stress. The larger the average hydrodynamic radius, the higher the liquid viscosity will be. As mentioned earlier, SPF30P contains multilamellar vesicles with larger average size compared to SPF30W. It is not surprising that SPF30P should have higher initial viscosity at the low shear rate. On the other hand, the shear-thinning rate is related to the deformation of the emulsion structure under the high shear rate. The faster the change in emulsion structure, the higher the shear-thinning rate should be. Since SPF30W does not have multilayer lamellar structure, the change in emulsion structure for SPF30W at high shear rate can be expected to be much less than the change for SPF30P at a similar shear rate. Therefore, the viscosity for SPF30W should be less affected by an increase in the shear rate.

Power Law and flow index: According to the Power Law,

$$\tau = KD^n \qquad [1]$$
$$\log \tau = \log K + n \log D \qquad [5]$$

where K, again, is the consistency index and D is the shear rate. Plots of logarithm of shear stress (log τ) versus logarithm of shear rate (log K) for the SPF30P and SPF30W sunscreen formulas are shown in Figure 7.

It can be seen that the rheological profiles of the two samples fit the Power Law very well. The flow index (n) is 0.2413 and 0.3745 for sunscreen formulas SPF30P and SPF30W, respectively. This result indicates that the sunscreen formulation containing phosphate emulsifier showed faster

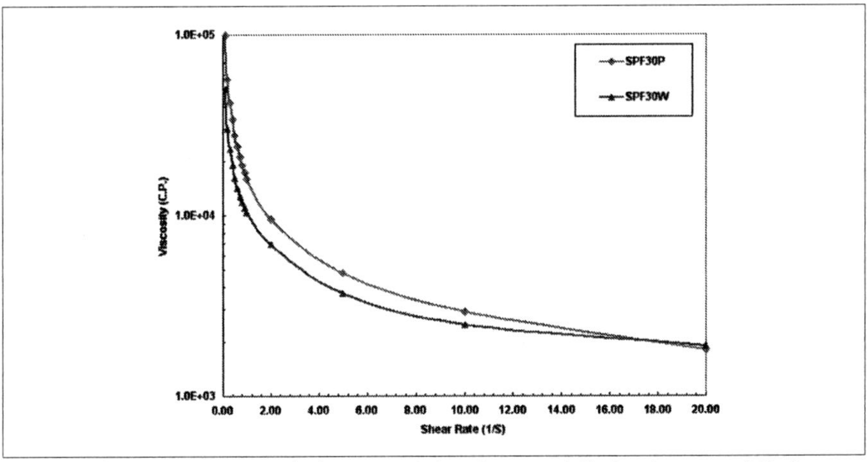

Figure 6. Change in viscosity with shear rate of phosphate-emulsified and wax-emulsified SPF 30 sunscreen formulas (SPF30P and SPF30W, respectively)

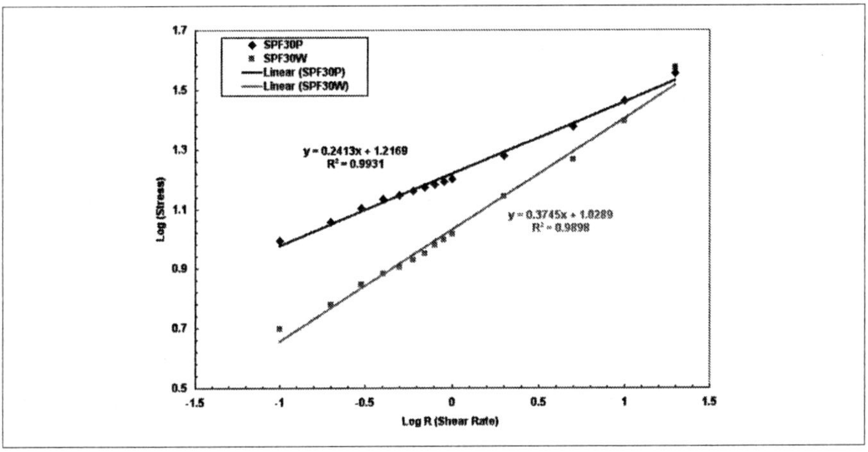

Figure 7. Power law and flow indexes of phosphate-emulsified and wax-emulsified SPF 30 sunscreen formulas (SPF30P and SPF30W, respectively)

decreasing rate (smaller n) in viscosity with an increase in shear rate than the corresponding wax-emulsified formulation. This is consistent with our predictions based upon the differences in their emulsion structure.

In practical applications, when applying the sunscreen product to skin, we always apply the shear stress to the product in order to evenly distribute the product on the skin. The less viscous after the applied shear stress the sunscreen product becomes, the easier it should be to flow and form a uniform layer of product on the skin. Therefore, it is expected that the phosphate-emulsified formula should generate a more uniform layer when applied on the skin.

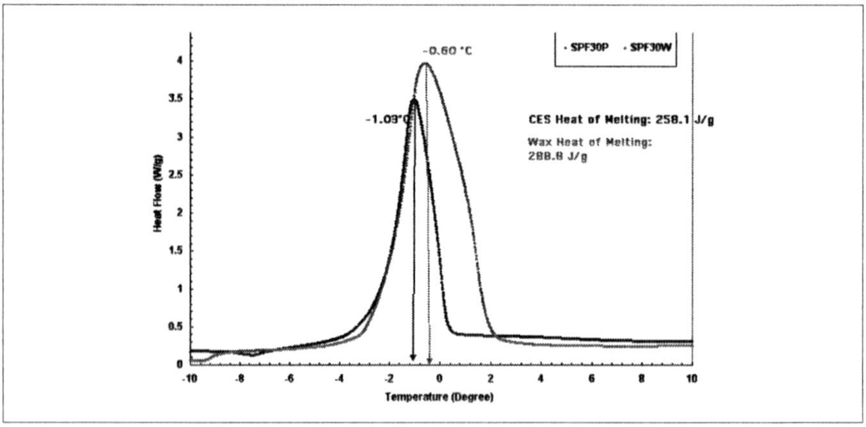

Figure 8. DSC curves of phosphate-emulsified and wax-emulsified SPF 30 sunscreen formulas (SPF30P and SPF30W, respectively)

Results on Structurally Bound Water

Differential scanning calorimetry melting curves of SPF30P and SPF30W emulsions are presented in Figure 8. It is observed that the phosphate-emulsified sunscreen formula showed lower melting point and smaller enthalpy of fusion of water compared to the corresponding wax-emulsified formula. The measurements suggest that the SPF30P emulsion contains more structurally bound water than the SPF30W does.

This experimental result correlates well with our observations of the multilamellar structure existing in the SPF30P sunscreen formula. Using equations [3] and [4], we are able to calculate the percent of structurally bound water in the two emulsions.

$$\text{PBW (phosphate)} = (1 - 258.1/333.6) \times 100\% = 22.6\%$$
$$\text{PBW (wax)} = (1 - 288.8/333.6) \times 100\% = 13.4\%$$

The calculated percentages of free water in the emulsion are 77.4% and 86.6%, respectively, for SPF30P and SPF30W sunscreen emulsions. In our previous study,[1] we reported the different water evaporation rates for phosphate and nonionic systems. The phosphate system showed higher water evaporation rate than the nonionic system. This result can be explained by the different content of structurally bound water in the two emulsion samples. The phosphate emulsion contains more structurally bound water and the structurally bound water molecules are in higher free energy state compared to those in the nonionic emulsion system. Therefore, these water molecules need less energy to evaporate from the emulsion and their water evaporation rate is greater than that for nonionic emulsion system.

Results on Water Wash Resistance

We determined the responses of fluorescence intensity versus the sunscreen concentration on the skin of three female panelists; the obtained calibration curves were

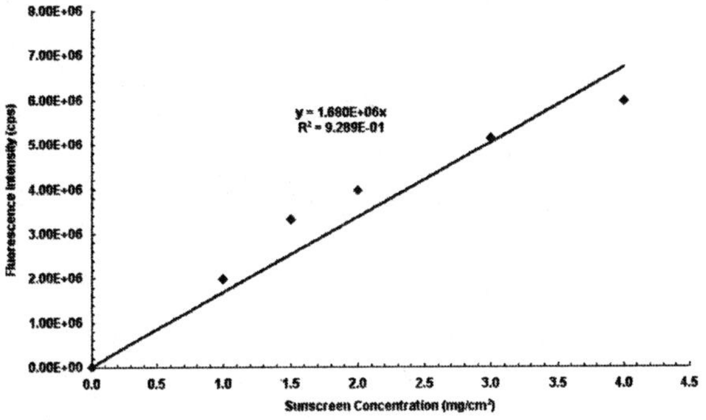

Figure 9. Response of fluorescence intensity vs. sunscreen concentration on skin

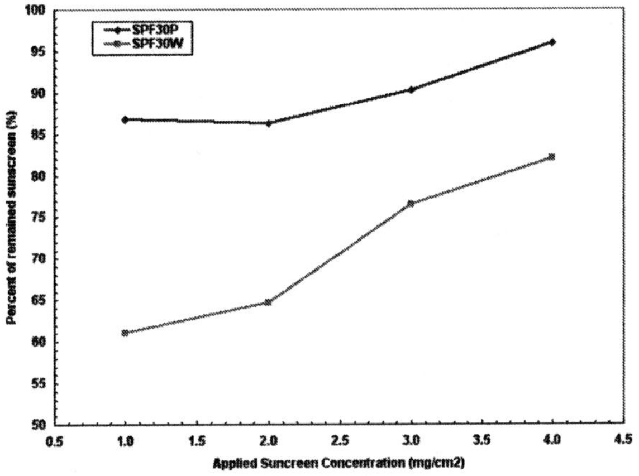

Figure 10. Water wash resistance of phosphate-emulsified and wax-emulsified SPF 30 sunscreen formulas (SPF30P and SPF30W, respectively)

very similar. A calibration curve of fluorescence intensity versus applied sunscreen concentration on skin is presented in Figure 9. We applied a known amount of sunscreen on the skin. After washing the skin with water, we measured the fluorescence intensity of sunscreen on the skin surface, and then used this calibration curve to obtain the concentration of sunscreen on the washed skin. From that we calculated the remained percent of the applied sunscreen on the skin surface.

Plots of the remaining percent of sunscreen concentration on skin after water washing for formulations SPF30P and SPF30W are shown in Figure 10. It can be seen that the phosphate-emulsified formula demonstrated better water wash resistance than the corresponding wax-emulsified formula.

These experimental results are consistent with SPF water wash resistance measurements obtained from panelist testing conducted by AMA Laboratories.[14] The results also strongly indicate the correlation between sunscreen water wash resistance and the emulsion structure.

As we mentioned before, the multilayer lamellar vesicles are very good delivery systems for active ingredients: they improve the dispersion of difficult-to-solubilize compounds, such as sunscreen oil and the adhesion on the skin surface and sustained release of the active ingredients. As a result of these improvements, the phosphate-emulsified SPF30P sunscreen formulation exhibited better water wash performance than the corresponding wax-emulsified formulation.

Conclusions

The phosphate-emulsified sunscreen emulsions formed multilayer lamella structure with large particle size, while the corresponding wax-emulsified emulsions demonstrated small particle size without multilayer lamella structure.

The phosphate-emulsified sunscreen emulsions had a higher amount of structurally bound water and exhibited a higher viscosity shear-thinning rate compared to the corresponding wax-emulsified emulsions.

The phosphate-emulsified sunscreen emulsions left a higher amount of sunscreen concentration on the skin surface where they were applied and demonstrated better water wash resistance compared to the corresponding wax-emulsified emulsions. This was attributed to the multilayer lamella structure in the formula.

—Timothy Gao, Ph.D., Jung-Mei Tien and Yoon-Hee Choi,
Croda Inc., North American Technical Center, Edison, New Jersey, USA

References

1. K Gallagher, A new phosphate emulsifier for sunscreens, *Cosmet Toil* 113(2) 73-80 (1998)
2. D Lasic, Lipsomes and niosomes, in *Surfactants in Cosmetics*, MM Rieger and LD Rhein, eds, 2nd edn, New York: Marcel Dekker (1997)
3. D Roux and F Gauffre, The onion phase and its potential use in chemistry, *ECC (European Chemistry Chronicles)* Research 17-24 (1999)
4. J Zipfel et al, Shear induced structures in lamellar phases of amphiphilic block copolymers, *Phys Chem Chem Phys* 1 3905-3910 (1999)
5. P Miner, Emulsion rheology: Creams and lotions, in *Rheological Properties of Cosmetics and Toiletries*, D Laba, ed, New York: Marcel Dekker (1993)
6. K Diec, W Meier and J Schreiber, Spontaneous formation of liposomes from lamellar liquid crystals, *Cosmet Toil* 117(3) 37-42 (2002)
7. J Lauger, R Weigel, K Berger, K Hiltrop and W Pichtering, Rheo-small-angle-light-scattering investigation of shear-induced structural changes in a lyotropic lamellar phase, *J Colloid & Interfaces Sci* 181 521 (1996)
8. S Motta and E Mignini, DSC: A new method for evaluating cosmetic emulsions and ingredients, in 15th *IFSCC International Congress*, vol A, London: IFSCC (1988)
9. *CRC Handbook of Chemistry and Physics*, D Lide, ed, 75th edn New York: CRC Press (1993-1995)
10. R Stokes and B Diffey, The feasibility of using fluorescence spectroscopy as a rapid, non-invasive method for evaluating sunscreen performance, *J Photochem Photobiol Biol* 50 137 (1999)
11. L Rhodes and B Diffey, Quantitative assessment of sunscreen application technique by in vivo fluorescence spectroscopy, *J Cosmet Chem* 47 109 (1996)
12. L Rhodes and B Diffey, Fluorescence spectroscopy: a rapid, noninvasive method for measurement of skin surface thickness of topical agents, *Brit J Dermatol* 136 12 (1997)
13. R Stokes and B Diffey, The water resistance of sunscreen and day-care products, *Brit J Dermatol* 140 259 (1999)

Film-formers Enhance Water Resistance and SPF in Sun Care Products

Keywords: olefin/MA copolymer, polyethylene, C20-40 alcohols, water resistance, viscosity, PVP/eicosene copolymer

At low formulating levels, film-forming polymers can increase water resistance and enhance SPF in sun care formulations, while also imparting improved aesthetics. In the case of C30-38 olefin/isopropyl maleate/MA copolymer, a synergistic SPF effect can be achieved with PVP/eicosene copolymer.

With ever higher levels of awareness, consumers are putting greater demand on sun care products. Performance is a key issue: products are expected to remain on skin for an extended time without the need for reapplication, even in the presence of water. Good aesthetics also are a must: consumers want lotions and creams that leave the skin feeling soft and moisturized, without the greasy, oily feel traditionally associated with sun care products.

By capitalizing on new technologies for improved aesthetics, greater resistance to wash-off and enhanced SPF, formulators can create innovative sun care products to meet specialized global requirements and the needs of individual skin types.

Polymers That Enhance Water Resistance and SPF

Among materials that impart water resistance, C30-38 olefin/isopropyl maleate/MA copolymer[a] is a low molecular weight, hydrophobic material that ensures a light feel. The MA in this INCI name refers to maleic anhydride. For convenience, this polymer will be referred to as the olefin/MA copolymer.

The maleic functionality of the olefin/MA copolymer makes it easy to disperse and helps it adhere to the skin. It does need to be neutralized for oil-in-water dispersions, but when neutralized, it disperses readily into water and also acts as an anionic emulsifier. The olefin/MA copolymer forms a water-resistant film that also enhances SPF. Because of the copolymer's highly efficient film-forming properties, it may be possible to use lower levels of potentially irritating active ingredients in formulation.

---

[a] PERFORMA V 1608 Polymer is a product of New Phase Technologies, Sugar Land, Texas, USA. PERFORMA V is a registered trademark of Baker Hughes Incorporated.

Polyethylene[b] and C20-40 alcohols[c] also have hydrophobic properties that make them useful ingredients for offering water resistance to formulations. As with the olefin/MA copolymer, the low molecular weight of these ingredients imparts a light feel on the skin. The linear polymer backbone of the polyethylene provides a foundation for thickening and structuring the oil phase of formulations, while giving a matte appearance and a dry, non-oily feel. In the case of long-chain linear alcohols, the alcohol functionality offers compatibility with silicones, while helping to stabilize these ingredients in sun care formulations.

The film-forming properties of the polyethylene and alcohols improve the water resistance of formulations and minimize the levels of active ingredients. Their superior ability to thicken oils makes these ingredients especially useful for enhancing SPF.

In Vitro Tests Assess SPF

The studies described in this article involved in vitro evaluations of water resistance for several materials. They were developed based on several protocols.[1,2] Water resistance was measured on the polymers at a use level of 2% by weight in an "easy-to-remove" prototype formula (Formula 1) that incorporates high levels of very hydrophilic surfactants.

[b] PERFORMALENE 400 Polyethylene is a product of New Phase Technologies. PERFORMALENE is a registered trademark of Baker Hughes Incorporated.
[c] PERFORMACOL 350 Alcohol is a product of New Phase Technologies. PERFORMACOL is a registered trademark of Baker Hughes Incorporated.

Formula 1. Sunscreen test formula

A. Glyceryl stearate (and) PEG-100 stearate	4.00% w/w
Polysorbate-20	2.00
Octyl methoxycinnamate	7.50
Benzophenone 3	4.00
Octyl salicylate	5.00
Cetearyl alcohol	0.75
C12-15 alkyl benzoate	5.00
Polymer chosen for evaluation	2.00
B. Water (*aqua*)	52.15
Disodium EDTA	0.10
Carbomer, 2%	12.00
Butylene glycol	4.00
C. Triethanolamine, 99%	0.50
D. Propylene glycol (and) diazolidinyl urea (and) methylparaben (and) propylparaben	1.00
	100.00

Table 1. In vitro SPF test conditions

Test parameter	Test condition or value
Substrate	Vitro-Skin Substrate[e] hydrated overnight in hydration chamber
Hydration chamber	Desiccator with 256 g water and 44 g glycerin in bottom of chamber
Amount of sunscreen on substrate	2 microliters/cm^2
Dry-down time	15 min
Measurements per sample	6 (at different positions on same piece of substrate)
Immersion time	80 min
Water bath temperature	37°C
Agitation rate	190 rpm
Delay time*	1 hr

* Interval between time of removal from bath to time of SPF measurement
[e] Vitro-Skin Substrate is a product of Innovative Measurement Solutions (IMS) Inc., Milford, Connecticut USA.

For the control condition, water was substituted for the polymer. Sunscreens were coated onto a clear substrate and analyzed with an SPF analyzer[d]. This instrument measures the amount of ultraviolet light coming through the sample compared to an uncoated substrate, then calculates the SPF value. Samples were then placed in a heated water bath with agitation. After immersion, a final SPF value was measured. Specific test conditions were as indicated in Table 1.

In vitro SPF measurements collected by this method appear to be higher than those obtained with in vivo tests, possibly because in vitro analyzers overestimate SPF by missing side scatter of UV rays.[3] However, the approach used in this study is a useful screening tool for making relative comparisons among materials. Figure 1 summarizes in vitro data for the test sunscreen containing 2% of some commercially available water-resistant polymers, before and after immersion.

Initial SPF was boosted versus the base formula for all the test polymers, particularly the polyethylene. However, results for olefin/MA copolymer show significantly better water resistance than the other polymers or the commercial benchmark, PVP/eicosene copolymer[f]. Not only is initial SPF enhanced with the olefin/MA copolymer, but it remains significantly higher after immersion compared to samples formulated with the other materials.

Although use levels of olefin/MA copolymer were 2% in the "easy to remove" prototype test formulation, lower levels may be sufficient in more optimized formulations. Figure 2 illustrates the water resistance that can be attained using 1% olefin/MA

[d] The SPF-290S Analyzer System is a product of Optometrics LLC, Ayer, Massachusetts USA
[f] Ganex V-220 alkylated polyvinylpyrrolidone is a product of International Specialty Products, Wayne, New Jersey USA

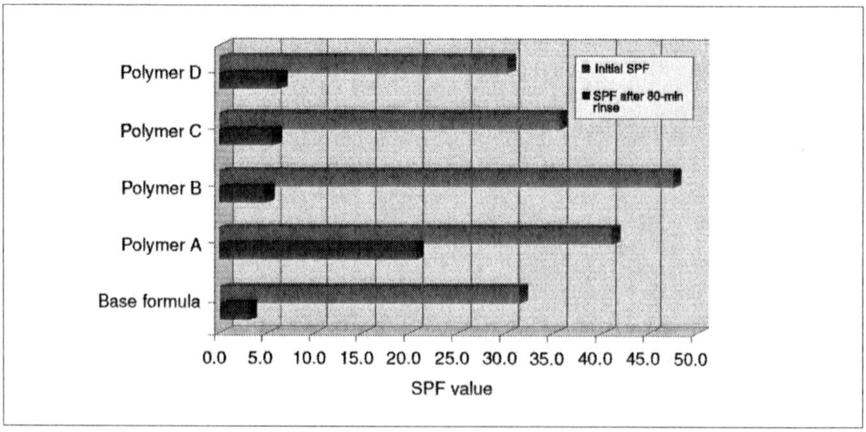

Figure 1. Comparison of water resistance and SPF; 2% polymer in test formula
Polymer A = C30-38 olefin/isopropyl maleate/MA copolymer
Polymer B = Polyethylene
Polymer C = C20-40 alcohols
Polymer D = PVP/eicosene copolymer

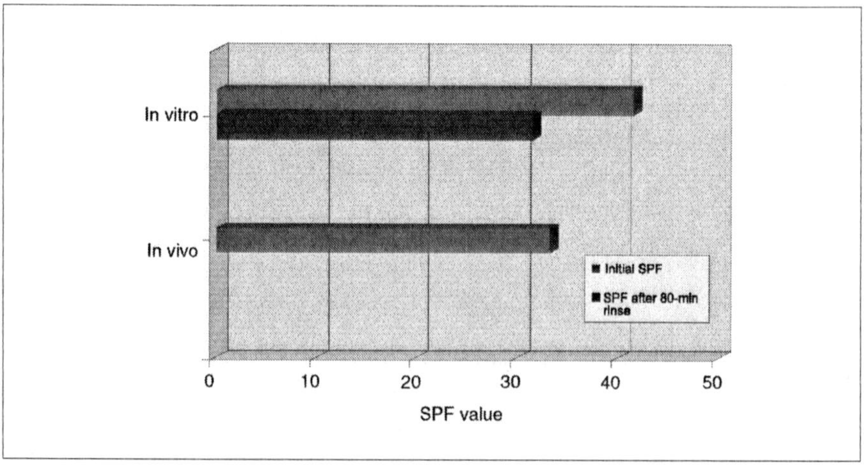

Figure 2. Water-resistant characteristics of prototype test formulation containing 1% C30-38 olefin/isopropyl maleate/MA copolymer

copolymer. Data in Figure 2 are based on Formula 2, a broad-spectrum sunscreen. This evaluation was based on a lower substrate coating level (1 μL/cm²) than previous study conditions, to obtain SPF values closer to in vivo results.

The in vivo test was performed at a level of 2 mg/cm² according to the U.S. Food and Drug Administration (FDA) Sunscreen Monograph.[4] Note that an in vivo water resistance test was not conducted in this portion of the study.

The olefin/MA copolymer can be used in oil-in-water systems (as discussed in this article) or water-in-oil systems. The polymer should be added to the oil

Formula 2. Water-resistant sunscreen (SPF 32)

A. Ethylhexyl salicylate	5.00% w/w
C12-15 alkyl benzoate (Finsolv TN, Finetex)	4.50
Isopropyl myristate	4.00
Diethylhexyl 2,6-naphthalate (Corapan TQ, Symrise)	5.00
PPG-2 myristyl ether propionate (Crodamol PMP, Croda)	0.50
Benzophenone-3	4.00
B. Butyl methoxydibenzoylmethane	3.00
C. Stearyl alcohol	0.30
Polyglyceryl-3 methylglucose distearate (Tego Care 450, Degussa)	3.00
C30-80 olefin/isopropyl maleate/MA copolymer (PERFORMA V 1608, New Phase Technologies)	1.00
Disodium EDTA	0.05
D. Butylene glycol	2.00
Glycerin (Glycerin, Procter & Gamble)	4.00
Phenoxyethanol (and) methylparaben (and) propylparaben (and) butylparaben (Phenonip, Clariant)	0.70
E. Water (*aqua*)	qs
F. Carbomer (Carbopol Ultrez 10, Noveon)	0.20
G. Triethanolamine, 99%	0.15

Procedure: Combine A and heat to 80°C with stirring. Add B and stir to dissolve. Increase heat to 90°C. Add C to AB, stirring after each addition until homogeneous. Heat water (less than 50 g) to 85°C. Preblend D and add to E. Predisperse F in 50 g water and set aside. With homogenization, add ABC to DE. Add F to batch. Maintain heat and homogenize for 10 min. Remove from heat. Stir with propeller while cooling. When temperature is below 40°C, slowly add G. Continue stirring to smooth, homogeneous lotion.

phase of the formula, which must be heated to 85–90°C to melt the polymer. To ensure proper incorporation, the water phase should also be heated to 85–90°C. An appropriate amount of base, such as triethanolamine or sodium hydroxide, must be added to oil-in-water formulations to neutralize the polymer.

The olefin/MA copolymer can be used alone or in combination with other polymers that impart water resistance. Figure 3 shows the synergistic effect when olefin/MA copolymer and PVP/eicosene copolymer are used together in the test sunscreen formula (Formula 1). In this formula, 2% PVP/eicosene copolymer did not enhance SPF or provide good water resistance. The olefin/MA copolymer alone performed very well in SPF enhancement and water resistance. However, the combination of 1% of each copolymer outperformed 2% of either used alone.

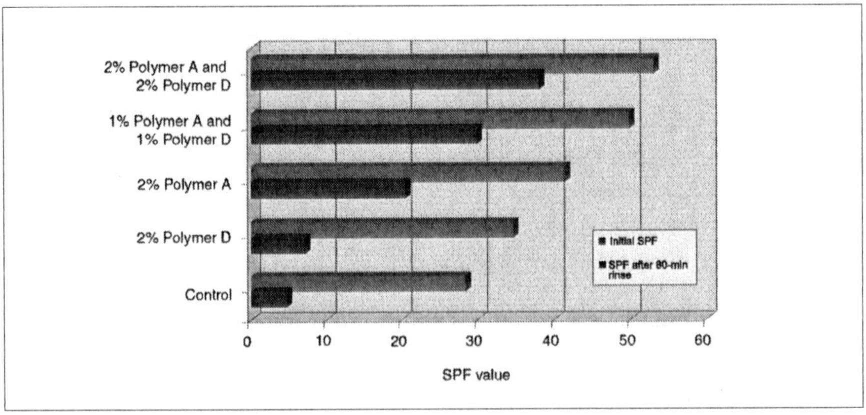

Figure 3. Synergistic water-resistance effect of C30-38 olefin/isopropyl maleate/MA copolymer (Polymer A) with PVP/eicosene copolymer (Polymer D)

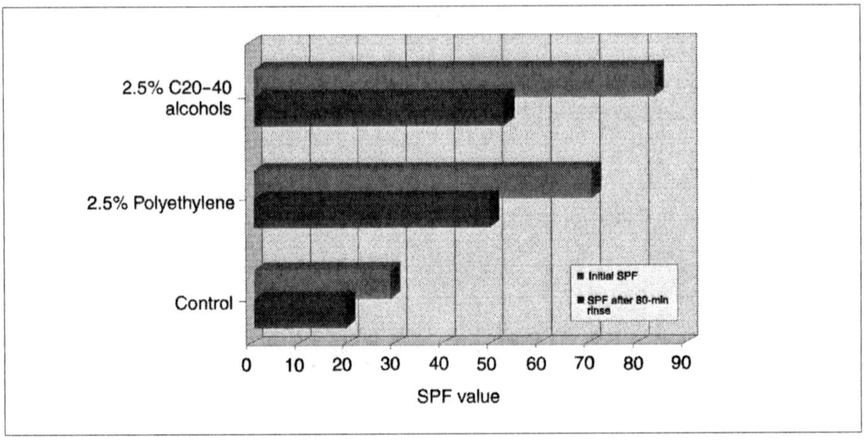

Figure 4. Water resistance of 2.5% polyethylene and 2.5% C20-40 alcohols

SPF was enhanced by more than 75% compared to the base formula. The combination also maintains a high SPF value even after an 80-minute warm water immersion. Doubling the level of each resulted in further improvement in both SPF enhancement and post-water immersion SPF value.

The Link Between Viscosity and SPF

Polyethylene and C20-40 alcohols can significantly improve the SPF of sunscreen formulas. Figure 4 shows that by adding 2.5% by weight — an amount selected so the oil phase will viscosify, but not solidify as it does at around 3%—of these film-forming polymers, the SPF of a prototype base formula used in the test more than doubled. Figure 4 also indicates the improvement in water resistance.

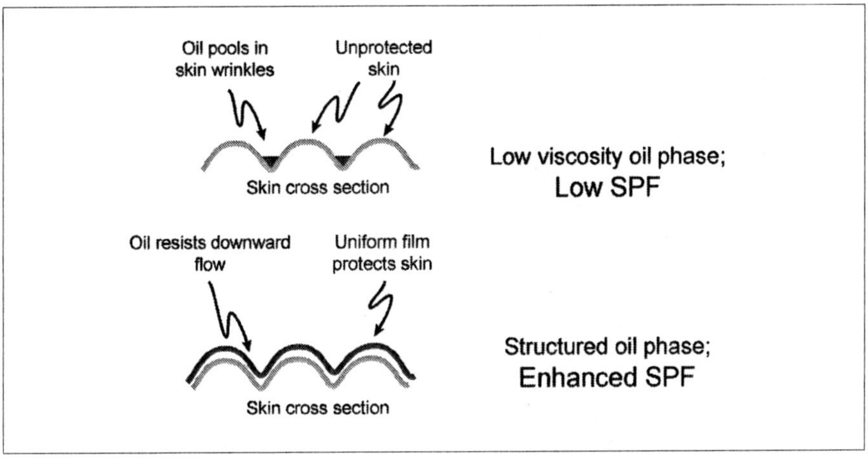

Figure 5. A structured oil phase of higher viscosity can optimize film formation to provide a uniform coating on the skin and enhanced SPF.

The property of enhanced SPF may be related to the viscosity-building properties of the material.[5] When added to the oil phase, waxes or other viscosity-building materials can enhance SPF through their ability to act as good film-formers and provide increased viscosity. Without their added benefit, the sunscreen formulation may flow downward into the wrinkles of the skin. The higher skin surface may be left without an adequate coating of sunscreen, and therefore more prone to damage from the sun. This proposed mechanism, illustrated in Figure 5, suggests why initial SPF may be higher with polyethylene and C20-40 alcohols, because they increase the viscosity of the oil phase.

Polyethylene and C20-40 alcohols can be used in oil-in-water or water-in-oil systems, and they should be added to the oil phase. Because these polymers are highly effective in building viscosity of the oil phase, use levels of 3% or less are recommended. The oil phase containing the polymers and the water phase should both be heated to 90-95°C to ensure complete incorporation.

If the resulting system appears grainy or if incompatibility is observed, addition of a small amount of C20-40 pareth-40[g] is recommended. The optimal ratio, on a weight basis, is 1 part C20-40 pareth-40 to 10 parts polyethylene or C20-40 alcohols. A higher proportion of C20-40 pareth-40 negatively impacts water resistance.

Figure 6 compares SPF values and stability in a prototype sunscreen using concentrations of 0.25% and 1.0% C20-40 pareth-40 and the same formula without the polymer. Notice that in both cases, the control formulation was grainy (in the case of polyethylene) or separated (when formulated with C20-40 alcohols), while the addition of C20-40 pareth-40 produced stable formulations with varying SPF levels.

[g] PERFORMATHOX 480 Ethoxylate is a product of New Phase Technologies. PERFORMATHOX is a registered trademark of Baker Hughes Incorporated.

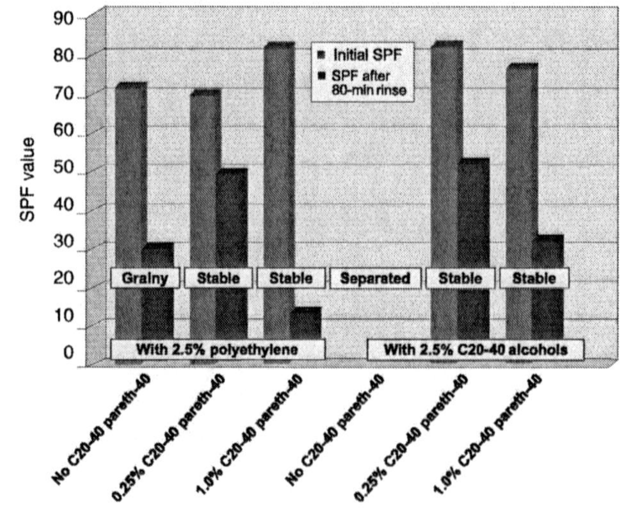

Figure 6. Optimal amount of C20-40 pareth-40 in sunscreen test formula with 2.5% polyethylene or 2.5% C20-40 alcohols

As Figure 6 indicates, just a small amount (0.25%) of C20-40 pareth-40 is effective in stabilizing the polyethylene and alcohols. Even though C20-40 pareth-40 is an emulsifier, which could be considered an ingredient that would encourage wash-off, at its optimum level in the formulation there is an increase in water resistance compared to a control formulation without this ingredient. By achieving a better emulsion at the optimal level, it is possible to get better SPF protection.

Formula 3 illustrates a nongreasy sport formula. Polyethylene provides water resistance, enhanced SPF and a dry feel. C20-40 pareth-40 acts as a secondary emulsifier to produce a smooth, creamy formula.

Conclusions

Tests of water resistance showed that C30-38 olefin/isopropyl maleate/MA copolymer provided significantly better water resistance than the competitive benchmark or the polyethylene or C20-40 alcohol polymers evaluated in this study. However, combining PVP/eicosene copolymer with C30-38 olefin/isopropyl maleate/MA copolymer had a synergistic effect that provided much better water resistance than either polymer alone. Other products were evaluated together and no synergistic results were seen, so the phenomenon appears distinctive for this particular combination.

Because of their viscosity-building properties, polyethylenes and long-chain linear alcohols showed the best SPF enhancing characteristics in this study. These ingredients as well as the C30-38 olefin/isopropyl maleate/MA copolymer have highly efficient film-forming properties, which can allow formulators to use lower levels of potentially irritating active ingredients in sun care products.

Formula 3. Water-resistant sport lotion (SPF 30)

A. Polyethylene (PERFORMALENE 400 Polyethylene, New Phase Technologies)	2.50% w/w
C20-40 pareth-40 (PERFORMATHOX 480 Ethoxylate, New Phase Technologies)	0.25
Stearic acid	1.30
Cetearyl alcohol (Crodacol CS-50, Croda)	2.00
Diisodecyl adipate (DIDA, Trivent)	8.50
Triethanolamine, 99%	0.25
B. Benzophenone-3	6.00
C. Octocrylene	10.00
D. Ethylhexyl methoxycinnamate	7.50
E. Water (*aqua*)	qs
Carbomer, 2% (Carbopol 940, Noveon)	7.50
Acrylates C10-30 alkyl acrylate crosspolymer, 2% (Pemulen TR-1, Noveon)	7.50
Propylene glycol	1.00
F. Phenoxyethanol (and) methylparaben (and) propylparaben and butylparaben (Phenonip, Clariant)	1.00
Fragrance (*parfum*)	qs

Procedure: Heat A to 85-90°C while propeller mixing. When ingredients are fully dispersed, reduce heat to 80-85°C. Add B, C and D in order, dispersing each phase before adding next. Combine ingredients of E and heat to 80-85°C while propeller mixing. Emulsify, adding first mixture to E with high speed mixing. Continue mixing for 10-15 min. Remove from heat and continue propeller mixing until cooled to 70-75°C. Change to sweep mix and allow mixture to cool to 25-30°C. Add F with mixing.

These versatile and multifunctional properties, combined with excellent aesthetics, present formulators with broader choices for creating high performance, distinctive sun care products to meet the needs of today's global consumers.

—Allison Hunter and Melanie Trevino,
New Phase Technologies, A Division of Baker Petrolite Corporation, Sugar Land, Texas USA

References

1. SPF-290S *Users Manual*, Version 2.1, Optometrics LLC (Revised October 2002)
2. Developing next generation suncare products: A unified model of sunscreen performance

on skin, a presentation by IMS Testing Group at the 2001 SCC Florida Chapter Sunscreen Symposium (Apr 25, 2002)
3. K Klein, Sun products: background, delivery and formulation, a presentation at the Advanced Technology Conference in Milan, Italy
4. US Food and Drug Administration, *Code of Federal Regulations*, Title 21, Vol 5, part 352: Sunscreen products for over-the-counter human use, (Revised Apr 1, 2003)
5. G Dahms, Efficacy of sunscreen emulsions on skin, a presentation at the 2001 SCC Florida Chapter Sunscreen Symposium (Apr 25, 2002)

A New Phosphate Emulsifier for Sunscreens

Keywords: sunscreen, ionic phosphate esters, phosphate emulsifier, viscosity, skin spreading, SPF, water resistance

This article describes and evaluates strategies for improving SPF water resistance through emulsion design.

During the development of a new emulsifying system to be used in caustic-based hair straighteners, we found that it was possible to design an emulsion that deposits high levels of oil phase onto the keratin surface of the hair.[1] We believe that the principles of this discovery can be used to help optimize the oil deposition — and hence deposition of oil-soluble UV absorbers — onto the keratinized surface of the stratum corneum (SC).

A number of parameters were found to be important in improving the oil phase deposition of an emulsion onto the hair or skin. These parameters include the following:

- Chemical nature of the emulsifier (including ionicity and amphiphillic character)
- Composition of the oil phase (especially relative hydrophobicity of the components)
- Ratio of oil phase to water phase
- Viscosity response (especially shear thinning of the emulsion)
- Structure of the emulsion (especially the presence of larger, possibly liquid crystal, structures).

This article describes the application of the principles of this new emulsifying system to sunscreen emulsions. It also interprets the resulting clinical SPF studies.

Emulsifier Description

Our new emulsifier system[a] has the INCI name cetearyl alcohol (and) dicetyl phosphate (and) ceteth-10 phosphate. It belongs to the class of ionic phosphate esters that have several important chemical characteristics (see sidebar) that affect their properties as emulsifiers.

---

[a]Crodafos CES, Croda Inc., Parsippany, New Jersey

Chemical Properties of Phosphate Esters

- The ionic phosphate group constitutes a powerful o/w emulsifying agent.
- The anionic phosphate group is shielded by the alkyl chains, hence the "crypto" in the term "cryptoanionic."
- Unlike carboxylic acid esters, the phosphate ester link is very stable at high and low pH.
- The emulsification characteristics of the ester depend on the degree of neutralization of the free-acid groups.
- The lipophilic character is critically dependent upon the alkyl chain-length distribution of the hydrophobe.
- Fatty alcohols work well in combination with phosphate esters made from these alcohols.

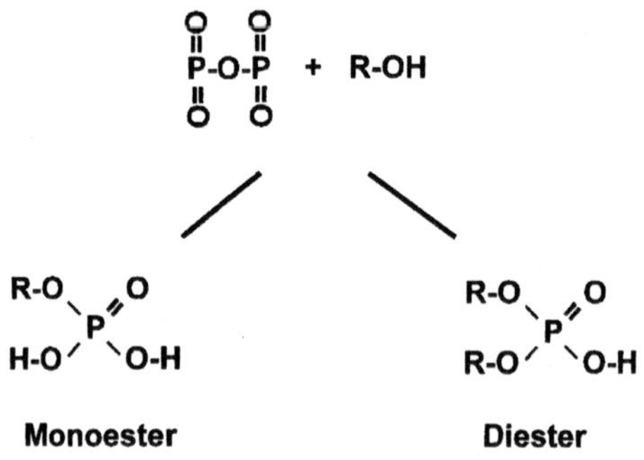

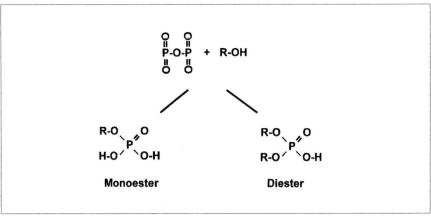

Figure 1. Phosphate ester used in the emulsifier cetearyl alcohol (and) dicetyl phosphate (and) ceteth-10 phosphate

Figure 1 gives a generalized description of the type of phosphate ester used in this new emulsifier as both an emulsifier and a conditioning agent. In this case, R is a C_{16} or C_{18} alkyl group. This type of phosphate ester is a very powerful o/w emulsifying agent. We have previously shown that it is substantive to hair.[2] One can reasonably assume that it would also be substantive to the SC.

An interesting and useful property of ionic emulsifiers is that the ionics appear to bind water to themselves much less strongly than do the repeated ether groups typically present in an ethoxylated nonionic emulsifier. An example of this is shown in Figure 2, where we have measured the relative evaporation rates from identical emulsions prepared using a nonionic emulsifier (emulsifying

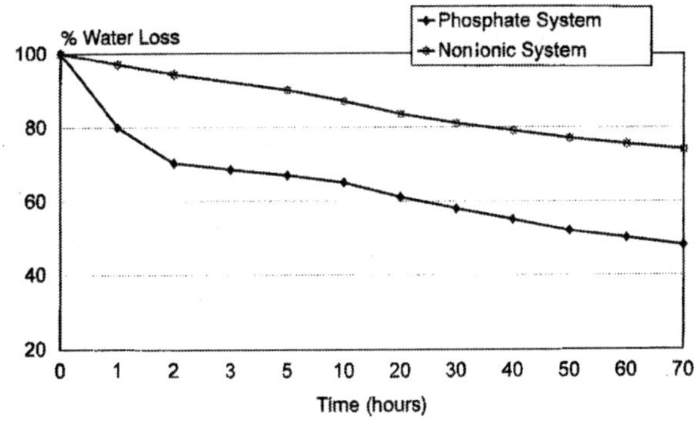

Figure 2. Water evaporation rates

Formula 1. Sunscreen Lotion With and Without the Emulsifier Cetearyl Alcohol (and) Dicetyl Phosphate (and) Ceteth-10 Phosphate

	5% Active		7.5% Active	
A. Cetearyl alcoyol (and) dicetyl phosphate (and) ceteth-10 phosphate (Crodafos CES, Croda Inc)	6.0%	0.0%	6.0%	0.0%
Emulsifying wax NF	0.0	5.2	0.0	5.2
Cetyl alcohol	0.5	0.5	0.5	0.5
Steareth-2	0.5	0.5	0.5	0.5
Steareth-10	1.0	1.0	1.0	1.0
Petrolatum	5.0	5.0	4.5	4.5
Mineral oil (*paraffinum liquidum*), 70 ssu	10.0	10.0	8.0	8.0
Octyl methoxycinnamate	5.0	5.0	7.5	7.5
B. Water (*aqua*) deionized	70.7	71.8	70.7	71.8
Triethanolamine	0.3	0.0	0.3	0.0
C. Propylene glycol (and) diazolidinyl urea (and) methylparaben (and) propylparaben	1.0	1.0	1.0	1.0
	100.0	100.0	100.0	100.0
pH (±0.5)	4.5	4.0	4.5	5.0
Viscosity (±10%, RVT spindle #5, 10 rpm, 25°C)	18,500	31,500	12,500	20,500

Procedure: Combine A with mixing and heat to 75-80°C. Combine ingredients of B with mixing and heat to 75-80°C. Add B to A with mixing and cool to 40°C. Add C with mixing and cool to desired fill temperature.

wax NF) and the phosphate emulsifier. The graph shows the greater evaporation from the phosphate emulsifier system. We believe that this may be an important characteristic in designing an emulsion that provides a separation of the oil and water phase during application to the hair or skin.

Formula Description

Formula 1 displays the sunscreen formulations used to compare the performance of the nonionic emulsions with that of the phosphate-based systems. We endeavored to keep the emulsifier levels exactly the same, although some compromise was necessary to achieve similar viscosities. We felt that these formulations, although somewhat simplistic for a sunscreen product, still offer a direct comparison between the two different emulsifier systems.

We examined these comparison formulations using two different levels of octyl methoxycinnamate, since we were not certain how different levels of UV absorber might affect the relative SPF performance of these emulsions.

Viscosity Response

In designing an emulsion system that will separate upon application, one has to consider not only the chemical characteristics of the emulsifier, but also the viscosity response of the emulsion. Specifically, shear-thinning of the emulsion appears to be strongly associated with the ability of the emulsion to "break" on application. In fact, emulsion formulators often speak of the "viscosity break" to describe the phenomenon in which the product is sheared during application, producing a drop in viscosity. Note that while these emulsions display this desired shear-thinning, they are stable under static conditions and therefore have adequate shelf life.

Figure 3 illustrates the viscosity vs. time measurements at constant shear for the emulsions containing 5% active sunscreen (octyl methoxycinnamate). Although viscosity in the nonionic system decreases slightly due to shear-thinning, one sees a more dramatic decrease (approximately 25%) due to shear-thinning in the phosphate-based system. This effect is even more apparent for the 7.5% sunscreen formulation shown in Figure 4, where the viscosity decrease for the phosphate system is closer to 30%.

It is our understanding that the shear-thinning effect probably indicates the disruption of the fatty alcohol gel network that extends from the dispersed phase through the external phase of the emulsion,[3] where it provides some stabilizing effect by joining emulsion particles. It is the disruption of this network that is responsible not only for shear-thinning, but also for the release of dispersed-phase particles (oil) onto the skin, as well as the possible release of water-soluble actives that would otherwise be trapped within this gel network.

Emulsion Character

Microscopic examination of the emulsions revealed that in the case of the 5% sunscreen systems, the gross structure of the emulsions (as observed under visual

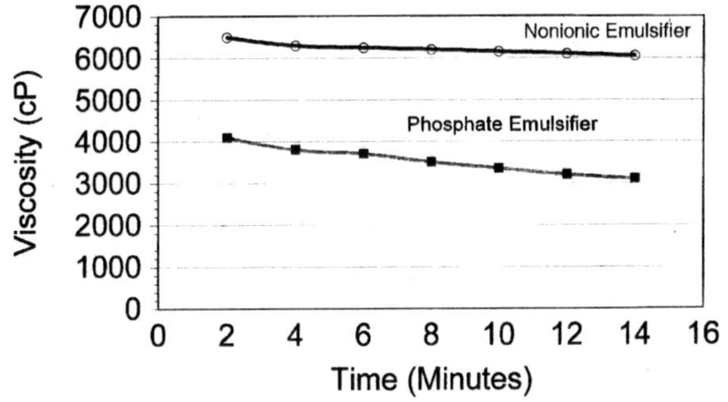

Figure 3. Viscosity at constant shear for a nonionic emulsifier and a phosphate emulsifier containing a 5.0% sunscreen active

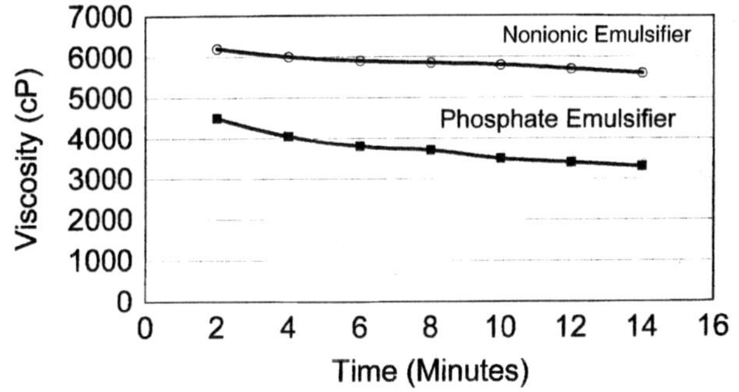

Figure 4. Viscosity at constant shear for a nonionic emulsifier and a phosphate emulsifier containing a 7.5% sunscreen active

light at 400x) are very similar for both the nonionic and the phosphate-based systems. For the 7.5% sunscreen systems, the emulsions appear significantly different, with the phosphate-based system displaying what appears to be large-scale structures in addition to the dispersed-phase particles.

SPF Performance

We asked an independent laboratory[b] using standard SPF protocols[4] to perform clinical SPF screening measurements on the four sunscreen emulsions described in Formula 1. Five-person panels measured SPFs initially and again after water immersions of 10, 20, 40 and 80 min. Figures 5 and 6 summarize the results.

[b]AMA Laboratories, New City, New York

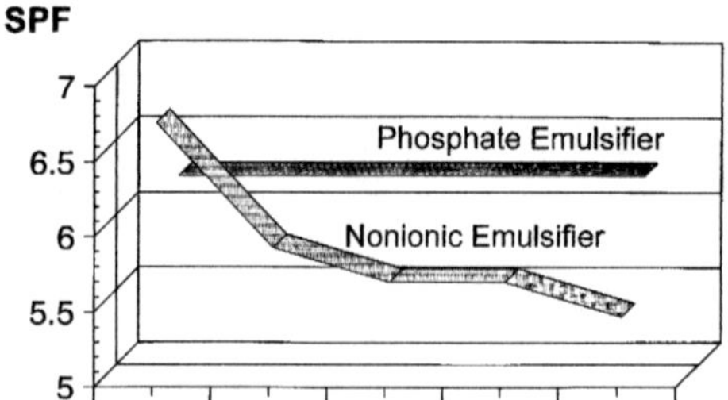

	0	10	20	40	80
Nonionic	6.73	5.9	5.67	5.67	5.44
Phosphate	6.23	6.23	6.23	6.23	6.23

Immersion Time in Water (mins).

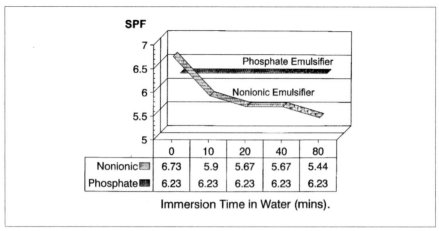

Figure 5. Effect of water immersion on the SPF of a nonionic emulsifier and a phosphate emulsifier containing a 5.0% sunscreen active

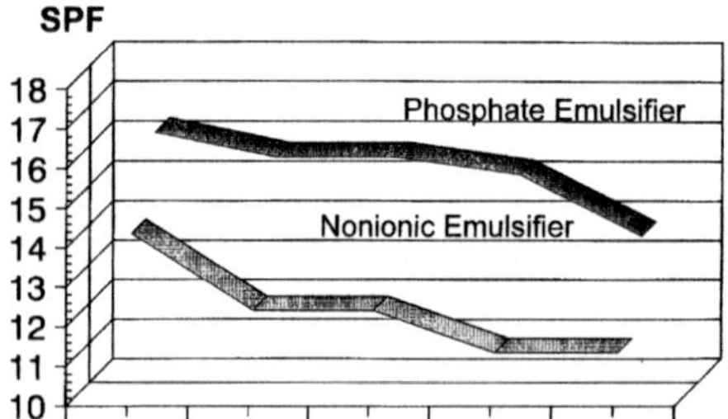

	0	10	20	40	80
Nonionic	14.25	12.3	12.3	11.24	11.24
Phosphate	16.2	15.58	15.58	15.15	13.6

Immersion Time in Water (mins).

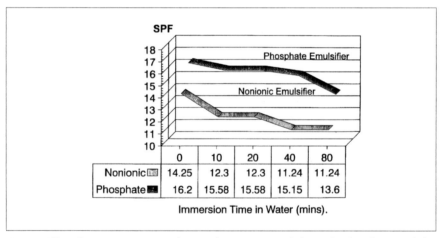

Figure 6. Effect of water immersion on the SPF of a nonionic emulsifier and a phosphate emulsifier containing a 7.5% sunscreen active

Skin Spreading

Once we had the SPF measurements, we attempted to correlate them with skin-spreading measurements on five sites on the lower forearm. We applied the test emulsions to small circular areas and used a black marking pen to outline the sites. We measured the spreading of the sunscreen emulsions by observing the migration of components of the black ink 2 h after application. Table 1 summarizes the results.

Table 1. Skin spreading factors of the sunscreen emulsions

Emulsifier	Sunscreen Level	Spreading Factor*
Cetearyl alcohol (and) dicetyl phosphate (and) ceteth-10 phosphate	5%	1.1
Emulsifying wax NF	5%	1.1
Cetearyl alcohol (and) dicetyl phosphate (and) ceteth-10 phosphate	7.5%	1.0
Emulsifying wax NF	7.5%	1.5

$$\text{*Spreading factor} = \frac{\text{Outer surface area of spreading}}{\text{Original inner surface area of site}}$$

Formula 2. Wash-off Resistant Sunscreen Lotion

A. Cetearyl alcohol (and) dicetyl phosphate (and) ceteth-10 phosphate) (Crodafos CES, Croda Inc)	6.0%
Cetyl alcohol	0.5
Steareth-2	0.5
Steareth-10	1.0
Lignoceryl erucate	15.0
Octyl methoxycinnamate	5.0
B. Water (*aqua*), deionized	70.7
Triethanolamine, 99%	0.3
C. Propylene glycol (and) diazolidinyl urea (and) methyparaben (and) propylparaben	1.0
	100.0

pH = 4.6±0.5
Viscosity = 18,000 cps ± 10% (RVT spindle #5, 10 rpm @ 25°C)

Procedure: Combine A with mixing and heat to 75-80°C. Combine B with mixing and heat to 75-80°C. Add B to A with mixing and cool to 40°C. Add C with mixing and cool to desired fill temperature.

Discussion

An examination of the SPF results indicates a substantial difference in performance for the phosphate-based emulsion compared to the nonionic system, although they contain identical levels of UV absorber. At 5% levels of sunscreen, the emulsions yielded similar initial SPFs, although the nonionic system appears to undergo a

Formula 3. Waterproof SPF 30 Sunscreen

Water (*aqua*), deionized	60.73%
Carbomer (Carbopol 981, BFGoodrich)	0.13
Cetearyl alcohol (and) dicetyl phosphate (and) ceteth-10 phosphate	6.50
Benzophenone-3	5.00
Octyl methoxycinnamate	7.50
Octyl salicylate	5.00
Menthyl anthranilate	5.00
Octyl stearate	5.00
Avocado oil unsaponifiables	2.50
NaOH, 10% soln	1.54
BHT	0.10
Propylene glycol (and) diazolidinyl urea (and) methyparaben (and) propylparaben	1.00
	100.00

pH = 7.0±0.5

Viscosity = 9,500 cps ± 10% (RVT spindle TB, 10 rpm @ 25°C)

Static SPF = 31.66 Waterproof SPF = 30.31

Procedure: Dust carbomer into the deionized water while stirring rapidly. Mix well for good hydration. Begin heating to 75-80°C. Add cetearyl alcohol (and) diecetyl phosphate (and) ceteth-10 phosphate and mix well until all is melted and homogeneous. Add benzophenone-3, octyl methoxy-cinnamate, octyl salicylate, menthyl anthranilate, octyl stearate and avocado oil unsaponifiables individually and with good mixing. Continue mixing at 75-80°C until homogeneous. Begin slow cooling and at 60°C, add NaOH solution. Cool to 45°C and add BHT and preservatives.

significant drop in SPF values during water immersion, whereas the phosphate-based system does not.

The results at the 7.5% sunscreen level are somewhat different, with the phosphate-based emulsion exhibiting a significantly higher initial SPF, despite the fact that both emulsions contain the same sunscreen levels. Although both emulsions appear to experience a drop in SPF performance during the test, the phosphate-based emulsion still has a much higher SPF at the end of the immersion test, probably due to its greater initial SPF values.

Additional Formulation Work

Since the original test emulsions are simplistic by the standards of current sunscreen formulations, we decided to conduct additional formulation work to help confirm our earlier results. For instance, mineral oil is not typically used in sunscreen formulations because of its relatively poor feel properties. It is also not the best solvent for oil-soluble UV absorbers.

We decided to measure the influence of the oily vehicle on the water resistance of a sunscreen emulsion. We compared the sunscreen lotion with 5% octyl methoxycinnamate from the original study (Formula 1) with a similar formulation (Formula 2) in which the mineral oil has been replaced by a high-molecular-weight ester: lingoceryl erucate.[c] We then compared the SPF values (static and waterproof) for both emulsions. The resulting SPF values for the sunscreen with the oily vehicle were higher — both initially (SPF 17.7) and after the 80 min of water immersion called for by the waterproof test (SPF 16.5) — than the corresponding values (SPF 16.2 and 13.6) for the sunscreen with mineral oil.

It is significant to note that this system provides a waterproof SPF above 15, whereas the same formulation with mineral oil does not (SPF 13.6). Clearly, the choice of an oily vehicle will also influence the SPF results.

We also wanted to demonstrate the effectiveness of the new phosphate emulsifier system with a high-SPF (30) waterproof sunscreen (Formula 3). The resulting SPF values (static SPF 31.7, waterproof SPF 30.3) show that this formula can provide for a waterproof SPF above 30 (waterproof SPF = 30.3), indicating the applicability of the new phosphate emulsifier system for high-SPF formulations.

Conclusions

It is apparent that significant differences in SPF performance exist for the four test emulsions in Formula 1. The results suggest that greater SPF performance can be achieved through the use of the phosphate emulsifying system than through the use of a similar nonionic emulsifier.

In the case of 5% octyl methoxycinnamate, although we did not observe a difference in initial SPF values (suggesting similar levels of oil phase deposition), we did observe that the phosphate-based system underwent no loss in SPF performance. This could be due to the greater substantivity of the phosphate-based emulsion, which could act to hold the sunscreen active more strongly to the skin surface.

In the case of the 7.5% octyl methoxycinnamate, we observed a greater initial SPF with the phosphate-based emulsion. This could indicate greater oil deposition on the skin, although one should not rule out the possibility that this higher initial SPF is due to the lack of spreading of this emulsion on the skin. It may also be important that the difference in viscosity "break" was greater with the phosphate-based formulation. It is possible that the larger-scale structures seen in the microphotographs could be evidence of liquid crystal structure. If so, this liquid crystal character could help prevent the spreading on skin of the emulsion, while possibly making it easier for these formulations to re-emulsify off the skin than is the case with the 5% active sunscreen emulsion.

We may conclude that the new phosphate emulsifier can be expected to improve the SPF performance of a sunscreen formulation, although other factors such as the structural characteristics of the emulsion are also very important.

[c]Crodamol LGE, Croda Inc., Parsippany, New Jersey

—**Kevin F. Gallagher,** *Croda Inc., Parsippany, New Jersey USA*

References

1. P Obukowho et al, A new approach to formulating hair relaxers (Part I), *Happi* 32 62-70 (1995)
2. Electron Spectroscopy for Chemical Analysis Study on Crodafos SG, Parsippany, NJ: Croda, Inc (1978)
3. GM Eccleston et al, *J Soc Cosmet Chem* 41 1-22 (1990)
4. AMA Laboratories, private communications (1995)

Formulating Water-Resistant TiO_2 Sunscreens

Keywords: sunscreens, water-resistance, physical UV-filters, liquid crystal gel networks, silicone w/o emulsions

This article describes and evaluates strategies for achieving water-resistance in sun-protection formulations containing physical sunscreen actives.

Water-resistance is now regarded as indispensable for most sun-protection products. Consumers demand products that will maintain their protective efficacy after swimming or water sports.

Another trend in sun protection is the increasing use of physical UV filters (titanium dioxide and zinc oxide), especially in products for children and persons with sensitive skin. This increased use of physical sunscreens is due partly to their low potential for producing irritant reactions and their resultant excellent safety record, but also to the efficacy of these materials. With titanium dioxide (TiO_2), cosmetic scientists can formulate high SPF products using just this single active. Zinc oxide (ZnO) is less effective at providing a high SPF, but provides a broad spectrum of both UVB and UVA protection. Both TiO_2 and ZnO have demonstrated impressive synergistic effects when used in combination with organic sunscreens.[1,2]

Achieving water-resistance with physical sunscreens requires some thought. Uncoated inorganic oxides are inherently hydrophilic and hence are not resistant to wash-off by water. All UV-attenuating grades of TiO_2 are, of course, coated; but in many cases the coating consists of other inorganic materials such as silica and alumina. The problem can be overcome, at least in part, by coating the TiO_2 with a hydrophobic organic material such as a stearate, or by using a suitable dispersant to disperse the oxide in the oil phase of an emulsion. However, as with organic sunscreens, whether the product is actually water-resistant depends more on the properties of the formulation into which the active is incorporated.

This paper will discuss the strategies available to the formulator for making sunscreen formulations water-resistant, and review how these strategies have been applied using TiO_2 as the physical sunscreen. Based on the fundamental requirements for making the product film water-resistant, a general approach to achieving water-resistance will be outlined, with specific examples given for TiO_2. Theoretically, the same principles should apply for ZnO as for TiO_2. However, because it provides the higher SPF per unit weight, TiO_2 has found more widespread use than ZnO in 'beach' products. for this reason, the work reported in this paper concentrates on TiO_2.

Waterproofing Strategies

A number of strategies have been developed for waterproofing sunscreen emulsions. These strategies are applicable to both organic and inorganic actives. A brief review of their use with inorganic sunscreens follows.

w/o emulsions: These emulsions are expected to be water-resistant for two reasons: because oil is the external phase of the emulsion; and because they contain predominantly hydrophobic emulsifiers. The film deposited on skin, therefore, can be expected to resist re-emulsification. The TiO_2 used in w/o systems is typically in a hydrophobic or oil-dispersed form, which also helps to enhance the product's water resistance. A recent paper included an example of the use of this strategy for water resistance.[3] There are also sample formulations available in the literature.[4]

Silicones: Silicone oils aid water resistance in two ways. First, the oils themselves are inherently hydrophobic. They also have very good spreading properties, which assist in forming a coherent, continuous film. Many different silicone fluids can be incorporated into formulations alongside TiO_2, although excessive quantities of silicone can cause agglomeration of the TiO_2, resulting in reduced efficacy.[5]

Specialized emulsifiers: Several specialized emulsifiers now available use different technologies to impart water resistance.

- Phospholipid emulsifiers, such as potassium cetyl phosphate, have chemical structures similar to normal skin lipids. This quality is claimed to facilitate increased water resistance.
- Acrylates/$C_{10\text{-}30}$ alkyl acrylate crosspolymer stabilizes emulsions by electrostatic means, forming an aqueous gel structure within which the oil droplets are suspended. By using this technology, surfactant-free (so-called "emulsifier-free") emulsions can be made. With no surfactant present, re-emulsification is prevented. Sunscreen formulations based on this crosspolymer and using TiO_2 as the active have been published.[4,6,7]

Liquid-crystal gel networks: The use of liquid-crystalline structures to stabilize o/w emulsions is now well-recognized and frequently used.[8,9] Theoretically, the so-called "liquid-crystal gel-network" systems have an inherent advantage for creating water-resistant sunscreen formulations. They are based on hydrophobic, lipid emulsifiers, which makes them difficult to re-emulsify after dry-down. Additionally, as far as physical sunscreens are concerned, such liquid-crystalline systems have been found to be especially suitable for use with aqueous TiO_2 dispersions.[10,11]

Film-forming polymers: Perhaps the most common approach is to incorporate a waterproofing agent. This is typically a film-forming polymer, which is used to make the product film more coherent and hence more substantive. Suppliers' literature and the trade journals give sample formulations in which these polymers are used with physical sunscreens.[4,7,12]

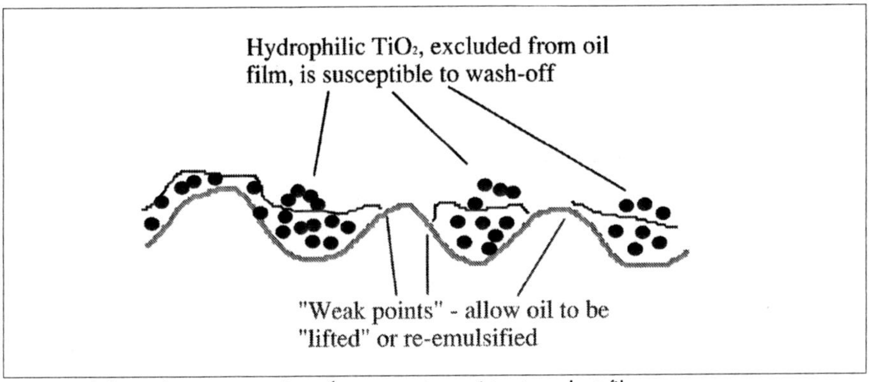

Figure 1. Schematic representation of a non-water-resistant product film

An additional aspect of this last approach was investigated in detail by Angelinetta and Barzaghi.[13] They studied the effects of film-forming polymers on static SPF in formulations containing either aqueous or oil-based dispersions of TiO_2. They observed significant increases in SPF when the film-forming polymers were incorporated into the formulation. However, the formulations used by Angelinetta and Barzaghi were based on high-HLB, hydrophilic emulsifier systems, which are readily re-emulsified. Despite the improvement found in static SPF testing, subsequent in vivo SPF tests demonstrated that the formulations were not water-resistant.

This last observation is an example of a common feature we have noted in evaluating the various waterproofing strategies with physical sunscreens: while all of these strategies have been used successfully, none provides a guarantee of success. With each strategy, we have found examples in which the strategy failed to yield a water-resistant formula.

To develop a general model for water resistance, it is necessary to go back to first principles. We must consider the fundamental requirements for an emulsion to be water-resistant.

Fundamental Principles

When a sunscreen emulsion has been spread on skin, it should maintain a constant level of UV attenuation, and hence a constant SPF, provided that the UV filters used are photostable, and provided that they remain on the skin. There are three processes by which UV filters may be removed on contact with water:

- Re-emulsification of the product film
- Removal of the product film by mechanical action of water moving over the skin surface
- Removal of active UV filters that are excluded from the product film

The first of these occurs where a significant concentration of hydrophilic emulsifiers is present in the product film. This facilitates easy re-emulsification of the oils, including any dissolved or suspended UV filters, by the water moving over the skin.

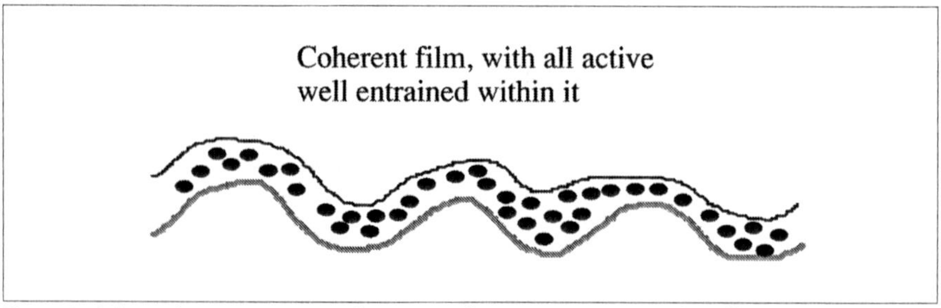

Figure 2. Schematic representation of a potentially water-resistant product film

The conditions necessary for the other two processes are illustrated schematically in Figure 1.

Mechanical removal can occur where the product film is discontinuous or has poor adhesion to the skin. This allows water to penetrate under the product film and "lift" it from the skin surface. This can be observed in a simple in vitro experiment. A film of an emulsion is applied onto a glass slide, left to dry, and then placed in a beaker of water which is gently agitated. If the product re-emulsifies, the water becomes turbid or cloudy; such products are unlikely to be water-resistant. However, some products that are not water-resistant do not re-emulsify; rather, they peel away from the slide in one piece.

Exclusion of actives from the film, the third condition, is best explained by considering a specific example. If hydrophilic TiO_2 is incorporated into an o/w emulsion, it will naturally be located in the aqueous phase. When this emulsion is spread on the skin, the water evaporates within a relatively short time, leaving a film which consists of oils, emulsifiers and other non-volatile (predominantly hydrophobic) components. At this point, the hydrophilic TiO_2 will be largely excluded from the oil film, and hence not protected by it. While the film itself may be water-resistant, the TiO_2 is readily re-dispersed and removed by water. (This helps explain why the water-resistant physical sunscreen formulas found in the literature tend to be based on hydrophobic grades of TiO_2.)

This situation is less likely to arise in a typical w/o emulsion. In such a case, the evaporation of water after spreading is slow, so the product film consists essentially of the emulsion itself. Hence, the active should remain within the film, regardless of the phase in which it is dispersed.

The fundamental requirements for water resistance, illustrated in Figure 2, are therefore:

- A low concentration of hydrophilic emulsifiers, to avoid re-emulsification
- A continuous, coherent product film after application and dry-down
- TiO_2 effectively dispersed within the product film

The Dual-Strategy Approach

The preceding discussion reveals why the waterproofing strategies discussed earlier are not always successful. For example, use of a film-forming polymer, which

Formula 1. Sunscreen Emulsion Based on High Silicone Content

A. Dimethicone, trimethylsiloxysilicate	5.00% w/w
Mineral oil (*paraffinum liquidum*)	5.00
Cyclomethicone pentamer	8.00
Laurylmethicone copolyol	3.00
B. Glycerol	4.00
Sodium chloride	1.00
Water (*aqua*), demineralized	58.60
C. Titanium dioxide, caprylic/capric triglyceride, 2-ethylhexyl palmitate, alumina, polyhydroxystearic acid silica (TiO_2 dispersion, 40% solids)	15.00
D. Butylparaben, ethylparaben, isobutylparaben, phenoxyethanol and propylparaben	0.40

Table 1. In vivo SPF data for Formula 1 sunscreen emulsion

Subject	SPF Before Immersion	SPF After Immersion
1	20.00	20.00
2	16.00	12.80
3	12.80	12.80
4	12.80	12.80
5	20.00	16.00
Mean	16.32	14.88

should help to ensure a coherent film, is pointless if the film contains hydrophilic emulsifiers and can be re-emulsified. Conversely, the inherent resistance of a w/o emulsion to re-emulsification does not ensure water resistance if its spreading properties are poor, and it forms a discontinuous film.

In order to build a more reliable approach for creating water-resistant products, we must consider all of the fundamental requirements. This means invoking at least two basic strategies in the same formula. The two chosen strategies should complement each other. For instance, one should address the re-emulsification issue, while the other should take care of forming a homogeneous film. This dual-strategy approach is illustrated by the following two examples.

Silicones and w/o emulsion: As discussed earlier, the hydrophobic nature of the emulsifiers used in w/o emulsions makes them inherently resistant to re-emulsification. However, if the emulsion has poor spreadability, resulting in a discontinuous film on the skin, it can still be removed by the mechanical action of water moving across it. Certain silicone fluids can be used to optimize a formulation's spreading and film-forming properties. Combined with the inherent hydrophobicity of the silicones themselves, this means that a w/o emulsion with a high content of silicones in the oil phase should show good water resistance. To investigate this we developed Formula 1.

In vitro SPF tests using the technique of Diffey and Robson indicated that Formula 1 has an SPF between 14 and 17.[14] An in vivo pilot study (using five human subjects) was carried out according to the methods described in the US Food and Drug Administration's Tentative Final Monograph on Sunscreens.[15] The results of these in vivo SPF tests are shown in Table 1.

The results demonstrate that the formulation is indeed water-resistant; it maintains more than 90% of the static SPF after two 20 min immersions. The testing was not extended to four immersions. However, the work of Stokes and Diffey suggests that, in most cases, loss of SPF during water-resistance testing occurs during the first one or two immersions.[16] Therefore, it is likely that this formula would also pass the test for a "very water-resistant" label (previously "waterproof").

Formula 1 contains TiO_2 with a hydrophilic coating (silica/alumina). However, it was incorporated as an oil-based pre-dispersion by using a suitable dispersing agent. This, of course, facilitates incorporation of the TiO_2 into the oil phase of the emulsion and thence into the product film on skin. Theoretically, use of a hydrophobic grade of TiO_2, such as in the dispersions described by Dransfield et al.,[17] would be expected to yield even better water resistance. This has yet to be evaluated experimentally.

Liquid-crystal gel networks: Of course, one would expect that w/o emulsions with oil-based dispersions and/or hydrophobic grades of TiO_2 would hold advantages for water-resistant products. It remains a commercial truth, however, that more consumers prefer the skin-feel of o/w emulsions, particularly in hot and humid climates. Aqueous dispersions have also been established as the easiest way to incorporate TiO_2 into o/w systems.[18] But, as explained earlier, exclusion of hydrophilic TiO_2 from the oil film after dry-down means that such systems are expected to have poor water resistance. So how does one make such a formula water-resistant?

The answer lies in liquid-crystalline or lamellar-gel networks.[19] As previously discussed, these structures are resistant to re-emulsification because they are based primarily on hydrophobic lipid emulsifiers. They are also particularly well suited to incorporation of aqueous dispersions of TiO_2.

Freeze-fracture transmission electron microscope photographs indicate that the TiO_2 is preferentially located within the lamellar liquid-crystalline structure. This helps to maintain the dispersion of the TiO_2 within the emulsion, and also provides a vehicle by which the hydrophilic TiO_2 remains incorporated in the oil film on the skin after application and dry-down. This mechanism also helps the product's water resistance: if the TiO_2 can be effectively dispersed within the lamellar structure, it can be protected from wash-off. Thus, in theory, a water-resistant formula can even be made using hydrophilic TiO_2 as an active.

By combining this strategy with that of using a film-forming polymer to ensure a continuous film, we developed Formula 2. This incorporates a combination of oil-dispersed and water-dispersed TiO_2, and uses tricontanyl PVP as the waterproofing agent. The in vivo SPF data for this formulation are shown in Table 2.

Once again, the data demonstrate a high degree of water resistance for the formula, despite the fact that 60% of the active was incorporated as an aqueous

Formula 2. Sunscreen Emulsion (o/w) Based on a Lamellar Gel Network and an Oil-Soluble, Film-Forming Polymer

A. Polyglyceryl-10 pentastearate/ behenyl alcohol/ sodium stearoyl lactylate	2.50% w/w
Mineral oil (*paraffinum liquidum*)	4.50
2-Ethylhexyl palmitate	6.00
Jojoba (*Buxus chinensis*) oil	2.50
Myristyl myristate	2.00
Dimethicone copolyol	0.50
Tricontanyl PVP	3.00
Titanium dioxide, $C_{12\text{-}15}$ alkyl benzoate, phenyl trimethicone, alumina, polyhydroxystearic acid, silica (Titanium Dioxide Oil Dispersion)	5.00
B. Water (*aqua*), demineralized	61.50
Xanthan gum	0.20
Magnesium aluminum silicate	0.80
Propylene glyc ol	4.00
Water (*aqua*), titanium dioxide, alumina, silica and sodium polyacrylate (Titanium Dioxide Aqueous Dispersion)	7.50
C. Preservative	qs

Table 2. In vivo SPF data for Formula 2 sunscreen emulsion

Subject	SPF Before Immersion	SPF After Immersion
1	18.00	14.40
2	14.40	11.52
3	18.00	18.00
4	18.00	14.40
5	18.00	18.00
Mean	17.28	15.26

dispersion. This provides further validation of the dual-strategy approach. However, further work in this area demonstrated that use of a water-dispersed or water-soluble polymer failed to impart water resistance.[18] Apparently, the polymer needs to be oil-soluble to be effective in this regard.

Conclusion

Sunscreen emulsions containing physical UV filters can be waterproofed by using the same strategies as have been previously used with organic UV filters. However, use of only one strategy provides no guarantee of success. Consideration of the fundamental requirements for the product to be water-resistant explains why this is the case, and also points the way to a more reliable approach. This approach involves the use of two complementary strategies in combination, to ensure a coherent product film on skin and to prevent re-emulsification of this film on contact with water. Two examples that use this approach with TiO_2 have been shown to be effective in generating water-resistant sunscreen emulsions:

- Using a w/o emulsion with a high silicone content
- Using an o/w emulsion based on a lamellar gel network, with an oil-soluble film-forming polymer

—**Julian P. Hewitt,** *Uniqema Solaveil, Peterlee, County Durham, UK*

References

1. JP Hewitt, Novel formulation strategies for high SPF and broad spectrum sunscreen products, *Parf Kosm* 80(4) 36-39 (1999)
2. SR Spruce, The Synergy Between Titanium Dioxide and other Materials for use in Sunscreen Formulations, Florida Sunscreen Symposium (1995)
3. DT Floyd et al, Formulation of sun protection emulsions with enhanced SPF response, *Cosm Toil* 112(6) 55-64 (1997)
4. Sun products formulary, *Cosm Toil* 109(11) 71-94 (1994)
5. JP Hewitt, The Influence of Emollients on Dispersion of Physical Sunscreens, In-Cosmetics Conference, Milan (February 1996)
6. HAPPI 33(12) 20 (1996)
7. Sun products formulary, *Cosm Toil* 107(10) 133-152 (1992)
8. GH Dahms, Properties of oil-in-water emulsions with anisotropic lamellar phases, *Cosm Toil* 101(11) 113-115 (1986)
9. P Loll, Liquid crystals in cosmetic emulsions, *Cosm Toil Manufacture Worldwide* (Aston Publishing Group), 108-120 (1994)
10. GH Dahms, Formulating with a physical sun block, *Cosm Toil* 107(10) 87-92 (1992)
11. UK Patent 2 264 703 (Tioxide Specialties Ltd)
12. *HAPPI*, 30(4) 16 (1993)
13. C Angelinetta and G Barzaghi, Influence of oil polarity on SPF in liquid crystal emulsions with ultrafine TiO_2 pre-dispersed in oil, and cross-linking polymers, *Cosmetic News* (Italy) 100 20-24 (Jan/Feb 1995)
14. BL Diffey and J Robson, A new substrate to measure sunscreen protection factors throughout the ultraviolet spectrum, *J Soc Cosmet Chem* 40 127-133 (May/Jun 1989)
15. 21 CFR Part 352 et al, Sunscreen Drug Products for Over-the-Counter Human Use; Tentative Final Monograph; Proposed Rule; in *Federal Register* 58 (90) pp 28298-28301 (May 12,1993)
16. RP Stokes and BL Diffey, The water resistance of sunscreen and day care products, *Brit J Dermatol* 140 259-263 (1999)
17. GP Dransfield et al, Advances in Titanium Dioxide Technology, In-Cosmetics Conference/ SCS Spring Conference, London (April 1998)
18. MW Anderson et al, Broad spectrum physical sunscreens, in *Sunscreens: Development, Evaluation, and Regulatory Aspects*, 2nd ed, Lowe, Shaath and Pathak, eds, Marcel Dekker, New York (1997) pp 353-397
19. JP Hewitt, Effects of film-forming polymers on efficacy and water-resistancy of physical sunscreens, *Cosmetics and Toiletries Manufacture Worldwide* (Aston Publishing Group) 135-141 (1999)

Sunscreen Interactions in Formulations

Keywords: UV filters, photostability, solubility, packaging, microfine pigments, pH

Avoid the incompatibility pitfalls that await sunscreen formulators and use favorable interactions, including synergistic effects, to improve sunscreen performance.

Sunscreen formulators today have many technical constraints imposed upon them that limit the choice of raw materials to be used in the finished product. These include such factors as the product profile dictated by a fast moving market, legislative issues dictated by governments, and environmental issues promoted by pressure groups and governments. At the same time, the cosmetic formulator must also be aware of how key ingredients, such as UV filters, interact with each other and with their environment to formulate cost-effective products acceptable to consumers.

This article will review such interactions in sunscreen formulation. It consists of two main sections: the first dealing with the bad news of unfavorable interactions, and the second dealing with the good news of favorable interactions.

Unfavorable Interactions

In an ideal world, no ingredient used in the formulation of cosmetic products would have any unfavorable interactions. However, such reactions can occur. These must be documented and acted upon to prevent surprises, which can result in expensive product reformulation and delayed R&D programs. Raw material suppliers are the preferred source for such information.

Fortunately for today's sunscreen manufacturer, the unfavorable interactions in finished sunscreen products that manifest themselves to the consumer are few and far between. It is extremely rare to remove a product from the market because of such unfavorable interactions. Experience in the United States can help quantify this. Manufacturers of cosmetics in the United States follow a voluntary recall program of products that are deemed to be potentially unsafe. This program is monitored by the U.S. Food & Drug Administration (FDA). There are 3 classes of recalls:

Class I: Product poses reasonable probability of serious adverse health consequences or death.

Class II: Product may cause temporary or medically reversible adverse health consequences; probability of serious consequences is remote.

Class III: Product is not likely to cause adverse health consequences.

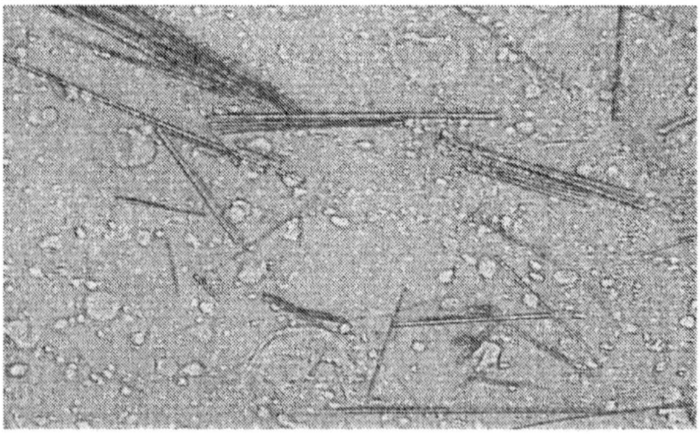

Figure 1. Crystals of phenylbenzimidazole sulfonic acid (pH 6.1)

The number of product recalls per year is published by the FDA and also in the weekly newsletter, *The Rose Sheet* (published by FDC Reports). In 1997[1] and 1998,[2] there were only two Class III recalls of a sunscreen product (out of 12 cosmetic product recalls in all) from the U.S. market, one product per year. These were due to mislabeling of the sun protection factor (SPF) and microbial contamination.

From this we can extrapolate that, in the cosmetic industry and, in particular, in sunscreen formulation, products with unfavorable reactions that have a potentially detrimental effect rarely reach the consumer.

Sunscreen formulators are, in general, well informed about most of the potential interactions between the individual ingredients in a sunscreen product. These are taken into account during the product development phase of the formulation. Therefore, the consumer is not aware of them.

The following section discusses incompatibilities that formulators should know about.

UV Filter Incompatibilities

Para-methoxycinnamates: Para-methoxycinnamates, such as octyl methoxycinnamate (OMC) or isoamyl methoxycinnamate, undergo a photochemical reaction to form minute traces (in ppm) of an intensely yellow compound on the surface exposed to daylight. Therefore, white sun care products containing these materials should not be packaged in colorless, transparent glass containers. The discoloration can be slowed by adding the UV filters benzophenone-3 or methylbenzylidene camphor, or by adding a pH-neutral form of microfine zinc oxide. The degree of discoloration must then be verified by stability testing under the appropriate conditions to determine whether it is acceptably small.

Recrystallization of UV filters: The SPF-boosting effect of the water-soluble filter phenylbenzimidazole sulfonic acid when it is used in combination with other organic UV filters or with inorganic microfine pigments is well known.[3,4] Indeed, this shall be discussed in more detail later. However, this synergistic effect will not work in sun care formulations containing α- or β-hydroxy acids, which require a low pH to be effective. Instead, an unexpectedly low SPF results and the product feels grainy on the skin.

Table 1. Solubility of solid UV filters in selected solvents

Solvent	%OT	%BMDM	%B-3
OMC	15	17	24
C_{12-15} Alkylbenzoate	4	15	10
Octocrylene	1	27	32

Key:
OMC = octyl methoxycinnamate
OT = octyl triazone
BMDM = butyl methoxydibenzoylmethane or avobenzone
B-3 = benzophenone-3

Phenylbenzimidazole sulfonic acid is virtually insoluble in water; it has to be neutralized with a suitable base, typically sodium hydroxide, for it to become soluble. The pH of the emulsion containing phenylbenzimidazole sulfonic acid must therefore remain in the region of 7.0 or above; the use of a suitable buffer system to keep the pH in this range is recommended.[5] If the pH falls below 7.0, the neutralized sodium salt may then rehydrolyze. The free acid then precipitates out, forming the long needles responsible for the grainy feel on the skin (Figure 1, see page 326). Because the crystals are not evenly distributed over the skin's surface, the protective effect of the phenylbenzimidazole sulfonic acid is greatly reduced. Hence, formulators should not use phenylbenzimidazole sulfonic acid in low-pH formulations. Another UV filter should be chosen, such as methylbenzylidene camphor or octocrylene, which will also boost the SPF but which do not recrystallize at low pH.

The crystallization problem is not unique to phenylbenzimidazole sulfonic acid. It can also occur when solid organic UV filters such as octyl triazone, butyl methoxydibenzoylmethane (BMDM), benzophenone-3 and methylbenzylidene camphor are incorporated into a formula if insufficient amounts of polar fatty acid esters are used to solubilize them. This becomes very important when mixtures of solid UV filters are used, considering that most o/w emulsions typically contain less than 35% oil phase. If 6% of this is taken up by solid UV filters, then, as a general guide, this amount has to be soluble in the balance of the 29% of liquid oils used to make up the formula, if all that 29% is liquid. Fortunately, liquid UVB filters such as cinnamates, salicylates, anthranilates, and octocrylene are also excellent solvents (Table 1, see page 327).

Inappropriate choice of packaging materials: Many of the better polar oils that are used both for their emolliency properties and as co-solvents are more expensive than nonpolar vegetable or mineral oils. This can become important in emerging economies where formulation costs must be kept to a minimum: such esters may seem to be an overly expensive luxury unless the importance of their co-solvent role is understood.

The co-solvent role of liquid UVB filters and polar emollients also affects the type of packaging chosen for the finished product. In addition to preventing the surface discoloration of cinnamates, packaging choices must take into account the solubility parameters (SPs) of the ingredients in a formulation.

The SPs of most liquid UV filters are in the range of 9 to 10.3, which is quite close to that of certain polymers used in packaging. For example, polystyrene has an SP of 8.9, low density polyethylene 8.5, and polyethylene terephthalate 10.3.[6] If these packaging materials are used, we can expect the UV filters to dissolve into the packaging. This can cause decomposition of polystyrene inserts used in the caps of the bottle or unacceptable discoloration of white packaging materials over time.

Other unwanted consequences include fracturing of the packaging and assay problems. The latter may have important legislative consequences in countries such as Canada, where the active ingredient contents must remain within a certain limit of what is claimed on the packaging. For this reason, high density polyethylene or high density polypropylene should be chosen for packaging sunscreen formulations. Vaughan's article on the importance of solubility parameters in cosmetic formulation discusses the topic in much more detail.[6]

Avobenzone: Over the past 19 years, sunscreen formulators outside of the United States have been able to provide broad UVA I and UVA II photoprotection by incorporating of BMDM, which in the United States is also called avobenzone. This excellent UVA filter has a number of formulation problems, however.[7] Most formulators have learned how to overcome these. In the US, where this filter is now listed as Category 1 in the FDA's Final Monograph for OTC Sunscreen Products,[8] formulators will now have to learn from this accumulated experience if they are able to incorporate the product successfully.

At temperatures above 30°C, the material reacts with active methylene groups such as those found in formaldehyde-releasing preservation systems. Hence, these preservation systems should be avoided. BMDM is an excellent chelator of iron, forming a highly colored red complex. Iron-chelating agents such as EDTA must therefore be added to the formula to prevent this colored complex from forming.

Because avobenzone is a solid, sufficient solvents must be used in the formulation to prevent crystallization.

Photostability: Photostability is a topic that has received much attention in the past five years or so. A controversial topic, it has had a marked influence upon the formulation of sunscreen products in Europe but not as much, so far, in other regions of the world. If a UV filter is not photostable, it could affect the efficacy of the formula to deliver its protection factor. For the protection factors determined in vivo (the SPF for erythema, or the UVA protection factors determined by Persistent Pigment Darkening or the Phototoxicity Protection Factor), any instability of the filter system would have been accounted for in the tests, because any photodegradation will have occurred during the measurement process. However, this is not the case for the in vitro techniques of measuring photoprotection unless a pre-irradiation step is incorporated into these measurements.

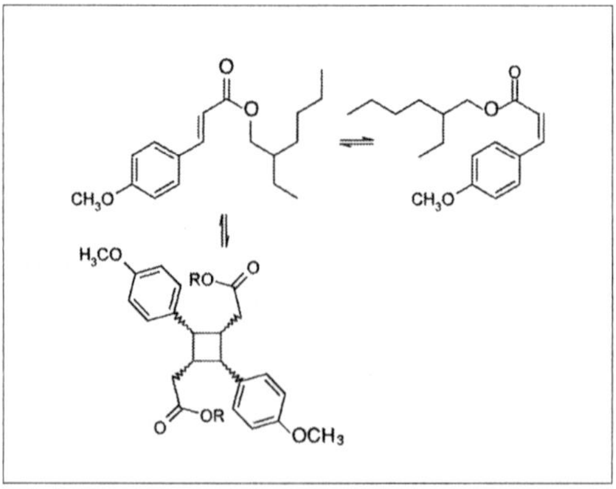

Figure 2. Octyl methoxycinnamate isomerizes and dimerizes with UV radiation

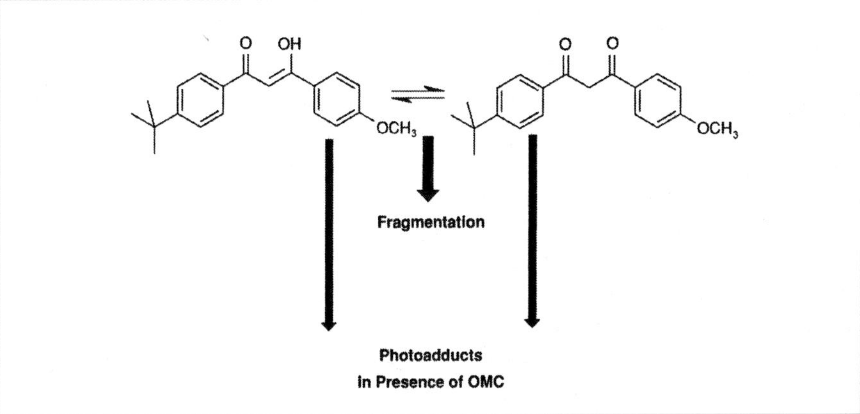

Figure 3. Butyl methoxydibenzoylmethane exposed to UV light tautomerizes, fragments and, in the presence of octyl methoxy-cinnamate, forms complex photoadducts

Many European sun care manufacturers consider photoinstability a seriously unfavorable interaction in a sunscreen formulation. As evidence of this, note that many patent applications associated with improving the photostability of sunscreen formulations have been published in the past seven years.

Of the UV filters commonly used today, only two are associated with some photoinstability.

- OMC undergoes cis/trans isomerization[9] and dimerization[10] (Figure 2, see page 328). Although cis/trans isomerization is not really considered a photoinstability, it is, in fact, a very efficient means of dispersing the absorbed energy. The loss of OMC due to photodimerization is generally low, and its effect upon the SPF as compared to a more photostable UVB filter is expected to be minimal, since any SPF variation will be lost in the standard errors of the SPF measuring method.
- BMDM undergoes photoinduced breakdown[11] and photoinduced interaction with OMC (Figure 3, see page 329).

This has led some manufacturers to remove OMC from formulations that contain BMDM and stabilize the latter with various materials such as octocrylene and methylbenzylidene camphor. However, the formulator has to be aware of recent patent restrictions that limit the choice of suitable combinations.[12]

An important point is that the issue of photostability measured by in vitro tests on artificial substrates, as described by the European Cosmetic, Toiletry, and Perfumery Association (COLIPA) work group on photostability,[13] does not translate into a safety issue. The combination of BMDM and OMC hasn't been proven to pose any hazard to human health in the many years they have been used together in sun care preparations.[14] In addition, there has been no documented evidence to support Sayre's postulate[15] that photoinstable combinations of these filters have resulted in consumers being excessively burned. In fact, we have contrary data from a study in Australia.[16] Participants in the study used an SPF 17 sunscreen containing 8% OMC and 2% BMDM during the summers of 1991 and 1992. Results showed that regular application of this sunscreen by a panel consisting of 210 skin Type I or II subjects reduced the number of solar keratoses—and by implication, possibly the risk of skin cancer in the long term—compared to a panel of 221 people using a placebo.

Other commonly used UV filters have a more optimal photostability. These include methylbenzylidene camphor, octocrylene, phenylbenzimidazole sulfonic acid, octyl salicylate, homomenthyl salicylate, benzophenone-3, microfine zinc oxide, and coated microfine titanium dioxide.

We have developed in our laboratories a number of very photostable formulations. In one example, we observed an SPF 50 containing 10% octocrylene, 3% methyl benzylidene camphor, 2.5% phenylbenzimidazole sulfonic acid, and 7% microfine zinc oxide having less than 5% loss of the organic UV filters after 4 h immersion in water. This loss was measured by HPLC after irradiation with 85 J/cm^2 of UV radiation from a solar simulator emitting 39.4 W/m^2 of UV radiation between 290-400 nm.

One result of this photostability issue is that any new UV filter being developed must possess an excellent photostability profile. In our own laboratories two different chromophors, which themselves were extremely photostable, had to be discarded because they interacted with OMC in an unacceptable way. One, in particular, caused very rapid decomposition of OMC such, was added to the large number of effective chromophors that did not pass the current standards of acceptable photostability.

Microfine pigments: Partly to avoid the photostability problems associated with organic UV filters, formulators have incorporated microfine zinc oxide and more commonly, microfine titanium dioxide into sunscreen preparations. However, the formulator has to be aware that there are a number of unfavorable reactions associated with their use.

- Microfine pigments have to be very well dispersed into the emulsion in order to be effective, because poor dispersion will reduce the performance of the product (see sidebar on dispersion).
- Microfine pigments must not only be well dispersed, but also must be kept in suspension so agglomeration does not occur. If the pigments start to coalesce and form larger aggregates with time, then the ultimate performance of the finished formulation will decrease as time goes on (see sidebar on agglomeration). This is where predictive stability testing becomes important. Also, if the pH of the emulsion equals the pH of the isoelectric point (the pH where the surface of the solid has zero charge), the particles of microfine pigment will coalesce. The point of zero charge for each microfine pigment will vary depending upon its surface treament.[19]
- The microfine pigments titanium dioxide and zinc oxide may, in some formulations, interact with BMDM. With titanium dioxide, this interaction may manifest itself as a yellow coloration; the interaction apparently depends very much on formulation and grade of titanium dioxide. With zinc oxide, the interaction causes the formation of a complex which may precipitate out of the emulsion.
- Formulators should not use carbomer thickeners with zinc oxide due to the slow formation of zinc acrylates. In time, this reduces the emulsion viscosity. This viscosity reduction can occur fairly rapidly, (within the three months of a stability test) or it may only show up at a much later date (one year of storage). Coating zinc oxide with dimethicone does not appear to protect products from this instability. In

a formula made in our laboratories, the viscosities decreased from an initial value of 27,500 mPa to 14,000 mPa after three months, and to 8,500 mPa, after six months storage at an ambient temperature for an uncoated grade of microfine zinc oxide. This compares to viscosities of 28,500 mPa, 9,000 mPa and 7,000 mPa measured on a formula using a dimethicone-coated zinc oxide over the same interval.

- If both titanium dioxide and zinc oxide are incorporated into an emulsion, the formulator must be careful that the pH value does not fall between the points of zero charge of the two pigments. If this happens, the pigments will have opposite surface charges and, therefore, will rapidly form aggregates. The pH must be below the point of zero charge of both pigments to prevent this. At the same time, the product pH must remain above 6 to prevent solubilization of the zinc oxide.[4]
- Incorporating microfine pigments into a formulation can also have an effect on the fragrance impact of the formula after storage. Due to the large surface area of the microfine particles, fragrance ingredients may adsorb onto the surface and thereby reduce the fragrance impact with time. This will be noticed during storage testing of the formula under development.

Despite the long list of unfavorable interactions that can occur between ingredients in sun-care formulations, one should remember that experienced formulators can still develop elegant and efficacious sun care products that are readily accepted by consumers.

Favorable Interactions

Moving on to the good news, a number of favorable interactions in sun care formulation also exist.

A number of means to obtain favorable interactions in a sunscreen formulation have been demonstrated by raw material suppliers and consultants to the cosmetics industry. For example, the way an emulsion is structured and the addition of a number of inactive ingredients can have a marked impact upon the SPF of a finished formulation. Here are more examples.

- Dow Corning has demonstrated that alkylmethylsiloxanes with a melting point close to the temperature of the skin can improve the performance of sun care formulations.[20]
- Dahms and Sottery have reported upon the importance of building in thixotropy to emulsions. This ensures that the emulsion, once it breaks on the skin, leaves behind an oily film that continues to flow and thus distributes itself over the total skin surface,[21] thereby obtaining more optimal performance of the UV filters present.
- Building a lamellar gel network into an emulsion by correct choice of the emulsifier has also been shown to improve the performance of sun care formulation for both organic and inorganic UV-attenuating systems.[4,21,22] The lamellar gel network is said to build a viscoelastic structure throughout the emulsion, thereby increasing its stability and spreading properties over the skin's surface.[23]
- The use of film-forming polymers such as PVP/eicosene, PVP/hexadecene, or acrylic/acrylate copolymers reportedly enables the oily film left on the skin to form an even layer, which improves SPF performance and water resistancy.[24]
- Water resistancy is improved by using cold-water-insoluble emulsifiers such as alkyl phosphates.[25]

Dispersion of Microfine Pigments

Poorly dispersed microfine pigments will reduce the performance of a sunscreen product, as shown by an experiment performed in 1998.[17] A simple o/w emulsion containing 7% microfine zinc oxide achieved 3 varying transmission profiles depending upon how the emulsions were treated (Figure 4). The smallest transmission resulted from the mixing process that incorporated a colloid mill. The product with intermediate transmission was manufactured using a rotary homogenization device to produce fine particle sizes. The greatest transmission resulted from the emulsion made using a turbine system. The latter emulsion did not pass the criteria that allow a broad-spectrum protection claim according to the Australian Standard.[18] This observation is relevant to all microfine insoluble particles that are dispersed in an emulsion and should not be discounted as a trivial matter.

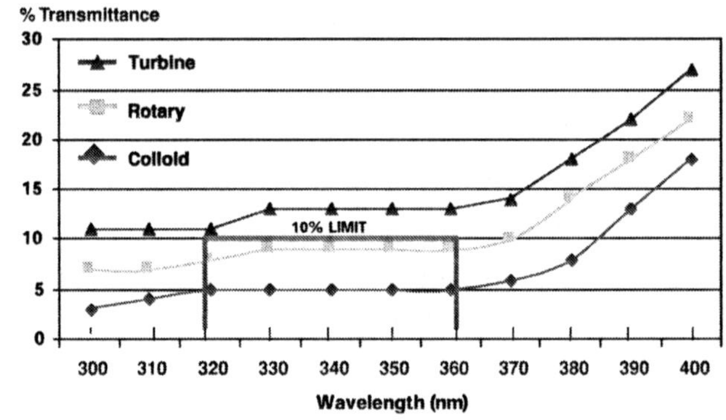

Figure 4. Effect of particle dispersion on transmittance curves of 7% ZnO in a w/o formula

Synergistic Effects to Boost the SPF

Once a good emulsification system is developed that contains all of the desired ingredients to promote an effective performance, the choice of UV-attenuating agents is crucial to obtaining the desired, favorable interactions to achieve the required SPF.

As mentioned at the beginning of this article, the addition of two UV filters together can give a synergistic increase in SPF. The combination said to give the greatest synergistic effect involves using water-soluble and oil-soluble UV filters together. The hypothesis behind this choice is that, by using oil- and water-soluble filters, the protective ingredients will be attracted to both the hydrophilic and lipophilic areas of the skin, thereby generating a better coverage of UV filters. A simple indication of what synergies may be obtained is shown by combining 3% OMC with 1% phenylbenzimidazole sulfonic acid for an SPF of 10. If the for-

Agglomeration of Microfine Pigments

Dispersed microfine pigments that allowed to agglomerate will reduce the performance of a sunscreen product.

In our laboratories we developed two formulas containing a mixture of organic UV filters along with 10% and 5% of microfine zinc oxide, respectively. We tested these formulas for their in vivo SPF. A modified COLIPA method, in which only six subjects were tested, was used to measure the SPFs on fresh samples. The tests were repeated on the same batches 27 months later along with fresh samples at the same test institute that undertook the first round of testing. The initial SPFs of 26 and 16 were found to have changed little on the aged and freshly remade samples (Figure 5).

This shows that if the microfine pigment is formulated properly, there is no negative effect on the SPF with time.

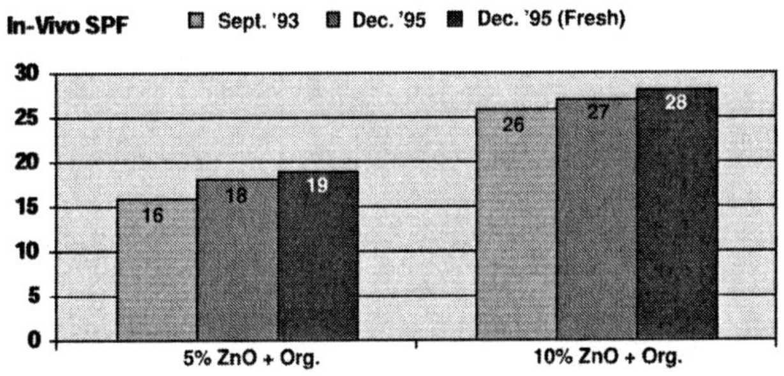

Figure 5. Effect of time on SPF of formulations containing organic UV filters and either 5% or 10% well dispersed microfine zinc oxide pigments

mula uses only OMC at 4%, an SPF of only 4 results. Other examples have been mentioned in the literature.[3]

The combination of 2 closely related UV filters, OMC and isoamyl methoxycinnamate, also shows the potential for synergistic effects, though the increases in SPF are not as large as those obtained by combining water- and oil-soluble filters. In tests, a product using 2.5% of each filter gave an in vivo SPF of 6.3; 3.5% of each filter gave an SPF of 7.4. In comparison, 7.5% of OMC gave an SPF of only 5 in the same excipient. In another study, 6% of OMC or 6% of isoamyl methoxycinnamate in separate formulations with 2% microfine zinc oxide gave in vivo SPFs of 11 and 12, respectively. However, when these two organic filters were combined with 2% of microfine zinc oxide in the same formulation, only 3% of each was needed to give an in vivo SPF of 16, an increase of 45% and 33%, respectively.

Formula 1. Daily protection facial lotion (SPF 25)

A. Cetyl phosphate (Crodafos MCA, Croda)	1.50% (w/w)
Glyceryl stearate (Cutina MD, Henkel)	2.00
Cetyl alcohol (Lanette 16, Henkel)	1.00
Cetearyl isononanoate (Cetiol SN, Henkel)	4.00
Caprylic/capric triglyceride (Myritol 318, Henkel)	5.00
$C_{12\text{-}15}$ Alkylbenzoate (Witconol TN, Witco)	5.00
Isostearic acid (Prisorine 3505, Unichema)	1.00
Tocopheryl acetate (Copherol 1250, Henkel)	0.50
Propylparaben (Solbrol P, Bayer)	0.10
Benzophenone-3 (Neo Heliopan BB, H&R)	1.50
Octocrylene (Neo Heliopan 303, H&R)	5.00
B. Water (*aqua*), demineralized	qs
Phenylbenzimidazole sulfonic acid (Neo Heliopan Hydro, H&R)	2.00
Sodium hydroxide solution	qs
Glycerin, 99%	3.00
Methylparaben, (Solbrol M, Bayer)	0.20
Carbomer (Carbopol ETD 2050, BFGoodrich)	0.20
Phenoxyethanol	0.70
C. Bisabolol (Alpha Bisabolol Nat, H&R)	0.10
Fragrance (*parfum*)	

In our laboratories, we recently developed a very light emulsion with an SPF of 15 for use as a daily protection product, aimed at the US and Asian markets (see Formula 1, page 334). The base emulsion uses cetyl phosphate as the emulsifier, into which we incorporated the following UV filters:

- octocrylene (5.0%);
- phenylbenzimidazole sulfonic acid (2.0%);
- benzophenone-3 (1.5%).

The in vivo SPF far exceeded our expectations, delivering individual results of >16, >25, 31, 31, 31, >31, >31, and 34.1 on the eight subjects tested using the guidelines set in the Australian standard for determining the SPF.[26] Note that these results are only indicative, because we did not use the full panel of 10 subjects demanded

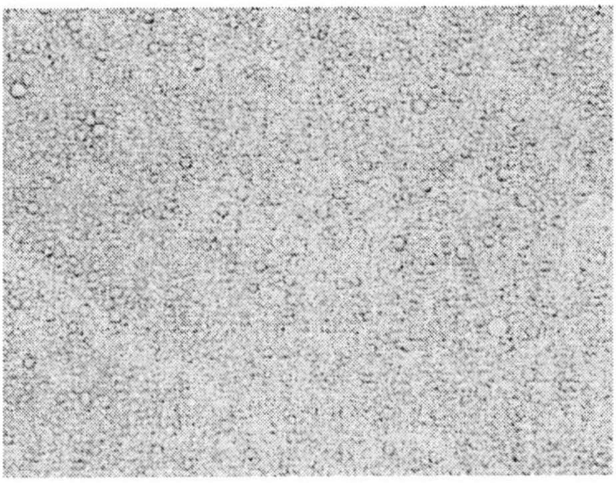

Figure 6. Microfine structure of daily protection lotion (400x magnification)

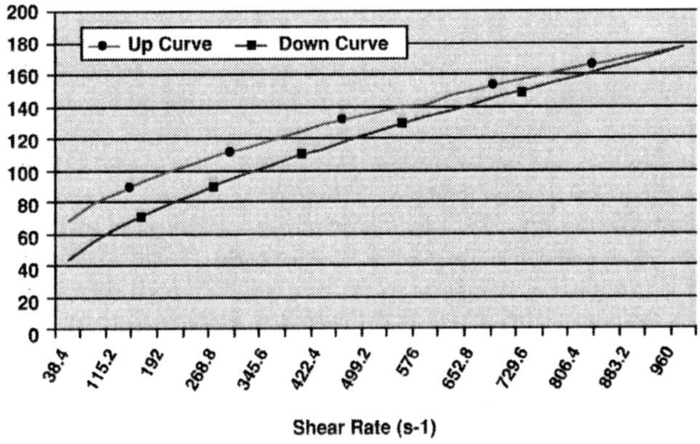

Figure 7. Rheology curve of daily protection lotion

by the standard. Nevertheless, we are confident that this formula is delivering a protection factor of 25+. These excellent results may be due to the following:

- an extremely fine (see Figure 6, page 335), slightly thixotropic (Figure 7) emulsion with liquid crystalline microstructure, all of which are due to the use of cetyl phosphate as primary emulsifier;
- sufficient UVA II protection offered by the combination of octocrylene and benzophenone-3;
- synergies of interaction between the three UV filters used.

Conclusion

Today's formulator of sunscreen products is aware of most of the unfavorable reactions that can occur in formulation development, as evidenced by the quality of the products which are found on the market. This shows that, by searching for the favorable interactions that can improve performance and sensory perception and by avoiding the unfavorable reactions associated with product formulation development, consumer-acceptable products can be formulated to achieve objectives that will build market share.

—**William Johncock ,** *Haarmann & Reimer GmbH, Holzminden, Germany*

References

1. *The Rose Sheet* (Jan 17, 1998)
2. *The Rose Sheet* (Jan 11, 1999)
3. W Johncock, *Euro Cosmetics* (7) 22-24 (1994)
4. A Anderson et al, Broad spectrum physical sunscreens, in *Sunscreens: Development, Evaluation & Regulatory Aspects*, 2nd ed, N Lowe, N Shaath and M Pathak, eds pp 353-398 (1997)
5. Haarmann & Reimer, *Neo Heliopan Technical Brochure*
6. C Vaughan, *Cosm Toil* 103(10) 47-69 (1988)
7. Givaudan-Roure, Parsol 1789 Technical Brochure (1993)
8. Sunscreen Drug Products For Over-The-Counter Human Use; Final Monograph. *Federal Register* 64(98) 27666-27693 (1999)
9. P Morlière, *Photochem Photobiol* 36 395-399 (1982)
10. A Schrader et al, *J Soc Cosmet Chem* 45 43-52 (1994)
11. W Schwack and T Rudolph, *Analytika von Kosmetika* 4 373-377 (1996)
12. EP 514491, L'Oréal
13. G Berset et al, *Int J Cosmet Sci* 18 167-177 (1996)
14. G Groves, *DCI* 37-39 (Aug 1994)
15. RM Sayre and JC Dowdy, *Cosm Toil* 114(5) 85-91 (1999)
16. R Marks et al, *New England J Med* 329(16) 1147-1151 (1993)
17. W Johncock, Formulating effective sunscreens with microfine zinc oxide and octocrylene, presented at the SCC Spring Symposium, London (1998)
18. Australian/New Zealand Standard 2604, Appendix C4 (1997)
19. D Fairhurst and MA Mitchnick, Particulates in sunblocks: General principles, in *Sunscreens: Development, Evaluation & Regulatory Aspects*, 2nd ed, N Lowe, N Shaath and M Pathak, eds pp 313-352 (1997)
20. I Van Reeth et al, Poster presented at the 19th IFSCC Congress, Sydney (1996)
21. GH Dahms and JP Sottery, Seminar on Development of High Efficiency Emulsions, London (June 1994)
22. GH Dahms, *Cosm Toil* 101(11) 113-115 (1986)
23. P Loll, *Cosmetic & Toiletries Manufacture Worldwide*, 108-120 (1994)
24. K Klein, Broad spectrum physical sunscreens, in *Sunscreens: Development, Evaluation & Regulatory Aspects*, 2nd ed, N Lowe, N Shaath and M Pathak, eds pp 285-311 (1997)
25. Croda Ltd, technical data on Crodaphos CES and MCA
26. Australian/New Zealand Standard 2604 (1997)

Cyclodextrins in Skin Delivery

Keywords: cyclodextrins, delivery systems, penetration enhancer, odor absorption, enhanced shelf life

The author discusses the physicochemical properties of cyclodextrins and their ability to camouflage various undesirable chemical and biological effects in cosmetic products.

Cyclodextrins are a group of structurally related natural products which are formed during bacterial digestion of cellulose. In 1891, Villiers announced his discovery of crystalline dextrin that possessed properties different from those known as saccharides. Villiers called this newly discovered dextrin cellulosine. Later it was shown that cellulosine contained a mixture of cyclic dextrins that were then named cyclodextrins or cycloamyloses. However, until about 30 years ago only small amounts of relatively impure cyclodextrin mixture (consisting mainly of α-, β- and γ-cyclodextrin) could be extracted from natural sources and high production costs prevented their industrial usage. Recent biotechnological advancements have resulted in dramatic improvements in the efficient manufacture of cyclodextrins lowering the cost of these materials and making highly purified cyclodextrins and cyclodextrin derivatives available.

In cosmetics and toiletry products they are mainly used as complexing agents to increase aqueous solubility of lipophilic water-insoluble compounds, and to increase both physical and chemical stability of active ingredients. In addition, cyclodextrins have been used to:

- Increase or decrease absorption of various compounds into skin
- Reduce or prevent skin irritation
- Reduce or eliminate unwanted body odor
- Control release of fragrances
- Stabilize emulsions and suspensions
- Inhibit foaming caused by surface active materials
- Prevent interactions between various formulation ingredients
- Remove stains
- Convert oils and liquids into microcrystalline or amorphous powders

Cyclodextrins in personal care products: Currently cyclodextrins can be found in a number of cosmetic and toiletry products such as creams, lotions, shampoos, toothpastes and perfumes. Following is a short review of cyclodextrins and their usage in skin care products. Cosmetic and toiletry applications of cyclodextrins are mainly described in the patent literature. However, Matsuda and Arima recently reviewed dermal and cosmetic applications of cyclodextrins.[1] For further information on cyclodextrins, their physicochemical properties and industrial applications the reader is referred to several books and reviews published in recent years.[2-11]

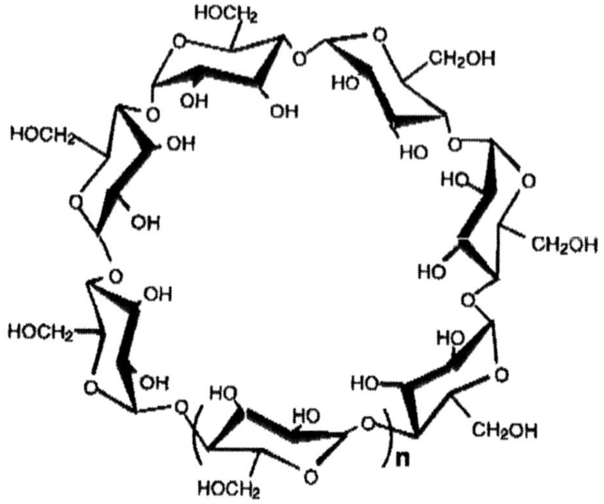

	αCD (n=0)	βCD (n=1)	γCD (n=2)
Molecular weight	972	1135	1297
Central cavity diameter (Å)	4.7-5.3	6.0-6.5	7.5-8.3
Height of torus (Å)	7.9	7.9	7.9
Approx. outer diameter (Å)	14.6	15.4	17.5
Solubility in water (w/v %)	14.2	1.85	23.2
Surface tension (mN/m)	71	71	71
Melting Range (°C)	255-260	255-265	240-245
Water of Crystallization (w/v %)	10	13-15	8-18
Number of water molecules in cavity	6	11	17

Figure 1. Structure and some physicochemical properties of α-cyclodextrin (αCD),β-cyclodextrin (βCD) and γ-cyclodextrin (γCD) at room temperature

Structure and Physicochemical Properties

Cyclodextrins are cyclic oligosaccharides, consisting of (α-1,4)-linked α-D-glucopyranose units, with a somewhat lipophilic central cavity and a hydrophilic outer surface. Due to the chair conformation of the glucopyranose units, the cyclodextrins are shaped like a truncated cone rather than perfectly cylindrical molecules. The hydroxyl functions are orientated to the cone exterior with the primary hydroxyl groups of the sugar residues at the narrow edge of the cone and the secondary hydroxyl groups at the wider edge. The central cavity is lined by the skeletal carbons and ethereal oxygens, which give it a lipophilic character. The natural product consists of a mixture of the various cyclodextrins, mainly α-cyclodextrin, β-cyclodextrin and γ-cyclodextrin, containing six, seven, and eight glucopyranose units, respectively (Figure 1).

Large ring cyclodextrins: While it is thought that, due to steric factors, cyclodextrins having fewer than five glucopyranose units cannot exist, cyclodextrins containing up to 21 glucosepyranose units have been isolated and characterized. However, due to their high production cost and unfavorable physicochemical properties these large ring cyclodextrins are presently of no industrial value. Cyclodextrin containing five glucopyranose units was recently described but its central cavity is closed and thus it is of no industrial value. The natural cyclodextrins, in particular β-cyclodextrin, have limited aqueous solubility, and their complex formation with lipophilic compounds frequently results in precipitation of solid cyclodextrin complexes.

In fact, the aqueous solubility of the natural cyclodextrins is much lower than that of comparable acyclic saccharides. This is thought to be due to relatively strong intermolecular hydrogen bonding in the crystal state. Substitution of any of the hydrogen bond forming hydroxyl groups, even by lipophilic methoxy functions, results in dramatic improvement in their solubility.

Cyclodextrin derivatives of industrial interest include the hydroxypropyl derivatives of β- and γ-cyclodextrin, the randomly methylated β-cyclodextrin, sulfobutylether β-cyclodextrin, the acetylated β- and γ-cyclodextrin, and the saccharide-conjugated cyclodextrins. The safety profiles of the various cyclodextrins, including their dermal safety, has been reviewed.[9]

Complex Formation

In an aqueous environment, cyclodextrins form inclusion complexes with many lipophilic molecules through a process in which water molecules located inside the central cavity are replaced by either a whole molecule, or more frequently, by some lipophilic structure of the molecule. The

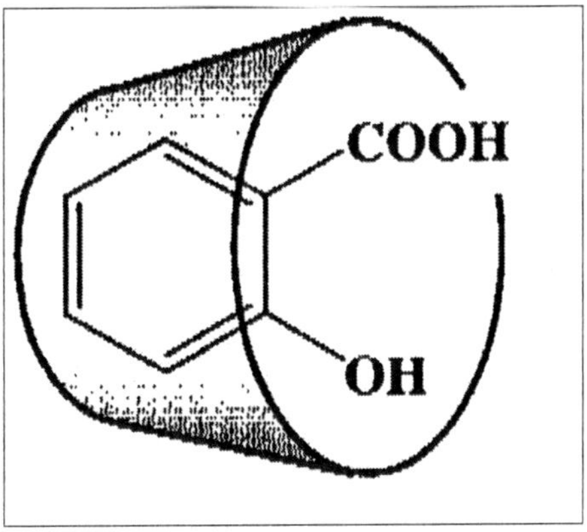

Figure 2. Salicylic acid/β-cyclodextrin complex

molecule located inside the cavity is then called the guest molecule and the cyclodextrin molecule the host molecule (Figure 2).

Since water molecules located inside the lipophilic cyclodextrin cavity cannot satisfy their hydrogen-bonding potential, they are of higher enthalpy than bulk water molecules located in the aqueous environment. The main driving force for complex formation, at least in the case of β-cyclodextrin and its derivatives, appears to be the release of these enthalpy-rich water molecules from the cavity, which lowers the energy of the system. However, other forces (e.g. van der Waals interactions, hydrogen bonding, hydrophobic interactions, release of structural strains and changes in surface tension) may also be important for the guest-host complex formation.[7]

Disassociation of guest molecules: Once included in the cyclodextrin cavity, the guest molecules may be disassociated from the host molecules through complex dilution. This occurs through replacing the included guest by some other suitable molecule (such as skin lipids) or, if the complex is located in close approximation to a lipophilic biological membrane (such as the skin surface), the guest may be transferred to the matrix for which it has the highest affinity. Importantly, since no covalent bonds are formed or broken during the guest-host complex formation, the complexes are in dynamic equilibrium with free guest and host molecules.

Preparing cyclodextrin complexes: Various methods have been applied to preparation of cyclodextrin complexes.[11-14] In solution, the complexes are usually prepared by addition of an excess amount of the guest to an aqueous cyclodextrin solution. The suspension formed is equilibrated (for periods of up to one week at the desired temperature) and then filtered or centrifuged to form a clear guest/cyclodextrin complex solution. For preparation of the solid complexes, the water is removed from the aqueous guest/cyclodextrin solutions by evaporation or sublimation, e.g. spray-drying or freeze-drying. Other methods can also be applied to prepare solid guest/cyclodextrin complexes including kneading and slurry methods, co-precipitation, neutralization and grinding techniques.[11,13]

In some cases the complexation efficiency is not very high, and therefore, relatively large amounts of cyclodextrins are needed to complex small amounts of a given guest compound. To add to this difficulty, various vehicle constituents, such as surfactants, lipids, organic solvents, buffer salts, and preservatives, often reduce the efficiency. However, it is possible to enhance the efficiency through formation of multicomponent complex systems (or co-complexes).

For example, the complexation efficiency of certain basic guest molecules can be enhanced through formation of guest-hydroxy acid-cyclodextrin ternary complex.[15] Water-soluble polymers are also known to enhance the complexation efficacy of a wide variety of guest molecules and to increase the aqueous solubility of the natural cyclodextrins.[11, 16]

Cyclodextrins as Penetration Enhancers

The effects of cyclodextrins on drug bioavailability and drug delivery through biological membranes has been investigated by a number of research groups.[1,6,8,17-20] Their findings can be summarized as follows. First, cyclodextrin molecules are relatively large molecular weight (ranging from almost 1000 to over 2000), with hydrated outer surface, and under normal conditions, cyclodextrin molecules and their complexes will only permeate lipophilic biological membranes with considerable difficulty. For example, only 0.02% of topically applied 2-hydroxypropyl-β-cyclodextrin was absorbed into hairless mouse skin under occlusive conditions.[21]

Second, cyclodextrins are able to extract various constituents from skin by taking them up into the central cavity. However, pretreatment of skin with aqueous cyclodextrin solution rarely results in enhanced drug permeability and reduced permeability is generally observed at relatively high cyclodextrin concentrations.

Third, cyclodextrins can be used to prevent absorption of drugs and other chemicals into skin. Fourth, hydrophilic cyclodextrins do not generally enhance skin permeability of water-soluble compounds.

Finally, hydrophilic cyclodextrins act as true carriers by keeping the hydrophobic drug molecules in solution and delivering them to the skin surface where they partition into the skin. The relatively lipophilic skin has low affinity for the hydrophilic cyclodextrin molecules or the hydrophilic drug/cyclodextrin complexes, which thus remain in the aqueous skin exterior, e.g. the aqueous vehicle system (such as o/w cream or hydrogel). Conventional skin penetration enhancers, such as alcohols and fatty acids, disrupt the skin barrier, whereas hydrophilic cyclodextrins enhance penetration by increasing drug availability at the skin surface (Figure 3).

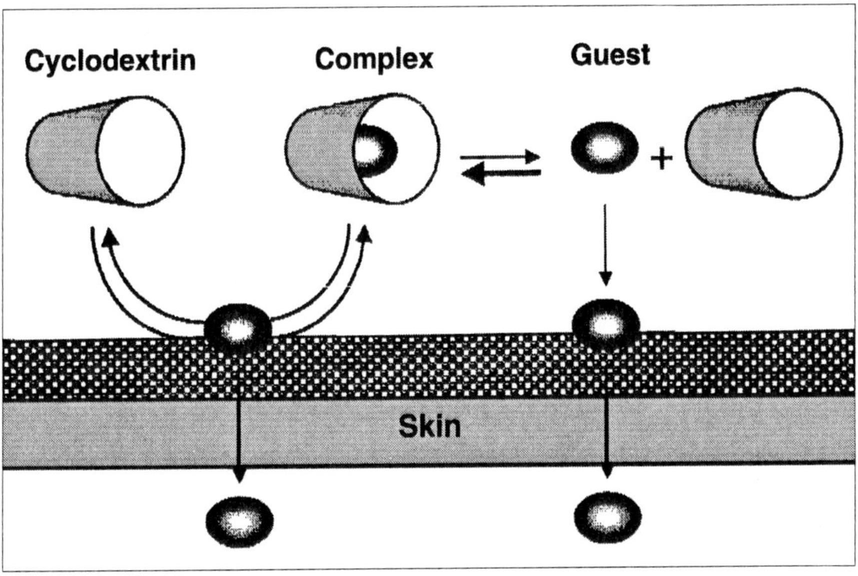

Figure 3. The apparent main mechanism of penetration enhancement

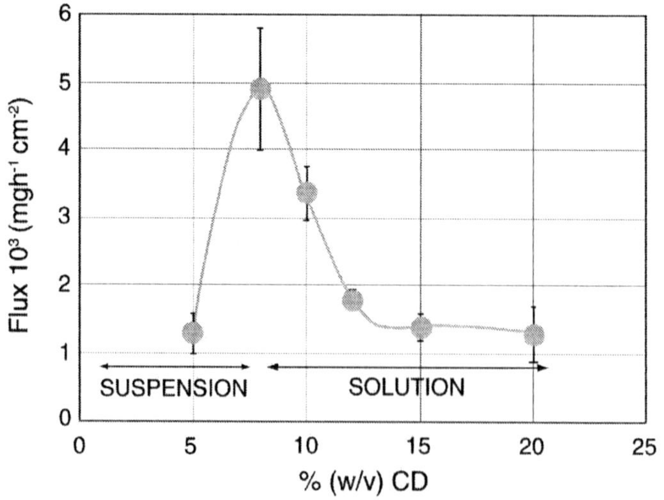

Figure 4. Permeability of hydrocortisone through hairless mouse skin in vitro. The vehicle consisted of 2.3% (w/v) hydrocortisone suspension or solution in water containing 5-20% (w/v) randomly methylated β-cyclodextrin and 0.25% (w/v) polyvinylpyrrolidone. The complexation efficiency was enhanced through heating in an autoclave (121°C for 40 min).

Cyclodextrins in Topical Formulations

Since neither cyclodextrins nor their complexes are absorbed into skin, cyclodextrins can both increase and decrease the host availability at the skin surface. For example, the effect of cyclodextrin concentration on the permeability of the lipophilic water-insoluble drug hydrocortisone through hairless mouse skin is shown in Figure 4. At low cyclodextrin concentrations, when the drug is in suspension, the flux of the drug increases with increasing cyclodextrin concentration.

At higher cyclodextrin concentrations, when all of the drug is in solution, the flux decreases with increasing cyclodextrin concentration. Maximum permeability is observed when just enough cyclodextrin is added to the vehicle to solubilize all of the drug. The effect of hydrocortisone/β-cyclodextrin molar ratio on the hydrocortisone availability in o/w cream has also been investigated (T. Loftsson: unpublished results).

The hydrocortisone/β-cyclodextrin complex was prepared in an autoclave by the slurry method. The slurry consisted of hydrocortisone and β-cyclodextrin in water or aqueous 0.25% (w/v) sodium carboxymethylcellulose solution. The resulting suspension was heated in a sealed container in an autoclave (121°C for 40 minutes). This was followed by equilibration at room temperature for three days and lyophilization. The molar rations of hydrocortisone/β-cyclodextrin in the dry complexes were 1:1, 1:1.5, 1:2, 1:2.5 and 1:3 in the different complex powders. The lyophilized hydrocortisone/β-cyclodextrin complexes were then incorporated into

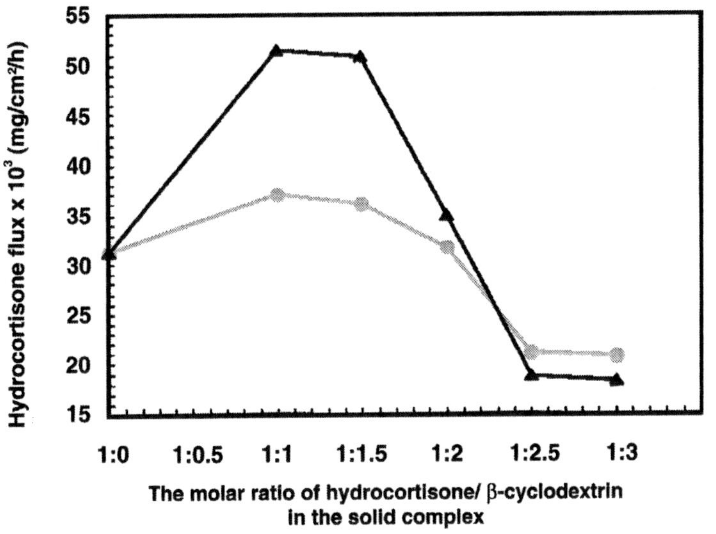

Figure 5. The effect of the hydrocortisone/β-cyclodextrin molar ratio on the hydrocortisone release from o/w cream through a cellophane membrane. The hydrocortisone concentration was kept constant at 2.8 mg/mL but the cyclodextrin concentration was varied. No polymer was added to the complexation media when the complex powder was prepared (O); 0.25% (w/v) sodium carboxymethylcellulose was present in the aqueous complexation media during preparation of the complex powder (Δ).

an o/w cream base. The availability of hydrocortisone in the cream was evaluated by determining the hydrocortisone release rate from the cream through a semi-permeable cellophane membrane (Figure 5).

The figure shows that it is very important to optimize the hydrocortisone release from the cream by adjusting the hydrocortisone/β-cyclodextrin ratio. Some of the ingredients of the cream will compete with hydrocortisone for a space in the cyclodextrin cavity reducing the solubilizing effect of β-cyclodextrin. At the same time, some of the ingredients will have solubilizing effect on hydrocortisone in the cream reducing the amount of β-cyclodextrin needed to solubilize hydrocortisone. Here the optimum hydrocortisone/β-cyclodextrin ratio is between 1:1 and 1:1.5. Reduced release was observed when the ratio was 1:2.5 and 1:3, compared to a cream containing no cyclodextrin (i.e. the 1:0 ratio).

The optimized formulations improved significantly the transdermal delivery of hydrocortisone.[11] Adding an optimal amount of the hydrophilic cyclodextrin to the cream base will increase the amount of hydrocortisone in the homogeneous aqueous phase, which enhances the hydrocortisone release from the cream, i.e. larger number of hydrocortisone molecules will be in contact with the skin surface. The large hydrated hydrocortisone/β-cyclodextrin complex does not permeate skin and, thus, excess cyclodextrin will reduce the permeability through the membrane. It is therefore possible to reduce or prevent absorption of compounds into and through skin by including cyclodextrins in their formulations.

Cyclodextrins and odor: Cyclodextrin complexation of volatile compounds will reduce or prevent their evaporation from the skin surface. For example, cyclodextrins can be used to prevent or reduce malodors on skin.[22] The odor absorbing composition will then consist of free (i.e. uncomplexed) cyclodextrin in an aqueous or alcoholic vehicle. When applied to skin, cyclodextrin will form nonvolatile complexes with the bad-smelling substances, thereby preventing their evaporation from the skin surface.

Likewise, through cyclodextrin complexation it is possible to obtain controlled release of fragrances. Only the free fraction of a fragrance material, which is in equilibrium with the complexed fraction, is released from the skin surface. At the same time cyclodextrin complexation of the fragrances will reduce their permeability into skin, thereby reducing their potential skin irritation and toxic side effects.[1,23]

Enhanced shelf life: Cyclodextrins are also able to enhance the shelf life of skin care products. Through complexation the cyclodextrin molecule shields, at least partly, the guest molecule against attack by various reactive molecules. In this way, cyclodextrins can prevent or reduce hydrolysis, oxidation, steric rearrangement, racemization, isomerization, polymerization and even enzymatic decomposition of the included molecule.[7,24] For example, the photodegradation of the sunscreen agent butyl-methoxydibenzoylmethane, can be reduced through cyclodextrin complexation. Cyclodextrin complexation enhanced the aqueous solubility of the sunscreen agent, increased its chemical stability and decreased its irritation potential at the skin surface.[25]

Another example is cyclodextrin stabilization of polyunsaturated fatty acids. Fish oils are rich in polyunsaturated fatty acids such as eicosapentanoic acid (EPA) and docosahexanoic acid (DHA). However, unsaturated fatty acids are sensitive towards oxidation and the oxidative products give the characteristic fishy odor. Even highly purified, odor-free fish oils rapidly undergo oxidation after topical application. The oxidative degradation of fish oils can be prevented, or at least hampered, though complexation with γ-cyclodextrin.[26]

Conclusions

Cyclodextrins are nontoxic complexing agents that are able to improve the various physicochemical properties of cosmetic ingredients, such as their aqueous solubility and chemical stability. Through cyclodextrin complexation it is possible to enhance and decrease skin delivery of cosmetic ingredients, as well as prevent or reduce their skin irritation and toxic side effects.

In addition, cyclodextrins can have a wide variety of other functions in skin care products. However, the preferred effects can only be obtained through careful optimization of the cyclodextrin containing product. Adding cyclodextrins, though, to already existing products rarely results in the desired outcome.

—**Thorsteinn Loftsson,** *Faculty of Pharmacy, University of Iceland, Reykjavik, Iceland*

References

1. H Matsuda and H Arima, Cyclodextrins in transdermal and rectal delivery, *Adv Drug Deliv Rev*, 36 81-99 (1999)
2. D Duchéne, *Cyclodextrins and their Industrial Uses*. Paris: Editions de Santé, 1987
3. J Szejtli, *Cyclodextrin Technology*. Dordrecht: Kluwer Academic Publisher, 1988
4. KH Frömming and J Szejtli, *Cyclodextrins in pharmacy*, vol. 5. Dordrecht: Kluwer Academic Publishers, 1994
5. VT D'Souza and KB Lipkowitz, Cyclodextrins, In *Chemical Reviews*, vol. 98(5). Washington DC: American Chemical Society (1998)
6. K Uekama, Cyclodextrins in drug delivery, In *Advanced Drug Delivery Reviews*, vol. 36(1). Amsterdam: Elsevier Science (1999)
7. T Loftsson and ME Brewster, Pharmaceutical applications of cyclodextrins. 1. Drug solubilization and stabilization, *J Pharm Sci*, 85 1017-1025 (1996)
8. RA Rajewski and VJ Stella, Pharmaceutical applications of cyclodextrins. 2. In vivo drug delivery, *J Pharm Sci*, 85 1142-1168 (1996)
9. T Irie and K Uekama, Pharmaceutical applications of cyclodextrins. III. Toxicological issues and safety evaluation., *J Pharm Sci*, 86 147-162 (1997)
10. VJ Stella and RA Rajewski, Cyclodextrins: their future in drug formulation and delivery, *Pharm Res*, 14 556-567 (1997)
11. T Loftsson, Pharmaceutical applications of β-cyclodextrin, *Pharm Technol*, 23(12) 40-50 (1999)
12. F Hirayama and K Uekama, Methods of investigating and preparing inclusion compounds, In *Cyclodextrins and their industrial uses*, D Duchêne (ed. Paris: Editions de Santé (1987), 131-172
13. AR Hedges, Industrial applications of cyclodextrins, *Chem Rev*, 98 2035-2044 (1998)
14. T Loftsson, M Másson, and JF Sigurjónsdóttir, Methods to enhance the complexation efficiency of cyclodextrins, *STP Pharma Sci*, 9 237-242 (1999)
15. E Redenti, L Szente, and J Szejtli, Drug/cyclodextrin/hydroxy acid multicomponent systems. Properties and pharmaceutical applications, *J Pharm Sci*, 89 1-8 (2000)
16. T Loftsson, Increasing the cyclodextrin complexation of drugs and drug bioavailability through addition of water-soluble polymers, *Pharmazie*, 53 733-740 (1998)
17. K Uekama, F Hirayama, and T Irie, Cyclodextrin drug carrier systems, *Chem Rev*, 98 2045-2076 (1998)
18. VJ Stella, VM Rao, EA Zannou, and V Zia, Mechanism of drug release from cyclodextrin complexes, *Adv Drug Deliv Rev*, 36 3-16 (1999)
19. T Loftsson and T Järvinen, Cyclodextrins in ophthalmic drug delivery, *Adv Drug Deliv Rev*, 36 59-79 (1999)
20. M Masson, T Loftsson, G Masson, and E Stefansson, Cyclodextrins as permeation enhancers: some theoretical evaluations and *in vitro* testing, *J Controlled Release*, 59 107-118 (1999)
21. M Tanaka, Y Iwata, Y Kouzuki, K Taniguchi, H Matsuda, H Arima, and S Tsuchiya, Effect of 2-hydroxypropyl-β-cyclodextrin on percutaneous absorption of methyl paraben, *J Pharm Pharmacol*, 47 897-900 (1995).
22. M McCoy, Cyclodextrins: great product seeks a market, *Chem Engin News*, 77(9) 25-27 (1999)
23. M Tanaka, H Matsuda, H Sumiyoshi, H Arima, F Hirayama, K Uekama, and S Tsuchiya, 2-Hydroxypropylated cyclodextrins as a sustained-release carrier for fragrance materials, *Chem Pharm Bull*, 44 416-420 (1996)
24. T Loftsson, Effects of cyclodextrins on chemical stability of drugs in aqueous solutions, *Drug Stability*, 1 22-33 (1995)

25. S Scalia, S Villani, A Scatturin, MA Vandelli, and F Forni, Complexation of the sunscreen agent, butyl-methoxydibenzoylmethane, with hydroxypropyl β-cyclodextrin, *J Pharm Sci*, 175 205-213 (1998)
26. H Reuscher, Stabilized PUFA triglycerides for nutraceuticals and functional foods using γ-cyclodextrin, presented at The 10th International Cyclodextrin Symposium, Ann Arbor, Michigan (2000)

A Laboratory Method for Measuring the Water Resistance of Sunscreens

Keywords: testing and instrumentation, sunscreens, water resistance, spectrophotometric analysis, in vitro skin model

A new laboratory in vitro method for the measurement of sunscreen water resistance uses spectrophotometric analysis of a model skin substitute before and after 80-minute immersion. It gives good correlation with the FDA's 80-minute immersion SPF results (very water-resistant). The method is especially good for screening new formulations or water-resistant technologies.

Normal summer activities put a great deal of stress upon sunscreen products, particularly water exposure through swimming and sweating. Most sunscreen products are therefore designed to be water resistant. Regulatory agencies around the world, including the US Food and Drug Administration (FDA) and the European Cosmetic, Toiletry and Perfumery Association (COLIPA), have defined protocols to assure consumers that a claim of water-resistance has been substantiated. Although protocols differ around the world, they all require testing on a significant number of people, which makes the testing expensive and time-consuming. The sunscreen formulator and chemist need a rapid and cost-effective method to determine at least a comparative level of water resistance in the laboratory in order to determine project progress.

A number of papers[1-3] have been published on in vitro methods for the measurement of water resistance. The first challenge for any in vitro method is to select a substrate that properly models at least the physical properties of skin. This requirement eliminates some convenient substrates such as polyethylene film and Transpore tape.[a] Although the latter is used for laboratory SPF determination,[4] its porous nature creates an artificial opportunity for sunscreen retention when immersed in water. Therefore, a number of methods have focused on the use of real skin from both human and animal sources.

For our purposes, actual skin does not meet the criteria for cost-effectiveness due to limited shelf life (decomposition), cost, and/or availability.

Model skin substrates must simulate the heterogeneous mixture of lipid and protein, the overall hydrophobic behavior of the skin, and its topology, which can

---

[a] Transpore is a registered trademark of 3M Co., St. Paul, Minnesota USA.

vary by ± 50 microns or more. One of the first to do so was from Charkoudian,[5] who described a composition and process to make a model skin surface. Although this material meets the requirements described above, it lacks the mechanical strength to be manipulated for water-resistance testing.

A material of similar composition is Vitro-Skin[b] from IMS, Inc. This product, made of cross-linked collagen with added lipids, does have mechanical integrity. Vitro-Skin simulates the properties of composition, wettability, pH, ionic strength and surface topography of human skin. Indeed, IMS has developed a water-resistance test based on their Vitro-Skin material using the Optometrics[c] SPF 290 to measure the SPF of the product on the skin before and after water immersion.[6]

We have found that the normally opaque Vitro-Skin becomes optically transparent upon hydration. This lets us use a normal UV-VIS spectrophotometer to measure the sunscreen

[b] Vitro-Skin is a trademark of IMS, Inc., Milford, Connecticut USA.

Table 1. Medium SPF sunscreens for in vivo and in vitro water-resistance measurements (percentages are wt %)

	A	B	C	D	E
A. Water (*aqua*), deionized	69.1%	67.1%	68.1%	68.1%	68.1%
Hexylene glycol	2.0	2.0	2.0	2.0	2.0
Carbomer	0.2	0.2	0.2	0.2	0.2
B. Octyl methoxycinnamate	7.5	7.5	7.5	7.5	7.5
Benzophenone-3	3.0	3.0	3.0	3.0	3.0
Octyl salicylate	3.0	3.0	3.0	3.0	3.0
Octyl palmitate	6.0	6.0	6.0	6.0	6.0
Cetearyl alcohol	0.2	0.2	0.2	0.2	0.2
Lamellar gel structurant*	3.0	5.0	3.0	3.0	3.0
PVP/hexadecene copolymer	-	-	1.0	-	-
PVP/eicosene copolymer	-	-	-	1.0	-
Tricontanyl PVP	-	-	-	-	1.0
C. Triethanolamine	0.2	0.2	0.2	0.2	0.2
Water (*aqua*), deionized	5.0	5.0	5.0	5.0	5.0
D. Diazolidinyl urea (and) iodopropynyl butylcarbamate	0.3	0.3	0.3	0.3	0.3
Phenoxyethanol (and) isopropylparaben (and) isobutylparaben (and) butylparaben	0.5	0.5	0.5	0.5	0.5
	100.0	100.0	100.0	100.0	100.0

* Stearic acid (and) behenyl alcohol (and) glyceryl stearate (and) lecithin (and) cetyl alcohol (and) myristyl alcohol (and) lauryl alcohol (and) palmitic acid

present before and after water immersion, and allows the use of smaller samples and measurements with greater precision than we experienced with the Optometrics SPF 290. Here we describe this new method and its validation with in vivo data as well as a comparison of some common water-resistant technologies for sunscreen formulations.

Materials and Methods

Sunscreen compositions: Medium sun protection factor (SPF) sunscreen formulations based on a lamellar gel structuring system (Table 1) were used for the

[c] Optometrics is a trademark of Optometrics USA, Inc., Ayer, MA USA.

Table 2. Medium SPF nonionic sunscreen for in vitro water-resistance measurements (percentages are wt %)

	F	G	H	I	J	K	L
A. Water (*aqua*), deionized	65.0%	64.0%	64.0%	64.0%	64.0%	64.0%	64.0%
Hexylene glycol	2.0	2.0	2.0	2.0	2.0	2.0	2.0
Carbomer	0.2	0.2	0.2	0.2	0.2	0.2	0.2
B. Octyl methoxycinnamate	7.5	7.5	7.5	7.5	7.5	7.5	7.5
Benzophenone-3	3.0	3.0	3.0	3.0	3.0	3.0	3.0
Octyl salicylate	3.0	3.0	3.0	3.0	3.0	3.0	3.0
Octyl palmitate	6.0	6.0	6.0	6.0	6.0	6.0	6.0
PEG-20 stearate	2.0	2.0	2.0	2.0	2.0	2.0	2.0
Glyceryl stearate (and) laureth-23	5.0	5.0	5.0	5.0	5.0	5.0	5.0
PVP/eicosene copolymer	-	1.0	-	-	-	-	-
Polybutene, hydrogenated	-	-	1.0	-	-	-	-
Adipic acid/diethylene glycol/ glycerin crosspolymer	-	-	-	1.0	-	-	-
Synthetic wax	-	-	-	-	1.0	-	-
C_{30-38} olefin/isopropyl maleate/ MA copolymer	-	-	-	-	-	1.0	-
Diglycol/CHDM/isophthalates/ SIP copolymer	-	-	-	-	-	-	1.0
C. Triethanolamine	0.2	0.2	0.2	0.2	0.2	0.2	0.2
Water (*aqua*), deionized	5.0	5.0	5.0	5.0	5.0	5.0	5.0
D. Diazolidinyl urea (and) iodopropynyl butylcarbamate	0.6	0.6	0.6	0.6	0.6	0.6	0.6
Phenoxyethanol (and) isopropylparaben (and) isobutylparaben (and) butylparaben	0.5	0.5	0.5	0.5	0.5	0.5	0.5
	100.0	100.0	100.0	100.0	100.0	100.0	100.0

validation of the in vitro water-resistance method described here. The method was then used to examine the effect of some commercial water-resistant technologies in a medium SPF nonionic formulation (Table 2). This formulation was chosen because nonionic-based emulsions are well known to be difficult to make water-resistant.

Sunscreen preparation: Sunscreen formulations were made using the following procedure. Water and hexylene glycol of Phase A were combined at room temperature. Carbomer was slowly sprinkled onto the surface while stirring. After incorporating all the carbomer, Phase A was heated to 70-75°C with stirring. Phase B was prepared separately, heated to 75-80°C and stirred until uniform. Phase B was slowly added to Phase A with homogenization at 70°C. When the mixture appeared uniform, Phase C was added with homogenization. After achieving uniformity again, the heat was turned off and mixing was switched to sweep at 60°C. Sweep mixing was continued throughout cool-down. Phase D was added with stirring at 45°C. Finally, water was added to make up for loss during heating and stirred to room temperature.

In Vitro Testing Methodology

Before the application of sunscreens, large strips (20x10 cm) of Vitro-Skin N-19 were hydrated following the manufacturer's recommendations (16-18 hours at 90-95% relative humidity at room temperature).

A measured amount (6-7 mg) of formulation was applied on pre-hydrated pieces of Vitro-Skin (28x38 mm) that were mounted in 35 mm slide mounts. The emulsion was carefully spread using a rubber-gloved finger with initial circular and then linear motion for approximately one minute. Samples were then placed in a humidity chamber (90-95% relative humidity at room temperature) for 20 minutes to allow for emulsion coalescence.

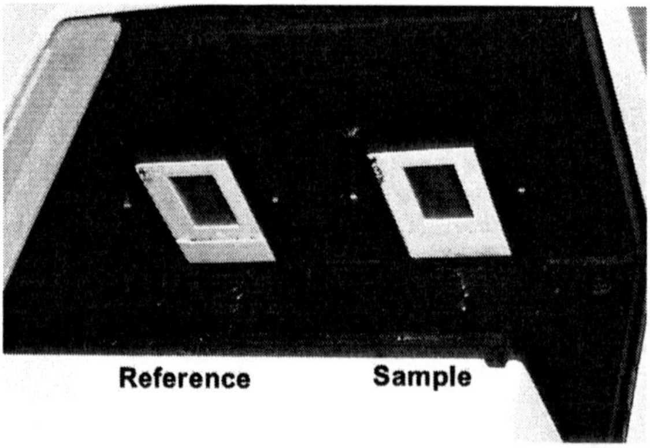

Figure 1. Placement of Vitro-Skin sample and reference in the spectrophotometer sample compartment

Four UV spectra were collected per sample in the wavelength range of 250-350 nm using a spectrophotometer.[d] Each spectrum was collected after a 90° rotation of the sample giving four separate area scans for each sample. Samples were scanned against the untreated reference sample in the reference beam of the two-beam spectrophotometer, as demonstrated in Figure 1. Absorbance readings were taken at 310 and 291 nm and were labeled as the initial absorbance, A_i.

Samples were then immersed in a temperature-controlled water bath (25° ± 0.2°C) for 80 minutes with constant mixing using a paddle type impeller at 50 rpm, as shown in Figure 2. The volume of the water bath was large enough (2000 ml) to prevent a high concentration of dispersed sunscreen and possibility of re-adsorption.

[d] Cary 1-E Spectrophotometer, Varian Corp., San Fernando, California USA

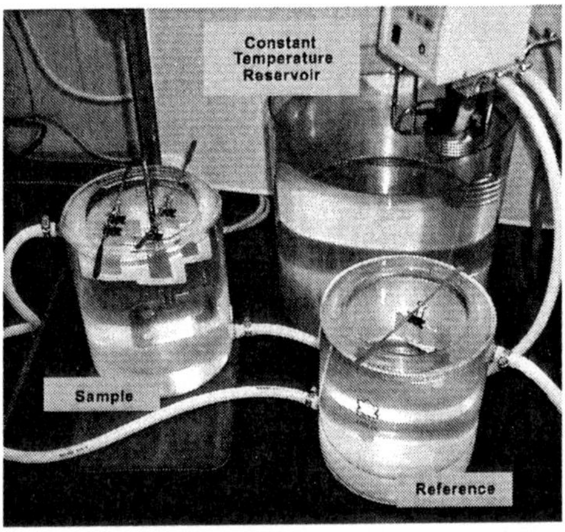

Figure 2. Immersion apparatus for the in vitro measurement of water resistance

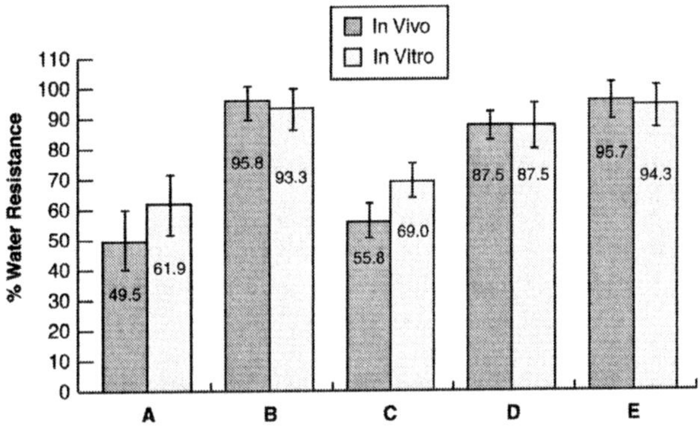

Figure 3. Comparison of FDA in vivo 80-minute immersion results with in vitro 80-minute immersion results for formulations shown in Table 1

After immersion, the samples were taken out of the water, lightly shaken to remove the largest water droplets and hung in the air in a climate-controlled room at 50% relative humidity for 30 minutes. Afterward, the samples were placed back into the humidity chamber for 120 minutes. Final absorbance readings, A_f, were taken in the same manner as the initial ones. The percentage of water resistance was calculated as $(A_f / A_I) \times 100$.

Samples were run in quadruplicates giving 32 readings (4 samples x 4 orientations x 2 wavelengths) for each formulation tested. It is important that the samples remain hydrated; if they lose too much moisture, the Vitro-Skin becomes opaque and unsuitable for spectrophotometric analysis. Each blank control sample (without sunscreen) was treated and measured in exactly the same way. The control sample was immersed in a separate water bath, as shown in Figure 2, to ensure no sunscreen transfer from formulation-treated samples.

Results and Discussion

Comparing in vivo and in vitro test methods: The in vivo water resistance of the series of sunscreen formulations in Table 1 was determined according to the FDA method using five human panelists of different skin types for the very water-resistant protocol (80-minute immersion). This series shows the water-resistance effect of a lamellar gel structuring system at 3% (A) and 5% (B), and when the 3% lamellar gel formulation is augmented with 1% of three alkylated PVP polymers: PVP/hexadecene copolymer (C), PVP/eicosene copolymer (D), and tricontanyl PVP (E). The in vivo performance is reported in Figure 3 as a percentage of SPF before and after the 80-minute immersion. For example, formulation C gave an average pre-immersion SPF of 16.5 and average post-immersion SPF of 9.2 giving an average retention of 55.8%. The initial SPF of all five formulations was between 16.5 and 18.0.

When the same formulations were tested in the new method described here, the agreement, as a percentage of retained material, was very good. Not only does the new method provide similar accuracy, it also gave standard deviations on

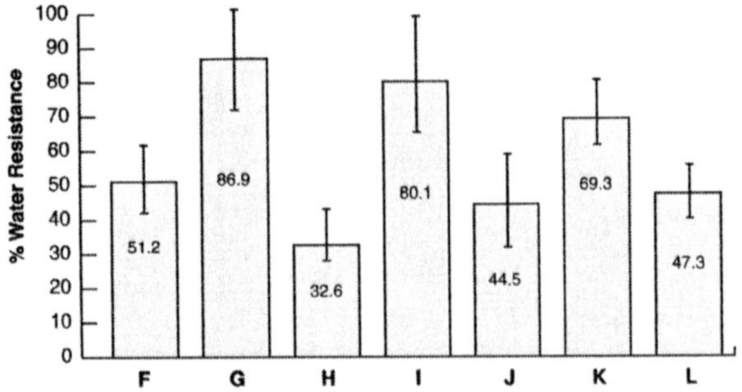

Figure 4. Comparison of water-resistant technologies in the medium SPF nonionic formulations shown in Table 2

the same order as the FDA in vivo method. It was also important that there was agreement between formulations that retained most of their activity (B, D and E) and those that lost about half of their activity (A and C). Best agreement was seen for the formulations that were very-water-resistant (B, D and E). There was more uncertainty with the formulations that lost significant amounts of activity (A, C), which can probably be attributed to the mechanisms expected for the loss of materials that are inhomogeneous in character.

Since this method is designed to rapidly screen formulations for water resistance, best agreement with the most water-resistant samples is most desirable. Those that are not water-resistant are selected away from further examination, as desired. The original study showed that for a medium SPF (~15) formulation, 3% of the lamellar gel structurant and 3% lamellar gel structurant with 1% PVP/hexadecene copolymer did not meet the requirements for very-water-resistant claims at an SPF of 15. Formulations with 3% of the lamellar gel structurant and 1% PVP/eicosene copolymer and 1% tricontanyl PVP, as well as a 5% level of the lamellar gel structurant, easily met the criteria for a very-water-resistant SPF 15 claim. The same conclusion could be reached much more quickly and inexpensively in the laboratory using the new method described here. Then only the final selected formulation need be tested in vivo for the final label claim.

In vitro testing of four commercial sprays: Sunscreen spray formulations have become very popular recently. We tested four commercial spray formulations that claimed to be waterproof or very water resistant. Table 3 summarizes the results of the testing of the four sprays in the new method described here.

Product 1 claimed a waterproof SPF of 8 and we found retention of 85.4% of the initial absorbance. Product 2 claimed a very-water-resistant SPF of 25 and we found 99.5% retention of the initial absorbance. Product 3 claimed a very-water-resistant SPF of 30 and we found 98.0% retention of the initial absorbance. Product 4 claimed a waterproof SPF of 30 and we found 96.2% retention of the initial absorbance. Since these products come under FDA regulation, they should deliver on their claims; therefore we believe this gives an independent confirmation of the correlation of this method with in vivo testing.

In vitro comparison of seven water-resistance technologies: Finally, we looked at a simple medium SPF nonionic formulation (Table 2) to screen the performance of a number of technologies that are marketed for sunscreen water resistance at a level of 1%. The level of 1% was selected to show differences, not to generate very water-resistant formulations. Nonionic formulations are known

Table 3. In vitro test results for four commercial spray products

Product	SPF	Claim	% retained
1	8	waterproof	85.4 ± 8.1
2	25	very water-resistant	99.5 ± 0.9
3	30	very water-resistant	98.0 ± 3.9
4	30	waterproof	96.2 ± 2.4

to be difficult to make water-resistant, probably due to the excellent detergency of the nonionic emulsifiers. This was confirmed in our study.

Formulation F, the control nonionic formulation, retained only 51.2% of the initial absorbance. Hydrogenated polybutene (H) retained less than the control at 32.6% of the initial absorbance. Synthetic wax (J) and diglycol/CHDM/isophthalates/SIP copolymer (L) were similar to the control at 44.5% and 47.3% retention of initial absorbance, respectively. Three technologies improved retention: $C_{30\text{-}38}$ olefin/isopropyl maleate/MA copolymer (K), adipic acid/diethylene glycol/glycerin crosspolymer (I), and PVP/eicosene copolymer (G). The PVP/eicosene copolymer gave the best retention at the 1 wt % level. Several other waterproofing technologies could not be evaluated in this model because they were incompatible with the system. Results are shown in Figure 4.

Conclusion

In this paper we have described a new laboratory in vitro test method for determining the water resistance potential of sunscreen formulations. This method is rapid, accurate and inexpensive, based on its results compared to in vivo results for the same formulations. The method is easily suited to most laboratory environments requiring only a thermostated water bath, glass-jacketed vessels, constant humidity chambers, and a UV-visible spectrophotometer. The substrate is readily available. We have found the method fast and reliable for screen both sunscreen formulations and water-resistant technologies.

—Berislav Markovic, Donna Laura and Mark Rerek,
International Specialty Products, Wayne, New Jersey, USA

References

1. RP Stokes, BL Diffey, LC Dawson and SP Barton, A novel in vitro technique for measuring the water resistance of sunscreens, *Int J Cosmet Sci* 20 235-240 (1998)
2. RP Stokes and BL Diffey, The water resistance of sunscreen and day-care products, *British J Dermatol* 140 259-263 (1999)
3. R Tarroux, MF Assalit, J Hemmerle and J Ginestar, Influence of applied quantity, water-immersion and air-drying on covering and microstructure of physical sunscreen films, *Int J Cosmet Sci* 22 447-458 (2000)
4. BL Diffey and J Robson, A new substrate to measure sunscreen protection factors throughout the ultraviolet spectrum, *J Soc Cosmet Chem* 40 127-133 (1989)
5. JC Charkoudian, A model skin surface for testing adhesion to skin, *J Soc Cosmet Chem* 39 225-234 (1988)
6. RL Sellers and FG Carpenter, An instrument for in vitro determination of SPF, *Cosmet Toil* 107 119-123 (1992)

Photoaging and Photodocumentation

Keywords: testing and instrumentation, photoaging, photodocumentation, clinical grading, light imaging techniques, video-microscopy

Techniques to photograph or image skin photodamage have reached new levels of sophistication. This survey discusses clinical grading, light imaging techniques, video-microscopy and three-dimensional in vivo measuring systems.

An image conveys an immediate and powerful message, both to the "untrained" consumer and to the expert grader. A comparison of two images, such as before and after treatment, creates an even stronger statement. Therefore, significant attention has been given to the standardization of imaging and photographic tools in order to use images as part of scientific evaluation methods in acne and photoaging.[1-7]

"Skin photodamage" refers to the changes in the skin tissue caused by acute or chronic ultraviolet (UV) exposure. In photoaging, the photodamage is superimposed on the changes occurring with intrinsic aging.[8] Clinical characteristics of cutaneous photoaging include fine lines, coarse wrinkles, crinkles, roughness, laxity, sallowness, dullness, telangiectasia, pebbly appearance, mottled pigmentation, disruption of microtopography, actinic keratoses and skin cancers (Table 1).[1,9-14]

Antiaging studies usually rely on the clinician to score the severity of photodamage parameters by visual and tactile inspection using a descriptive scale.[3,4,15] In acne studies, however, it has become routine to score acne severity by comparing the patients to standard photographic grades.[5] On this premise and to design a more reproducible evaluation system in photoaging, a few photonumeric scales have been published with the aim of anchoring the evaluator to fixed visual grades.[1,2]

Additionally, thanks to the recent advances and excellent standardization in photographic systems, clinicians have used sequential photographs to evaluate retrospectively the improvement of photodamage.[6,7] Finally, state-of-the-art technologies in the optical arena are offering new methodologies, such as PRIMOS, which merge three-dimensional imaging of the skin surface with quantification of macrostructures such as wrinkles.[16]

The aim of this review is to discuss various evaluation approaches and imaging methods based on the specific needs of the photoaging trial.

Clinical Grading of Photoaging

When designing a photonumeric scale for aging it is important to recognize that the face can age differently depending on the phototype, genetic background and sun-exposure history of the patient.

In Caucasoid subjects, severe photodamage on the face can appear with severe hyperpigmentation, actinic keratoses and skin cancers on an atrophic, telangiectatic and pinkish skin.[14] Alternatively, severe photodamage may present with only deep and coarse wrinkles all over the face on a yellowish, pebbly and thickened skin, while hyperpigmentation and keratoses are rare.

Additionally, it is important to note that different photoexposed body sites may present different photoaging characteristics. Wrinkles are, of course, studied on the face because it is in this location that they are of most concern and most prominent. On the other hand, evaluation of crinkling and microtopography is best achieved on the forearm.[11] The dorsum of hands is also often studied for mottled pigmentation.

Table 1. Typical parameters evaluated in photoaging studies

Graded Parameter*	Measurement Site	Description
overall severity[4]	face, dorsal forearm	overall general appearance
fine lines[1,21]	face	fine shallow wrinkles that usually disappear when the skin is slightly pulled
coarse wrinkles[1,21]	face	deeper and wider furrows than fine lines
mottled pigmentation[15,17,25]	face, forearms, hands	solar lentigines, hypopigmented and hyperpigmented spots
sallowness or yellowing[1]	face	appearance of the skin tone from pinkish to yellowish
brightness or clarity[15]	face	skin appearance is scored from bright to dull
roughness[17,21]	face, forearms	assessed by gently touching the skin
laxity[15,17]	face, forearms	inability of the skin to return back to normal after being pulled. It is sometimes indicated by the "pinch recoil time" where the skin lateral to the zygomatic arch is pulled and suddenly released and the time for the skin to return to normal is measured in hundredths of a second.[15]

*Other parameters such as lentigines[4] (independently from pigmentation) and telangiectasia are less frequently graded.

Wrinkles and hyperpigmentation (also called mottled pigmentation) are the two major parameters studied in photoaging.

Wrinkles: Kligman has described five types of wrinkles: crinkles, linear facial wrinkles, glyphic facial wrinkles, facial creases, and naso-labial folds.[13] Of these, usually the first two are the focus of topical treatments.

Linear facial wrinkles are the typical facial furrows, such as "crow's feet" in the eye region and "rhytids" on the upper lip. Their distribution is dictated by the insertion on the skin of the muscles of mimicry.

Crinkles, instead, are redundant fine folds of skin caused by dermal degradation and loss of skin elasticity. They are most noticeable on the arms and can be displaced by massaging the skin or by movements.

Mottled pigmentation: The second parameter, mottled pigmentation, refers to irregular pigmentation. Nevi are not included. Some authors have included only patchy hyperpigmentation, solar freckling and melasma[1] while others have also included lentigines and hypopigmentation.[17]

The mottling is caused by the heterogeneity in the activity of epidermal melanocytes, with some being very active while others are no longer producing melanin.[12] Since areas of decreased melanin (hypopigmentation) are also due to the damaging effects of solar radiation and contribute to the mottled appearance, they are usually included as part of the overall mottled pigmentation score.

Lentigines and melasma: At times, parameters such as lentigines and melasma are graded separately. Solar lentigines are sun-induced, well-defined, hyperpigmented lesions, which, unlike ephelides or freckles, do not fade during absence of sun exposure.

"Melasma" refers to an acquired hyperpigmentation that involves larger areas and it is usually related to hormonal variations. Melasma is graded using a specifically designed clinical index (MASI = Melasma Area and Severity Index).[18,19] Here the total face area is divided into four parts: forehead (F, 30%); right malar (MR, 30%); left malar (ML, 30%); and chin (C, 10%). The melasma in each part is graded according to the following:

- Percentage of site involvement (A) is graded 0 to 6, representing 0 to 90-100% site involvement
- Darkness (D) is graded 0 to 4, representing none to severe darkening
- Homogeneity (H) is graded 0 to 4, representing minimal to maximum homogeneity

The MASI is then calculated as follows:

$$MASI = 0.3(DF+HF)AF + 0.3(DMR+HMR)AMR + 0.3(DML+HML)AML + 0.1(DC+HC)AC.$$

Using standardized photography the MASI score could also be calculated from images.

Clinical photoaging scales: Several clinical photoaging scales are reported in the literature.[1,3,4,17,20] A peculiar scale designed by Glogau is sometimes used for a rough categorization of the overall facial photodamage.[20] The distinctiveness of this scale is the analysis of wrinkle appearance during facial movements, which denotes a problem difficult or impossible to manage with only topical products.

In Glogau's scale the face is categorized into four types. Type I patients (mild) have "no wrinkles." Type II patients (moderate) have "wrinkles in motions." Type III patients (advanced) present "wrinkles at rest." Type IV patients (severe) appear with "only wrinkles." Because of the parameters evaluated and of their limited range of severity, the Glogau scale is not usually used to study topical formulations.

To detect improvement in at least some of the photoaging features, grading should involve evaluation of various photodamage parameters, both visual and tactile (see Table 1). The grading is often designed on a 10-point descriptive scale (0-9), where 0 equals absence, 1-3 is mild, 4-6 is moderate and 7-9 is severe.[1,3,4]

Typical antiaging studies usually enroll subjects who fit into mild to moderate overall photodamage (4-6 or 3-7) (Figure 1). Extreme grades would not be able to show changes because either the damage is too little to be quantified or it is too severe to improve with topicals. Clinical evaluations can be conducted unanchored or anchored to baseline photographs or scores.[3,4]

Photographic Evaluation of Photodamage

Anchored evaluation and photonumeric scales: Anchored evaluation may indicate that the subject's baseline photograph is used by the evaluator at each grading visit as a reference to assess the subject's progression.[3,7] On the other hand, it may also refer to the use of standard images from a photonumeric scale to match the subject severity to that image grade.

Griffith et al. have published a photonumeric scale to anchor the clinical visual assessment to a photographic grade.[1] They have presented five anchor severity grades (0, 2, 4, 6 and 8), frontal and side view, of a nine-point scale. This scale, however, is based only on wrinkle severity while hyperpigmentation is not evaluated.

Since severity of photodamage is based both on wrinkles and dyspigmentation, Larnier et al. addressed this issue by publishing a second photonumeric scale.[2] Here, they depicted three photodamage variations within each grade. Although this scale is more refined than the previous one, its six-point range may be restrictive when used to detect subtle improvement in topical anti-aging studies.

Retrospective evaluation of photographs: Another photographic approach consists in the use of standardized sequential photography to evaluate the improvement of photodamage retrospectively. This method is believed to detect the more subtle within-patient improvements during antiaging treatments.[6,7]

Maddin et al. have used a system in which baseline and final photos were projected side-by-side in a random right/left manner, assuring a blinded analysis.[7] A 13-point balanced categorical scale was used by each evaluator to indicate that the photograph projected on the corresponding side was better. The center of the

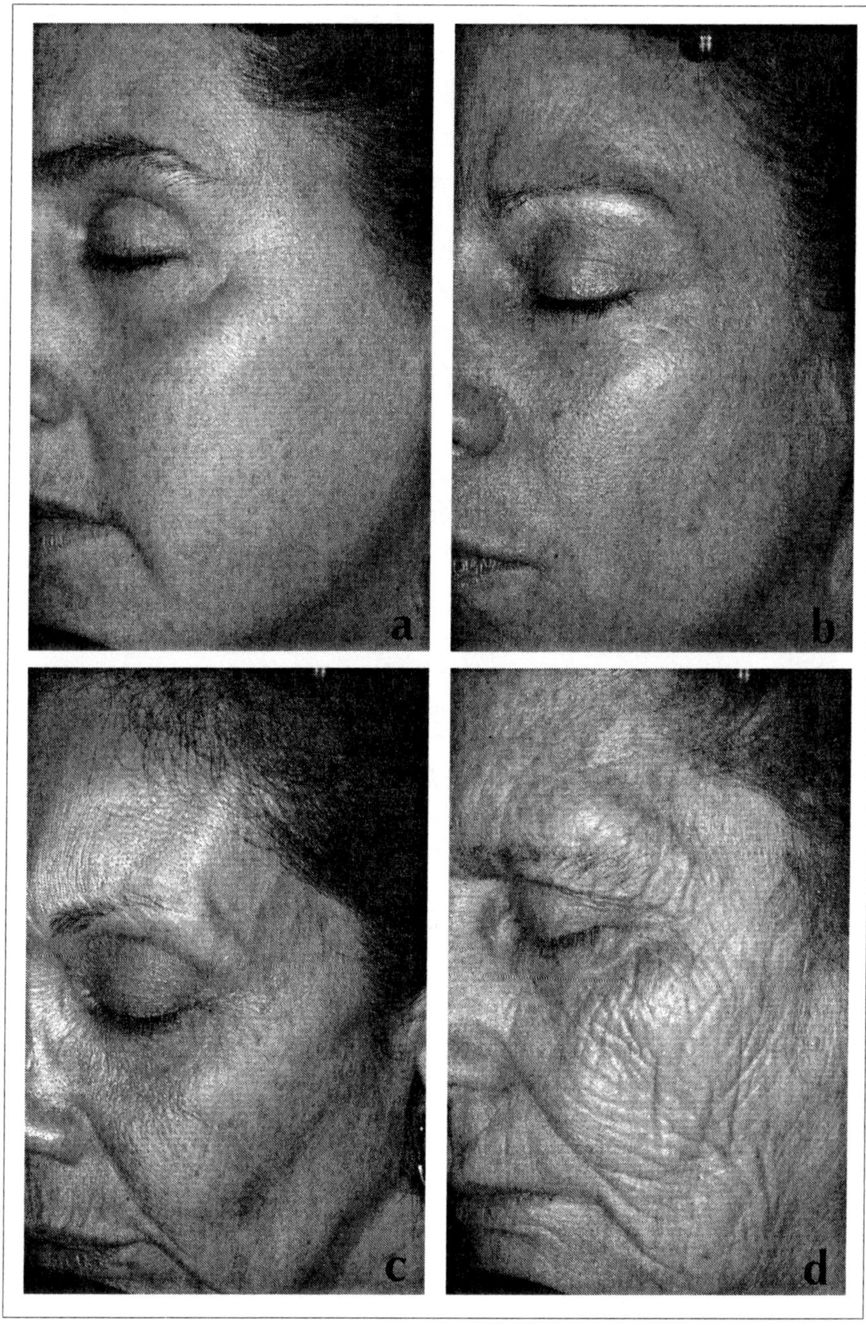

Figure 1. Examples of photoaging grades classified using a 0-9 descriptive scale for Overall Photodamage (OP), Coarse Wrinkles (CW) and Fine Lines (FL). a) OP=3, CW=0, FL=4; b) OP=4, CW=2, FL=5; c) OP=6, CW=4, FL=6; d) OP=8, CW=8, FL=0

scale corresponded to "no changes" between the two images, while six categories to the right and six to the left were used to quantify the change (much better <6>—better—slightly better—no change <0>—slightly better—better—much better <6>).

To aid the retrospective analysis of photographs, attention should be paid to the type of photography used. For example, to study the changes in wrinkles, a "parallel polarized" photography may be better than unfiltered photography.[22-24] A discussion of the different imaging methods may help in deciding which technique would be better for a particular study.

Light Imaging Techniques

Photography is becoming an important part of testing methods, owing to recent advances in photographic systems. Different standardized photographic and lighting techniques are today available that can best highlight specific skin characteristics.

Flash light versus polarized light: We define flash photography as an unfiltered system. It is used to document the general skin appearance (Figure 2a).

Polarized photography, on the other hand, is usually a filtered system.[22-24] It was developed to satisfy the need to enhance the visualization of either surface or sub-surface details. To achieve polarization, most photographic systems place a linear polaroid filter on each of the light sources (usually two) and a rotatable third linear polaroid filter on the camera lens.

In parallel polarization (Figure 2b), the light is reflected by the stratum corneum-air interface, allowing a better discrimination of the skin superficial structures (such as pores, wrinkles, fine lines and glyphics). In this case the filter on the lens is aligned parallel with the flash filters.

In cross (perpendicular) polarization (Figure 2c) the light is instead remitted after traversing the epidermis and papillary dermis, bringing with it information about the deeper skin components (information such as erythema, telangiectasia, hyper- or hypopigmentation). This lighting is achieved by aligning the filter on the lens perpendicularly to the flash filters.

Black-and-white UV/fluorescence photography: Fluorescence photography employs light emitting in the UVA range. As melanin strongly absorbs in the UVA, UVA photography can better detect hyperpigmentations such as freckles and melasma as well as hypopigmented lesions such as vitiligo or idiopathic guttate hypomelanosis. Sites with higher epidermal melanin appear darker compared to the surrounding area.

Particular care should be taken when photographing skin with both hyperpigmentation and inflammation. In fact, hemoglobin also attenuates UVA, and erythema may appear as dark as hyperpigmentation.[26] In subjects with inflammatory conditions, it may be impossible to distinguish on photographs what is inflammation (e.g., acne lesion) and what is hyperpigmentation.

Since UVA photography is monochromatic, its images are more suitable for quantification by image analysis than the color cross-polarized ones. Image analysis may be used to quantify both the darkness of the pigmented spot, as well as its size.[25,26]

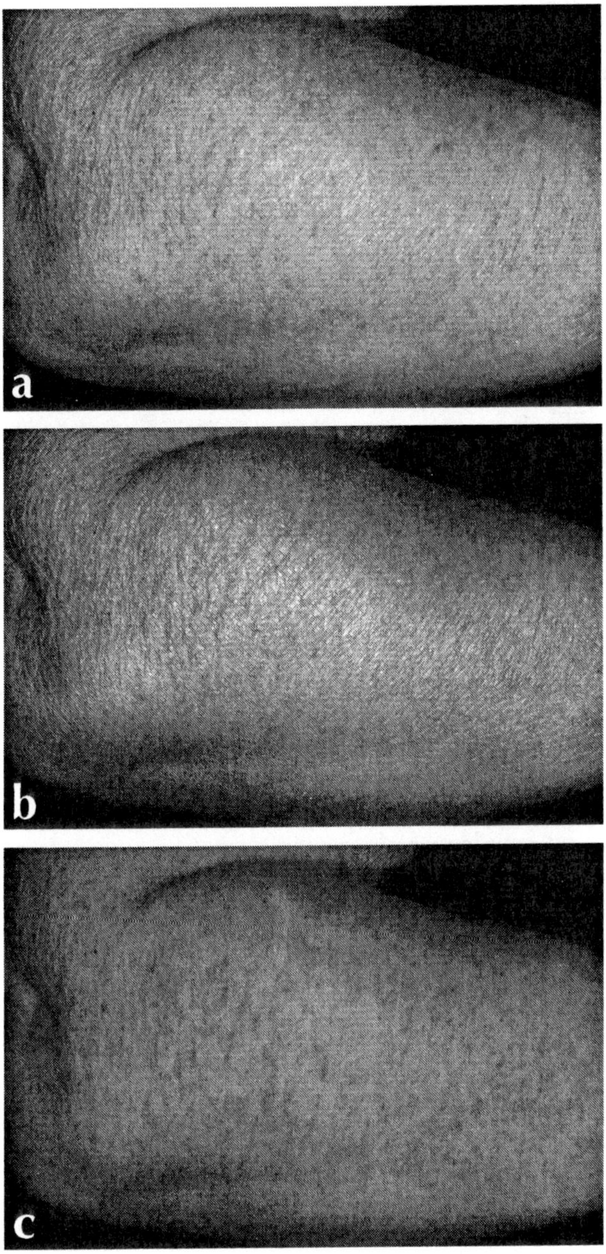

Figure 2. Examples of filtered photography (polarization) of a dorsal forearm. a) Unfiltered photograph; b) Parallel Polarization enhances surface information such as wrinkling and microtopography; c) Cross Polarization highlights subsurface details such as erythema and pigmented spots.

Usually, two filtered light sources (Schott UG1 filter[25]) emitting at a center wavelength of 365 nm (UVA) are positioned symmetrically at the side of the camera body. These can be substituted with Wood's light. The camera lens can be kept unfiltered or it can be fitted with a UVA-cutting filter. If UVA light reflected from the skin is allowed to pass into the camera (unfiltered), wrinkles, pores and other surface details are enhanced together with hyperpigmentation.[25] However, the intensity and delineation of hyperpigmentation is compromised by the surface reflection. Therefore, for capturing pure dyspigmentations such as lentigines, freckles and melasma, a UVA-cutting filter is placed in front of the camera lens.

Digital camera versus 35 mm camera: A question often arises concerning the best type of camera to use for a study: digital or 35 mm. Although professional 35 mm cameras still have overall a better resolution compared to professional digital cameras, the quality and the "risk/benefit ratio" of the latter is such that more and more studies are documented using digital photography. The resolution of two million pixels or higher in professional digital cameras is enough to achieve excellent information.

Advantages of a high-resolution digital camera system are multiple.

- Images are captured, downloaded and viewed within a couple of minutes, increasing significantly the efficiency and turnaround time. Bad images can be immediately retaken.
- Pictures are consistent in color because the development and differences in film lots are avoided.
- Repositioning of subjects, especially during high magnification, is easier because baseline images are pulled up on the computer screen to guide the subject set-up.
- Retrospective evaluation of images is easier and faster.

When using digital cameras, attention must also be paid to the type of software used for storage and database. More sophisticated software, such as the Mirror DPS[a], can have great features, one of which is the tracking of any manipulation that occurred to the stored original picture (image authentication).

Videomicroscopy: High-resolution video-microscopy is used to record ultra-fine in vivo changes in experimental and clinical dermatology.

An example of a UVA-videomicroscope is the Visioscan[b]. Visioscan is a hand-held black-and-white videocamera, which uses UVA light (350-400 nm, peak at 375 nm) to produce a sharp definition of surface morphology. The strong reflection of the UVA at the skin surface allows a clear definition of glyphics (Figure 3) and dry skin. Its best application in aging is in the documentation of changes in microtopography (Figure 3). The camera is easy to operate and to standardize since it works under fixed magnification on a 6 x 8 mm field of view.

[a] Mirror is a registered trademark of Canfield Clinical Systems, Fairfield, New Jersey, USA.
[b] Visioscan is a registered trademark of Courage+Khazaka Electronic GmbH, Cologne, Germany.

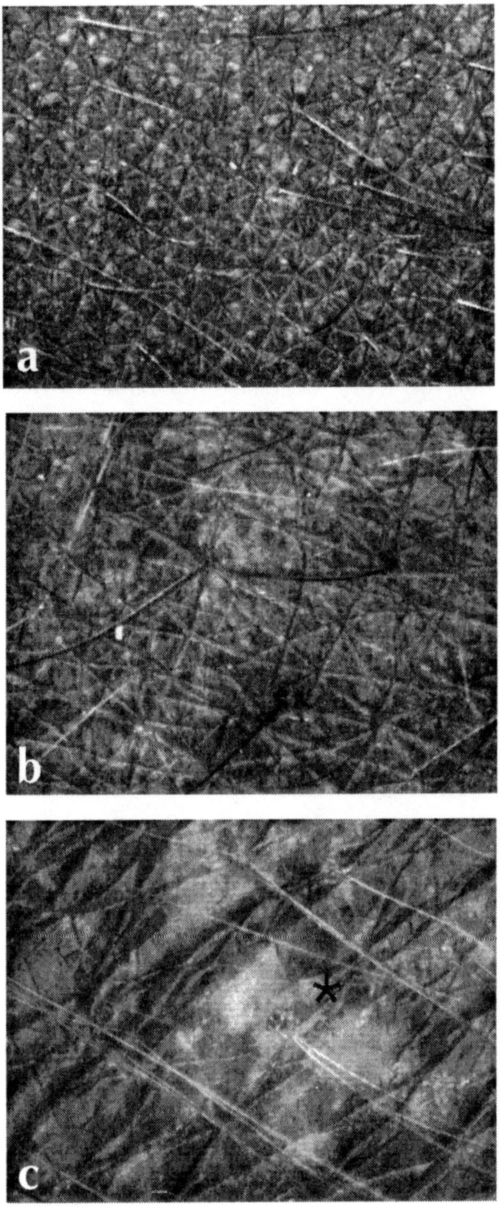

Figure 3. Visioscan images of photodamage changes in dorsal forearms of persons 6 years old (a), 30 years old (b) and 50 years old (c) under the same magnification.

In the young adult (b) only minor changes in glyphics are visible. In the photodamaged person (c) the microtopography is very disorganized, with some glyphics disappearing while others become more pronounced giving a "micro-wrinkled" look to the arms. Also, dyspigmentation (*) is visible.

The use of color videomicroscopy has brought new insights in the study of the skin under magnification in vivo.[27] Several types of videomicroscopes are today available with magnifications ranging from 1x to approximately 700x or more.

In selecting a videomicroscope it is important to consider the resolution, the polarization system and the clarity of the images. The most useful range of magnification is 20x-80x. In photoaging, videomicroscopy is used to study hyperpigmented lesions under cross polarization and microtopography under parallel polarization mode.

Three-dimensional in vivo measuring system: Among methodologies for the measurement of skin topography, a new 3-D in vivo scanning system seems very promising.[16] Quantification of fine lines and wrinkles has usually relied upon the silicon replica technique and its evaluation by "profilometry." The need for a quantification of surface structures in vivo and real-time led to the development of PRIMOS[c] (Phaseshift Rapid In vivo Measurement Of Skin).

PRIMOS allows "contact-free" analysis of different measuring areas of the skin with a resolution of 800 x 600 or 1024 x 768 micro mirrors. It is based on the principle of the stripe projection technique, where an optical parallel stripe pattern is projected from a digital micromirror projector onto the skin area to be measured and then displayed on the chip of a CCD-matrix camera. If the skin surface presents even small differences in height, this will cause deflection of the parallel stripes, which can be quantified for the respective height profile. Quantitative assessment of roughness is usually obtained from the 2D-profile cuts using dedicated software. Additionally, by enlargement or reduction of the projection area, it is possible to make 3D-measurements of smaller or larger areas (from 4 x 4 to 100 x 100 mm), for wounds, scars, nevi, wrinkles, limbs or even larger body parts.[16]

Conclusion

Documenting photodamage changes by photographic or imaging means has reached new levels of sophistication. These have greatly reduced the subjectiveness of photodamage grading systems and helped to identify true skin amelioration with topical treatments. A future goal will be the decrease in cost of these state-of-the-art imaging systems so that they will become standard methodology in antiaging trials.

—**A. Pagnoni,** *Hill Top Research, Inc., Milltown, New Jersey, USA*

[c] PRIMOS is a product of GF Messtechnik GmbH, Teltow, Germany.

References

1. CEM Griffith, TS Wang, TA Hamilton, JJ Voorhees and CN Ellis, A photonumeric scale for the assessment of cutaneous photodamage, *Arch Dermatol* 128 347-351 (1992)
2. C Larnier, JP Ortonne, A Venot, B Faivre, JC Beani, P Thomas, TC Brown and E Sendagorta, Evaluation of cutaneous photodamage using a photographic scale, *Br J Dermatol* 130 167-173 (1994)

3. D Piacquadio, M Dobry, S Hunt, C Andree, G Grove, KA Hollenbach, Short contact 70% glycolic acid peels as a treatment for photodamage skin. A pilot study, *Dermatol Surg* 22 449-452 (1996)
4. CEM Griffiths, S Kang, CN Ellis, KJ Kim, LJ Finkel, LC Ortiz-Ferrer, GM White, TA Hamilton and JJ Voorhees, Two concentrations of topical tretinoin (retinoic acid) cause similar improvement of photoaging but different degrees of irritation, *Arch Dermatol* 131 1037-1044 (1995)
5. BM Burke and WJ Cunliffe, The assessment of acne vulgaris: the Leeds technique, *Br J Dermatol* 111 83-92 (1983)
6. RB Armstrong et al, Clinical panel assessment of photodamage skin treated with isotretinoin using photographs, *Arch Dermatol*, 128 352-356 (1992)
7. S Maddin, J Lauharanta, P Agache, L Burrows, M Zultak, L Bulger, Isotretinoin improves the appearance of photodamaged skin: Results of a 36-week, multicenter, double-blind, placebo-controlled trial, *J Am Acad Dermatol* 42 56-63 (2000)
8. BA Gilchrest, A review of skin aging and its medical therapy, *Br J Dermatol* 135 867-975 (1996)
9. CR Taylor, RS Stern, JJ Leyden and BA Gilchrest, Photoaging/photodamage and photoprotection, *J Am Acad Dermatol* 22 1-15 (1990)
10. BA Gilchrest and M Yaar, Ageing and photoageing of the skin: observation at the cellular and molecular level, *Br J Dermatol* 127 25-30 (1992)
11. A Pagnoni, AM Kligman, I Sadiq and T Stoudemayer, Hypopigmented macules of photodamaged skin and their treatment with topical tretinoin, *Acta Dermato-Venereologica* 79 305-310 (1999)
12. JP Ortonne, Pigmentary disorders associated with sun exposure, *J Dermatol Treatment* Suppl 2 S7-S8 (1996)
13. AM Kligman, The classification and treatment of wrinkles, in *Cutaneous Aging*, AM Kligman and Y Takase, eds, Tokyo: University of Tokyo Press (1988) pp 547-555
14. R Marks, Aging and Photodamage, in *Sun-Damaged Skin*, R Marks, ed, London: Martin Dunitz (1992) pp 13-34
15. A six-month clinical study to evaluate the long-term efficacy and safety of an AHA lotion, *Cosmet Derm* 9 33-40 (1996)
16. S Jaspers, H Hopermann, G Sauermann, U Hoppe, R Lunderstadt and J Ennen, Rapid in vivo measurement of the topography of human skin by active image triangulation using a digital micromirror device, *Skin Res Technol* 5 195-207 (1999)
17. J Sefton, AK Kligman, SC Kopper, JC Lue and JR Gibson, Photodamage pilot study: A double-blind, vehicle-controlled study to assess the efficacy and safety of tazarotene 0.1% gel, *J Am Acad Dermatol* 43 656-63 (2000)
18. CK Kimbrough-Green et al, Topical retinoic acid (Tretinoin) for melasma in black patients, *Arch Dermatol* 130 727-733 (1994)
19. N Lawrence, SE Cox and HJ Brody, Treatment of melasma with Jessner's solution versus glycolic acid: a comparison of clinical efficacy and evaluation of the predictive ability of Wood's light examination, *J Am Acad Dermatol* 36 589-593 (1997)
20. RG Glogau, Aesthetic and anatomic analysis of the aging skin, *Semin Cutan Med Surg* 15(3) 134-138 (1996)
21. LM Harnisch, MK Raheja, LK Lockhart, A Lopez and A Gabbianelli, Substantiating antiaging product claims, *Cosmet Toil* 114(10) 33-47 (1999)
22. JA Muccini, N Kollias, SB Phillips, RR Anderson, AJ Sober, MJ Stiller and LA Drake, Polarized light photography in the evaluation of photoaging, *J Am Acad Dermatol* 33 765-769 (1995)
23. J Philp, NJ Carter and CP Lenn, Improved optical discrimination of skin with polarized light, *J Soc Cosmet Chem* 39 121-132 (1988)
24. RR Anderson, Polarized light examination and photography of the skin, *Arch Dermatol* 127(7) 1000-1005 (1991)

25. N Kollias, R Gillies, C Cohen-Goihman, SB Phillips, JA Muccini, MJ Stiller and LA Drake, Fluorescence photography in the evaluation of hyperpigmentation in photodamaged skin, *J Am Acad Dermatol* 36 226-230 (1997)
26. A Pagnoni, AM Kligman and T Stoudemayer, UVA photography to monitor the progression of freckles in young children, 55th meeting, American Academy of Dermatology, P 362, Mar 21-26, 1997
27. PL Dorogi and EM Jackson, In vivo video microscopy of human skin using polarized light, *J Toxicol - Cut & Ocular Toxicol* 13 97-107 (1994)

Determination of the In Vitro SPF

Keywords: testing and instrumentation, sunscreen, SPF, spectral transmittance, erythema-effective irradiance, polymethylmethacrylate, UVB

A ring test at six European test centers showed that SPF evaluations with good reproducibility and comparability could be obtained by an in vitro protocol measuring either transmittance or erythema-effective irradiance of product samples applied on PMMA plates.

Nowadays sun care products are highly technical and sophisticated cosmetic products. All basic ingredients, such as filter systems or additives, must be carefully selected not only for their performance in both the UVA and UVB spectrum, but also regarding their influence on photostability, on the substantivity on the skin (water resistance) and on the rheological behavior. The goal is to find synergies between the filter system and the vehicle after spreading on the skin. Then the technical requirements (such as good dispersion of inorganic powder, an optimized particle size and the solubility of organic filters) must be optimized in a sun care product that has nice-feeling properties and a homogenous film applied to the skin.[1-4]

Reproducibility of the batch and long-term stability are additional parameters of increasing interest concerning regulatory compliance, safety and communication media.

The UVB protection of a sun care product is given by the Sun Protection Factor (SPF). For economical, practical and ethical reasons a reliable in vitro measurement of the SPF would be very welcome as a supplement – not as a replacement – of the in vivo SPF measurement. The in vitro SPF is particularly useful for screening test during product development.[5-8]

This article is the result of a scientific collaboration between different industrial laboratories and testing institutes in order to optimize an in vitro SPF method. The method calls for measuring either the spectral transmittance or the erythema-effective irradiance of a sunscreen product sample applied on polymethylmethacrylate (PMMA) plates. The goal of the collaboration is to achieve a common procedure for a reproducible, reliable and sensitive method for a maximum accuracy in transmittance measurements.

The method was checked by using different instruments, commercial sun care products, emulsion types, filter systems and photostability behaviors.

Table 1. Colipa guidelines for in vivo SPF tests

Parameter	Colipa Test (1994)
UV source	Artificial lamp source. Output defined by distribution of erythemal effectiveness between 290 & 400 nm based on standard sun 40° latitude.
Product quantity and application	2 mg/cm², minimum area of 35 cm²; prescribed method of weighing by loss and application technique for uniform coverage. Product drying time 15 min.
Test subjects	Minimum number = 10; actual number determined by statistical variation (no more than 20).
	Skin types I, II, and III according to Fitzpatrick.
Test area	Region of the back between the scapula and the waistline
	Irradiation sites minimum area = 0.4 cm²; recommend 1 cm².
MED determination	Constant flux variable time
	Minimum of five sites
	MED read after 20 ± 4 h
	1.25x increments between irradiation doses.
	MED of protected skin determined simultaneously with unprotected skin.
Calculation of SPF	The calculated mean SPF is the arithmetic mean of the individual SPF values.
	95% confidence interval applied to the calculated mean ± 20%

Methods and Materials

A ring test was carried out in six European test centers using two SPF calculation approaches and four measuring devices on 10 commercial sunscreens whose SPF had already been independently determined by in vivo techniques. Different types of measuring devices were included in the ring test in order to examine the possible influences of measuring techniques. PMMA plates were used as a substrate to which we applied a homogenous amount of the sun care products to be tested. Two ring tests with different test product amounts were performed – one with 1.4 mg/cm² and one with 1.2 mg/cm². The results of these two ring tests were compared with the in vivo SPFs determined separately.

Principle of in vivo SPF measurement: The in vivo measurement of the SPF is described in a variety of national and international specifications.[9-16] The major elements of the SPF tests published by Colipa are shown in Table 1.

In this study, we selected different commercial products that had been tested previously following the Colipa Guidelines[9] in different European institutes. The in vivo mean values shown in Table 2 derive from several new readings at multiple laboratories.

It is now well known that the test results according to these guidelines can vary by 20% among test subjects (due to differences of skin biology of the test subjects) and by up to 40% when the test is performed at different labs (due to differences in the way the test procedures are executed at the various laboratories).[17]

Principle of in vitro SPF measurement: The in vitro SPF measurement considers the sunscreen as a filter, containing both organic and inorganic filter substances, that protects the skin against UV radiation. There are two methods of calculating SPF. One requires the measurement of the product's spectral transmittance, from which the attenuated erythema-effective irradiance can be calculated. The other directly measures the product's attenuated erythema-effective irradiance by using an integral detector.

In Method 1, the SPF by spectral transmittance, as reported by Diffey and Robson,[18] is defined by equation 1:

$$SPF = \frac{\int_{290}^{400} E_\lambda s_{\lambda,er} d\lambda}{\int_{290}^{400} E_\lambda \tau_\lambda s_{\lambda,er} d\lambda} \qquad (1)$$

where E_λ is the spectral irradiance of "standard sun" corresponding to COLIPA "SPF test method" or DIN 67501; $s_{\lambda,er}$ is the erythema action spectrum corresponding to CIE publication No. 90 (1991) or DIN 5031-10; and τ_λ is the spectral transmittance of the sunscreen.

E_λ and $s_{\lambda,er}$ are known and tabulated values. With the help of the measured spectral transmittance τ_λ, the SPF can be calculated via equation 1. Since the dimensions of E_λ in equation 1 are energy x area^{-1} x time^{-1} and those of $s_{\lambda,er}$ are erythema x energy^{-1} x area, and since τ_λ is the dimensionless ratio of transmitted to incident intensity, both the numerator and denominator have dimensions of erythema per unit of time (erythema effective irradiance).

Because most sunscreens do not only absorb but also scatter radiation, care must be taken that when the spectral transmittance τ_λ is measured, the radiation scattered by the sample is also included. Usually, this can be accomplished by using an integrating sphere.

In Method 2, the SPF by directly measuring erythema-effective irradiance, as reported by Diffey and Robson,[18] is defined by equation 2:

$$SPF = \frac{E_{er} \text{ without sunscreen}}{E_{er} \text{ with sunscreen}} \qquad (2)$$

where E_{er} is the erythema-effective irradiance. Equation 2 suggests to expose the sample to the "standard sun" E_λ and to measure the erythema-effective irradiance

$$E_{er} = \int_{290}^{400} E_\lambda s_{\lambda,er} d\lambda \qquad (3)$$

without and with the sunscreen. This is in line with the principle of the in vivo measurement of the SPF. The procedure requires an integral detector whose spectral sensitivity is well adjusted to the erythema action spectrum $s_{\lambda,er}$ of the skin.[19] This method makes it possible to observe any changes of the SPF during the radiation process. The sample is continuously irradiated by simulated solar radiation in the UV range and the erythema-effective irradiance behind the sample is continuously monitored.

Tested products: The 10 commercial products used in this ring test are listed in Table 2. One additional product, G, was initially included among the tested products, but it is no longer commercially available so its data is not presented here.

Sample preparation: Of utmost importance for both reproducibility and comparability of the results from any in vitro procedure for the determination of the SPF is the sample preparation.

For this ring test, UV-transmitting PMMA plates[a] (50 mm x 50 mm) with a 5 µm medium roughness were used as a substrate.[20] The plate's roughness differs from that of human skin. In order to ensure a reproducible and homogeneous product film and due to the physical limits of transmission measurement, we reduced the amount of applied cream.

The sun care product was spotted evenly on the plate's surface. The applied amount was weighed before spreading or evaporation occurred.

In the first ring test, amounts of 1.4 mg/cm^2 per test product were used. In the second ring test the applied amount was reduced to 1.2 mg/cm^2.

By using light pressure, the product was immediately spread over the whole plate surface until a homogeneous distribution was achieved.

The sample thus obtained was allowed to settle for 15 min at room temperature to ensure a self-leveling of formulation.

A roughened PMMA plate with homogeneous layer of glycerin was used as reference. The transmittance was measured in the UV range (290-400 nm) in 1 nm steps. Up to 9 UV transmittance spectra from 290 to 400 nm were taken from each substrate at different locations except for those apparatuses that measure the whole area of the plate.

Five different plates were used for each sunscreen tested (except Lab 6) for calculating average SPF data.

Labs and instruments: Table 3 lists the labs and the devices they used to measure transmittance (or the erythema-effective irradiance in the case of Lab 6). The instruments differ according to how they include the scattered radiation and what they use for detectors.

Results

First ring test: For the first ring test an application amount of 1.4 mg/cm^2 was dropped on the plate and allowed to spread. This amount enables homogeneous spreading of the product on this plate. The results obtained are outlined in Table 4.

[a] Helioplates, from Helioscience, Creil, France

Table 2. The tested sunscreen products

Product	In vivo SPF*	Filtering composition
A	6 ± 1.2	ethylhexyl methoxycinnamate butyl methoxydibenzoylmethane 4-methylbenzylidene camphor
B	7 ± 1.4	butyl methoxydibenzoylmethane methylene bis-benzotriazolyle tetramethylbutylphenol
D	12 ± 2.4	butyl methoxydibenzoylmethane ethylhexyl methoxycinnamate
F	17 ± 3.4	ethylhexyl methoxycinnamate benzophenone-3 ethylhexyl salicylate butyl methoxydibenzoylmethane
J	18 ± 3.6	ethylhexyl methoxycinnamate ethylhexyl salicylate butyl methoxydibenzoylmethane
I	37 ± 7.4	microtitanium dioxide ethylhexyl methoxycinnamate benzophenone-3
H	38 ± 7.6	benzophenone-3 butyl methoxydibenzoylmethane 4-methylbenzylidene camphor diethylhexyl butamido triazone phenylbenzimidazole sulfonic acid
K	40 ± 8	ethylhexyl methoxycinnamate isoamyl p-methoxycinnamate ethylhexyl salicylate benzophenone-3 phenylbenzimidazole sulfonic acid butyl methoxydibenzoylmethane methylene bis-benzotriazolyle tetramethylbutylphenol
E	58 ± 11.6	microtitanium dioxide ethylhexyl methoxycinnamate benzophenone-3 ethylhexyl salicylate
C	79 ± 15.8	butyl methoxydibenzoylmethane octyl triazone octocrylene ethylhexyl methoxycinnamate

* 20 volunteers ± 20%

The analysis of Table 4 shows that despite the large difference in the level of in vivo protection (SPF 6-79), the results obtained from one laboratory to another are generally close regardless of the apparatus used. For most of the products it can be noted that the values obtained by the laboratories show a good reproducibility except for products B, E, H.

Table 3. Labs and instruments

Instrument	Lab	Type	Integrating device	Detection system
OL754[a]	1	spectroradiometer	integrating sphere behind sample	double monochromator PMT
UVIKON 933[b]	3,4	spectrophotometer	integrating sphere behind sample	single monochromator PMT
UV 1000S[c]	2,5	spectrophotometer	integrating sphere before sample	Diode array
Sunscreen Tester[d]	6	photometer	scattering dome behind sample	integrating detector with spectral sensitivity adjusted to erythema action spectrum

[a] Optronic OL754, Optronics Laboratories Inc, Orlando, Florida, USA
[b] Kontron UVIKON, UVK-LAB Service, Trappes, France
[c] Labsphere UV 1000S, Labsphere, North Sutton, New Hampshire, USA
[d] Kockott UV technik, UV technik, Hanau-Steinheim, Germany

Table 4. Results of the in vivo SPF measurements and the in vitro SPF measurements for products A-K (excluding product G) at 1.4 mg/cm²

	A	B	C	D	E	F	H	I	J	K
In vivo	6	7	79	12	58	17	38	37	18	40
Lab 1	8	6	54	12	67	23	58	46	59	78
Lab 2	8	5	46	10	56	20	46	36	46	74
Lab 3	8	4	42	10	57	16	31	31	42	59
Lab 4	8	8	50	15	103	18	39	49	39	95
Lab 5	8	9	57	15						
Mean	8	6	50	12	71	19	44	40	47	77
SD%	1	32	12	21	31	16	26	21	19	19

Representations in transmittance (T) have a very small range. It is a better and common way to illustrate the results in absorbance (–logT) for this type of experiment. In that way it should also be noted that the absorption spectra are very close to this wavelength range even for high absorbing products, as shown in Figure 1.

Table 5. Results of the in vivo SPF measurements and the in vitro SPF measurements for products A-K (excluding product G) at 1.2 mg/cm²

	A	B	C	D	E	F	H	I	J	K
In vivo	6	7	79	12	58	17	38	37	18	40
Lab 2	6	4	28	7	37	16	20	21	19	37
Lab 3	4	3	32	5	31	12	23	22	21	34
Lab 4	6	4	33	5	69	17	24	28	21	37
Lab 5	6	6	40	10	59	17	36	40	31	41
Lab 6	6	5	43	8	50	13	30	44	25	28
Mean	6	4	35	7	49	15	27	31	24	36
SD%	17	29	17	34	32	16	23	34	20	14

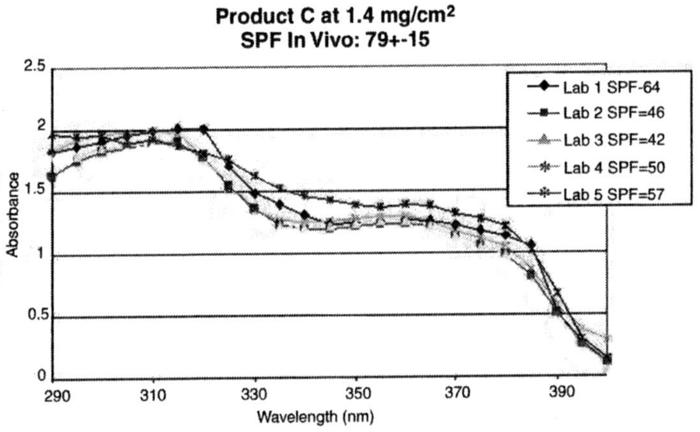

Figure 1. Absorption spectrum of product C (in vivo SPF 79 ± 15) at 1.4 mg/cm²

Second ring test: In the first ring test the results showed a good reproducibility between different laboratories. Nevertheless the in vitro SPF values were often higher than the in vivo SPFs of the tested sun care products. Therefore, in the second ring test an amount of 1.2 mg/cm² (instead of 1.4 mg/cm²) was used to achieve a better comparability with the in vivo SPF determinations. Additionally a new apparatus (Sunscreen Tester) at Lab 6 was included in the ring test, allowing continuous radiation of the products. Finally, for technical reasons, Lab 1 could not participate in the second ring test.

As shown in Table 5, the inter-laboratory reproducibility of the second ring test was comparable to that of the first ring test.

Figures 2 and 3 give more information about the inter-laboratory differences during the second ring test. Here the sun care products were divided into two

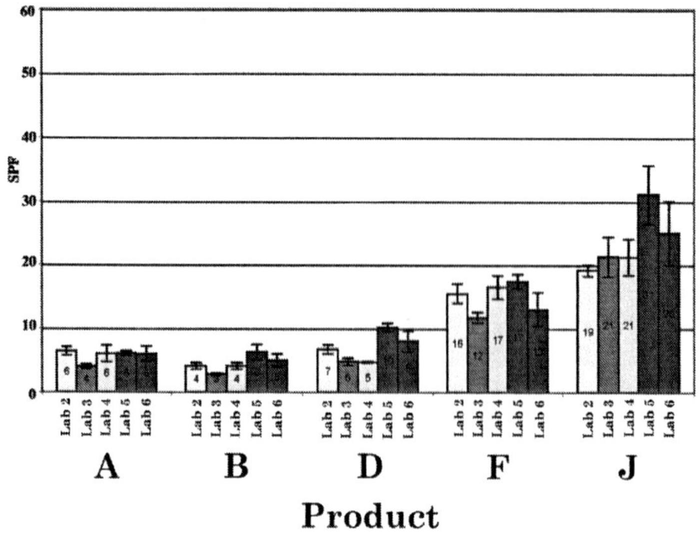

Figure 2. Inter-laboratory comparison of measured SPF values for low-SPF sunscreen products applied at 1.2 mg/cm²

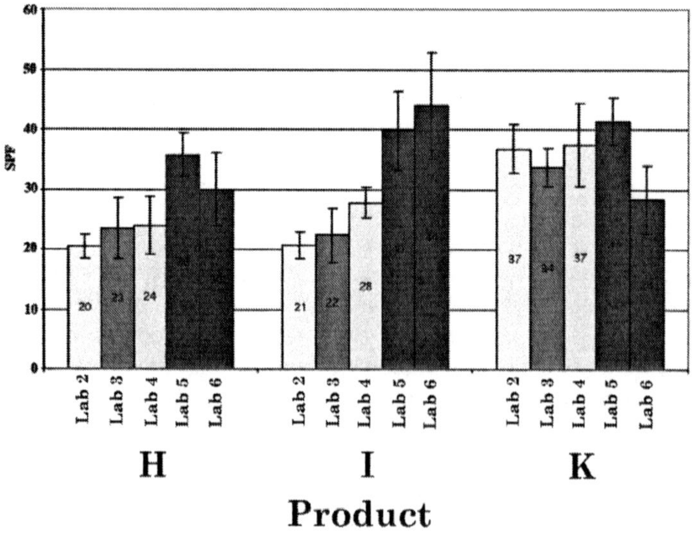

Figure 3. Inter-laboratory comparison of measured SPF values for high-SPF sunscreen products applied at 1.2 mg/cm²

groups, those of lower SPF levels (Figure 2) and those of higher SPF levels (Figure 3).

As shown in Figure 2, all laboratories have similar results for these lower SPF ranges. For in vitro determinations from SPF 20-30+ (Figure 3) the inter-laboratory differences increase, but remain in an acceptable range.

Comparison of in vitro and in vivo SPF results: In Figure 4 the means of the in vitro SPF results, obtained in the second ring test, are compared to the means of the in vivo SPF values. All in vivo SPF tests mentioned in this article were examined by different laboratories out of this ring test. The standard deviations for the in vivo results are related to the specific laboratory. By contrast the standard deviations for the in vitro results are based on the results of five different laboratories for each product.

During this ring test, two particular sun care products (Figure 5) showed a different behavior in the examination.

First, the product C in vitro SPF, which showed good inter-laboratory reproducibility, did not compare well to the in vivo SPF. This could be explained by the presence of special biological active ingredients in the formula, with "anti-inflammatory" anti-redness effects.

Second, product E in vitro SPF shows a low inter-laboratory reproducibility, but the mean value is close to the in vivo SPF. The analysis of this product showed a physical instability that could explain the large variability in these results.

Comparison of in vitro and in vivo SPF results: Another aim of the ring test was to study the correlation between the mean in vitro SPF measured by the different laboratories and the SPF values obtained in vivo. The results in Figure 6 show a clear advantage of the procedure chosen in the second ring test with the reduced amount of product application. The correlation to the in vivo SPFs could be raised from R^2=0.77 to R^2=0.93.

The method in this study achieved quite good correlation with the in vivo values. Nevertheless, no analysis of individual results compared to the accepted human SPF results could be investigated for the moment because precise standard in vivo values do not yet exist.

Discussion

The results given in this work demonstrate the possibility of in vitro determination of a sun care product's UVB protection.

The aim of the ring test was to find an appropriate protocol for the in vitro SPF determination. Special interest was given in the use of different devices (as shown in Table 2) and different test products (Table 1).

The PMMA plates are useful substrates for in vitro tests, as reported also from other groups.[21,22] The reduction of the applied amount of product from 1.4 mg/cm^2 to 1.2 mg/cm^2 led to an improvement of correlation to the in vivo measurements. According to guidelines, 2 mg/cm^2 are applied for in vivo determination on human skin. The PMMA plates and the human skin are chemically and physically different (mainly surface roughness is not the same); this can explain why 1.2 mg/cm^2 in vitro achieve about the same result as 2 mg/cm^2 in vivo.

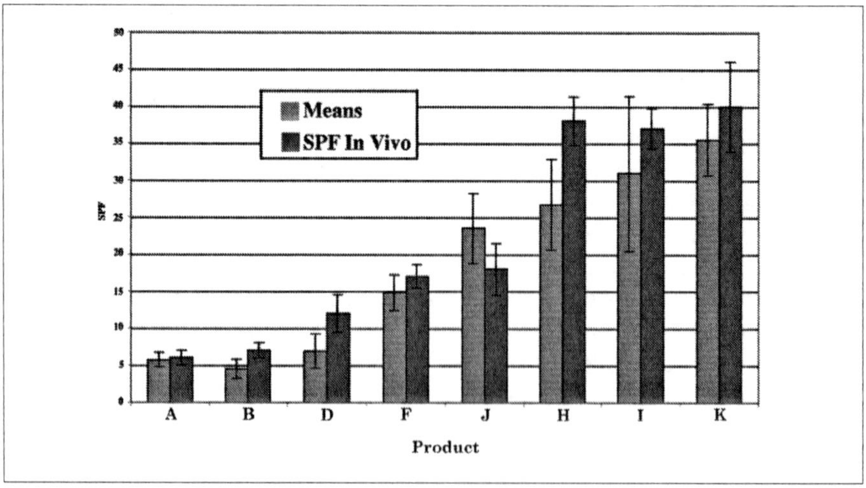

Figure 4. Comparison of the in vivo SPF measurements and the in vitro SPF measurements at 1.2 mg/cm²

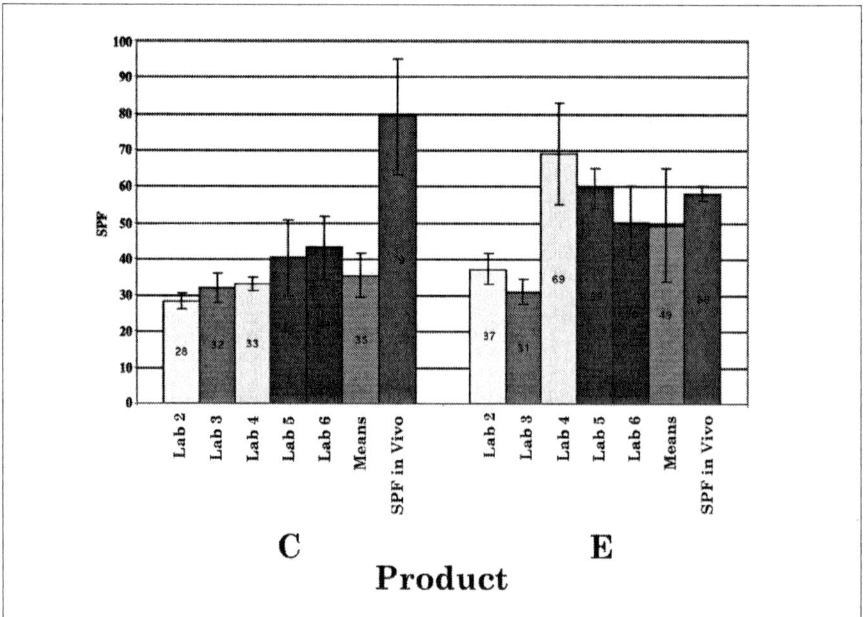

Figure 5. Particular in vivo SPF measurements and the in vitro SPF measurements at 1.2 mg.cm²

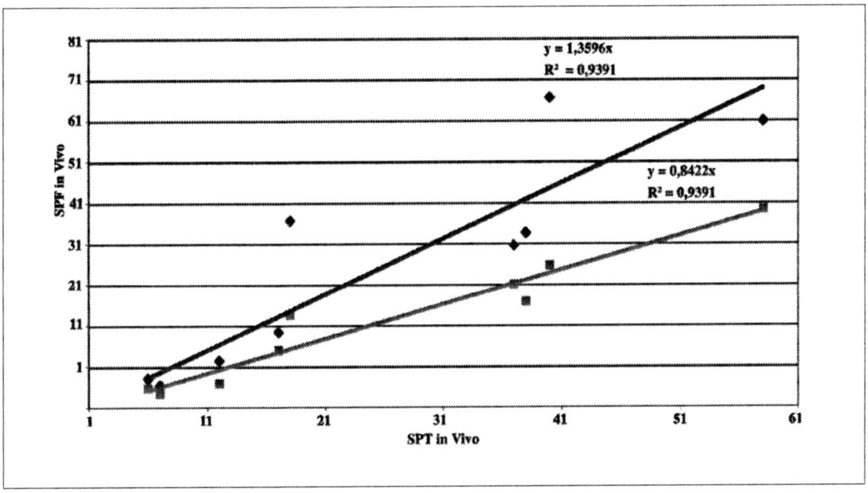

Figure 6. Correlation between in vitro and in vivo SPF measurements for the first ring test (1.4 mg/cm^2) (upper curve) and for the second ring test (1.2 mg/cm^2)

It should be mentioned here that in vivo examinations, especially for products with higher SPF values, can be characterized by a certain variability due to the biological variations of the volunteers. International standards, such as the Colipa SPF test method,[9] accept a deviation of 20% during the tests.

There is an advantage of the presented method for 1.2 mg/cm^2 of slightly underestimating the SPF in vivo, while remaining within the margin of error proposed by the Colipa (20%). This principle of precaution makes it possible to select products that should be tested in vivo before marketing them.

Conclusion

The presented method is a practical, convenient and relevant tool for in vitro SPF screening for sun care and skin care products. It also permits to characterize heterogeneities of chemical and/or physical filter dispersion of a product by analyzing the standard deviations. Additional information of product stability and photo stability can be obtained.

The reliability of this method allows good comparison between different products or batches, as well as different time stoking, to ensure the endurance of its properties. This method has been proved as a quick, inexpensive and easily applicable method for use as a convenient protocol in different laboratories regardless of which device has been used. Finally, the quite good correlation of this method – when 1.2 mg/cm^2 was applied – with in vivo values helps to enhance the quality and the efficacy of sun care products and therefore the safety for the consumers. To optimize this correlation, a new inter-laboratory ring test with standard products should be used to calibrate the uncertainty.

—M. Pissavini and L. Ferrero
Coty-Lancaster, International Research & Development Center, Monaco

—**V. Alard,** *LVMH Recherche, branche Parfums et Cosmétique, St. Jean de Baye, France*
—**U. Heinrich and H. Tronnier,** *Dermatronier, Institut für Experimentelle Dermatologie, Universität Witten/Herdecke, Witten, Germany*
—**D. Kockott,** *UV-Technik, Hanau-Steinheim, Germany*
—**D. Lutz,** *Helioscience, Marseille, France*
—**V. Tournier,** *Dipta, Aix-en-Provence Cedex, France*
—**M. Zambonin,** *Eurochem Ricerche, Mestrino Podova, Italy*
—**M. Meloni,** *VitroScreen, Milan, Italy*

References

1. FJ Moloney, S Collins and GM Murphy, Sunscreens: safety, efficacy and appropriate use, *Am J Clin Dermatol* 3(3) 185-191 (2002)
2. BL Diffey, Sources and measurement of ultraviolet radiation, *Methods* 28(1) 4-13 (2002)
3. FP Gasparro, M Mitchnick and JF Nash, A review of sunscreen safety and efficacy, *Photochem Photobiol* 68(3) 243-256 (1998)
4. L Scherschun and HW Lim, Photoprotection by sunscreens, *Am J Clin Dermatol* 2(3) 131-134 (2001)
5. RM Sayre, PP Agin, GJ LeVee and E Marlowe, A comparison of in vivo and in vitro testing of sunscreening formulas, *Photochem Photobiol* 29 559-66 (1979)
6. RM Sayre, Correlation of in vivo tests, in vitro SPF predictions, *Cosmet Toil* 108(2) 111-114 (1993)
7. GA Groves, PP Agin and RM Sayre, In vitro and in vivo methods to define sunscreen protection, *Aust J Dermatol* 20 112-119 (1979)
8. JC Hubaud, L Magnani, A Denis and S Magnan, Rapid in vitro screening method to quantify the protection of sunscreen against the UVA, *Proceedings of the European UV Sunfilters Conference, Paris, France* (1998)
9. COLIPA SPF test method (The European Standard) (Méthode Colipa de détermination du Facteur de protection solaire, version française, FIP) (1994)
10. Department of Health and Human Services, FDA (USA), Sunscreen drug products for over-the-counter human use, *Federal Register* 64(98) 27666-27693 (1999)
11. JCIA, Japan Cosmetic Association Standard Sun Protection Factor test method and Japan Cosmetic Industry Association measurement standards for UVA protection efficacy (1999)
12. Standards Australia, Standards New Zealand, Sunscreen products, evaluation and classification, AS/NZS 2604 (1998)
13. South African Bureau of Standards, Standard specifications, sunscreen products, SABS 1557(1992)
14. Commission Internationale de l'Eclairage (CIE), Sunscreen testing (UVB), Technical report number CIE 90 (1991)
15. Deutsches Institut für Normung (DIN), Experimental evaluation of the protection from erythema of external sunscreen products for the human skin, DIN 67501 (1999)
16. Önorm, Sonnenschutzmittel. Bestimmung des Lichtschutzfactors im Labor, Önorm S1130 (Jan 5, 1999)
17. MW Brown, The Sun Protection Factor: Test methods and legal aspects, SÖFW 128 10-18 (2002)
18. BL Diffey and J Robson, A new substrate to measure sunscreen protection factors throughout the ultraviolet spectrum, *J Soc Cosmet Chem* 40 127-33 (1989)
19. D Kockott, H Tronnier, B Hölzner and U Heinrich, Automatic in vitro evaluation of suncare products, 21st I*FSCC Congress Berlin, Proceedings* (2000)
20. D Lutz, Contribution to measuring in vitro protection; some keys for the reliability of sunscreen spectrophotometric evaluation, *Proceedings of the European UV Sunfilters Conference, Paris, France* (1999) 26-39
21. H Gers-Barlag et al, In vitro testing to assess the UVA protection performance of suncare products, *Int J Cosmet Sci* 23 3-14 (2001)
22. L Ferrero, M Pissavini, S Marguerie and L Zastrow, Sunscreen in vitro spectroscopy application to UVA protection assessment and correlation with in vivo persistent pigment darkening, *Int J Cosmet Sci* 24 63-70 (2002)

Quality Comparison of w/o and o/w Photoprotection Creams

Keywords: emulsions, photoprotection, melanin, microbial control, physical-chemical stability, sensory appeal

The authors determined the quality of a photoprotector w/o cream by evaluating its microbial control, physical-chemical stability and sensory appeal in comparison to an o/w photoprotector cream.

An emulsion used in photoprotection creams, like any emulsion, is a heterogeneous system made up of at least an immiscible liquid dispersed in another liquid in the form of small drops. These systems have only minimal stability that can be increased by adding substances such as surfactants or finely divided solids.

It is possible to classify the emulsions into two distinct types—water-in-oil (w/o) and oil-in-water (o/w)—according to the nature of the respective dispersed phase because water and oil, as well as a lipo-soluble substance, are the two basic components of an emulsion.[1]

Sensorial tests allow the measurement of how much volunteers like or dislike a certain product. They also enable the identification of the presence or absence of perceptible sensorial differences in characteristics such as the flavor, the texture, the color and the global impression, the spreading and the hydration sensation.[2]

In this article we describe how we determined the quality of a photoprotector w/o cream by evaluating its microbial control, physical-chemical stability and sensory appeal in comparison to an o/w photoprotector cream.

Photoprotection

The melanin distribution inside the epidermis is one of the most important factors in protecting the skin against sun light-induced chronic damages such as skin aging and cancer. Nevertheless, melanin is a very poor classical solar filter, because it presents very little sun protection factor (SPF) in concentrations that can be considered biologically useful.

Melanin is a bio-chemically non-reactive free radical. It is unique in its capacity to neutralize the free radicals produced in the skin when skin is exposed to sunlight.[3] Because it is stable thermally and photo chemically, melanin is not degraded by enzymes. When topically applied, it neutralizes and removes the free radicals of the skin, besides working as an antioxidant. Its antioxidant properties are comparable to those of tocopherol because it is a water-soluble polyphenol and consequently stable under light and heat. However, melanin cannot be used as a replacement to traditional antioxidants, but only together with them.[3]

Formulas 1 and 2. Proposed w/o photoprotective cream (Formula 1) and an o/w photoprotective cream (Formula 2)

Raw material	Formula 1 w/o	Formula 2 o/w
Cetearyl glucoside/cetearyl OH	7.50%	3.00%
Ceteareth 20	-	3.00
Coco caprate /caprylate	10.00	3.00
Cetearyl alcohol	-	6.00
Dicaprylyl ether	4.00	3.00
Cetostearyl palmitate	-	3.00
Caprylic and capric acid triglyceride	1.00	3.00
Isopropyl myristate	-	4.00
Silicon fluid	0.50	1.00
Propylene glycol	3.00	3.00
Imidazolidinyl urea (Germall 115, Sutton)	qs	qs
Water dispersion at 2% of carboxyvinylpolymer	6.00	3.00
Lanolin ethoxy	-	2.00
pH Adjuster, adjust to pH = 6.0	qs	qs
Octyl salicylate	4.50	4.50
Camphor benzalkonium	7.50	7.50
Butyl methoxydibenzoylmethane	2.00	2.00
Water (*aqua*), distilled	qs 100.00	qs 100.00

The solar filter efficacy is determined in vivo through the SPF that is a numeric value. It is the ratio of the time required for a certain dose of UV exposure to provoke the appearance of perceptible erythema in protected skin of a given person to the time required for the same dose to provoke the same response in unprotected skin of the same person. An elevation of the SPF is observed in the following order, once the quantities of UV filters are kept constant in the different types of preparation: hydro-alcoholic lotion (one phase); ointments (one phase); o/w type emulsion (two phases); w/o type emulsion (two phases).[4]

One aim of this study was the quality determination of a photoprotector w/o cream, by determining the microbial contamination, the physical-chemical stability and a sensorial analysis of flavor, texture, color, spreading and global impression. Another aim was to determine how much consumers liked the product and whether they observed significant differences between the proposed w/o cream (Formula 1) and an o/w cream (Formula 2).

Methods

Microbial control: The total number of microorganisms and the presence of pathogenic ones—such as Salmonella sp, Escherichia coli, Staphylococcus aureus and Pseudomonas aeruginosa—were analyzed according to methods described in the U.S.P.5 and the British Pharmacopeia.[6]

Acceptance analysis: A team of 30 volunteers was used for the acceptance test. The team analyzed the two cream samples (Formulas 1 and 2) presented in six combinations of two occurrences of one sample and one occurrence of the other sample, coded each time with three-digit numbers. The volunteers used a nine-point structured hedonic scale[7] to report their degree of acceptance of the two samples based on five sensory attributes.

Data statistical analysis: Computer software[a] was used to obtain statistical analysis of the data by univariant variance analysis (ANOVA), Tukey average tests and histogram analysis of the volunteers' hedonic scale ratings for the five sensory attributes for each of the two samples.

Difference tests: The two cream samples (Formula 1 and Formula 2) were submitted to the difference triangular tests[9] performed by 30 volunteers. The x2 test10 was used for the triangular test data analysis.

Spectrum-photometric analysis: The spectrum-photometric analysis was performed in a spectrophotometer[b] with quartz cubes of 1 cm optical path. Chloroform and isopropanol (V/V) were used as solving mixture. The base cream (Formula 1 without the analyzed filters) was used to zero the instrument. To verify the stability, the absorbance determination and absorption wavelength band of the filters (separately and mixed) were evaluated as well.

Results and Discussion

Microbial control: The results showed that no *Salmonella sp, E. coli, S. aureus* and *P. aeruginosa* microbial growth occurred in the analyzed samples. The total microorganism count was less than 10 CFU/g of product, which is within permissible limits according Brazilian requirements.[11]

The microbiologic control has the aim of assuring that consumer products are of good quality, free of any pathogenic or potentially harmful microorganisms, permitting a limited number of acceptable microorganisms. In this case, the testing for microorganisms used a series of cosmetic treatment agents that are good nutrient media.

The important point of exposing the microbial contamination of creams must be solved with allowance for their physicochemical properties and component composition that customarily hamper the isolation of microorganisms. This microbial contamination test also is used in pharmaceutical quality control that meets the ANVISA (Agência Nacional de Vigilância Sanitária) requirements[12] of GMP (Good Manufacturing Practices).

Physical-chemical stability: Spectrophotometric analysis was carried out 24 hours after the cream was prepared and again after the cream was submitted to 28 days of thermal stress (45°C), yielding similar spectrophotometric profiles. This analysis suggests that these filters did not degrade after 28 days of thermal stress (Figure 1) and shows the stability of filters in the studied preparation under these conditions.

Sensorial analysis: Sensorial analysis consists of a complete and real characterization of the sensorial and tactile properties of cosmetic products. The volunteers must be able to detect and describe all the sensations related to product characteristics and properties. However, some precautions must be taken to avoid fatigue, mainly in smelling.

Table 1 shows the sample dispersal in the difference triangular tests and the number of people who were able to identify the sample that was different from the other two.

[a] SAS User's Guide: Statistics, Cary, North Carolina: SAS Institute (1993)
[b] Hitachi U-2001 Spectrophotometer, Hitachi, Japan

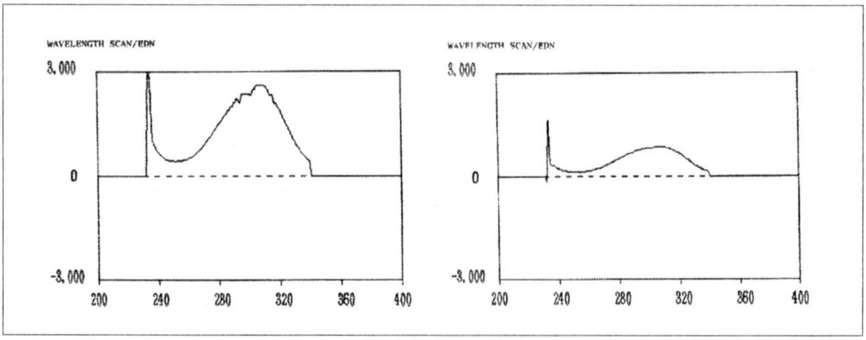

Figure 1. Spectrophotometric analysis of w/o cream (left) and o/w emulsion (right) following 28 days of thermal stress (45°C)

Table 1. Sample dispersal in the difference triangular tests and the number of people who were able to identify the sample that was different from the other two.
A = w/o emulsion B = o/w emulsion

Sample dispersal	Number of people who identified a different sample
A A B	19
A B A	26
B A A	25
B B A	24
B A B	23
A B B	25
Total	142

Table 2 shows the volunteers' average ratings of five attributes for the w/o emulsion and the o/w emulsion. Through the significant minimum difference (SMD), which was obtained by the Tukey average test ($p \leq 0.05$), a comparison was performed among the averages, and again no significant difference between them was found.

When the SPF for UVB using identical chemical filters in o/w and w/o emulsions was compared, the w/o emulsions showed the higher SPF. This may be explained by the fact that the silicone fluid in the emulsion enables a higher water resistance value because of the formation of a hydrophobic film in the skin. The silicone fluid also improves the product's thixotropic properties.

However, this type of emulsion rarely has been used because of its low stability and sticky skin feel. For this reason some agents specific for w/o emulsions and with rheological properties were developed. Among them are the following:

- Cetearyl glucoside and cetearyl alcohol[c], an emulsifying agent that gives viscosity for cosmetic preparation without oily sensations
- Dicaprylyl ether[d], an emollient product
- Coco caprate/caprylate[e], an emollient and skin conditioning agent

These agents are indicated for the preparation of photo-filters in w/o emulsions because they increase the overall stability, product retention time on the skin and resist washoff. They also avoid a greasy sensation. So at least one source has concluded that the w/o emulsions are the best vehicles in preparing sun protection creams.[4]

The acceptance analysis (Table 3 and Figure 2) showed that 60% of the volunteers said they definitely or probably would buy the w/o product, compared to 53% for the o/w product. Usually, consumers prefer o/w emulsions because they give a dry sensation and a softer skin feel. The authors sought to develop a w/o emulsion that does not cause an oily skin feel after use. The sensorial analysis showed the w/o emulsion is more acceptable than the o/w emulsion. So it is possible to develop a w/o emulsion that does not give the oily skin sensation that is characteristic of w/o emulsions.

Conclusion

This study determined the quality of a photoprotector w/o cream by evaluating its microbial control, physical-chemical stability and sensory appeal in comparison

[c] Emulgade PL68/50, Henkel KgaA, Düsseldorf, Germany
[d] Cetiol OE, Henkel KgaA
[e] Cetiol LC, Henkel KgaA

Table 2. Volunteer acceptance average of Samples 1, 2 for the studied attributes

Samples	Color	Flavor	Texture	Global Impression	Spreading
1 (A/O)	7.6000	5.5667	7.5000	7.1667	7.5667
2 (O/A)	7.7667	4.8000	7.3000	6.7000	6.9000

Averages in each column do not differ significantly between samples according to the Tukey average test ($p \leq 0.05$).

Table 3. Acceptance analysis (% acceptance) for w/o emulsion and o/w emulsion

Purchasing attitude	w/o	o/w
I definitely would not buy this product	0	0
I probably would not buy this product	10	16.66
I am not sure if I would buy this product	30	30
I probably would buy this product	46.66	40
I definitely would buy this product	13.33	13.33
Total = 30		

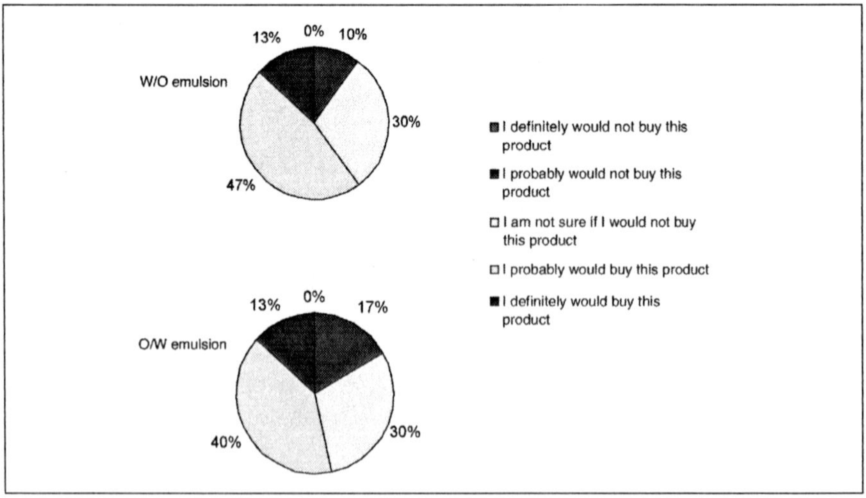

Figure 2. Acceptance analysis

to an o/w photoprotector cream. Microbial challenge tests in cultures of the w/o cream showed that no microbial growth occurred. Spectrum-photometric analysis indicated the stability of the w/o cream by showing that there was no degradation in its solar filters. Sensorial analysis including a triangular test comparing the w/o cream and an o/w cream on five sensory characteristics indicated that there was no significant difference ($p < 0.05$) between the two creams, and 60% of the volunteers said that they probably or certainly would buy the w/o product.

—Tatiana Maria de Almeida Silva, Ketylin Fernanda Migliato,
—Hérida Regina Nunes Salgado and Vera Lucia Borges Isaac Rangel
Faculty of Pharmaceutical Sciences, UNESP, Araraquara, Brazil

References

1. LN Prista, AC Alves and R Morgado, Emulsões, *Tecnologia Farmacêutica*, 5 ed, 1(8) 597–669 (1995)
2. AM Muñoz, GV Civille and BT Carr, *Sensory Evaluation in Quality Control*, New York: Van Nostrand Reinhold (1993) 240
3. MR Chedekel, A melanina pode melhorar os filtros solares, *Cosmet Toil* 10(5) 60–62 (1998)
4. SC Azzelini, Agentes potencializantes de fotoprotetores, *Cosmet Toil* 7(4) 34–37 (1995)
5. USP 26, *The United States Pharmacopeia*, Rockville, Maryland: The United States Pharmacopeial Convention (2003)
6. *British Pharmacopea*, London: Her Majesty's Stationary Office (2001)
7. H Stone and J Sidel, *Sensory Evaluation Practices*, 2nd ed, New York: Academic Press (1993) 338
8. *SAS User's Guide: Statistics*, Cary, USA: SAS Institute (1993)
9. M Meilgaard, GV Civille and BT Carr, *Sensory Evaluation Techniques*, Boca Raton, Florida: CRC Press (1987) 281
10. EB Roessler and HL Alder, *Introduction to Probability and Statistics*, 5th ed, San Francisco: WC Freeman (1972) 195–220
11. Brazil, DIÁRIO Oficial da União. DOU. Brasília (08 Oct 1999)
12. Brazil, Resolução RDC no 33 (19 Apr 2000)

Quantifying Benzophenone-3 and Octyl Methoxycinnamate in Sunscreen Emulsions

Keywords: benzophenone-3, octyl methoxycinnamate, sunscreen, emulsions, quantitative determination, HPLC

The authors have validated a high performance liquid chromatographic (HPLC) method for the quantitative determination of benzophenone-3 and octyl methoxycinnamate present in sunscreen emulsions

Editor's note: Octyl methoxycinnamate is the previous INCI name and drug name for what has been more recently renamed ethylhexyl methoxycinnamate (INCI/EU name) and octinoxate (U.S. drug name).

In order to minimize the adverse effects of UV radiation, the use of sunscreens in cosmetic preparations has been increasing. Many new cosmetic products containing sunscreens are being developed and are commercially available. Sunscreens are considered drugs in the United States and several other countries, thus the FDA and other regulatory agencies mandate the actives be present at the labeled level by a validated method. Consequently there is a need for development and validation of analytical methods for quantitative determination of sunscreen agents in cosmetic products. High performance liquid chromatography (HPLC) is the most used chromatographic method for qualitative and quantitative determination of sunscreen agents in cosmetic products.[1-6]

The objective of this study was to develop a rapid, selective and accurate method by reversed-phase HPLC for quantitative simultaneous determination of benzophenone-3 (B-3) and octyl methoxycinnamate (OM) contained in sunscreen emulsions.

Material and Methods

Sunscreens: B-3 presents its maximum UV absorption between 288 and 289 nm and absorbs UVA as well as UVB radiation. The maximum concentration used in solar protector formulations is 6%. It is normally used in a concentration range from 2% to 6% in combination with other chemical filters.[7-9]

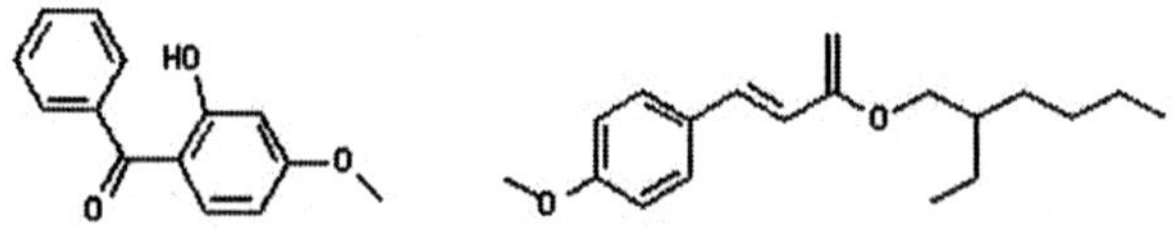

Figure 1. Chemical structures of benzophenone-3 (left) and octyl methoxycinnamate (right)

OM presents its maximum UV absorption between 289 and 311 nm and absorbs UVB radiation and to a lesser extent UVA radiation. The maximum concentration used in solar protector formulations is 10%. Normally it is used in a concentration range from 2% to 7.5%.[7-9] The chemical structures of these substances are shown in Figure 1.

Materials: All reagents and solvents were of analytical grade or of HPLC grade: methanol[a]; water from a water purification system[b]; sunscreen filters[c] B-3 and OM.

Sample: The simulated sample used in this research was prepared in the laboratory according to Formula 1.

Apparatus: A liquid chromatographic system[d] equipped with a variable UV detector was connected to an electronic integrator[e] and manual injection valve fitted with a 20 mL sample loop[f].

Methods

HPLC operating conditions: LiChrospher 100 RP-18 column[g]; particle size 5 mm, 125x4 mm id; mobile phase methanol:water (87:13 v/v); flow rate of 1.0 mL min^{-1}; UV detection at 290 nm; detector sensitivity 0.32 a.u.f.s.; room temperature (25 ± 1°C). All solutions and solvents were filtered through a filter membrane[h], pore size 0.45 mm and vacuum degassed by sonication before use.

Preparation of standard solution: A 5 mL aliquot of standard stock solution containing 50 mg mL^{-1} of B-3 and 100 mg mL^{-1} of OM in methanol was transferred to a 25 mL amber volumetric flask and diluted to volume with methanol (final concentration 10 mg mL^{-1} of B-3 and 20 mg mL^{-1} of OM).

Sample solution: Sample solution was also prepared in methanol. An amount of sample containing 5 mg of B-3 and 10 mg of OM was weighed and transferred to a 100 mL volumetric flask; 50 mL of methanol was added and the mixture was sonitcated for 20 min. The volume was completed with methanol and the solution was filtered. After adequate dilutions a final solution containing 10 mg of B-3 mL^{-1} and 20 mg of OM mL^{-1} was obtained.

[a] Merck, Darmstadt, Germany
[b] Milli-Q Plus Compact Water System, Millipore, São Paulo, SP, Brazil. Milli-Q is a registered trademark of Millipore Corporation, Billerica, Massachusetts USA.
[c] Filters were obtained from Galena, São Paulo, Brazil.
[d] CG Model 480C, Instrumentos Científicos CG Ltda. São Paulo, Brazil
[e] Model CG-200, Instrumentos Científicos CG Ltda. São Paulo, Brazil
[f] Instrumentos Científicos CG Ltda, São Paulo, Brazil

Formula 1. Simulated sample

Butylhydroxy toluene (BHT)	**0.05% w/w**
Silicone oil	3.00
Glyceryl monostearate	2.00
Cetostearyl alcohol	2.00
Mineral oil (*Paraffinum liquidum*)	3.00
Emulsifying Wax NF (Polawax, Chemyunion Química Ltda)	2.50
2-Bromo-2-nitropropane-1,3-diol (Chemynol, Chemyunion Química Ltda)	0.80
Benzophenone-3	2.00
Octyl methoxycinnamate (Neo Heliopan AV, Chemyunion Química Ltda)	4.00
Titanium dioxide and capric/caprilic triglyceride (Tiosorb TG, Chemyunion Química Ltda)	2.00
Corallina officinalis extract (Phycocorail, Chemyunion Química Ltda)	0.50
Propyleneglycol	5.00
Magnesium aluminum silicate (Veegum ultra, R.T. Vanderbilt)	0.50
Xanthan gum	0.30
Imidazolidinyl urea	0.50
Water (*aqua*), distilled	qs 100.00

Table 1. Statistical results obtained in the validation of the HPLC method

Statistical Data	Benzophenone-3	Octyl Methoxycinnamate
Concentration range (mg mL^{-1})	4.0-18.0	8.0-36.0
Number of injections	8	8
Slope	2120.60	3202.66
Intercept	-612.35	821.50
Coefficient of correlation	0.9999	0.9999
Limit of detection (mg mL^{-1})	0.11	0.59
Limit of quantitation (mg mL^{-1})	0.38	1.98
Confidence limit*	99.88 ± 0.42	100.68 ± 0.32
Precision (RSD)*	0.58	0.45
Accuracy (Average recovery %)	101.88	98.28

*Average of 10 determinations

Placebo solution: 250 mg of placebo was weighed and a solution was prepared as described above.

Results and Discussion

Validation of the analytical method: The method was validated according to the routine procedures used to validate a chromatographic method in order to obtain reliable results.[10-14] The separation of a mixture of standards is shown in Figure 2(c). The peaks are very well resolved within 7 minutes.

Linearity: The linearity of an analytical method is the ability of producing results proportionate to the concentration of the analyte in the sample within a certain interval.[11,14,15] Standard curves were obtained using solutions of eight different levels of concentration ranging from 4 to 18 µg mL^{-1} for B-3 and from 8 to 36 µg mL^{-1} for OM. The solutions were prepared by diluting a standard solution containing 50 µg mL^{-1} of B-3 and 100 µg mL^{-1} of OM in methanol and each solution was injected three times in the chromatographic system. The calibration curves were constructed by plotting the peak area against concentration (µg mL^{-1}). The linear regressions for each sunscreen are shown in Table 1. The correlation coefficients are in accordance of those established in the scientific literature.[16,17]

Limit of detection (LD) and limit of quantitation (LQ): The LD and LQ were determined according to the literature.[12,14] Five determinations were made at five levels of concentration ranging from 2 to 10 µg mL^{-1} for B-3 and from 4 to 20 µg mL^{-1} for OM. These solutions were obtained by diluting a standard solution containing 50 µg mL^{-1} of B-3 and 100 µg mL^{-1} of OM. LD and LQ were calculated based on the standard deviation of the responses and the slope of a calibration curve constructed at the lower end of the linearity range. Results are presented in Table 1. LD was defined as three times the signal-to-noise ratio and LQ as 10 times the same signal.

Precision: The precision was defined as the ability of the proposed method to repeat responses at given analytical conditions, expressed as relative standard deviation (RSD) and was obtained by injecting each sample solution 10 times and standard solution three times into the chromatographic system. In pharmaceutical analysis a method is precise if the RSD is less than 1%.[10,15,16,18] Results are presented in Table 1.

Accuracy: The accuracy was determined by recovery tests, obtained by adding standard solution to the sample being analyzed.[11,13-15] The test was performed according to AOAC International guidelines.[19] Known amounts of B-3 (6 to 10 µg mL^{-1}) and OM (12 to 20 µg mL^{-1}) standard solutions were added to samples and the resulting spiked samples were submitted to the assay procedure. All samples were injected three times. Table 2 shows the results of recovery tests. For pharmaceutical preparations the percentage of recovery must be within the interval 98% and 102%.[15] As the sample used in this research was a cosmetic preparation, the experimental results can be considered satisfactory.[20]

Specificity: The specificity of a method is its ability of producing reliable results in the presence of the excipients of the formulation and other components, such as

[g] Merck, Darmstadt, Germany
[h] Millipore, São Paulo, Brazil

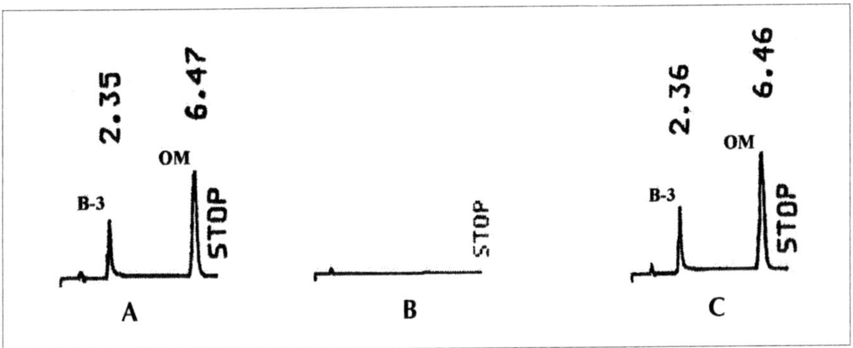

Figure 2. Chromatograms
A = simulated samples of benzophenone-3 and octyl methoxycinnamate
B = placebo formulation
C = standard solutions of benzophenone-3 and octyl methoxycinnamate

Table 2. Recovery of standard sunscreen solution of benzophenone-3 (B-3) or octyl methoxycinnamate (OM) added to sample analyzed by HPLC

Amount added (μg mL⁻¹)		Amount found* (μg mL⁻¹)		Average recovery (%)	
B-3	OM	B-3	OM	B-3	OM
6.0	12.0	6.08	11.41	101.33	95.08
8.0	16.0	8.20	16.08	102.50	100.50
10.0	20.0	10.18	19.85a	101.80	99.25

* Average of 10 determinations

impurities.[11] The specificity was determined by comparing the results obtained in the analysis of placebo formulation containing excipients only with those obtained in the analysis of a simulated sample and a standard solution containing only sunscreens. It can be observed that the placebo (Figure 2b) does not present any peak that can interfere in the sample analysis (Figure 2a).

Conclusions

The proposed HPLC method enables the simultaneous quantitative determination of sunscreens B-3 and OM contained in cosmetic emulsions. The preparation of sample is easy and efficient. Both compounds were separated with high efficiency in a short period of time (less than 7 min). There was no excipient interference in the method. Linearity of calibration curves for B-3 and OM was obtained in a concentration range from 4 to 18 μg mL^{-1} and 8 to 36 μg mL^{-1}, respectively, with a good correlation coefficient. The relative standard deviations for both compounds were satisfactory. The accuracy of the method was confirmed by the obtained recovery test results, ranging from 101.33% to 102.5% for B-3 and from 95.08% to 100.5% for OM.

—**M. Inês R. M. Santoro, D.A.G.C. Oliveira, E.R.M. Kedor-Hackmann, A. K. Singh,**
Department of Pharmacy, Faculty of Pharmaceutical Sciences, University of São Paulo, Brazil

References

1. EA Dutra, ERM Kedor-Hackmann and MIRM Santoro, Validation of a high performance liquid chromatography method for sunscreen determination in cosmetics, *Int J Cos Sci* 24 97-102 (2002)
2. SC Rastogi and GH Jensen, Identification of UV filters in sunscreen products by high-performance liquid chromatography-diode array detection, *J Chromatogr A* 828 311-316 (1998)
3. V Vanquerp, C Rodriguez, C Coiffard, LJM Coiffard and Y De Roeck-Holtzhauer, High-performance liquid chromatographic method for the comparison of the photo stability of five sunscreen agents, *J Chromatogr A* 832 273-277 (1999)
4. G Portad, C Laugel, A Baillet, H Schaefer and JP Marty, Quantitative HPLC analysis of sunscreens and caffeine during in vitro percutaneous penetration studies, *Int J Pharm* 189 249-260 (1999)
5. S Scalia, Determination of sunscreen agents in cosmetic products by supercritical fluid extraction and high-performance liquid chromatography, *J Chromatogr A* 870 199-205 (2000)
6. MIRM Santoro, FCF Silva and ERM Kedor-Hackmann, Stability analysis of emulsions containing UV and IR filters, *Cosmet Toil* 115 55-62 (2000)
7. *The Martindale, The Extra Pharmacopeia,* 30th edn (1993) 1233-1235
8. P Romanowski and R Schueller, Introdução aos produtos fotoprotetores, *CosmetToilet* edn Port 12 60-67 (2000) (In Portuguese)
9. NA Shaath, Enciclopédia de absorvedores de UV para produtos com filtro solar, *Cosmet Toilet* edn Port **7** 47-58 (1995) (In Portuguese)
10. P Bruce, P Minkkinen and ML Riekkola, Practical method validation: validation sufficient for an analysis, *Mikrochim Acta* 128 93-106 (1998)
11. International Conference on Harmonization: Text on validation of analytical procedures (ICH Q2A), Geneva (Mar 1995)
12. International Conference on Harmonization: Validation of analytical procedures: methodology (ICH Q2B), Geneva (Nov 1996)
13. ME Swartz and IS Krull, Validação de métodos cromatográficos, *Pharm Technol* Braz edn 2 12-20 (1998) (In Portuguese)
14. *United States Pharmacopeia*, 25th edn, Rockville: United States Pharmacopeial Convention (2002)
15. DR Jenke, Chromatographic method validation: a review of current practices and procedures. II. Guidelines for primary validation parameters, *J Liq Chrom & Rel Tech* 19 737-757 (1996)
16. *Quality Assurance Principles for Analytical Laboratories,* FM Garfield, ed, Arlington: Association of Official Analytical Chemists Internacional (1991) 196
17. *Métodos instrumentais para análise de soluções–Análise quantitativa: espectrofotometria no visível e ultravioleta: erros em química analítica.* MLSS Gonçalves, ed, Lisboa: Fundação Calouste Gulbenkian (1983) 13-82; 609-640 (In Portuguese)
18. IR Krull and M Swartz, Analytical method development and validation for the academic research, *Anal Lett* 32 1067-1080 (1999)
19. *Official Methods of Analysis of the Association of Official Analytical Chemists*, 15th edn, Arlington: Association of Official Analytical Chemists International (1990) 15-17
20. JS Azevedo, NS Viana Jr and CDV Soares, UVA/UVB sunscreen determination by second-order derivative ultraviolet spectrophotometry, *Il Farmaco* 54 573-578 (1999)

Photostability Testing of Avobenzone

Keywords: photostability, avobenzone, sunscreens, testing, photodegradation

This study investigates the photostability of avobenzone by measuring various sunscreen products. The authors suggest some areas to overcome in formulating.

Photostability studies usually involve the examination of the photochemical degradation of a specific chromophore, and most work generally focuses on issues regarding the detection and analytical measurement of the degradation of sunscreen agents.[1-16] Work on photostability in the 1970s and 1980s focused on cinnamates, benzoates and benzilidine camphor sunscreen derivatives.[1-10] More recently, sunscreen photostability studies have emphasized di-benzoyl methane derivatives but have not systematically examined films of products.[11-16] In this study, we tried to duplicate as closely as possible real sun exposure to dynamically assess the photostability of sunscreen drug products.[17,18] Recently, Maier et al. investigated similar sunscreen products and reported comparable findings.[19]

This study focuses on some implications of photostability relative to the efficacy of sunscreen products. Photo-safety testing is used to ensure that formulations, including common photo-breakdown products, are safe. However, no practical amount of testing can ensure that an undesirable photointeraction will not result when other topical products are used in conjunction with a photoreactive sunscreen formula.

Methods

For this study, we devised an apparatus and an analysis technique designed to dynamically assess the photostability of sunscreen drug products exposed to a solar-like radiation source. The experimental setup simultaneously irradiates and monitors changes in the transmittance of films of sunscreen products in place without disturbing the sample in the beam (Figure 1).

Initially the solar simulator, using a 1 mm WG-320 filter,[a] was measured using the spectroradiometer calibrated in irradiance mode to determine the MED. For this study, a MED is defined as 20 mJ/cm^2 of erythemically effective exposure. The same spectroradiometer was then calibrated against the solar simulator source for transmittance measurements with an ultraviolet

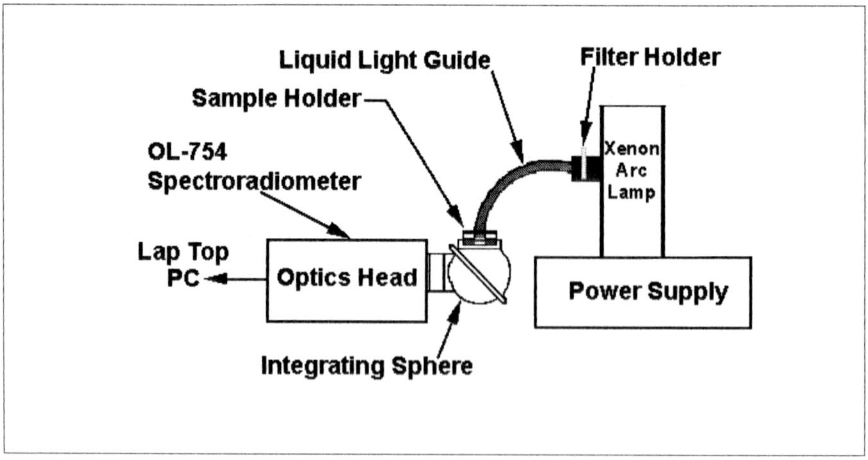

Figure 1. Schematic of photostability test system.
The OL-754 spectroradiometer is used in two modes. It is used in irradiance mode to measure the spectrum of the filtered solar simulator, and it is used in transmittance mode to measure the UV transmittance of the sunscreen film being irradiated. A 150-watt xenon arc is filtered to achieve a solar-like spectrum using a Schott 1 mm WG-320 filter. Radiation is conveyed to the sunscreen applied to a Teflon film mounted in the integrating sphere via the fiber optic light guide. All transmittance measurements are made automatically without moving the Teflon membrane/product film mounted in the integrating sphere.

(UV) transparent Teflon membrane mounted in the entrance aperture of the integrating sphere.

Thin films of sunscreen products, about 1 to 2 mg/cm^2, were then applied to the Teflon membranes and positioned in the entrance to the integrating sphere. Exposures were administered through the liquid light guide coupled to the integrating sphere, so that transmittance measurements of the sunscreen films could be made while continuously exposing the film to full-spectrum, solar-simulated radiation.

We programmed the spectroradiometer to measure the initial product transmittance and then automatically remeasure the transmittance of the film at 1-MED intervals throughout the exposure period. For each specimen, an array of transmittance measurements resulted, beginning with the unexposed product and proceeding through a series of multiple MED exposures.

Products tested: A series of sunscreen products currently marketed in North America were examined:

A. SPF-15 (US), 3% avobenzone, 7% padimate O
B. SPF-15 (US), 3% avobenzone, 7.5% octyl methoxycinnamate, 3% oxybenzone
C. SPF-30 (US), avobenzone, ethylhexyl-*p*-methoxycinnamate, oxybenzone, 2-ethyl hexyl salicylate, homosalte (percentages not indicated on label)

[a]Schott WG-320, Schott Glass Technologies Inc., Duryea, Pennsylvania, USA

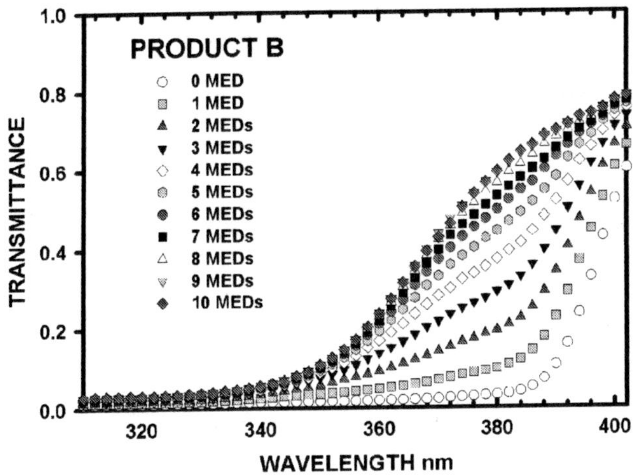

Figure 2. Transmittance of avobenzone-containing product.
The transmittance of the sunscreen applied to a Teflon film is measured periodically during UV exposure at each MED increment. Note the UVA transmittance increases with increasing exposure. While changes in the UVB protection are evident, the magnitude is not apparent.

D. SPF-30 (Canadian), 2.5% avobenzone, 7.5% octyl methoxycinnamate, 6% benzophenone-3, 2% titanium dioxide
E. SPF-30 (US), 3% avobenzone, 7.5% octyl methoxycinnamate, 5% octylsalicylate, 6% oxybenzone
F. SPF-30 (US), (no avobenzone) ethylhexyl-*p*-methoxycinnamate, homosalate, oxybenzone (percentages not indicated on label)

Results

The results of the data analysis were viewed in three formats:

1. Film transmittance for each MED exposure interval (Figure 2)
2. Monochromatic protection factors (MPF) (Figure 3): The MPF is the reciprocal of the transmittance
3. Changes in product performance, shown as the remaining fraction of the initial average MPF (Figure 4)

Figure 2 shows how the transmittance spectrum of an avobenzone-containing formula changes with UV exposure. In this representation, we discovered increased transmittance as the products photodegrade. When this data was expressed as MPF (Figure 3), we found that a considerable loss of protection occurred within the first two MED exposures. While most of this degradation occurred in the UVA spectrum, loss throughout the entire spectrum was also observed. Changes are apparent in

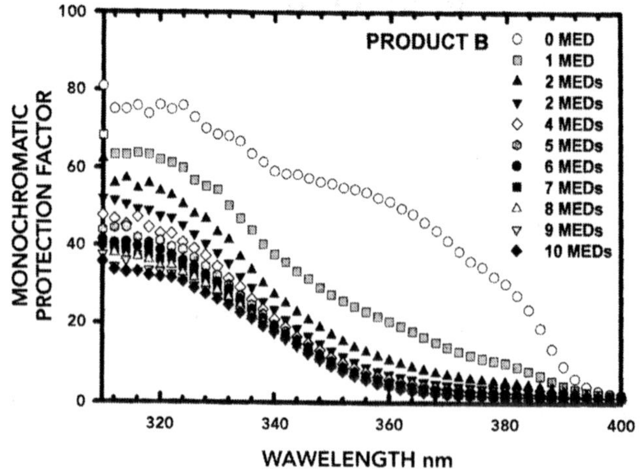

Figure 3. Change in monochromatic protection factor (MPF) with UV exposure.
Many laboratories use monochromatic protection factor representation to analyze sunscreen efficacy. The monochromatic protection factor MPF = 1/T, where T is the transmittance of the sunscreen formula. Note the loss of MPF in the UVA-1 ($\lambda > 340$ nm). After 2 or 3 MEDs exposure, UVA-1 protection provided is trivial. Also note the significant decrease in the monochromatic protection in the UVB. Avobenzone contributes less than 10% of the monochromatic protection in the UVB, but the decrease in monochromatic protection is 30% after 2 MEDs and approaches 50% by 10 MEDs. The UVB sunscreens are being changed. The potential protection provided by avobenzone is clearly shown in this plot. At 360 nm before irradiation, the MPF due to 3% avobenzone is 75% of the UVB MPF (300-320 nm) of the other 10.5% UVB sunscreens. After 10 MEDs exposure, the MPF at 360 nm is about 10% that of the remaining UVB sunscreens at 300-320 nm.

the UVB portion of this plot. The degree of these changes, in light that they do not occur when avobenzone is absent in the formulation, suggests that the UVB sunscreen(s) may be degraded by an avobenzone-photosensitized mechanism.

Our work, encompassing an examination of products containing avobenzone that we obtained on the U.S. and Canadian markets, found that this ingredient showed photoinstability. Figure 4 shows how the fraction protection remaining of the initial average MPF changes with increased exposure. Note that Product F, which does not contain avobenzone, appears to be photostable throughout the duration of exposure.

Discussion

These results show that sunscreen products containing avobenzone may experience photodegradation from UV exposures. Our study also suggests that octyl methoxycinnamate and padimate O, normally shown to be photostable UVB sunscreens, did not show such effectiveness in the products containing avobenzone (Figure 3).

SPF 30 (US), (no avobenzone) ethylhexyl-p-methoxycinnamate, homosalate, oxybenzone (percentages not indicated on label)

SPF 30 (Canadian), 2.5% avobenzone, 7.5% octyl methoxycinnamate, 6% benzophenone-3, 2% titanium dioxide

SPF 30 (US), avobenzone, ethylhexyl-p-methoxycinnamate, oxybenzone, 2-ethyl hexyl salicylate, homosalte (percentages not indicated on label)

SPF 30 (US), 3% avobenzone, 7.5% octyl methoxycinnamate, 5% octylsalicylate, 6% oxybenzone

SPF 15 (US), 3% avobenzone, 7.5% octyl methoxycinnamate, 3% oxybenzone

SPF 15 (US), 3% avobenzone, 7% padimate O

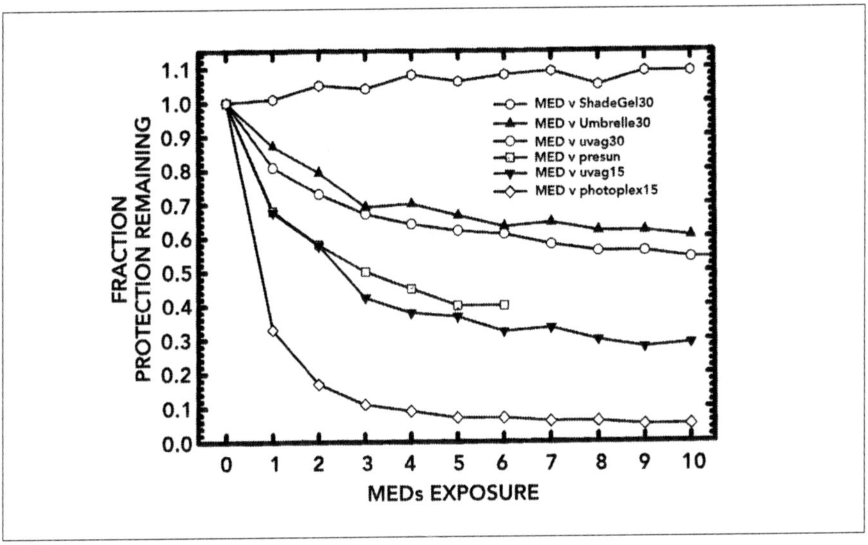

Figure 4. Change in average monochromatic protection factor with UV exposure.
The average monochromatic protection factor is calculated for each formula (Average MPF = Sum MPFλ/Sumλ). Each point represents the average MPF after the exposure to that point. It is shown as fraction of the protection remaining. Note that, by 2 or 3 MEDs exposure, all available avobenzone has been photodegraded. At this point, only UV-stable sunscreens remain available in the product. This figure shows the data for a selection of formulas tested. Note the SPF 30 product which does not contain avobenzone whose protection remains constant throughout exposure.

Further studies of this are warranted to determine if avobenzone (the ingredient) or the formulations (the methods of using avobenzone) can cause photosensitized degradation of other sunscreen ingredients, as products containing either padimate O or octyl methoxycinnamate without avobenzone appear photostable in this study.

UVB unrelated: When we removed the UVB radiation from our system using a 2 mm WG-360 filter, the photodegradation of avobenzone occurred with as little as 0.2 MED of UVA exposure, suggesting that UVB is not required to cause photolysis of avobenzone.[17,18] As the amount of UVB increased in the spectrum, the total amount of UVA decreases for each MED administered. Sources with more UVA-1 (340 to 400 nm) radiation could destroy avobenzone

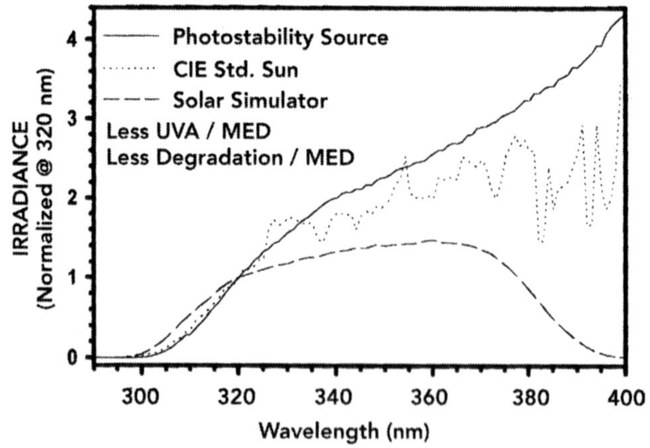

Figure 5. Comparison of sunlight, our solar simulator and standard human test solar simulator. The typical solar simulator used to determine the SPF of sunscreen products meets existing FDA and COLIPA standards. This solar simulator has more short UVB radiation than is present in sunlight and has less than half of the UVA-1 (340-400 nm) intensity available in sunlight.

faster than sources with less UVA-1. Therefore, we believe the precise number of MEDs required to destroy the available avobenzone depends on the amount of UVA-1 in the source.

Sunlight vs. solar simulator: Figure 4 shows the fraction of protection remaining based on the change in average MPF during exposure of a product film. For each product we tested, the majority of loss of protection appears to have occurred by a 2 or 3 MED exposure. We setup our exposure system to spectrally resemble sunlight (Figure 5) in that it has the full complement of UVA-1 (340-400 nm) radiation that is available in sunlight. While the intensity of our system is similar to sunlight, it is lower than the intensity of other solar simulators commonly used in SPF testing. In sunlight, a fair-skinned individual's MED might be 15 to 20 min. The estimated MED for our source is 10 to 15 min, rather than 10 to 15 sec commonly used for in vivo sunscreen testing.

As Figure 5 shows, our solar simulator, like sunlight, has considerably less short UVB radiation per MED than solar simulators commonly used in sunscreen testing. Regarding UVA photostability, this means that, in sunlight, an individual is exposed to 6 to 15 J/cm^2 of broad-spectrum UV per MED, mostly UVA. Conversely, when using conventionally filtered solar simulators to test SPF, a volunteer is exposed to only 1 or 2 J/cm^2 of UV per MED, mostly UVB with little UVA. Because sunlight is so rich in UVA, compared to solar simulators used in SPF testing, sunscreen UVA photostability may not be adequately accounted for in the current SPF test in the US.[21-22]

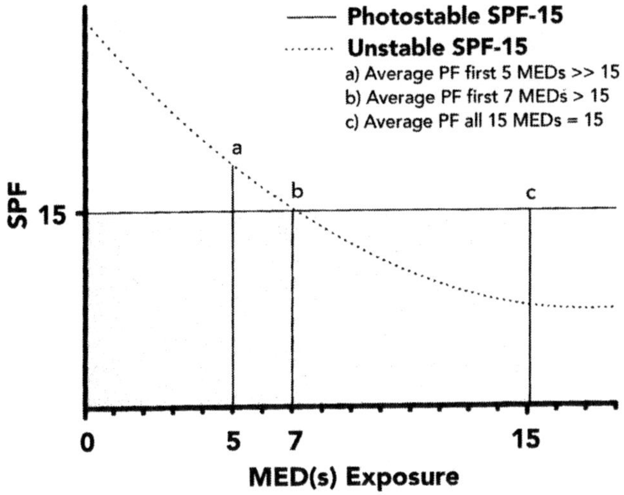

Figure 6. Typical photodegradation curve. (Modified from Petit and Gonzenbach)
This figure shows how photounstable products initially provide better protection than claimed. As photodegradation proceeds, unstable products lose protective capability. By the time the SPF exposure is reached, considerably less protection than claimed still remains. If overexposed, consumers will sunburn worse "through" unstable products than stable ones of the same SPF. This also explains why, in certain studies, avobenzone products appeared to be so effective. In one such study, mice treated with SPF-15 products were exposed to 5 to 7 MEDs. Because the unstable product was only partially degraded, the average SPF was higher than claimed, and the formula appeared more protective compared to a photostable product with the same SPF.

In clinical SPF testing, it is customary to have the expected SPF spaced as the center exposure in a sequence. For instance, to test an SPF 15 product, the center exposure is 15 MEDs, the next exposure is 19 MEDs and the last exposure is 23 MEDs. For a photostable SPF 15 product, the 19 MED exposure causes a 1.27 MED burn and the full 23 MED exposure produces a sunburn of 1.53 MED. Sunburns less than 2 MEDs are not significantly serious. For a photounstable SPF 15 product whose protection has degraded to the extent shown by product A (Figure 4), a 19 MED exposure will result in a severe 4 or 5 MED sunburn.

Other studies: Parts of this work have been presented at two recent scientific meetings: the 1998 Photomedicine Society meeting[17] and the 1998 American Society for Photobiology meeting.[18] At both events, participants from pharmaceutical and personal-care manufacturers asked essentially the same question: Why did the work of Dr. L. Kligman and colleagues[23] indicate that an avobenzone-containing product was more effective than an avobenzone-free product of the same SPF after UV exposures were administered? This is a valid question. The Kligman et al. study shows that an SPF 15 with avobenzone provided superior protection to non-avobenzone SPF 15 products, compared the products using 5 to 7 MED exposures.

We understand the SPF of a product is defined by an endpoint of a 1 MED burn delivered through the product. Stable products provide the same amount of protection throughout the entire duration of the exposure, while photounstable products provide a decreasing amount of protection over the exposure. Since photounstable products have a defined SPF, when the test begins, the product must initially be substantially more protective than the SPF indicates. This dynamic, illustrated in Figure 6 (a modification of an earlier figure proposed by Gonzenbach and Pittet[24]), shows that when fractional exposures are administered using less than the SPF indicated, the protection provided is significantly greater. Therefore, the comparison in the study made at 5 to 7 MED was not relevant for consumers requiring a full SPF 15 protection. The practical difference for consumers is that, with photounstable products, once the SPF is exceeded, more severe sunburns may result than would occur with photostable products of the same SPF.

Formulators' challenge: Formulators have a challenge facing them. Based on this study, we advise awareness to these results when formulating products that include avobenzone in combination with UVB sunscreens. In Figure 3, one can observe that, at 360 nm, 3% avobenzone in a product has 75% of the maximum MPF that the other 10.5% of UVB sunscreen agents exhibit at 310 nm. Clearly, a small amount of avobenzone has the potential to provide a significant amount of protection and could be extraordinarily effective and beneficial for consumers.

While other studies examining the stability of avobenzone are needed, based on our results, we believe two hurdles need to be addressed to develop photostable avobenzone-containing products. First, formulators need to be aware that avobenzone could make other UVB sunscreens photounstable; simply adding avobenzone to an established product may potentially decrease the SPF due to this loss of UVB protection. There may be combinations of UVB sunscreens and avobenzone that minimize or prevent this interaction. Additionally there may be stabilizer molecules which, if added to formulations, could alleviate this potential difficulty and better stabilize the product. A second potential hurdle is to prevent photodecomposition of avobenzone itself; again there may be stabilizer or quencher ingredients that could block or retard this possible effect.

There are good reasons why formulators should address these challenges head-on. Clearly, any formula that is photostable will have a significant advantage over other competitive products; because avobenzone is a very potent-absorbing compound, a formulator should be able to produce very high SPFs with less active ingredients. The resulting stable formula, using fewer sunscreens at low concentrations, should result in competitive pricing. Also, because avobenzone is such a potent UVA sunscreen, a photostable product containing it should be more effective in clinical tests, especially involving UVA-rich sunlight.

—**Robert M. Sayre,** *Rapid Precision Testing Laboratories, Cordova, Tennessee USA*
—**John C. Dowdy,** *The University of Memphis, Department of Microbiology and Molecular Cell Sciences, Memphis, Tennessee USA*

References

1. CDM Ten Berge, CHP Bruins and JS Faber, Die Photochemie van Sonneschtzmitteln. I Uber die Photochiemie von mMethyl-p-dimethylaminobenzoat, *J Soc Cos Chem* 23 289–299 (1972)
2. CDM Ten Berge and CHP Bruins, Die Photochemie van Sonneschtzmitteln., II Uber die Photochiemie von methyl-p-dimethylaminobenzoat, *Soc Cos Chem* 2255 263–269 (1972)
3. I Beck, A Deflandre, G Lang, R Arnaud and J Lemaire, Study of the photochemical behaviour of sunscreens: benzylidene camphor and derivatives, *Int J Cosm Sci* 3 139–152 (1981)
4. P Morliere, O Avice, T Sa E Melo, L Dubertret, M Giraud and R Santus, A study of the photochemical properties of some cinnamate sunscreens by steady state and laser flash photolysis, *Photochem Photobiol* 36 395–399 (1982)
5. H Sunal, JP Laget and H Delonca, Evaluation de la photostabilite d'une filter solaire, *Parfums Cosmeticques Arones* 48 49–55 (1982)
6. I Beck, A DeFlandre, G Lang, R Arnaud and J Lemaire, Study of the photochemical behaviour of sunscreens:Benzlidene camphor and derivatives II: Photosensitized isomerization by aromatic ketones and deactivation of 8-methoxy psoralen triplet state. *J Photochemistry* 30 221–227 (1985)
7. AA Shaw, LA Wainschel and MD Shetlar, The photochemistry of p-aminobenzoicacid, *Photodermatology* 2 151–157 (1985)
8. FP Gasparo, UV-induced photoproducts of para-aminobenzoic acid, *Photodermatology 2* 151–157 (1985)
9. A Kammeyer, W Westerhof, PA Boluis, AJ Ris, EA Hische, The spectral stability of several sunscreen agents on stratum corneum sheets Int *J Cosm Sci* 9 125–136 (1987)
10. A Deflandre and G Lang, Photostability assessment of sunscreens: benzylidene camphor and dibenzolmethane derivatives, *Int J Cosm Sci* 10 53–62 (1988)
11. P Yankov, S Sultiel and I Petkov, Photoketonization and excited state relaxation of dibenzoyl methane in non-polar solvents, *J Photochem Photobiol A* 41 205–214 (1988)
12. NA Shaath and HM Fares, Photodyegradation of sunscreen chemicals: Solvent considerations, *Cosm Toil* 105 41–44 (1990).
13. H Gonzenbach, JJ Hill, TG Truscott, The triplet energy levels of UVA and UVB sunscreens, *J Photochem Photobiol B: Biol* 16 377–379 (1992)
14. NM Roscher, MKD Lindemann, SB Kong, CG Cho and P Jiang, Photodecomposition of several compounds commonly used as sunscreen agents, *J Photochem Photobiol A* 80 417–421 (1994)
15. W Schwack and T Rudolph, Photochemistry of dibenzoyl methane-UVA filters part 1, *J Photochem Photobiol B: Biology* 28 225–234 (1995)
16. JM Allen, CJ Gossett and SK Allen, Photochemical formation of singlet molecular oxygen (1O2) in illuminated aqueous solutions of p-aminobenzoic acid (PABA), *J Photochem Photobiol B: Biology* 32 33–37 (1996)
17. RM Sayre and JC Dowdy, Avobenzone and the Photostability of sunscreen products, Photomedicine Society Annual Meeting, Orlando FL, *Photoderm Photomed Photoimmun* 14 38 (1998)
18. RM Sayre, JC Dowdy and DL Sayre, Photoinstability of Avobenzone-containing sunscreen products, American Society for Photobiology, 26th Annual Meeting, July 12th 1998, Snowbird, Utah, Symposium VI: *Sunscreen Photobiology, Photochemistry and Photobiology* 67S:20S (1998)
19. H Maier, K Brunnhofer, G Schauberger, H Honigsmann, Photoinactivation of sun protection products, *Archives of Dermatological Research* 34 290 (1998)
20. Fed Reg 62(83) (Apr 30, 1997)
21. Sunscreen SPF Testing is Accurate Reflection of Ingredient Photostability. *The Rose Sheet*, F-D-C Reports Inc. (5–7) (June 8, 1998)
22. CTFA Sunscreen Comments Highlight Photostability of Combination Formulas, *The Rose Sheet*, F-D-C Reports Inc. (5) (September 28, 1998)
23. LH Kligman, H Lorraine, PP Agin and RM Sayre, Broad-spectrum sunscreens with UVA I and

UVA II absorbers provide increased protection against solar-simulating radiation-induced dermal damage in hairless mice, *J Soc Cosm Chem* 42 139–155 (1996)
24. H Gonzenbach, and G Pittet, *Photostability, a must?* in: Symposium proceedings, Broad Spectrum Sun Protection: The Issues & Status, Commonwealth Institute, London, UK 11–12 (1997)

Treatment of Photoaged Hands

Keywords: hand skin, antiaging, photoaging, α-hydroxy acid, polyhydroxy acid

A hand cream containing a combination of α-hydroxy acids and polyhydroxy acids has antiaging benefits on the hands.

Facial antiaging has received significant attention through topical application of cosmetics and drugs, in-office skin care procedures (such as ablative and nonablative laser resurfacing, chemical peels and microdermabrasion) and spa techniques (such as facials and exfoliation treatments). Those who care about the quality of their skin are diligent users of broad spectrum sunscreens on the face year round to prevent additional photoaging.

By comparison, little effort is taken to maintain the health and youthfulness on the back of the hands, which often provide strong clues in determining one's age. While the face is often covered with some level of sunscreen from makeup and/or moisturizers, the hands are rarely protected with sunscreen. Consider that while driving during daylight, for example, one hand is nearly always exposed to UV from sunlight and the damaging UVA rays that penetrate glass windows. Chronic exposure to sun causes noticeable pigmentation irregularities and epidermal and dermal atrophy leading to altered skin texture and laxity.[1-3] On facial skin, the development of wrinkles is an obvious hallmark of aging. On the back of hands, the condition is often described as crepeness (i.e., like crepe paper), a term that encompasses both skin looseness and lack of smoothness, but not necessarily roughness.

An effective antiaging product targeted for use on hands should provide demonstrable reductions in the visual signs of aging that are prominent on hands, namely uneven skin tone and diminished skin firmness.

A Formulation Approach

An antiaging hand cream as an oil-in-water emulsion containing a blend of α-hydroxy acids (AHAs) and polyhydroxy acids (PHAs) was formulated. This formula has been commercialized[a] and cannot be presented here. The beneficial ingredients are highlighted in Table 1. Here we describe our thinking behind the selection of these ingredients.

These ingredients were selected to provide skin smoothing and plumping effects as well as enhanced exfoliation and improved skin clarity. Multiple emollients were added to moisturize and condition. Also added were pro-vitamin A (to augment the antiaging benefits of the product) and sunscreens (to protect against further sun damage).

Table 1. Benefit ingredients in antiaging hand cream

Ingredient	Description
Ethylhexyl methoxycinnamate	organic sunscreen
Titanium dioxide	inorganic sunscreen
Lactobionic acid	polyhydroxy acid
Gluconolactone	polyhydroxy acid
Mandelic acid	α-hydroxy acid
Citric acid	α-hydroxy acid
Retinyl acetate	vitamin A acetate
Arginine	amphoteric pH adjuster

AHAs and PHAs: The blend of AHAs and PHAs was selected to provide skin smoothing and plumping effects, as well as enhanced exfoliation and improved skin clarity.

The PHAs were lactobionic acid and gluconolactone (total 7%). The AHAs were mandelic acid and citric acid (total 3%). The formulation pH was adjusted to 3.6 using an amphoteric agent (arginine) to help impart gentleness to the AHA/PHA formulation[4,5] in combination with a small amount of ammonium hydroxide.

Sunscreens: Both organic sunscreens and inorganic sunscreens were added to provide UV protection and help prevent further sun damage. The SPF rating is 15.

Inorganic sunscreens absorb and scatter UV energy, while organic sunscreens act purely by absorbing UV light.[6] A high solids content along with a surface-treated titanium dioxide dispersion (an inorganic sunscreen agent) and ethylhexyl methoxycinnamate/octinoxate (an organic sunscreen agent) were selected for use in this formulation.

The titanium dioxide is surface-treated with aluminum stearate and is dispersed in a vehicle selected for ease of handling and incorporation during manufacturing by providing better wetting and dispersion characteristics with a reduced risk of agglomeration.

Ethylhexyl methoxycinnamate is a widely used UVB absorber with maximum absorption occurring at 290–320 nm.[6]

The combination of organic and inorganic sunscreens provided a sun protection factor (SPF) of 15 for this formulation when tested according to the United States Food and Drug Administration (FDA) Monograph test method.

Pro-vitamin A: A significant amount of vitamin A acetate was incorporated to augment the antiaging benefits of the product. The ester forms of vitamin A are significantly more stable in cosmetic formulations than retinol and have similar effects in skin.[7,8]

Emollients: A combination of emollients was used to moisturize and condition, and it provided a non-greasy, cushiony feel to the skin.

To avoid the quick hydrolysis of short-chain fatty alcohol and fatty acid ester emollients that occurs under the acidic conditions needed to maximize bioavailability of AHAs and PHAs, long-chain and branched esters and ether-type emollients were used.

[a] The product is a currently marketed formulation that is covered by patents for AHAs + sunscreens, and AHAs + retinoids, and others. A quantitative formulation is not available for use in the publication, but the technology is available for licensing from NeoStrata Company.

Stabilizers: Long chain alcohols and acids (e.g., cetyl alcohol, stearyl alcohol and stearic acid) were used to help stabilize the emulsion and increase the viscosity of the cream. An emulsion stabilizer (magnesium aluminum silicate) and an aqueous phase thickener (xanthan gum) were added to ensure formula stability and to modify viscosity.

Testing for Antiaging Effects

The formulation passed routine stability testing and microbial challenge testing and was found to be non-irritating and non-sensitizing in a repeat insult patch test (RIPT) conducted on humans without occlusive conditions.[9] The formulation for specific antiaging benefits then was tested.

Method: A clinical study was conducted at an independent testing facility to evaluate the antiaging effects of the test hand cream on women with mild to moderate photodamage–specifically mottling, crepeness and roughness–on the dorsal aspect of hands.

Twenty-nine women (aged 30-60 years, Fitzpatrick types I-IV) participated in the 12-week study. Following a three-day wash-out period, each subject applied the treatment cream twice daily. Efficacy assessments were performed at each study visit corresponding to weeks zero, six and 12 and included the following:

- Irritation: We obtained clinical grading of objective irritation (erythema, edema, dryness) using a trained clinical evaluator, and made inquiries regarding subjective irritation (itching, stinging, burning, tightness, tingling) using the four-point scale: 0 = none, 1 = mild, 2 = moderate, 3 = severe.
- Photoaging: We obtained clinical grading of photoaging by a trained clinical evaluator for these parameters: crepeness (textural wrinkling of the hands), roughness, mottled pigmentation, clarity, and firmness. Evaluations were made visually using a 10 cm analog grading scale (Table 2) divided into 0.25 cm increments with descriptors on the terminal anchors.
- Smoothness: We took silicone replicas at baseline and endpoint. The replicas were analyzed using computerized image analysis to measure skin smoothness.
- Pinch recoil: We measured the time it takes for the skin to recover (recoil) after mechanical distortion (pinching). Measurements were generated by pinching the dorsum of the hand and recording the time with a stopwatch in seconds to full recovery of the skin. The measurements were performed in triplicate, and the average score was reported. Pinch recoil is a recognized indicator of skin resiliency and firmness.[10]
- Self-assessment: Via a questionnaire.

Results: No subjects discontinued the study due to product-related effects. All parameters were analyzed in comparison to baseline scores using a paired t-test.

- Irritation: There was a statistically significant decrease ($p< 0.05$) in dryness at weeks six and 12 and there were no other changes in irritation parameters, demonstrating product tolerance on skin (Figure 1).
- Photoaging: Statistically significant ($p< 0.05$) improvements in the signs of photoaging were evident at weeks six and 12 for all signs of photoaging. Week

Table 2. Analog grading scale of clinical conditions of aging in dorsal hand skin

Condition	Complete absence of the condition	Complete presence of the condition
Crepeness texture	0 = Smooth, uniform texture	10 = Loose, crepe paper
Roughness	0 = Soft, smooth	10 = Rough, coarse
Mottled pigmentation	0 = Even tone	10 = Mottled, uneven tone
Clarity	0 = Dull, matte	10 = Clear, radiant
Firmness	0 = Loose, pliable	10 = Firm, unpliable

12 showed a 33% improvement in skin clarity and 55% improvement in firmness on average across the participants, compared to baseline (Figure 2).

- Smoothness: In further support of the clinical grading, there was a statistically significant improvement ($p < 0.05$) in skin smoothness via silicone replica analysis at week 12. Skin roughness and line depth improved by 14%.
- Pinch recoil: Pinch recoil was significantly improved at weeks six and 12 ($p < 0.05$) with a mean relative percentage improvement of 12% at week six and 22% at week 12. This finding directly supports the improvement observed in visually assessed skin firmness.
- Self-assessment: Results from the self-assessment parameters were favorable and in agreement with the clinical findings (Figure 3). More than 60% of participants noticed an antiaging effect on their skin after two weeks of use, and 86% noticed an improvement within four weeks of product use (Figure 4).

Conclusion: The AHA/PHA antiaging hand cream exhibited significant improvements in all of the signs of photoaging, while being gentle to the skin. Pinch recoil and silicone replicas further confirmed the benefits of this product for increased resiliency and smoothness. Product users concluded that their skin looked and felt younger, smoother and firmer.

Summary

Different approaches for the use of AHAs and PHAs in the treatment of photoaged skin have been described.[11-14] PHAs offer gentleness as well as a high degree of moisturization and antioxidant effects, compared to AHAs. These qualities may make them preferable to some users.[15] For some applications, it may be best to combine both AHAs and PHAs to achieve optimal effects based upon the attributes of individual ingredients.

The current study demonstrates the antiaging benefits of using a combination AHA/PHA product in reducing the visible signs of photoaging and improving skin firmness on the back of hands. The results verify that antiaging treatments can extend to uses beyond wrinkles on the face. The subtle textural changes of aged skin on hands and prominent pigmentation irregularities can be minimized through continued antiaging treatment with AHAs and PHAs.

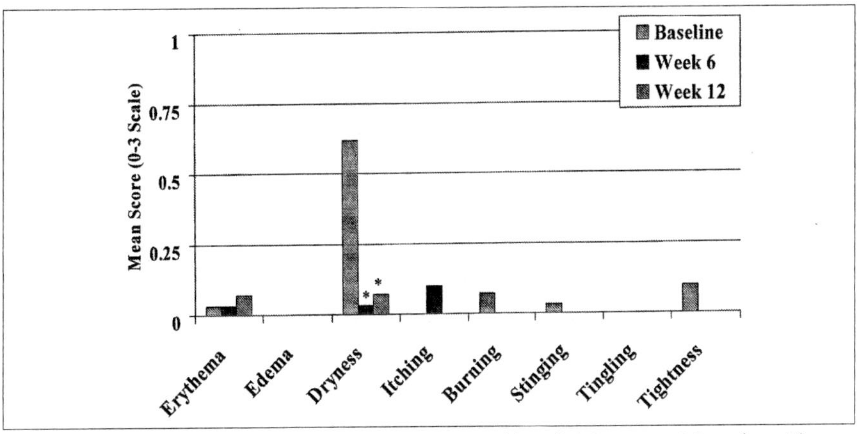

Figure 1. Irritation grading mean score (n = 29) assessed by clinician on a 0-3 scale, with asterisk indicating significant improvement from baseline (p<0.05)

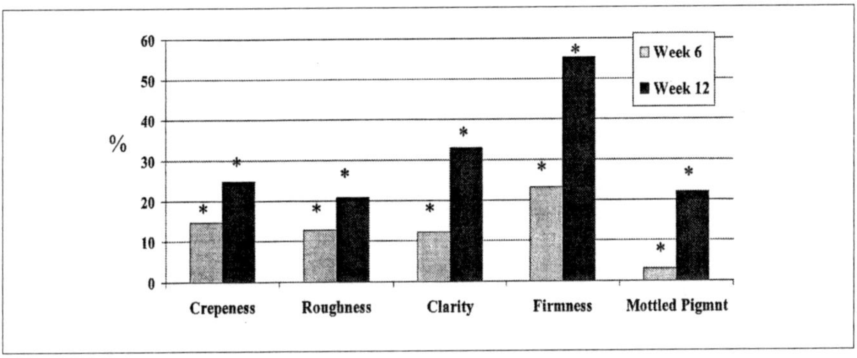

Figure 2. Photoaging mean relative % improvement (n = 29) in grading assessed by clinician, compared to baseline. All attributes significantly improved at weeks 6 and 12 (p<0.05) as shown by asterisks.

—**Barbara A. Green, Brenda L. Edison and Yaling Lee**
NeoStrata Company, Inc., Princeton, New Jersey, USA

References

1. EF Bernstein, Dermal effects of alpha hydroxy acids, Chapter 7 in *Glycolic Acid Peels*, R Moy, D Luftman and L Kakita, eds, New York: Marcel Dekker Inc (2002) 71–114
2. CM Ditre, TD Griffin, GF Murphy, H Sueki, B Telegan, WC Johnson, RJ Yu and EJ Van Scott, Effects of α-hydroxy acids on photoaged skin: A pilot clinical, histologic, and ultrastructural study, *J Am Acad Dermatol* 34 187–195 (1996)
3. EF Bernstein, CB Underhill, J Lakkakorpi, CM Ditre, J Uitto, RJ Yu and EJ Van Scott, Citric acid increases viable epidermal thickness and glycosaminoglycan content of sun-damaged skin, *Dermatol Surg* 23 689–694 (1997)
4. RJ Yu and EJ Van Scott, A discussion of control-release formulations of AHAs, *Cosmetic Dermatology* 10 15–18 (Oct 2001)
5. HU Kraechter, JA McCaulley, B Edison, B Green and DJ Milora, Amphoteric hydroxy complexes: AHAs with reduced stinging and irritation, *Cosmet Toil* 116(1) 47–52 (2001)

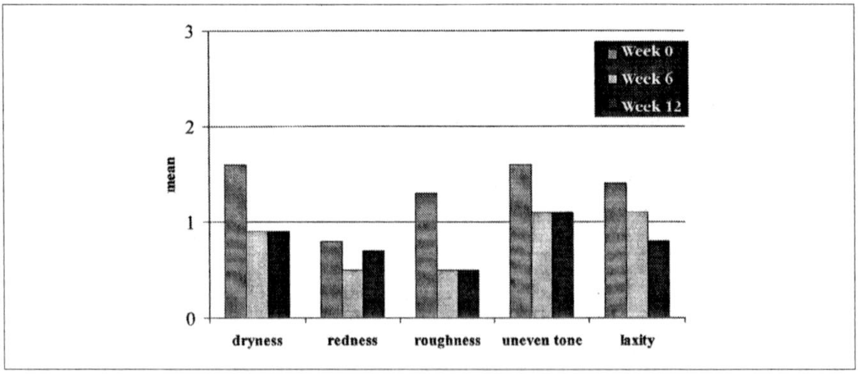

Figure 3. Irritation and photoaging mean score (n = 29) determined by self-assessment on a 0-3 scale

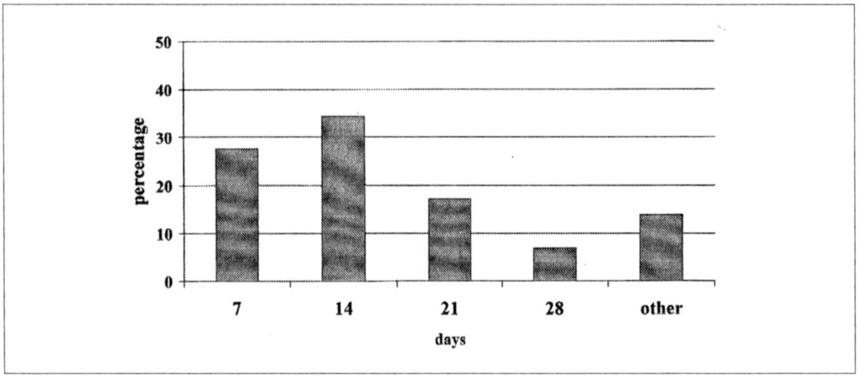

Figure 4. Percentage of subjects (n = 29) reporting improvement in youthful look and feel of the skin at selected time intervals

6. NP Patel, A Highton and RL Moy, Properties of topical sunscreen formulations: A review, *J Dermatol Surg Oncol* 18 316–320 (1992)
7. KL Keller and NA Fenske, Uses of vitamins A, C and E and related compounds in dermatology: A review, *J Amer Acad Dermatol* 39 611–625 (1998)
8. CC Zouboulis, Retinoids: Is there a new approach? *IFSCC Magazine* 3(3) 9–17 (2000)
9. Data on file, NeoStrata Company, Inc
10. Y Appa, BS Asuncion, TJ Stephens, RL Rizer, DL Miller and JH Herndon, A six month clinical study to evaluate the long term efficacy and safety of an α-hydroxy acid lotion, *Amer Acad Dermatol Poster Exhibit* (1996)
11. EJ Van Scott, CM Ditre and RJ Yu, α-hydroxyacids in the treatment of signs of photoaging, *Clinics in Dermatol* 14 217–226 (1996)
12. MI Rendon and G Okan, The use of α-hydroxy acids in xerosis and photoaging, in Chapter 8, *Glycolic Acid Peels,* R Moy, D Luftman and L Kakita, eds, New York: Marcel Dekker Inc (2002) pp 115–139
13. P Grimes, BL Edison, BA Green and R Wildnauer, Evaluation of inherent differences in ethnic skin types and response to topical polyhydroxy acid (PHA) use, American Academy of Dermatology Poster Exhibit, Washington DC, March 2001
14. BA Green, BL Edison, RH Wildnauer and ML Sigler, Lactobionic acid and gluconolactone: PHAs for photoaged skin, *Cosmetic Dermatology* 9 24–28 (Sep 2001)
15. EF Bernstein, BA Green, B Edison and RH Wildnauer, Poly hydroxy acids (PHAs): Clinical uses for the next generation of hydroxy acids, *Skin & Aging* Supplement 9 (Sep 2001)

Defending Against Photoaging: A New Perspective for Retinol

Keywords: skin, photoaging, matrix metalloproteinases, UV filter, antioxidants

Based on the understanding that photoaging of the skin is a multifactorial process with the major influence coming from UV rays, the author proposes a three-line defense strategy.

Photoaging of the skin is a multifactorial process with the major influence coming from ultraviolet (UV) rays. UVA and UVB both contribute to this phenomenon, but in different ways. While UVB directly affects the cellular and extracellular components to a certain extent, UVA exerts its detrimental effect predominantly via reactive oxygen species (ROS). The ROS break down lipids and proteins of the skin cells as well as the extracellular matrix of the skin. They also trigger different signal transduction cascades within the cells, thus modifying gene expression and giving rise to further detrimental processes such as inflammation and enzymatic matrix degradation through so-called matrix metalloproteinases (MMPs).

Based on this understanding of the processes of photoaging, a three-line defense strategy is proposed.

First, the amount of UV reaching the living layers of the epidermis has to be minimized. This can be achieved through a combination of UVA and UVB filters. Obviously, this can never afford full protection, and some ROS are still generated.

Therefore, at the second level, the use of antioxidants–particularly a combination of water-soluble (e.g., ascorbyl phosphate) and fat-soluble (e.g., tocopheryl acetate) compounds–reduces damage by oxygen radicals. Again, this type of defense is not absolute. Cellular responses such as the inflammatory reaction still occur, leading to the release of matrix-degrading MMPs.

The third and final line of defense modulates and, at best, normalizes these detrimental cellular processes. It comprises compounds such as bisabolol or panthenol, which help sooth the inflammatory response of the skin.

Finally, retinol has been shown to normalize the activation of MMPs after UV irradiation. This gives a new perspective on retinol. Apart from being an agent effective in treating wrinkles, it also helps to prevent the formation of wrinkles and should therefore be used at adequate levels in daily care products.

Combining all of the mentioned compounds would optimize the everyday protection of the skin against photoaging.

The Three Lines of Defense Against Photoaging

Retinol is an established antiwrinkle active ingredient for high-end, up-to-date skin care products. It increases the mitotic activity and normalizes the enzyme activity of keratinocytes, thus improving the epithelization of the skin. As a result, retinol smoothes fine lines and wrinkles, improves the structure and normalizes the physiology of the epidermis. This also leads to an improved barrier function and helps retain skin moisture and it has a normalizing effect on skin pigmentation leading to a healthier skin color.

The best known effect of retinol is its high efficiency in treating skin wrinkles. Until now it is used mostly in skin care products for mature skin to diminish fine lines and other signs of premature skin aging. But retinol has additional effects. Recent studies have shown that retinoids also counteract the physiological processes that actually lead to the formation of wrinkles.

It is well accepted that a major cause for skin aging is UV irradiation, which leads to certain physiological changes within the skin. This is an immediate response that is invisible to the eye but, after accumulating over the years, it leads to prematurely aged skin. Oxidative processes are commonly believed to be involved in these processes.[1] A variety of factors contribute to an increased occurrence of highly detrimental ROS. Apart from UV irradiation, these factors include various xenobiotics such as ozone, cigarette smoke or nitrogen oxides. Yet, the main factor of the extrinsic, premature aging is sun exposure. Therefore, it is often termed photoaging.[2-7]

Although a crucial part of the aging process, ROS and the oxidative stress do not represent more than a small slice of the whole process. Other physiological processes are equally important: Figure 1 depicts some crucial aspects of these events: UVA and UVB generate ROS both within the skin cells and in the intercellular space.[8] They induce different signal transduction cascades, either directly or via unknown mediators. The activation of AP-1 alters the gene expression of the skin cells. Interleukins, the mediators of inflammation, are released and initiate the inflammatory response in the skin. As this process proceeds, further ROS are generated, keeping the vicious circle going.

The altered gene expression also leads to an overexpression of matrix degrading proteins. These so-called MMPs are responsible for the degradation of collagen and elastin[9] which are important matrix proteins that give the skin its mechanical strength and elasticity. Figure 1 also shows the different targets for anti-photoaging products, which are discussed in more detail below.

Level 1: Sunscreens: Within the spectrum of sunlight, UV light is the major factor leading to skin damage. The exposed parts of the skin are permanently subjected to UV radiation. When UV filters are applied to the skin, e.g. in the form of a sunscreen, the amount of UV radiation that reaches the skin is markedly reduced.[10,11]

However, sunscreens are not usually applied on a routine basis. Only a certain percentage of the Western population uses sunscreens at all,[12] and most of those who

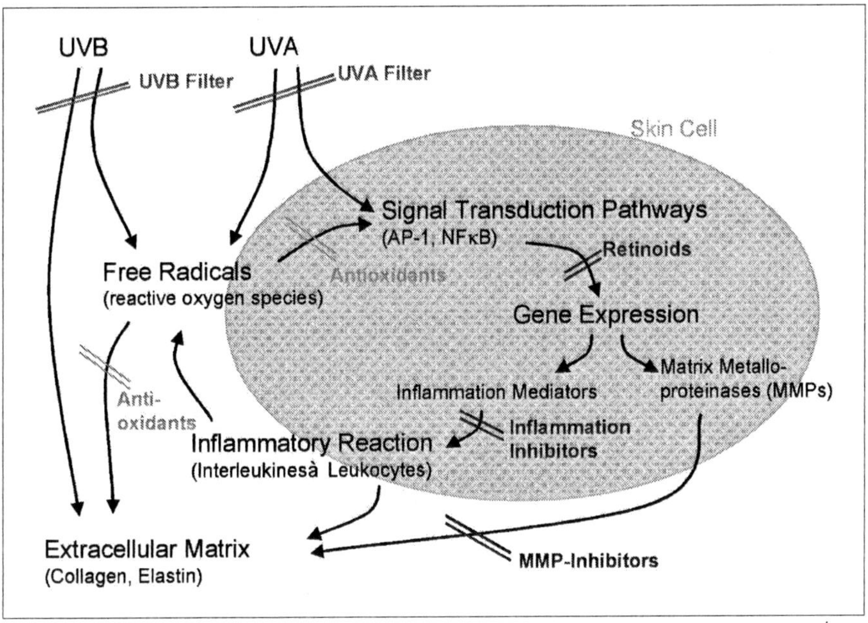

Figure 1. UV-mediated alteration of physiological processes within the skin

do only do so upon extensive sun exposure. The protection afforded is often well below the theoretical SPF value due to insufficiencies in both dose applied and mode of application.[13]

It can be estimated that for more than 95% of a year, when people are not explicitly sunbathing, they usually remain totally unprotected. This lack of protection can be overcome by the moderate use of UV filters in daily cosmetic products such as face lotions, hand creams, body milk and so forth. The SPF achieved should not be too high, so that the light-dependent synthesis of vitamin D will not be impeded.

Both UVA and UVB filters should be incorporated into daily care products. Whether this is through organic filters or inorganic pigments depends on the product to be formulated. Both UVA and UVB protective filter types are nowadays available in stable form (See Formula 1, page 410).

Level 2: Antioxidants: UV filters can never offer total protection. The remaining radiation penetrating into the skin–especially in the UVB (280-320 nm) and UVA (320-400 nm) range–leads to the formation of reactive species. These species, which are often oxygen-centered free radicals, are probably mediated via intracellular chromophores or photosensitizers.[14,15] They have a detrimental effect on all molecular components of the skin cells and on the extracellular matrix. Thus DNA, proteins and lipids are all affected.

Damaged DNA bases could cause gene mutations,[16] leading to skin cancer in the worst case. Lipid peroxidation, as well as damage to certain enzymes, impairs cellular functions, leading to decreased cell turnover. This results in dry, flaky, dull-looking skin with decreased barrier properties.

Formula 1. Sun Care Lotion (o/w) with UVA, UVB protection, SPF 28	
A. Dibutyl adipate	8.0 %wt
C12-15 alkyl benzoate	8.0
Cocoglycerides	12.0
Sodium cetearyl sulfate	1.0
Diethylamino hydroxybenzoyl hexyl benzoate (Uvinul A Plus, BASF Aktiengesellschaft)	2.0
Lauryl glucoside (and) polyglyceryl-2 dipolyhydroxystearate (and) glycerin	4.0
Cetearyl alcohol	2.0
Ethylhexyl triazone (Uvinul T 150, BASF Aktiengesellschaft)	3.0
Vitamin E acetate	1.0
B. Zinc oxide (Z-Cote, BASF Aktiengesellschaft)	4.0
C. Glycerin	3.0
Disodium EDTA (Edeta BD, BASF Aktiengesellschaft)	0.05
Allantoin	0.2
Xanthan gum	0.3
Magnesium aluminum silicate	1.5
Water (*aqua*)	48.45
D. Citric acid	0.5
Phenxoxyethanol (and) methylparaben (and) ethylparaben (and) butylparaben (and) propylparaben (and) isobutylparaben	1.0
Fragrance (*parfum*)	qs

Extracellular matrix proteins, such as collagen and elastin, are partly destroyed by the direct action of free radicals. Crosslinks between collagen fibrils occur more frequently in photo-aged skin than in normal skin. Interestingly, the type of crosslinking is different in photo-aged skin. Elastosis, a typical sign of photo-aged skin, occurs when elastin fibrils are degraded to a certain extent and clump together. Accumulated damage of matrix proteins results in structural changes with impaired mechanical functionality and leads to a loss of skin elasticity with wrinkles and sagging.

The use of antioxidants in cosmetic products is therefore highly recommended.[17,18] To achieve a good antioxidant activity, two requisites must be met. First, the compounds have to be stable for the storage life of the product. Second, they have to penetrate into the skin to exert their antioxidant activity at the site where the oxidative stress occurs.

Tocopherol and ascorbic acid are well-known antioxidants, but they should only be used as technical antioxidants to prevent oxidation of the cosmetic product. They are inadequate for protecting the skin, so more stable compounds should be

Formula 2. Daily Care Aerosol according to the "three lines of defense" concept

A. Ethylhexyl methoxycinnamate (Uvinul MC 80, BASF Aktiengesellschaft)	4.0 %wt
Octocrylene (Uvinul N539, BASF Aktiengesellschaft)	1.5
Caprylic/capric triglyceride	9.0
Buxus chinensis (jojoba) oil	5.0
Cyclomethicone	1.5
Hydrogenated coco-glycerides	3.0
VP/hexadecene copolymer	1.0
Ceteareth-6 (and) stearyl alcohol (Cremophor A 6, BASF Aktiengesellschaft)	1.0
B. Zinc oxide (Z-Cote, BASF Aktiengesellschaft)	5.0
C. Ceteareth-25 (Cremophor A 25, BASF Aktiengesellschaft)	2.0
Panthenol (D-Panthenol 75W, BASF Aktiengesellschaft)	1.2
Sodium ascorbyl phosphate (BASF Aktiengesellschaft)	0.2
Imidazolidinyl urea	0.3
Disodium EDTA (Edeta BD, BASF Aktiengesellschaft)	0.1
Water (*aqua*)	64.17
D. Tocopheryl acetate (Vitamin E-Acetate, BASF Aktiengesellschaft)	0.5
Bisabolol (Bisabolol rac., BASF Aktiengesellschaft)	0.2
Fragrance (*parfum*)	qs
Caprylic/capric triglyceride (and) retinol (Retinol 15D, BASF Aktiengesellschaft)	0.33

Fill in suitable cans and pressurize with liquid pressure gas (e.g. 10 % Propan/Butan 25/75).

chosen. Derivatives are often the choice, and both tocopheryl acetate and ascorbyl 2-phosphate salts are stable enough to survive long storage times. They are also able to penetrate into the skin and within the skin the ester bonds are cleaved liberating the respective antioxidants.

By applying a combination of these two antioxidants, a synergistic effect that gives maximum protection should be achieved. It was shown in a recent study on human keratinocytes that ascorbyl phosphate and tocopheryl acetate represent a highly efficient combination for the protection of skin cells against ROS.[19] Adequate use levels should be between 0.5% and 5% for tocopheryl acetate and around 0.2% and 2% for ascorbyl phosphate (see Formula 2, page 411).

Level 3: Modulators of cellular responses: In addition to the direct action of UV-generated free radicals, a vast variety of cellular responses occur. Mediated directly by UV irradiation or indirectly by signals coming from cell and tissue damage, the skin reacts with an inflammatory process. When skin cells mistake UV-mediated cell damage for a mechanical insult, bearing the risk of an infection, the skin cells release pro-inflammatory cytokines like interleukin (IL)-1alpha, IL-1beta, and IL-6, as well as prostaglandins.[20,21] Subsequently, white blood cells infiltrate the skin, generating additional radical species in order to combat a non-existing infection, thus augmenting the detrimental process in a vicious circle.

This inflammatory reaction of the skin is an immediate response that, in extreme cases, manifests itself as an erythema or sunburn. But, even if a response is not visible, the underlying mechanisms still occur to a certain extent, and their detrimental effects accumulate over time. Few compounds are available to the cosmetic chemist seeking to alleviate these temporary processes. One such active compound is bisabolol, which is the active principle of chamomile and a traditional medical herb. A recent study has shown its activity against UV irradiation-induced erythema.[22] Another active compound for this purpose is panthenol, long known for its effectiveness at healing wounds.[23]

On the same level of cellular responses, the UV-induced gene products c-jun and c-fos form AP-1 (activator protein-1). AP-1 binds to a family of retinoid receptors, the so-called retinoic acid receptors (RAR), giving rise to an overexpression of MMPs.[24] These MMPs, in turn, are responsible for the degradation of extracellular matrix-proteins, most importantly collagen and elastin.[25]

Retinoids have been shown to moderate this overexpression and, thus, norma-lize the cellular function.[26-27] Therefore, retinol is not only a highly potent active against existing fine lines and wrinkles, which various studies have shown previously, but also acts directly on UV-induced cellular processes, thus guarding the skin against premature aging.[28,29]

While panthenol and bisabolol are easy to handle, a few words should be said on retinol. The major drawback of this highly potent skin active for the formulator is its inherent instability, which has—until now—not been solved to satisfaction.

To achieve an adequate stability of retinol in a cosmetic product, the manufacturer has follow two prerequisites: first is the use of an inert gas atmosphere during production and packaging, and the second is the use of oxygen-impermeable packaging such as aluminum-lined tubes. These packaging types should also prevent the re-entry of air when samples are taken under use conditions.

With these two prerequisites, the stability of retinol in a typical cream formulation would be approximately 90–95% after three months at 40°C. Even with the currently available stabilized forms of retinol, one would probably have to follow these precautions to achieve this stability—and consequently the desired physiological effect in the skin.

Conclusion

Based on the previously described understanding of UV-mediated skin damage, there are three lines of defense against photoaging:

- Minimize UV irradiation either by permanently using adequate UVB and UVA filters applied to the exposed areas of the skin or by not exposing the skin to the sun or similar light sources.

- Strengthen the skin's antioxidative network to minimize direct insult from free radicals by supplying adequate amounts of synergistic antioxidants such as ascorbyl phosphate and tocopheryl acetate.
- Soothe the inflammatory processes with active compounds such as bisabolol and panthenol and normalize overall cellular responses by using adequate concentrations of retinol.

—Axel Jentzsch, Harald Streicher and Valérie André
BASF Aktiengesellschaft, Ludwigshafen, Germany

References

1. I Emerit, Free radicals and aging of the skin, *EXS* 62 328–341 (1992)
2. S Beissert and RD Granstein, UV-induced cutaneous photobiology, *Crit Rev Biochem Mol Biol* 31 381–404 (1996)
3. LH Kaminester, Current concepts: Photoprotection, *Arch Fam Med* 5 289–295 (1996)
4. Y Miyachi, Photoaging from an oxidative standpoint, *J Dermatol Sci* 9 79–86 (1995)
5. NH Nicol and NA Fenske, Photodamage: Cause, clinical manifestations, and prevention, *Dermatol Nurs* 5 263–275 (1993)
6. K Scharffetter-Kochanek, M Wlaschek, P Brenneisen, M Schauen, R Blaudschun and J Wenk, UV-induced reactive oxygen species in photocarcinogenesis and photoaging, *Biol Chem* 378 1247–1257 (1997)
7. CR Taylor and AJ Sober, Sun exposure and skin disease, *Annu Rev Med* 47 181–191 (1996)
8. R Kohen, Skin antioxidants: Their role in aging and in oxidative stress—new approaches for their evaluation, *Biomed Pharmacother* 53 181–192 (1999)
9. A Thibodeau, Metalloproteinase inhibitors, *Cosm Toil* 115 75–82 (2000)
10. H Schaefer et al., Photoprotection of skin against ultraviolet A damage, *Met. Enzymol* 319 445–465 (2000)
11. HV Debuys et al., Modern approaches to photoprotection, *Dermatol Clin*, 18 577–590 (2000)
12. B Diffey, Sun protection, *Nederl Tijdschrift Dermatol Venerol*, 9 333–334 (1999)
13. B Diffey, Personal Protection: The Way Forward, *Radiation protection dosimetry* 91 293–296 (2000)
14. K Scharffetter-Kochanek et al. UV-induced reactive oxygen species in photocarcinogenesis and photoaging, *Biol Chem* 378 1247–1257 (1997)
15. M Dalle Carbonare, MA Pathak, Skin photosensitizing agents and the role of reactive oxygen species in photoaging, *J Photochem Photobio. B: Biol*, 14 105–124 (1992)
16. J Lehmann et al., Kinetics of DNA strand breaks and protection by antioxidants in UVA- or UVB-irradiated HaCaT keratinocytes using the single cell gel electrophoresis assay, *Mut Res*, 407 97–108 (1998)
17. I Savini et al., Ascorbic acid maintenance in HaCaT cells prevents radical formation and apoptosis by UV-B, *Free Radic Biol Med* 26 1172–1180 (1999)
18. L Vaillant et al., Long term topical antioxidant treatment provides protection against clinical signs of photoaging, *Trace Elem Man Anim* 10 437–439 (2000)
19. A Jentzsch, H Streicher, K Engelhart, The Synergistic Antioxidant Effect of Ascorbyl 2-Phosphate and Alpha-Tocopheryl Acetate, *Cosmetics & Toiletries* 116 55–64
20. T Schwarz, UV radiation and cytokines, In: *Skin cancer and UV radiation* edited by P Altmeyer and K Hoffmann 219–226 (1997)
21. K Isoherranen et al., Ultraviolet irradiation induces cyclooxygenase-2 expression in keratinocytes, *Brit J. Dermatol* 140 1017–1022 (1999)
22. H Streicher, A Jentzsch, Alpha-Bisabolol, a versatile active ingredient for cosmetics, *PCIA*

Bangkok 8.–10.3. (2000)
23. U Wollina, Zur klinischen Wirksamkeit von DexPanthenol, *Kosmetische Medizin* 4 180–186 (2001)
24. GJ Fisher, JJ Voorhees, Molecular mechanisms of photoaging and its prevention by retinoic acid: Ultraviolet irradiation induces MAP kinase signal transduction cascades that induce Ap-1-regulated matrix metalloproteinases that degrade human skin in vivo, *J Invest Dermatol Symp Proc* 3 61–68 (1998)
25. AN Malak, E Perrier, TIMP-1 like: A new strategy for anti-aging cosmetic formulations, *20th IFSCC congress Cannes* 1 79–89 (1998)
26. J Varani et al., Inhibition of type I procollagen synthesis by damaged collagen in photoaged skin and by collagenase-degraded collagen in vitro, *Am J Pathol*, 158 931–942 (2001)
27. J Varani et al., Molecular mechanisms of intrinsic skin aging and retinoid-induced repair and reversal, *J Invest Dermatol Symp Proc* 3 57–60 (1998)
28. J Varani et al., Vitamin A antagonizes decreased cell growth and elevated collagen-degrading matrix metalloproteinases and stimulates collagen accumulation in naturally aged human skin *J Invest Dermatol* 114 480–486 (2000)
29. JJ Voorhees, Retinoid repair/prevention of photoaging and natural aging in human skin in vivo, *Clin Exp Dermatol*, 25 161–162 (2000)

Participation of Metalloproteinases in Photoaging

Keywords: Matrix metalloproteinases, tissue inhibitors, dermatoheliosis, photoaging

This review is intended to acquaint readers with a novel explanation for photoaging. It starts with a short introduction to the basic biology of metalloproteinases and then describes a novel approach to our understanding of photoaging.

Matrix Metalloproteinases

Extracellular matrix metalloproteinases (MMP) have been studied for many years. The best known of this ever-increasing number of proteolytic enzymes are the collagenases.[1]

The body's requirements for the presence of MMPs arise from the need for degradation of essentially insoluble polymeric ground substance components. Proteolysis of fibrillar macromolecules requires fragmentation (or partial solubilization) to allow the better known phagocytic pathways to complete the task of removing unwanted or damaged polypeptides. The role played by MMPs in dermal homeostasis is complex and not fully understood.

MMPs degrade extracellular matrices during wound repair and remove damaged tissue. They play a part in the remodeling processes during tissue repair. The number of identified MMPs is quite large (about 15), and many of them are capable of degrading the same substrate. This redundancy, shown in Table 1, is unexplained, and many unknowns remain. It is generally agreed[1,2] that plasmin converts the proenzymes to the active MMPs. Once formed, their proteolytic activity must be controlled. Removal of zinc atoms from a MMP by chelation, for example with ethylenediamine tetraacetic acid, destroys the activity. In vivo, macroglobulins and the so-called tissue inhibitors of MMPs, TIMPs, control the activity of the MMPs. TIMPs, especially, have attracted the interest of skin scientists. TIMPs form non-covalent bi-molecular complexes with active MMPs and inhibit their proteolytic activity. The amino acid sequences of some TIMPs are known, but the mechanism for regulating TIMP expression is not fully understood.

Table 1 Partial Listing of Metalloproteinases

Number	Enzyme	Selected Substrates
MMP-1	Matrix Collagenase (Fibroblast Collagenase)	Collagens I, II, III, VII, and X
MMP-8	Neutrophil Collagenase	Collagens I, II, and III, Link Protein, Aggrecan
MMP-13	Collagenase 3	Collagens I, II, and III
MMP-18	Collagenase 4	Collagen I
MMP-2 (Mol. Wt. 72K)	Gelatinase A	Gelatins, Collagens I, IV, VII, and XI, Fibronectin, Laminin, Elastin
MMP-9 (Mol. Wt. 92K)	Gelatinase B	Gelatins, Collagens IV, V, and XIV, Aggrecan, Elastin
MMP-3	Stromelysin 1	Aggrecan, Gelatin, Fibronection, Laminin, Collagens III, IV, IX, and X
MMP-10	Stromelysin 2	Aggrecan
MMP-14	(membrane type)	Collagens I, II, and III, Laminin
MMP-15	—— ?	
MMP-17	—— ?	
MMP-7	Matrilysin	Aggrecan, Fibronectin
MMP-11	Stromelysin 3	Fibronectin
MMP-12	Metalloelastase (Macrophage)	Elastin

Proteolytic action of MMPs: Despite some uncertainties about the precise mechanism, the proteolytic action of MMPs in healthy skin is evidently necessary. In fibrotic skin diseases, such as scleroderma, excessive TIMP-3 levels are suspected.[3] The expression of mRNA for TIMP-3 is enhanced by diverse mitogenic stimuli. High levels of TIMP-3 have been shown to be accompanied by increased amounts of collagen 1 in cultured fibroblasts.[4]

TIMP

A specific TIMP, TIMP-1, has also been implicated in the accumulation of collagen in scleroderma.[5] Additional reports on the inhibition of MMPs have appeared in recent years with an emphasis on wound healing.[6-8] In addition, TIMP-2 reportedly has a somewhat different beneficial effect because it inhibits tumor cell metastasis in basal cell carcinoma.[9] Growth stimulating effects on keratinocytes of TIMP have also been reported.[10] These observations are pertinent to this discussion because thickened epidermis has been associated with photoadaptation of exposed skin.

Therefore, it is safe to assume that various TIMPs have different effects on skin and that the expression of TIMP genes plays an important role in skin health. By analogy, the activation of MMPs in the extracellular environment must also contribute to skin health. The precise balance between proteinases and their inhibitors in skin health is a topic of serious research throughout the world. For example, Boelsma et al.[11] reported that a proteinase inhibitor, Skin Derived Antileukoproteinase (SKALP), was elevated in skin exposed to SLS or oleic acid and that this skin-derived inhibitor is a marker for skin irritancy. The expression of SKALP was induced by serum, epidermal growth factor, and fibroblasts but was inhibited by 1,2-dihydroxyvitamin D_3 and retinoic acid in vitro; it was unaffected by tocopherol and ascorbic acid. There is no evidence at this time that SKALP also inhibits MMPs. In contrast, McIlrath et al.[12] reported that the MMPs in pigskin cultures could be modified differentially. MMP-9 was increased by retinoic acid, while MMP-2 was slightly decreased. Retinoic acid appears to modify dermal proteinase activity. On the other hand, these results suggest that the repair of photoaged skin by retinoic acid is not simply due to enhanced proteolytic activity of the extracellular matrix.

It is becoming increasingly apparent that MMP activity and its induction is significant in a variety of human skin conditions.[13-18] MMP activity is initiated by a plasminogen activator commonly expressed by basal keratinocytes. One unexpected phenomena is the observation that TIMP stimulates the growth of human keratinocytes in tissue culture but has no effect on fibroblasts.[10]

Proteolysis in Photoaging

Most readers are familiar with Kligman's work on photoaging and its reversal in the hairless mouse.[21-23] Her findings can be summarized by the statement that UVB irradiation can cause damage to elastic fibers, collagen and glycosaminoglycans (GAG) in the extracellular matrix (ECM) within the dermis. Two aspects of her reports are especially pertinent: The damage was attributed to overproduction of dermal constituents, and reversal was enhanced by retinoic acid. Such repair in the absence of UVB exposure is a relatively slow reconstruction process.

The generally accepted mechanisms for photoaging include direct excitation of DNA bases by UVB and dimeric photo additions, type II photoreaction (singlet oxygen), type I photoreactions (electron abstraction, for example, free radical formation), or superoxide generation (followed by a Fenton-type reaction). The end result of these reactions is attack on the bases in the DNA.

Data on photoaging of human skin under normal conditions are not easily available. Thus, Kligman relied on Skh-1 and Skh-2 mice. More recently, results on photoaging obtained with cultured human fibroblasts were reported from Japan.[24] The fibroblast culture was exposed to externally generated reactive oxygen species (ROS), and the end-point parameter was synthetic activity of the fibroblasts. Elastin and GAG were increased in this experiment, while collagen production was decreased. Interestingly, the investigator reports that the level of TIMP-1 was increased by ROS. These results support the suspicion that ROS may effect biological (metabolic) changes in the dermis.

The approach by Fisher et al. provides a different explanation for the mechanism of photoaging with participation by MMPs.[19] In their first paper, they report on their investigation of UVB radiation damage on human buttock skin. They noted that messenger ribonucleic acids (RNAs) for MMP and subsequent MMP activation occur very rapidly and at doses "well below those that cause skin reddening." Transcription factors of MMP genes are activated by very low doses of UVB (0.5 mW/cm^2) in subjects exhibiting minimal erythemal doses (MEDs) of 30-50 mJ/m^2. The reported observations are of considerable importance. Most research on the impact of UV light on skin depends on erythema for quantification. If Fisher's data are valid, the MED may provide a false sense of security with regard to photodamage.

The investigators specifically assayed for a 54K collagenase (MMP-1) and for a 92K gelatinase (MMP-9) after exposures below one MED and report a sharp and rapid increase in these two proteinases. The authors also noted that retinoic acid effects transrepression of the transcription factor. The striking feature of these biological events after irradiation is the rapidity with which the system responds to low doses of UVB. The investigations described are impressive and evidently have not been challenged.

No response to this paper, nor a follow-up by other scientists, appeared for some time. The publication presented an explanation for dermatoheliosis and photoaging that did not call for the involvement of reactive oxygen species or free radicals. The authors did not provide a mechanism for the induction of the MMP messenger RNA, and the terms free radical or ROS cannot be found in their papers. They described effects in the dermis after irradiation as repair processes requiring substantial destruction of damaged ECM components by specialized enzymes. The only contribution to this idea was provided by Tanaka,[24] who observed upregulation of TIMP-1.

Another feature that made the scientific community uncomfortable was the speed with which the repair process (MMP formation) was initiated, and that it occurred at levels below the MED. The latter must be distressing to those who assess skin damage and its potential prevention on the basis of erythema. Instead, Fisher's data suggest that insolation, at less than one MED, triggers photoactivation of MMPs in the dermis.

The second shoe fell about a year later in a second publication[20] by the Fisher group. In this case, too, the authors exposed buttock skin of Caucasians to low doses of UVB. Skin samples for analysis were obtained by biopsy or dermatotome from irradiated and non-irradiated sites.

Collagenase (MMP-1, MMP-9, and MMP-3; cf Table I) m-RNAs appeared to be almost absent in non-irradiated skin, but rapidly elevated after irradiation. The m-RNA for these enzymes were found predominantly in the epidermis. The enzymatic activity was found in the epidermis and dermis, but stromelysin (MMP-3) activity was located primarily in the dermis. Addition of TIMP reduced hydrolytic activity to normal. Continued (once a day) exposure to UV maintained the increased levels of MMPs for at least four to five days.

Pretreatment of the skin with all transretinoic acid before UV exposure reduced the induction of the three studied MMPs. A significant portion of the total MMPs is synthesized in the epidermis and then transported to the dermis. Retinoic acid appears to inhibit MMPs without elevating levels of TIMPs.

UV-Induced Photoaging

These observations prompted the authors to provide a hypothetical model for UV-induced photoaging.[20] Levels of UV light, causing no detectable sunburn, induce formation of MMPs in epidermal keratinocytes and dermal fibroblasts, with resulting degradation of the ECM. Keratinocytes also produce TIMPs, reducing the activity of MMPs. This process is followed by formation of imperfectly repaired collagen. During multiple intermittent UV exposure, this process, if repeated, results in the formation of more severely damaged collagen, for example, the solar scars visualized by Kligman.

In a third paper by Fisher et al., the authors began the arduous task of sorting out sequences of biological reactions which, after UV irradiation, trigger the induction of MMPs in human skin.[25] It is noteworthy that another group of investigators recently reported on the upregulation of MMPs and of structural protein formation by cytokines and integrins after UV irradiation of hairless mice.[26] It seems safe to predict that current research will clarify the relationship between UV light, MMP and photoaging.

The following needs to be done to clarify the mysteries of MMP and photoaging:

- Research for a sound reason for the poor repair process described in the Fisher model. One would expect that the repair process, or the formation of new collagen, should be "perfect."
- Research into agents that are totally unreactive with ROS or free radicals, but still may prevent photoaging.
- Research leading to a decision as to whether inhibition of MMPs by diverse approaches would be appropriate because the enzymatic destruction of damaged ECM proteins or GAGs is likely to be required.

Clearly, direct confirmation of the ideas presented by Fisher and co-workers at the University of Michigan would be welcome. In the meantime, cosmetic formulators might start the search for actives having the potential of controlling the repair process following the action of MMPs.

Protection Against Photoaging

Full recognition of the significance of the Fisher et al. findings has been slow. However, publicity in the May/June 1998 *R$_x$emedy* is likely to alert professionals, as well as laymen, to the hazards of sun bathing. If the very short exposures to UVB studied by Fisher can trigger a rapid cascade of events in human skin, even a partial UV blocker is unlikely to afford full protection against skin photoaging.

Much current thinking about photoaging relies on the oncogenic activity of insolation. There is considerable debate on the merits of sunscreens and the optimal sun protection factor in the prevention of skin cancers. This debate obviously impacts the benefits of sunscreens in the prevention of photoaging. Despite regulatory uncertainties, and in the absence of sound clinical data, a rational argument can be made for the use of sunscreens for reducing the damage of sun (UV) exposure. Treatment options for photodamaged skins are currently limited to retinoids and facial peeling.

It would appear premature to abandon the use of agents acting as antioxidants or free radical scavengers. There is sufficient clinical evidence to support the use of compounds that can inactivate ROS and similar UV-created substances to reduce pathological consequences.

For future control of photoaging, it would appear wise to examine a battery of diverse ingredients that can inhibit the formation of this and other pathological skin conditions. Retinoids, as a group, seem to be able to interfere with photoaging, but the mechanism for their action remains obscure. Some recent patents (PCT Int. Appl. WO 98 113,017 and PCT Int. Appl. WO 98 113,018) extol the benefits of retinal and retinyl esters in skin care products. There is no reason to assume, however, that retinoids are the only chemicals able to alleviate photoaging. Even if the mechanism proposed by Fisher et al. is not fully confirmed, the findings of these investigators suggest that there may be other cosmetic ingredients that can modulate photoaging by controlling some of the biochemical or transcriptional events in the skin.

Summary

Investigations published in recent years relate the stigmata of photoaging to improperly repaired ECM proteins after low levels of insolation. At this time, it is not clear how the repair process, requiring proteolytic activity in the dermis as a first step, occurs and why it should fail to restore the ECM to its original state. Following the hydrolytic action of MMPs, the proteolytic potential of MMPs must be reduced to allow synthesis of replacement proteins. The solution to this conundrum can be expected to provide formulators and clinicians with unexpected tools for preventing or repairing photoaged skin.

—**Martin Rieger,** *M&A Rieger Associates, Morris Plains, New Jersey USA*

References

1. H Birkedal-Hansen, WGIMoore, MKBodden, LJ Windsor, B Birkedal-Hansen, A DeCarlo and JA Engler, Matrix metalloproteinases: A review, Critical Reviews, Oral Biology and Medicine 4 (2) 197-250 (1993)
2. G Murphy, R Ward, J Gavrilovic and S Atkinson, Physiological mechanism for metalloproteinase activitation, Matrix Special Suppl 1 224-230 (1992)
3. L Mattila, K Airola, M Ahonen, M Hietarinta, C Black, USaarialho-Kere and VM Kähäri, Activation of tissue inhibitor of metalloproteinases-3 (TIMP-3) mRNA expression in scleroderma skin fibroblasts, *J Invest Dermatol*, 110 416-421 (1998)
4. EC C LeRoy, M McGuire and N Chen, Increased collagen synthesis by scleroderma skin fibroblasts in vitro, *J Clin Invest*, 54 880-889 (1974)
5. K Kikuchi, T Kadomo, M Furue, and K Tamaki, Tissue inhibitor of metalloproteinase 1 (TIMP-1) may be an autocrine growth factor in scleroderma fibroblasts, *J Invest Dermatol*, 108 281-284 (1997)

6. F Grinnell and M Zhu, Fibronectin degradation in chronic wounds depends on the relative levels of elastase, 1-proteinase inhibitor, and macroglobulin, *J Invest Dermatol*, 106 335-341 (1996)
7. PW Park, K Biedermann, L Mecham, DL Bissett and RP Mecham, Lysozyyme binds to elastin and protects elastin from elastase-mediated degradation, *J Invest Dermatol*, 106 1075-1080 (1996)
8. F Grinnell, M Zhu, and WC Parks, Collagenase-1 complexes with 2-macrogobulin in the acute and chronic would environments, *J Invest Dermatol*, 110 771-776 (1998)
9. SN Wagner, HM Ockenfels, C Wagner, HP Soyer, and M Goos, Differential expression of tissue inhibitor of metalloproteinases-2 by cutaneous squamous and basal cell carcinomas, *J Invest Dermatol*, 106 321-326 (1996)
10. B Bertaux, W Hornebeck, AZ Eisen, and L Dubertret, Growth stimulation of human keratinocytes by tissue inhibitor of metalloproteinases, *J Invest Dermatol*, 97 679-685 (1991)
11. E Boelsma, S Gibbs and M Ponec, Expression of skin-derived antileukoproteinase (SKALP) in reconstructed human epidermis and its value as a marker for skin irritation, *Acta Derm Venereol* (Stockh), 78 107-113 (1998)
12. EM McIlrath, U Santhanam, and MR Greene, The effect of retinoic acid treatment on matrix metalloproteinase (MMP) and plasminogen activator (PA) production by pig skin organ culture, *B J Dermatol*, 138 752 (1998)
13. R DeCastro, Y Zhang, H Guo, H Kataoka, MK Gordon, BP Toole and C Biswas, Human keratinocytes express EMMPRIN, an extracellular matrix metalloproteinase inducer, *J Invest Dermatol*, 106 1260-1265 (1996)
14. M Weckroth, A Vaheri, J Lauharanta, T Sorsa and YT Konttinen, Matrix metalloproteinases, gelatinase and collagenase, in chronic leg ulcers, *J Invest Dermatol*, 1061119-1124 (1996)
15. A Oikarinen, M Kylmäniemi, H Autio-Harmainen, P Autio and T Salo, Demonstration of 72-kDa and 92-kDa forms of type IV collagenase in human skin: Variable expression in various blistering diseases, induction during re-epithelialization, and decrease by topical glucocorticoids, *J Invest Dermatol*, 101 205-210 (1993)
16. DR Yager, L-Y Zhang, H-X Liang, RF Diegelmann and IK Cohen, Wound fluids from human pressure ulcers contain elevated matric metalloproteinase levels and activity compared to surgical wound fluids, *J Invest Dermatol*, 107 743-748 (1996)
17. K Airola, T Reunala, S Salo, and U Saarialho-Kere, Urokinase plasminogen activator is expressed by basal keratinocytes before interstitial collagenase, stromelysin-1, and laminin-5 in experimentally induced dermatitis herpetiformis lesions, *J Invest Dermatol*, 108 7-11 (1997)
18. L Vaalamo, Mattila, N Johansson, A-L Kariniemi, M-L Karjalainen-Lindsberg, V-M Kähäri, and U Saarialho-Kere, Distinct populations of stromal cells express collagenase-3 (MMP13) and collagenase-1 (MMP-1) in chronic ulcers but not in normally healing wounds, *J Invest Dermatol*,109 96-101(1997)
19. GJ Fisher, SC Datta, HS Talwar, Z-Q Wang, J Varani, S Kang and JJ Voorhees, Molecular basis of sun-induced premature skin ageing and retinoid antagonism, *Nature*, 379 335 (1996)
20. GJ Fisher, Z-Q Wang, SC Datta, HS Talwar, J Varani, S Kang, and JJ Voorhees, Pathophysiology of premature skin aging induced by ultraviolet light, *N Engl J Med*, 337 1419-28 (1997)
21. L Kligman, Connective tissue photodamage in the hairless mouse is partially reversible, *J Invest Dermatol*, 88 21s-17s (1987)
22. L Kligman, Skin changes in photoaging: characteristics, prevention, and repair, Aging and the Skin, AK Balin and AM Kligman (eds) Raven Press, New York (1989)
23. L Kligman, The hairless mouse and photoaging, *Photochem Photobiol*, 54 1109-18 (1991)
24. H Tanaka, Alterations in metabolism of elastin, collagen and glycosaminoglycan induced by reactive oxygen specis. Analysis of photoaging mechanisms using cultured human dermal fibroblasts as an experimental model, Kyoto-furitsu Ika Daigaku Zasshi, 107 (2) 207-222 (1998) through Chem Abstr, 128 268632m (1998)

25. GF Fisher, HS Talwar, J Lin, P Lin, F McPhillips, Z Wang, X Li, Y Wan, S Kang and JJ Voorhees, Retinoic acid inhibits induction of c-Jun protein by ultraviolet radiation that occurs subsequent to activation of mitogen-activated protein kinase pathways in human skin in vivo, *J Clin Invest*, 101(6), 1432-1440(1998)
26. E Schwartz, AN Sapadin and LH Kligman, Ultraviolet B radiation increases steady-state mRNA levels for cytokines and integrins in hairless mouse skin; modulation by topical tretinoin, *Arch Dermatol Res*, 290(3) 137-144 (1998)

Index

Z